AF443504

Turbomachinery – Volume A

Conference Organizing Committee

D Bohn
Aachen, Germany

G Bois
Ecully, France

M Davies
Limerick, Ireland

R Decuypere
Brussels, Belgium

G Dibelius
Aachen, Germany

R Dunker
EC

T Fransson
Stockholm, Sweden

G Gyarmathy
Zurich, Switzerland

B Haller
Lincoln, UK
(Local Organizer)

E G Hencke
Dusseldorf, Germany

H Jericha
Graz, Austria

J Krzyanowski
Gdansk, Poland

F Martelli
Firenze, Italy

V Molnar
Bratislava, Slovakia

M Moore
Guildford, UK
(Local Organizer)

K Papailiou
Athens, Greece

W Riess
Hannover, Germany

C T J Scrivener
Derby, UK
(Local Organizer)

C H Sieverding
St Genesius Rode, Belgium

M Stastny
Plzen, Czech Republic

J P van Buijtenen
Delft, The Netherlands

A Wilson
Bergen, Norway

Associated Organizations

Associazione Termotecnica
Italy

**Belgian Society of Mechanical and
Environmental Engineering**
Begium

**Koninklijke Vlaamse
Ingenieursvereiniging**
Italy

Institution of Engineers in Ireland
Ireland

Société Française des Mécaniciens
France

VDI–Gesellschaft Energietechnik
Germany

Norwegian Society of Engineers
Norway

Komitet Problemow Energetyki P.A.N.
Poland

**Association of Czech Mechanical
Engineers**
Czech Republic

Slovenska Asociacia Strojnych Inzinierov
Slovakia

Svenska Mekanisters Riskforbund
Sweden

**Royal Institution of Engineers in the
Netherlands (Klvl)**
The Netherlands

IMechE
Conference Transactions

I MECH E

Third European Conference on

Turbomachinery – Volume A
Fluid Dynamics and Thermodynamics

2–5 March 1999
Royal National Hotel, London, UK

Organized by the Energy Transfer and Thermofluid Mechanics Group of the
Institution of Mechanical Engineers (IMechE)

With support and sponsorship from

European Commission
Rolls-Royce plc
ALSTOM Energy Limited
ALSTOM Technology Centre
ALSTOM Gas Turbines Limited

IMechE Conference Transactions 1999–1A

**Professional
Engineering
Publishing**

Published by Professional Engineering Publishing Limited for the Institution of
Mechanical Engineers, Bury St Edmunds and London, UK.

First Published 1999

ISSN 1356–1448
ISBN 1 86058 196 X

A CIP catalogue record for this book is available from the British Library.

Printed and bound in Great Britain by Antony Rowe Limited, Chippenham, Wiltshire, UK.

Contents – Volume 1

Blade–Row Interaction and Aeroelasticity

Erosion

Secondary Flows and Tip Clearance Flows

Additional Paper

Contents – Volume 2

Compressors

Rotating Stall and Instability

Performance

Authors' Index

Turbine Aerodynamics

C557/063/99

Aerodynamic performance of two isolated turbine stators in transonic annular cascade flow

K FREUDENREICH, M JÖCKER, W HÖHN, and **T H FRANSSON**
Royal Institute of Technology, Stockholm, Sweden
H-J REHDER
German Aerospace Center, DLR, Industrial Aerodynamics Section, Göttingen, Germany

ABSTRACT

The objectives of the present work are to investigate experimentally the steady aerodynamic characteristics of two isolated turbine stators and the development of the wakes at transonic flow conditions in an annular turbine facility. The steady three-dimensional flow field was measured at several axial downstream positions and three radial positions using a three-dimensional Laser-Two-Focus anemometer and a total pressure probe. Steady vane surface pressure distributions were recorded on the vane surfaces. A trailing edge shock propagating in the stator downstream flow field was found to have a significant influence on the velocity and turbulence intensity distribution. In the wake strong radial velocity components were obtained. Close to the trailing edge they are directed towards the hub, further downstream the flow is directed towards the tip.

At midspan the experimental vane surface data and wake flow data are compared with predictions from two different numerical models: Firstly a 2D/Q3D code solving the viscous terms only in the boundary layer region of the vanes (k-l turbulence model) and elsewhere the Euler equations, secondly a fully viscous 3D code using a Baldwin-Lomax type turbulence model. Fairly good agreement between numerical results and experimental data were achieved. Differences, which are mainly observed in the wake flow predictions, are assumed to be due to turbulence modeling and a possible grid resolution influence. Furthermore, three dimensional flow contributions were observed in the 3D predictions, especially in the wake region, which could partly lead to the differences between VOLSOL and UNSFLO results. Different trends for the 3D effects have been found in the experiments and for VOLSOL.

The data will serve as part of a database for further experimental and numerical investigations of stator-rotor interactions in a turbine configuration. It is assumed that the presented data describe key parameters determining the forced response of a downstream rotor blade row.

INTRODUCTION

The flow in turbine stators is complex, including three-dimensional secondary flow effects and wake induced flow disturbances. This leads to loss, noise, unsteady flow and rotor blade vibrations due to stator-rotor interactions. The origin, decay and interaction phenomena of stator wakes and shocks are essential for the prediction of the aerodynamical and structural performance of the following blade rows. Numerous publications have been presented on annular stator flows. Binder and Romey [1982] presented subsonic results. Vortices were identified to contribute strongly to losses and mixing of the wake with the core flow, even at positions far downstream of the trailing edge. Radial components of the flow velocity towards the hub were obtained especially in wake regions. Sieverding et al. [1983] came to similar conclusions, testing a turbine nozzle guide vane in a low-speed annular facility. Total pressure loss, flow angle and static pressure measurements revealed a radial migration of the passage vortices in the passage flow field and at the outlet towards the hub under the influence of the radial pressure gradient. Both experiments showed total pressure distributions in the wake close to the trailing edges resembling Gaussian curves in the pitchwise direction. A vane geometry almost identical to the present one, but with coolant ejection slots close to the trailing edges, was tested in transonic flow in a straight cascade by Colantuoni et al. [1995] and in an annular facility by Kapteijn [1995]. The wake was found to resemble Gaussian curves with separate standard deviations for pressure and suction side of the jet. A weak shock structure at the suction side trailing edge was found.

The pressure and velocity distributions, created by wakes and shock structures, affect the steady and unsteady loading and pressure excitation of downstream positioned rotor blades. These phenomena were studied numerically by several authors, eg. Korakianitis [1992], who used wake velocity deficit data from several experimental investigations as rotor inlet conditions. For his examined subsonic cases the computed unsteady rotor passage flow was found to be sensitive to the amplitudes of the potential flow interactions, but not to changes in the viscous wake amplitudes. Kielb and Chiang [1992] and Izsak and Chiang [1993] modeled velocity and pressure distributions downstream of a turbine stator as incoming flow distortions into a rotor row in their forced response analysis. They reported problems in modeling stator wakes due to lack of appropriate turbulence models and lack of high speed stator wake data for verification.

OBJECTIVES

The objectives of this investigation are to gain a better understanding of the aerodynamic effects of turbine stator induced disturbances on the downstream flow field. Experiments and computations were carried out with two isolated stator configurations in a transonic annular wind tunnel. The wake characteristics and its interaction with shocks downstream of the cascades were examined. The aim was to gain a detailed understanding of the wake flow with the perspective to study stator-rotor interactions, which is planned as future work.

It was of interest how accurate the wake flow can be predicted with different numerical tools, because this is expected to be essential for the foreseen unsteady stator-rotor interaction simulations. Therefore, two numerical methods, which differ in turbulence modeling and flow field solution strategy, were applied to predict the flow around and behind the stator.

TEST FACILITY AND TEST CONFIGURATION

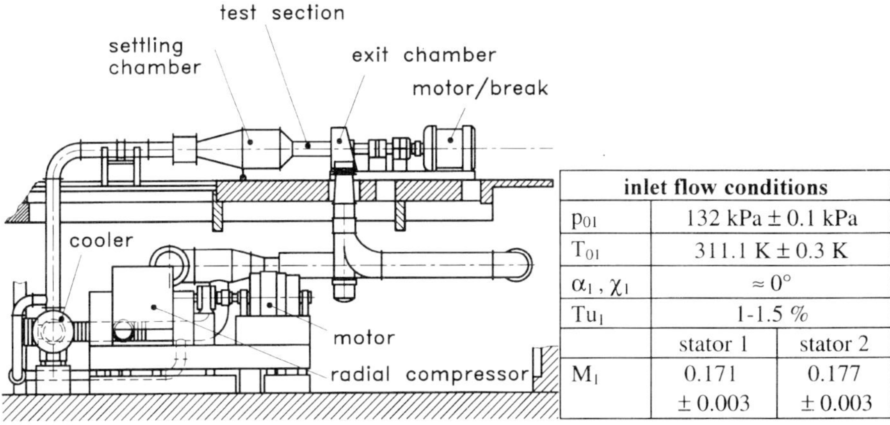

Fig. 1: RGG test facility and inlet flow conditions

The experiments were performed in the wind tunnel for rotating cascades (RGG), a high pressure cold flow test turbine at the German Aerospace Center (DLR) in Göttingen, Germany. The RGG is a closed facility operating in continuous mode, offering the opportunity to vary independently both Mach and Reynolds number. Figure 1 gives a schematic sketch, including the test section inlet flow conditions. Details about the test facility can be found in Amecke and Kost [1993]. The investigated stators have been designed with cylindrical geometries and identical exit Mach number and exit flow angle at midspan, but with 43 vanes for stator 1 and 70 vanes for stator 2 in order to change the disturbance frequency on a downstream rotor. Both stators were tested in transonic flow regime with isentropic outlet Mach numbers of M_{2is}=1.05 ± 0.002. The hub contour, where for rotor or stage tests the rotor is located, was closed by a conical ring. An impression of the stator 1 cascade being installed in the test section can be obtained from Fig. 2. Table 1 includes the measurement position for the 3D-L2F and the total pressure probe. The stator cascades are presented in Fig. 3.

Table 1: Measurement positions for 3D-L2F (L) and total pressure probe (P)

axial measurement positions						
mm from trailing edge	stator 1			stator 2		
	x/c_{ax}	L	P	x/c_{ax}	L	P
0.8	1.03	x		1.04	x	
4.3	1.14	x		1.21	x	
8.3	1.28	x	x	1.40	x	x
11.3	1.38	x		1.54	x	
15.3	1.51	x	x	1.74	x	x
27.4	1.92	x	x	2.32	x	x
radial measurement positions						
span		20%	50%	90%		

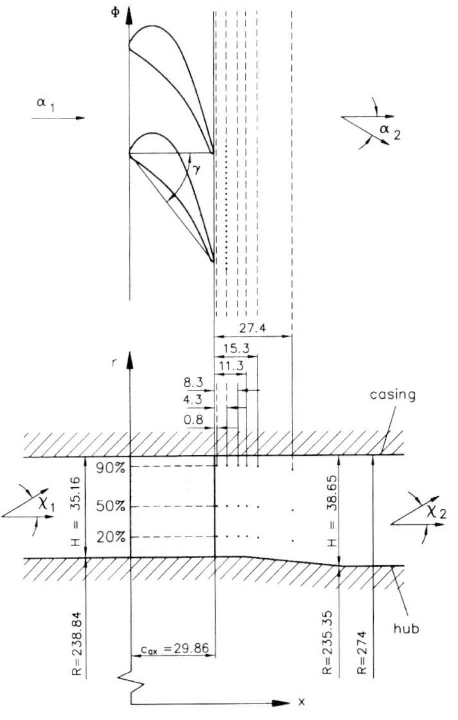

	stator 1	stator 2
No. of vanes	43	70
axial chord	29.86mm	20.79 mm
γ	51.9°	54.16°
M_1	0.174	0.174
M_{2is}	1.05	1.05
α_2	73°	73°

Fig 2: *Test section with stator 1 installed*

Fig. 3: *Stator 1 and 2 cascade geometries and design data*

MEASUREMENT TECHNIQUE

A three-dimensional Laser-Two-Focus Anemometer (3D-L2F) was used to measure the steady three-dimensional mean velocity vector and the turbulence intensity parallel and vertical to the main flow direction. The system was described by Maass et al. [1994]. The measurement volume has a size of 0.3x0.3x1.0 mm in the axial/circumferential/radial direction. The measurement errors were calculated for a 95% confidence interval and are given in Table 3, being subdivided into core flow and wake measurements. Wake flow measurements have a lower accuracy due to a higher turbulence level and a lower data rate. Percentage errors are based on the local measured values, while degree errors are absolute. Details about the error calculations can be found in Freudenreich [1998].

Inlet boundary layer total pressure traverses at hub and tip were performed with a flattened Pitot probe of 0.4 mm thickness. For total pressure wake traverse measurements a wedge type probe was used. Wake traversing were done by indexing the stator in pitchwise direction.

Table 3: 3D-L2F measurement errors

	core flow	wake flow
M	±0.3%	±2%
α	±0.3°	±1°
χ	±1.4°	±(1.4°+13%)
Tu_u , Tu_v	±2%	±4%

All pressures were recorded with an Pressure System Inc. unit with an accuracy of approximately ±0.1% based on the local measured value [Rehder [1997]. Wake traverse measurement positions for both 3D-L2F and probe are given in Fig. 2 and Table 1. Steady vane surface pressure distributions were recorded at 42 positions on stator 1 and 32 positions on stator 2, each at 20%, 50% and 90% span. The pressure tappings on the vanes are small orifices with a diameter of 0.3 mm. Due to mechanical constraints the pressure orifices were distributed over 5 to 7 vanes per vane height section.

COMPUTATIONAL APPROACH

Two different numerical methods were applied to predict the steady flow around and behind the investigated stators. UNSFLO [Giles, 1991] is a 2D/Q3D method solving explicitly the Euler equations on an H-type grid. It takes into account viscosity with a thin layer approach using an implicit method on an O-type grid containing the boundary layer around the vanes. The inviscid algorithm is of Lax-Wendroff type extended for unstructured grids, the viscous algorithm is based on an ADI procedure by Beam and Warming [1978]. For the present results a one equation k-l turbulence model [Bailey, 1993] was used. Quasi 3D effects are modeled by a stream tube specification.

The 3D multiblock Navier-Stokes solver for reacting flow VOLSOL 2.6 [Groth et al., 1996] solves the Reynolds averaged Navier-Stokes equations using various turbulence models in a finite volume fashion [Eriksson, 1995] based on structured H-type or O-type meshes. The turbulence model used in the present study is the Baldwin-Lomax turbulence model [Baldwin and Lomax, 1978]. The governing equations are marched in time by an explicit three stage Runge-Kutta scheme. The code is speeded up applying parallel computing as well as preconditioning and multi-grid.

Table 4: Mesh data and CPU times

	UNSFLO Q3D coarse mesh	UNSFLO Q3D fine mesh	VOLSOL 3D coarse mesh	VOLSOL 3D fine mesh
nodes	8018	13855	129600	388800
nodes/ 2D radial section	4483 (H) + 3535 (O)	7855 (H) + 6000 (O)	120x30=3600	180x60=10800
vane surface nodes /2D radial section	101	150	100	200
no of radial sections	1	1	36	36
streamtube expansion behind stator at hub	linear Q3D option	linear Q3D option	linear 3D model	linear 3D model
turbulence model	k-l	k-l	Baldwin Lomax	Baldwin Lomax
approx. CPU times	2 node hours on CRAY J90*	3.4 node hours on CRAY J90*	90 node-hours on IBM SP2**	100 node-hours on IBM SP2**

*   *max. performance 200 MFLOPS /processor*

** *max. performance approx. 640 MFLOPS /processor*

The UNSFLO code was applied on both stators, whereas the VOLSOL code was only applied on stator 2 in order to compare the UNSFLO results and the experimental data to 3D computational results. The limitation of VOLSOL to the stator 2 was mainly done due to space limits and computational costs. Table 4 includes details about mesh data and CPU times. The presented results are computed on refined meshes. The UNSFLO grid refinement lead to comparably small local changes of the Q3D results in the wake flow and thus these results are judged to be sufficiently mesh independent. The refinement of the VOLSOL 3D grid resulted in a significant improvement of the wake flow predictions but only in small changes of the vane surface pressure distribution. Hence, the presented VOLSOL solution is mesh independent on the vane surface but shows a mesh dependence in the wake flow prediction. Further mesh studies will be done in future.

RESULTS

Inlet and Outlet Flow Conditions

Velocity profiles of the inlet hub and tip boundary layers are presented in Fig. 5. Total pressure measurements were performed with the flattened Pitot probe at $x/c_{ax}=-0.6$ (18mm) upstream of the stator 1 leading edge. The static pressure was measured at the hub and tip walls. The profiles and shape factors show typical turbulent boundary layers with relative thickness of about 9% of the flow channel height at hub and about 20% at tip respectively. A comparison of the numerically obtained inlet and outlet flow conditions with experimental data is made in Table 5. This shows that the predictions compare well in the boundary conditions but with a deviation from the experimental inlet Mach number of about 0.007 to 0.011. It is also seen that the numerically obtained losses are higher than the measured losses. Furthermore, UNSFLO gives even higher losses than VOLSOL.

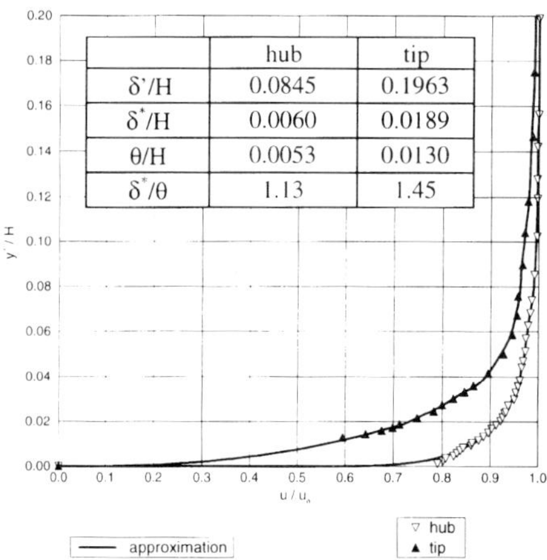

approximation — ▽ hub — ▲ tip

For the approximation a cubic interpolation of the data points was performed and the "gap" between the last measurement point and the hub or tip wall was filled by applying a power law profile, where the exponent n was calculated from the gradient at the last data point nearest to the wall.

Fig. 5: *Inlet hub and tip boundary layer*

Table 5: *Comparison of predicted and experimental flow conditions flux averaged*

	stator 1					stator 2				
	M_1	α_1	M_{2iso}	α_2	ζ (%)	M_1	α_1	M_{2iso}	α_2	ζ (%)
exp	0.171	0.0	1.05	72.2	4.35	0.177	0.0	1.05	71.9	4.68
unsflo	0.163	0.0	1.052	74.66	7.15	0.166	0.0	1.052	73.70	8.15
volsol						0.167	0.0	1.054	73.51	6.24

exp, unsflo: flux averaged pitchwise at midspan
volsol: flux averaged pitchwise and spanwise

$\zeta = 1 - v_{2is}^2 / v_2^2$

<u>Vane Surface Pressures</u>

Results from vane surface pressure measurements and numerical calculations are given in Fig. 6 for both stators. In the diagrams the local static pressure normalized with the inlet total pressure is plotted versus the relative axial chordwise coordinate. Both stators show a similar flow behavior with a strong acceleration on the suction side up to the supersonic region. A shock, propagating from the trailing edge of the neighboring vane, acts with the profile boundary layer. This interaction is clearly visible at 20% span, x/c=0.5 for stator 1 and x/c=0.55 for stator 2, but then reduces with increasing span. The pressure side curves indicate a continuous acceleration between leading and trailing edge.

As expected for the steady stator flow calculations the numerical results compare well to the experimental vane surface pressure data (Fig. 6). The results obtained with UNSFLO show fairly good agreement on both stators beside an under-prediction of the pressure in the suction side aft part of the vane. Also the small re-compression at midspan at about 45% chord is predicted well on both stators with UNSFLO. The 3D prediction on stator 2 by VOLSOL shows a significant difference: A stronger expansion of the flow followed by a shock-like compression at around 50% chord length on suction side can be seen, even though the inlet flow and the back-pressure of this case are similar in all presented numerical results (see Table 5). A look at the Mach number contours in the passage (not presented here due to the limited space) indicates a larger trailing edge shock predicted by VOLSOL than by UNSFLO, which is impinging on the suction side followed by the re-compression. In general, the averaged effective Mach number level at the midspan outlet is a little bit higher in VOLSOL than in UNSFLO, which is also confirmed by the mixed out losses in Table 5: VOLSOL gives smaller losses than UNSFLO indicating a higher averaged effective outlet Mach number.

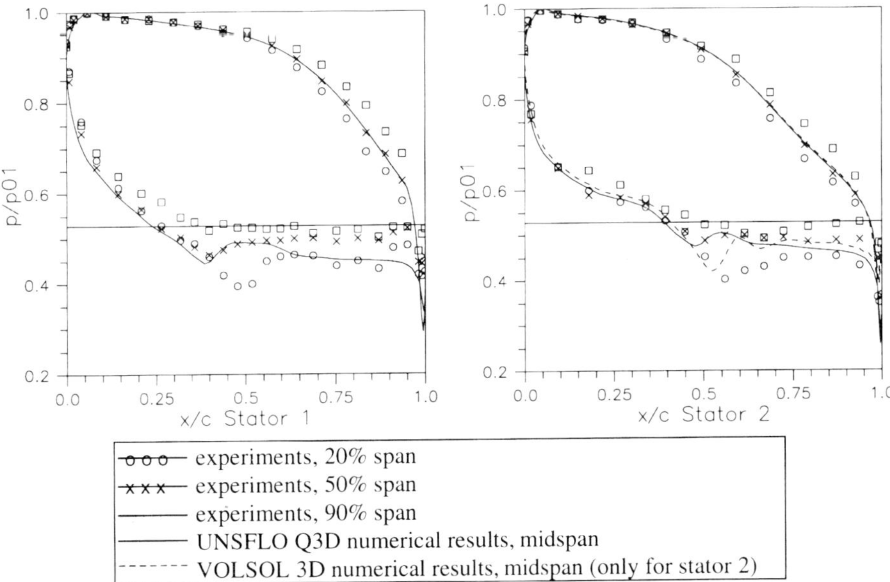

Fig. 6: *Vane surface pressure distribution: stator 1 and 2, experiments vs. numerical midspan results, refined meshes*

These discrepancies between the UNSFLO and VOLSOL results on stator 2 could still be due to the different mesh topologies applied for the UNSFLO respectively VOLSOL, the differences in the used turbulence model or due to 3D effects predicted with VOLSOL. The observation of the 3D flow prediction by VOLSOL as well as the observation of the experimental data indicate the flow to be rather two dimensional at midspan, beside the wake flow (see discussion on page 13). Still, the radial flow angles up to about 5 degrees in the core flow passage, which were found in the VOLSOL results, can have an influence on the surface pressure distribution.

<u>Wake traversing:</u>

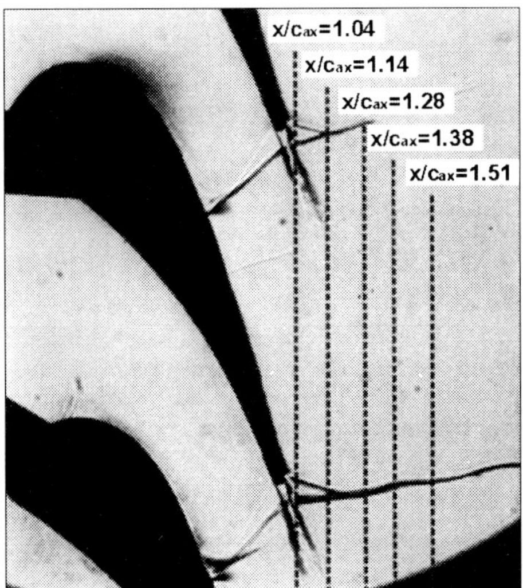

Fig. 7 Schlieren picture of cascade measured by Kapteijn, see Colantuoni et al. [1995], at M_{2is}=1.05, including first five axial measuring positions of stator 1 in present investigation

A Schlieren photo illustrates the flow structure of the 43 vane stator in Fig. 7. This photo, taken from Colantuoni et al. [1995], shows an almost identical vane geometry to stator 1 at identical flow conditions (M_{2is}=1.05) in a linear cascade wind tunnel. In difference to the present stator 1 geometry the trailing edge has a coolant ejection slot but the picture was taken with no ejected coolant mass flow. The first five axial measurement planes for wake traverses of stator 1 in the present investigation are added to the figure. The complete measurement locations can be seen in Fig. 2. For wake traverse measurements, Fig. 8-10, 12-13, about 1.2 pitches were covered. The circumferential coordinate Φ was non-dimensionalized with the stator pitch angle φ_t. The three-dimensional velocity magnitude $|\bar{c}_{3D}|$ was measured with the 3D-L2F, the total temperature T_0 was measured in the settling chamber and assumed to be constant over the vane row. The midspan Mach number wake traverses (M_{2w}) for stator 1 are presented in Fig. 8. The graph being closest to the trailing edge, x/c$_{ax}$=1.03, presents a strong velocity deficit at Φ/φ_t=1.95. This deficit is generated by low momentum fluid from the suction side (SS) and pressure side (PS) boundary layers sweeping around the trailing edge. Further on in the text these velocity deficits are referred to as wakes. The wake center of each graph is defined to be the location with the strongest velocity deficit, U_w. This measurement position is indicated with a square mark (■) for all wake traverse figures. Fluid on the right side of the wake center (high Φ/φ_t values) originates from the SS boundary layer, while fluid on the left side of the wake center (low Φ/φ_t values) originates from the PS boundary layer. SS and PS are indicated for the lowest x/c$_{ax}$ graphs of each figure. The core flow on the other hand is referred to be fluid not influenced by wake or shock effects and traveling with the free stream velocity. The measurement point with the highest velocity, U_m, is indicated in figures with Mach number distributions, Fig. 8 and 9, using a circle mark (●).

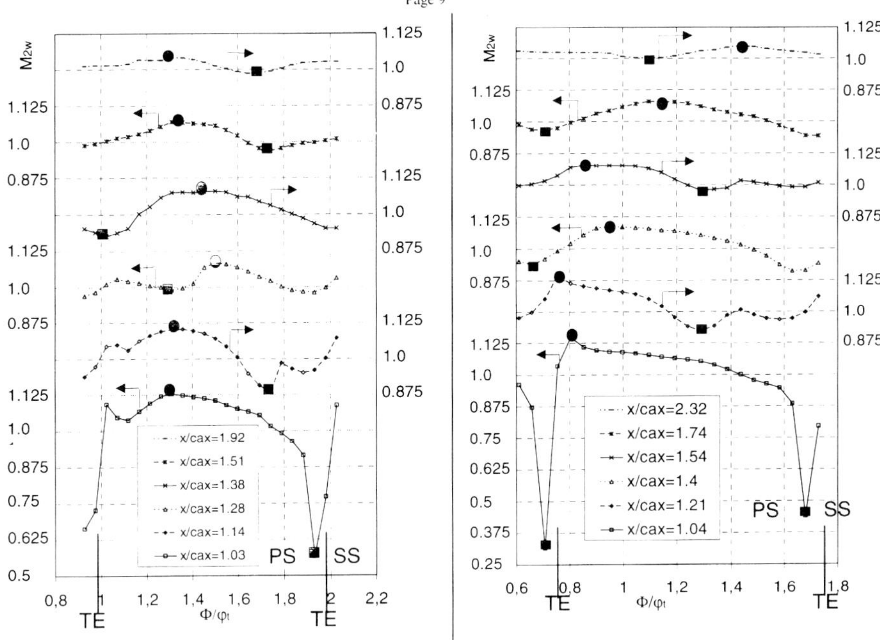

Fig. 8: *M₂ᵥ distribution downstream of stator 1 at midspan, measured with 3D-L2F*

Fig. 9: *M₂ᵥ distribution downstream of stator 2 at midspan, measured with 3D-L2F*

The M_{2w} graph $x/c_{ax}=1.03$ in Fig 8 shows a velocity maximum in the core flow at $\Phi/\varphi_t=1.3$. The velocity drops slower towards the PS of the wake center than towards its SS. This is generated by the lower velocity on the PS of the vane that is still felt downstream of the trailing edge. An additional velocity deficit, located on the SS of the wake centers, can be detected at $x/c_{ax}=1.03$ and $\Psi/\varphi_t=1.12$. This additional deficit is probably created by a shock structure. In the Schlieren picture, Fig. 7, one shock is located between the adjacent vane SS at the throat position and the wake formation location slightly downstream of the trailing edge PS. As this shock propagates through the wake it is directed almost axially. A second shock originates at the SS trailing edge and propagates also in almost axial direction. Both shocks merge at about $x/c_{ax}=1.15$ and form a λ-like structure. The λ-foot is crossed by the traverse at $x/c_{ax}=1.03$ for stator 1 and at $x/c_{ax}=1.04$ for stator 2. The shock within this λ-structure is smeared in circumferential direction, the induced velocity defect is small compared to the wake induced velocity defect. For $x/c_{ax}=1.03$ this additional velocity deficit is clearly separated from the wake center (wake: $\Phi/\varphi_t=0.95/1.95$, shock: $\Phi/\varphi_t=1.12$ at the SS of the wake at $\Phi/\varphi_t=0.95$). The wake-shock separation can be followed also for $x/c_{ax}=1.14$ (wake: $\Phi/\varphi_t=1.73$, shock: $\Phi/\varphi_t=1.9$) and $x/c_{ax}=1.28$ (wake: $\Phi/\varphi_t=1.28$, shock: $\Phi/\varphi_t=1.92$). Compared to the shock induced velocity deficit the wake velocity deficit decays faster in axial direction. The wake fluid streamlines are inclined to the axial direction, while the shocks propagate axially. Moreover, an increased mixing of the wake fluid is generated by its vortical structure.

Merging between both velocity deficits occurs at x/c_{ax}=1.38. The three most downstream velocity distributions show sinusoidal shapes.

Figure 9 presents the M_{2w} distributions for stator 2 at h/H=0.5, corresponding to Fig. 8 for stator 1. In general, similar distributions are obtained. The wake centers of the x/c_{ax}=1.04 graph at Φ/φ_t=0.7 and 1.68 appear to be more symmetrical than the ones of stator 1 at x/c_{ax}=1.03 and Φ/φ_t=1.95. This might be caused by the measurement points not being close enough to resolve the exact distributions. The wake center width of stator 2 at x/c_{ax}=1.04 seems also to be a bit narrower than for stator 1 at x/c_{ax}=1.03. A second velocity deficit on the wake center SS for stator 2 at x/c_{ax}=1.04 can not be identified clearly. The differences compared to stator 1 might originate from differences in machining the small trailing edge radii of stator 1 and 2, leading to slightly altered wake and trailing edge shock performances. Separated velocity deficits from wake and shock can be identified for stator 2 at x/c_{ax}=1.21 (wake at Φ/φ_t=1.3, shock at Φ/φ_t=1.59) and x/c_{ax}=1.54 (wake at Φ/φ_t=1.3, shock at Φ/φ_t=1.63) but not at x/c_{ax}=1.4 (wake at Φ/φ_t=0.67 and 1.65). Merging of both velocity deficits occurs at x/c_{ax}=1.74, further downstream than for stator 1.

The turbulence intensities parallel to the main flow direction, Tu_u, and normal (vertical) to the main flow direction (in the x-Φ plane), Tu_v, are presented in Fig. 10 for stator 1 at midspan. For clarity only the graphs of the first four downstream positions are given. In the wake flow high values of Tu_u and Tu_v of approximately the same order of magnitude are obtained. They decay in axial direction, resulting from mixing with the core flow fluid. The x/c_{ax}=1.03 graphs shows a higher intensity of the parallel intensity Tu_u, while further downstream slightly higher values for the vertical intensity Tu_v are obtained. This might result from trailing edge vortex shedding creating locally non-isotropic turbulence intensity structures in streamwise direction. For some Tu_u graphs a small maximum can be recognized on the SS of the wake centers, decaying in axial direction. The graph at x/c_{ax}=1.03 shows Tu_u maxima at Φ/φ_t=1.15 and, less clear, further downstream at x/c_{ax}=1.14 and Φ/φ_t=2.05 and at x/c_{ax}=1.28 and Φ/φ_t=2.05. These Tu_u maxima are most probably caused by the shock originating at the SS trailing edge discussed above. Normally, a shock can not be recognized in the turbulence intensity measured by a L2F. But the back pressure in the RGG fluctuates slightly due to a not perfectly steady compressor performance. This results in a slight oscillation of the shock over the measurement volume in main flow direction. The L2F detects pre and aft shock flow velocities, appearing as

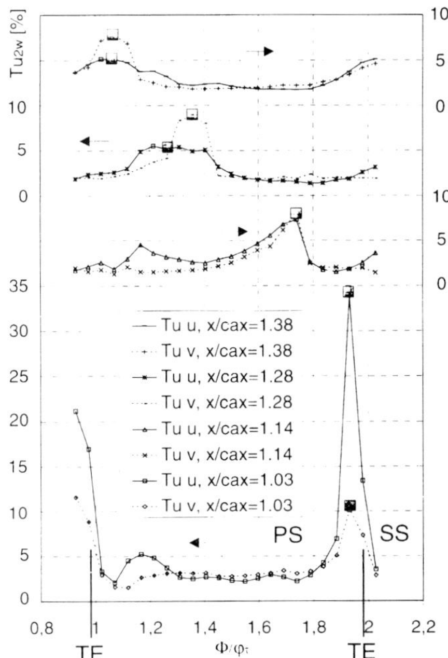

Fig. 10 *Turbulence intensities downstream of stator 1 at midspan, measured with 3D-L2F*

C557/063 © IMechE 1999

a local maximum in the turbulence intensity in main flow direction. The local Tu_u maxima correspond fairly well to the shock positions in the velocity distribution, Fig. 8. Another Tu_u maximum occurs at $x/c_{ax}=1.14$ and $\Phi/\varphi_t=1.12$, corresponding to a velocity minimum at $x/c_{ax}=1.14$ and $\Phi/\varphi_t=1.12$ in Fig. 8. This remains unexplained.

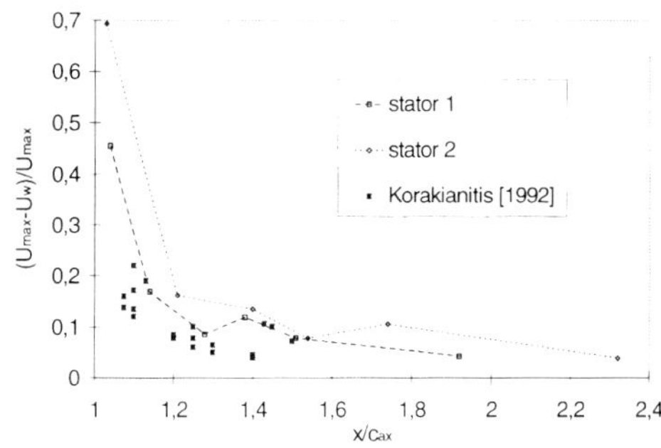

Fig. 11: *Wake center velocity deficit at midspan for stator 1 and 2 measured with 3D-L2F and experimental data, taken from Korakianitis [1992]*

Figure 11 displays the development of the wake center velocity deficit in axial direction for stator 1 and 2. The wake center velocity deficit is defined as the difference of U_w (■) and U_m (●) being non-dimensionalized with U_m. Both graphs show a rapid decrease due to mixing of wake fluid with the core flow. Close to the trailing edges the wake center velocity deficit of stator 2 is stronger than that of stator 1. Experimental turbine stator data are shown as well, taken from Korakianitis [1992] who collected these data exclusively from subsonic flow cases. A comparison shows that the stator 1 and 2 transonic velocity deficits are in general stronger than the subsonic velocity deficits. Korakianitis [1992] computed the unsteady flow and pressure distribution in the rotor passage of a turbine stator-rotor configurations with the aim of forced response prediction. His calculations applied either a sinusoidal velocity defect with an amplitude of 6.5% or an asymmetric Gaussian distribution with a 10% velocity defect at the rotor inlet for subsonic flow cases. In difference to that for the presented transonic cases velocity distributions with two deficits (wake and shock) were obtained, having neither Gaussian nor sinusoidal shape. The velocity deficits are as high as 15% at axial up to $x/c_{ax}=1.4$. Izsak and Chiang [1993] compared predictions of the wake generated pressure deficit in a high speed turbine with measured data for their forced response analysis. The empirically modeled first harmonic velocity distribution overpredicted the deficit and underpredicted the wake width compared to experimental results. Further numerical approaches might consider experimental data as presented here to model transonic rotor inlet conditions.

The flow angle α_{2w} distribution at midspan of stator 1, as presented in Fig. 12, shows strong gradients at the wake center locations (marked with ■). A sweeping of the flow around the trailing edge can be observed for the distribution being closest to the trailing edge $x/c_{ax}=1.03$. On the SS of the trailing edge an overturning occurs (higher local flow angle), while on the PS a corresponding under-turning (lower local flow angle) is detected. A significant influence of the SS shock can not be seen in this first graph ($x/c_{ax}=1.03$), but for $x/c_{ax}=1.14$ at $\Phi/\varphi_t=1.82$ and $x/c_{ax}=1.28$ at $\Phi/\varphi_t=1.35$, both showing a local minimum on the SS of the wake centers. Mixing of the wake fluid with the core flow results in an almost sinusoidal variations at about the same axial positions as the velocity distribution from $x/c_{ax}=1.38$ on downstream.

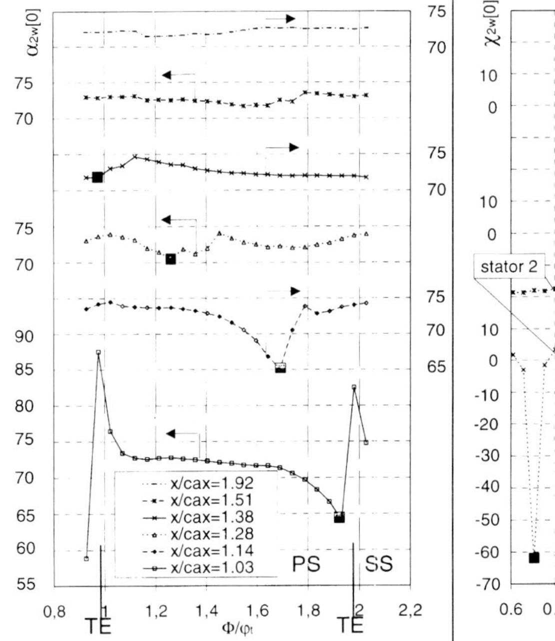

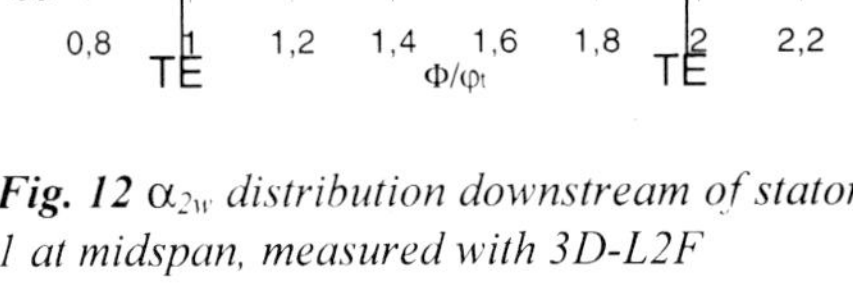

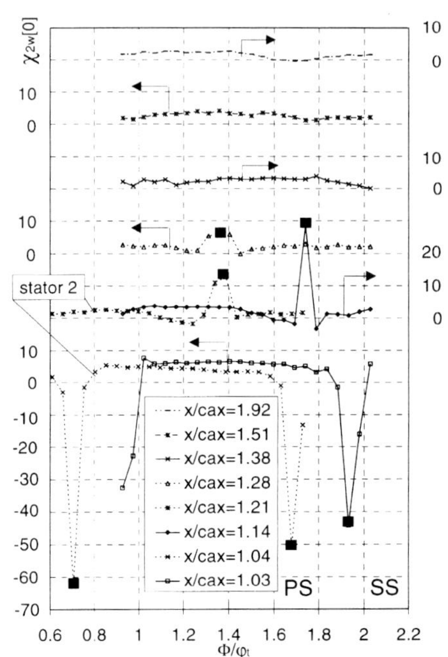

Fig. 12 α_{2w} *distribution downstream of stator 1 at midspan, measured with 3D-L2F*

Fig. 13: χ_{2w} *distribution downstream of stator 1 and 2 (only at x/c_{ax}=1.04 and 1.21) at midspan, measured with 3D-L2F*

In difference to linear cascades annular cascade test facilities create a radial static pressure gradient (higher pressure on the casing). As investigated by eg. Binder and Romey [1983] and Sieverding et al. [1983], this radial pressure drives low momentum fluid from tip to hub, as in wakes and secondary flow regions. Figure 13 presents the radial flow angle χ_{2w} distribution for stator 1 at midspan. For comparison, graphs from two axial planes of stator 2 (x/c_{ax}=1.04 and 1.21) are superposed on the stator 1 graphs at x/c_{ax}=1.03 and 1.14. χ is positive being directed from hub to tip. In agreement to the above mentioned experiments the distributions of stator 1 at x/c_{ax}=1.03 and stator 2 at x/c_{ax}=1.04 show a strong velocity component towards the hub in the wake centers. The stator 2 negative component is stronger. A change in the radial flow direction from negative to positive occurs for both stators at the next axial locations, where the stator 1 peak at x/c_{ax}=1.14 is higher than the stator 2 peak at x/c_{ax}=1.21. Comparisons with Fig. 8 and 9 show that these positive peaks are also located in the wake centers. Further downstream the positive peaks decrease until eventually an almost sinusoidal distribution is obtained. Sieverding et al. [1983] obtained slightly positive velocity components on the PS of the wake center at midspan for an axial position of x/c_{ax}=1.3. In the wake center a strong fluid migration towards the hub was observed, what is not detected in the present cases. A final explanation for the "switch" of the radial angle distribution in the wake center can not be given. Considering the core flow regions of the graphs, a positive radial angle χ_{2w} is obtained for all axial measurement locations, see also the homogeneous values in

Fig. 16. These results stay in contrast to the above referred comparable investigations, who attributed negative directed radial velocity distributions in the core flow to the radial pressure gradient. Zaccaria and Lakshminarayana [1995] investigated the nozzle guide vane exit flow field of a low speed turbine. They concluded that their also negative directed radial flow field over most of the passage flow was caused by the radial inward lean of the nozzle trailing edge. Richards and Johnson [1988] tested a stator with outward lean of the trailing edge in a subsonic single stage turbine. The rotor influence was limited through a large axial gap. As for all ohter experiments, strong migration of low momentum fluid occurred from tip to hub, especially in wake regions, while a radial outward directed velocity component was present in the core flow. The present stator geometries have also trailing edges with outward lean. Thus, the conclusion might be drawn that the radial trailing edge lean has an influence on the radial velocity in the core flow region, while the strong migration of wake fluid is caused by the radial pressure gradient. The observation of VOLSOL contour plots of radial flow angle at midspan (not presented due to limited space) indicate some additional differences in the prediction to the experimental data. The predicted shock regions are characterized by outward flow (hub to tip), whereas the wake flow is characterized by inward flow. Only immediately behind the trailing edge the predicted wake flow is towards the tip. Otherwise, the core flow is directed slightly towards the hub (flow angles up to 5 degrees). This contradicts the measured trends in these flow regions. Possible reasons for these discrepancies (for example numerical resolution in radial direction) have to be investigated in the continuation of the work.

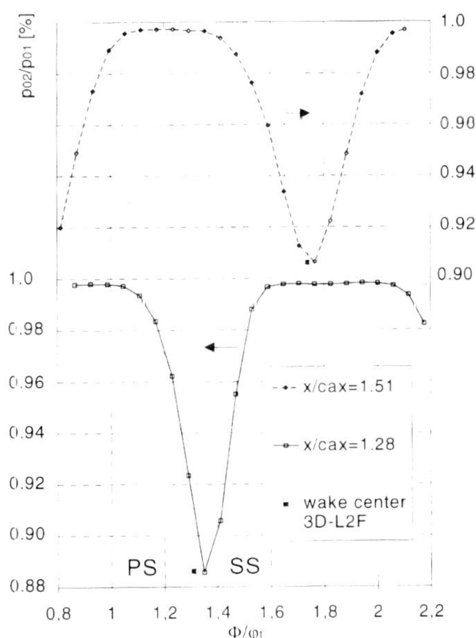

Fig. 14: p_{02w}/p_{01} *for stator 1 at midspan, measured with wedge type total pressure probe and wake center position, measured with 3D-L2F*

The total pressure loss distribution at midspan of stator 1 is shown in Fig. 14. The wake induced total pressure loss decays in axial direction from $x/c_{ax}=1.28$ to 1.51. A distinction between wake and shock losses can not be made. The wake and shock effects have possibly already merged at the first investigated axial plane (compare Fig. 8). Additionally, the shock is oblique and fairly weak, contributing not much to the losses. The wake centers measured with the 3D-L2F were copied from Fig. 8. The data taken with the probe and the laser anemometer almost coincidence. The slight deviations might result from a limited spatial resolution of the measurement points of the probe and 3D-L2F and from flow distortions created by the presence of the probe head.

Figure 15 compares the predicted wake flows of stator 2 with the traverse measurements at 5 axial midspan positions in terms of effective Mach number. The numerical results show some differences to the measured data. Neither of the numerical methods reflect the diminution of the wake

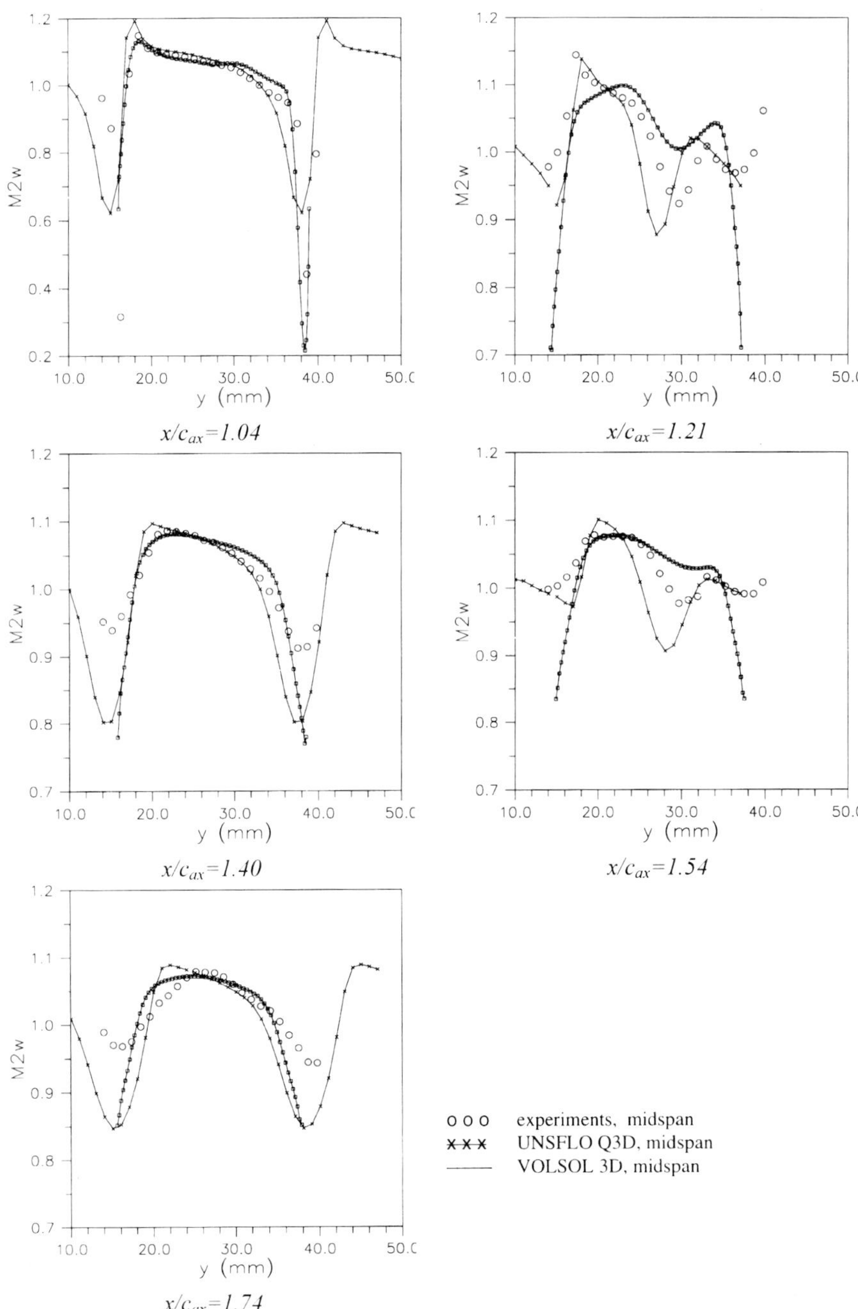

Fig. 15: *Comparisons M₂ᵥᵥ, numerical and experimental results at midspan, stator 2*

deficit with axial distance. UNSFLO gives a too small wake deficit shortly behind the vane (x/c_{ax}=1.04), which then does not decay in axial direction as fast as the measured wake deficit. VOLSOL predicts the wake deficit at x/c_{ax}=1.04 too high, but ends up with a similar wake decay as UNSFLO at x/c_{ax}=1.74. This difference in diffusion in flow direction might be due to the inviscid approach of UNSFLO outside the boundary layer region of the vane. The earlier stated observation of a higher effective outlet Mach number level at the midspan outlet in the VOLSOL prediction is not obvious in Fig. 15. However, the VOLSOL trailing edge shock is stronger and that dominates the Mach number distribution behind the stator: The strong Mach number drop predicted with VOLSOL is due to the trailing edge shock and the wake. In the plots, where the wake and the shock are at approximately the same positions UNSFLO and VOLSOL show comparable predictions. In the regions where the wake is circumferentially in between the trailing edge shocks (x/c_{ax}=1.21 and 1.54) deviations between the predictions can be seen. These indicate that the wake is predicted weaker with VOLSOL than with UNSFLO. Some possible reasons for the deviation could be the differences in mesh resolution as well as turbulence model. This has already been discussed in the context of the vane surface pressure analysis. From the discussion of the measured and predicted radial flow angle with VOLSOL it is obvious that the flow is three dimensional in the wake at midspan and the predicted trends do not coincide with the measured ones. This could also contribute to the discussed deviations of the VOLSOL prediction seen in Fig. 15. The observed wake flow predictions will be helpful to analyze the unsteady stage calculations planned to be done with these numerical methods on this configuration.

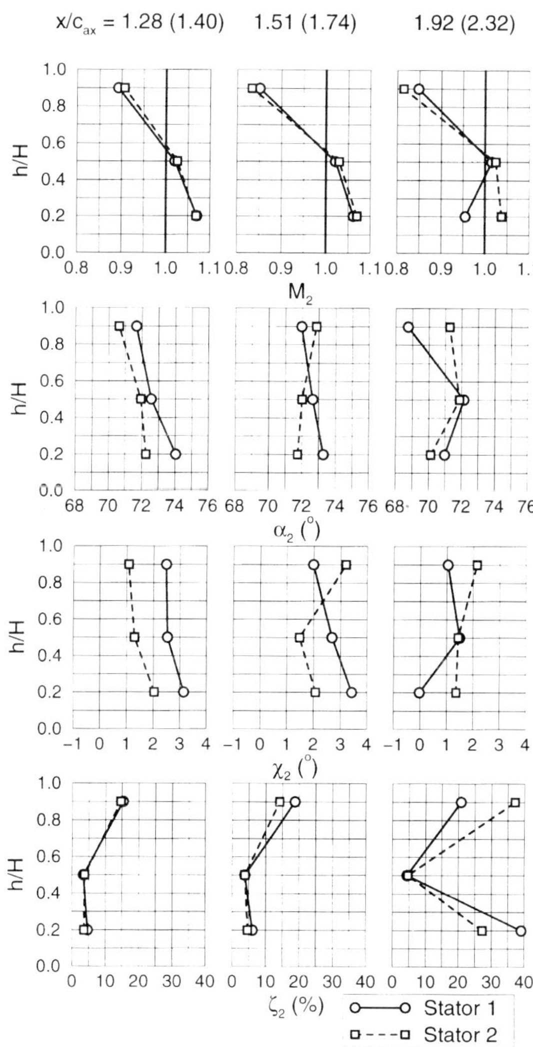

Fig. 16: *Homogeneous values of M_2, α_2, χ_2 (3D-L2F) and ζ_2 (wedge type total pressure probe) for stator 1 and 2, shown for increasing span (h/H)*

<u>Homogeneous Values</u>
Homogeneous (mixed-out) values from wake measurements are presented in Fig. 16. These values were calculated at three axial planes by using total pressures from probe measurements, flow angles and velocities from 3D-L2F measurements and taking into account the continuity laws of mass, momentum and energy between the inhomogeneous and the homogeneous flow field. The procedure was derived from a two-dimensional method, presented by Amecke and Safarik [1995], and is only applicable on thin annular stream surfaces at constant radii. A mixing in radial direction has not been taken into consideration. The homogeneous values of M_2, α_2, χ_2 and the kinetic energy loss coefficient ζ_2 are presented as functions of axial position x/c_{ax} and spanwise position. ζ_2 is based on the kinetic energy in the real flow to the maximum kinetic energy in the corresponding isentropic flow field. Due to the increasing throat width from hub to tip the Mach number decreases from the supersonic to the subsonic range. The flow is always directed towards the tip, indicated by a positive χ, while deviations from a mean value of 2-3 degree occur close to hub and tip, possibly caused by secondary flow effects. The losses show a minimum at midspan. Near the hub and tip wall the losses are higher due to interactions of the end wall boundary layers with the wakes and shocks from the stator row and due to secondary flow effects at tip. These effects are reinforced with increasing axial distance. A comparison between stator 1 and 2 indicates a relative good agreement at midspan, whereas some larger differences occur near hub and tip.

CONCLUSIONS

Two isolated stator rows with different vane number, but same design Mach numbers and flow angles, have been tested under annular transonic flow conditions. The midspan results show a strong wake induced velocity deficit decaying fast in downstream direction. A shock could be identified originating at the SS trailing edge, causing a second velocity deficit that exists separated from the wake velocity deficit until about $x/c_{ax}=1.3$ for stator 1 and $x/c_{ax}=1.5$ for stator 2. Further downstream both deficits merge and almost sinusoidal velocity distributions are obtained in pitchwise direction.

Both stators show similar aerodynamic behavior, just close to the trailing edge the wake velocity deficit of stator 2, with the higher vane number, is stronger. The stator 1 wake decays faster than the stator 2 wake. Compared with subsonic cases from the open literature the obtained transonic wake velocity deficits are stronger and the decay rates are slower.

Strong radial wake flow components were obtained for both stators. Close to the trailing edge the radial static pressure gradient drives the low momentum fluid towards the hub, corresponding to cases from the open literature. Further downstream the wake flow is directed towards the tip, what can presently not be explained sufficiently.

Comparisons of the experimental data with numerical results on stator 2 show fairly good agreement of the vane surface pressures but some deviations in wake flow effective Mach numbers. The codes can not correctly predict the measured decrease of wake deficit with axial distance behind the stator. Also the interaction with the trailing edge shock is captured differently both in position and in strength. These differences can be due to the turbulence models as well as the grid resolution. Also the prediction of the three-dimensional flow with VOLSOL shows some significant differences to the experiments in the radial flow. Even though 3D effects at midspan can be judged to be small the differences due to the 3D prediction can be a reason for the differences in the trailing edge shock strength. It is assumed that the differences in the stator outlet flow calculations will have an impact on the results of stator-rotor interaction predictions. These aspects will be investigated in the continuation of the project.

ACKNOWLEDGEMENT

This work was initiated and supported by the European Community, Brite/Euram project "Aeromechanical Design of Turbine Blades" (ADTurB), contract number BEPR-CT95-0124. The 3D-L2F system used in this investigation was acquired with funds provided by "Johnsonstiftelsen" and "Knut och Alice Wallenbergs Stiftelse", both in Sweden.
The authors wishes to thank for the support by the Industrial Aerodynamics Section at DLR, especially Mr. Schüpferling and Mr. Tappe for running the test facility and Dr. Kost for many helpful discussions about the measurement techniques.
Thanks go also to Mr. Christian Trapp, who performed the VOLSOL calculations during his M.Sc. thesis work.

REFERENCES

Amecke, J.; Kost, F.; 1993
"Rotating Annular Cascades.", AGARD-AG-328 on "Advanced Methods for Cascade Testing", 1993

Amecke, J.; Safarik, P.; 1995
"Data Reduction of Wake Flow Measurements with Injection of an Other Gas.", Institute Report DLR-FB 95-32, German Aerospace Center, Göttingen, Germany, 1995

Bailey, R. H.; 1993
"UNSFLO Version 6.0 - User's Manual", Rolls Royce plc , Aerothermal Methods Group, Report No. TSG0717 Baldwin, B. S., 1993

Lomax, H.; 1978
"Thin Layer Approximation and Algebraic Model for Separated Turbulent Flow.", AIAA Paper 78-257

Beam, R. M.; Warming, R.F., 1978
"An Implicit Factored Scheme for the Compressible Navier Stokes Equations.", AIAA Journal, 16, pp. 393-403, 1978

Binder, A.; Romey, R.; 1982
"Secondary Flow Effects and Mixing of the Wake Behind a Turbine Stator.", ASME Paper 82-GT-46

Colantuoni, S.; Colella, A.; Santoriello, S.; Kapteijn, C.; 1995
"Aerodesign and Performance Analysis of a Transonic Turbine Inlet Guide Vane with Trailing Edge Coolant Flow Ejection.", VDI Berichte Nr. 1185, 1995

Eriksson, L. E.; 1995
"Development and Validation of Highly Modular Flow Solver Versions in G2DFLOW and G3DFLOW Series for Compressible Viscous Reacting Flow.", Technical Report 9970-1162 Volvo Aero Corporation, Sweden, 1995

Freudenreich, K.; 1998
"Velocity and Turbulence Intensity Measurements in Transonic Turbomachinery Flow using 3D-L2F.", Institute Report KTH/HPT 98/16, 1998

Giles, M.; 1991
"UNSFLO: A Numerical Method for the Calculation of Unsteady Flow in Turbomachinery." GTL Report 205, 1991

Groth, J. P.; Mårtensson, H.; Eriksson, L. E.; 1996
"Validation of a 4D Finite Volume Method for Blade Flutter.", ASME Paper No. 96-GT-429

Izsak, M. S.; Chiang, H. W. D.; 1993
"Turbine and Compressor Wake Modeling for Blade Forced Response.", ASME Paper 93-GT-236

Kapteijn, C.; 1995
"Wake Development Downstream of a Transonic Turbine Inlet Guide Vane with Trailing Edge Ejection.", AGARD CP-571 on "Loss Mechanisms and Unsteady Flow in Turbomachines.", 1995

Kielb, R. E.; Chiang, H. D.; 1992
"Recent Advancements in Turbomachinery Forced Response Analyses.", AIAA Paper 92-0012

Korakianitis, T. P.; 1992
"On the Prediction of Unsteady Forces on Gas Turbine Blades: Part 1 and 2.", J. Turbomachinery, Vol. 114, pp. 114-131, 1992

Maass, M.; Förster, W.; Thiele, P.; 1994
"Unsteady Flow Experiments in the Exit of a Ducted Propfan Rotor.", AIAA Paper No. 94-2970

Rehder, H.-J.; 1997
"Stator 1 only Tests at DLR: Blade Surface Pressure Distribution and Probe Measurements.", ADTurB Report ADTB-DLR-4006, 1997

Richards, P. H.; Johnson, C. G.; 1988
"Development of Secondary Flows in the Stator of a Model Turbine.", Experiments in Fluids 6, pp. 2-10, June 1988

Sieverding, C. H.; van Hove, W.; Boletis, E.; 1983
"Experimental Study of the Three-Dimensional Flow Field in an Annular Turbine Nozzle Guide Vane.", ASME Paper 83-GT-120

Zaccaria, M.; Lakshminarayana, B.; 1995
"Investigation of Three-Dimensional Flowfield at the Exit of a Turbine Nozzle.", J. Propulsion and Power, Vol. 11, pp. 55-63, 1995

Numerical simulation of three-dimensional inlet guide vanes

V MICHELASSI and **E BELARDINI**
Energetics Department, University of Florence, Italy

1 - ABSTRACT

The fluid dynamics of three dimensional inlet guide vanes is studied by numerical simulation. Two inlet guide vanes are simulated by using an implicit Navier-Stokes solver which accounts for the effects of turbulence on the mean flow field by introducing two extra transport equations for turbulence quantities. For the HP IGV the computed solution is compared positively with experiments in terms of pitchwise averaged quantities downstream of the trailing edge. In the second, and LP, IGV, the solver needs to be modified to be able to cope with the very low Mach number by introducing a preconditioning matrix. The computed results are compared with a set of experimental data including the pressure distribution around the blade at different spanwise positions which indicate the large impact of secondary flows.

Nomenclature

D^2, D^4	second and fourth order artificial damping
F	convective flux vector
F^v	viscous flux vector
J	coordinate transformation Jacobian
Q	vector of unknowns (conservative variables)
Q_v	vector of unknowns (primitive variables)
t	time
T	static temperature
T, P, N	convective eigenvector matrices
U	contravariant velocity vector
x	Cartesian coordinate

Greek symbols

β, β'	preconditioning factors
β	turbulence model coefficient
γ	specific heat ratio (=1,4)

δ	centred difference operator
ξ, η, ζ	curvilinear coordinates
λ	spectral radius
Λ	convective eigenvalue vector
Λ^v	diffusive eigenvalue vector
σ	spectral radius scaling
ω^2, ω^4	second and fourth order artificial damping weight
Ω	artificial damping scaling factor

2 - INTRODUCTION

Gas turbine inlet guide vanes (IGV) are normally designed on the basis of correlations, experimental testing and with a continuos increasing help of computer simulations. The flow in modern gas turbines can be at very low speeds or transonic, the latter involving complex phenomena such as shocks interacting with the boundary layers, and shock induced separation and transition. For the design of such turbines, and IGV in particular, reliable calculation methods are needed. The computational tools must include models for simulating realistically the turbulence and can cope with any value of the Mach number while still maintaining an acceptable degree of accuracy, which should be verified by testing against experimental data. The present paper reports on the effort made to compute the flow in three-dimensional inlet guide vanes. In most calculations accounting for viscous effects, algebraic turbulence models have been used which do not alter the robustness and speed of the Navier-Stokes solvers. However these models often give poor accuracy (Lakshminarayana(1), Mayle (2)). At the other end of the spectrum, second moment closures simulate the complex turbulence phenomena more realistically and promise more accuracy, but they require large computational effort and their superiority could not always be confirmed (Rodi et al.(3), Michelassi et al. (4)). In the present work, the two-equation approach is adopted as a compromise between accuracy and computational efficiency. Two-equation models are generally able to yield good accuracy in terms of pressure distribution and head losses, which are of primary importance for the evaluation of the blade load and efficiency.

The model is applied to the computation of two IGV the first of which is transonic and the second low subsonic. This choice is motivated by the need to evaluate the performances of a solver in a relatively wide operational range and compare, when possible, with experiments.

3 - DESCRIPTION OF THE SOLVER

In the implicit time marching code the transported, density, momentum, and total specific energy, and related quantities are made non dimensional with respect to the inlet total pressure P_0 and inlet total temperature T_0. Viscosities and diffusion coefficients are made non dimensional with respect to the inlet laminar viscosity. The equations can be written in a compact vector form as:

$$\frac{\partial Q}{\partial t} + \frac{\partial F_k}{\partial x_k} = \frac{\partial F_k^v}{\partial x_k} \quad , \qquad k = 1, 2, 3 \tag{1}$$

The equations are discretized by centred finite volumes in curvilinear non orthogonal co-ordinate systems. The FLOS3D(5) program is based on the scalar implicit algorithm proposed by Pulliam and Chaussée (6). The program solves three-dimensional compressible flows with complex boundaries in inviscid and turbulent regime. The implicit solver needs the solution of scalar tridiagonal matrices because of the centred space discretization. In order to enforce stability artificial damping terms are introduced as proposed by Jameson et al. (7) in the manner suggested by Michelassi (8) on both the implicit and the explicit sides of the operator. Equation (1) is then solved by an ADI step as follows:

$$
\begin{cases}
T_\xi \cdot \left\{ I + \theta\Delta t \left[\delta_\xi \Lambda_\xi - \left(\delta_\xi^2 \Lambda_\xi^v + \delta_\xi \left(\Omega^\xi \omega_\xi^2 \delta_\xi (J) \right) \right) + \delta_\xi \left(\Omega^\xi \omega_\xi^4 \delta_\xi^3 (J) \right) \right] \right\} \cdot \Delta Q^* = \Delta t \cdot RHS \\[6pt]
N \cdot \left\{ I + \theta\Delta t \left[\delta_\eta \Lambda_\eta - \left(\delta_\eta^2 \Lambda_\eta^v + \delta_\eta \left(\Omega^\eta \omega_\eta^2 \delta_\eta (J) \right) \right) + \delta_\eta \left(\Omega^\eta \omega_\eta^4 \delta_\eta^3 (J) \right) \right] \right\} \cdot \Delta Q^{**} = \Delta Q^* \qquad (2) \\[6pt]
P \cdot \left\{ I + \theta\Delta t \left[\delta_\zeta \Lambda_\zeta - \left(\delta_\xi^2 \Lambda_\zeta^v + \delta_\zeta \left(\Omega^\zeta \omega_\xi^2 \delta_\zeta (J) \right) \right) + \delta_\zeta \left(\Omega^\zeta \omega_\zeta^4 \delta_\zeta^3 (J) \right) \right] \right\} \cdot T_\zeta^{-1} \cdot \Delta Q = \Delta Q^{**}
\end{cases}
$$

in which

$$
RHS = \Delta t \left(-\frac{\partial F_k}{\partial x_k} + \frac{\partial F_k^v}{\partial x_k} + D_k^2 - D_k^4 \right)^n
$$

$\Delta Q = Q^{n+1} - Q^n$, Q is the solution vector, the matrices $N = T_\xi^{-1} T_\eta$ and $P = T_\eta^{-1} T_\zeta$ are solution independent and θ allows weighting of the explicit-implicit nature of the space operator in round brackets. The terms D_k^2 and D_k^4 are second and fourth order differences of the transported quantities given by vector Q computed and scaled according to the directional weights Ω and ω as indicated in Michelassi and Martelli (9).

3.1 - Low Mach number preconditioning scheme

When the flow reference Mach number (i.e. in the exit plane) is low, for instance below 0.3, the scheme fails to give converged and accurate solutions. This is due to the large differences in the eigenvalues used in equation (2). Aim of the preconditioning scheme is to equilibrate the eigenvalues of the jacobian matrices. According to Pulliam and Jespersen (10) the preconditioning scheme for the convective terms in equation (1) can be obtained by a linearization of the operators and passing from conservative Q to primitive variable Q_v through matrix M defined as:

$$
M = \frac{\partial Q}{\partial Q_v}
$$

The convective part of equation (1) is modified introducing the convective jacobians A_k in terms of Q and subsequently in terms of $A_{v,k}$ observing that $A_{v,k} = A_k \cdot M$:

$$
M \cdot \frac{\partial Q_v}{\partial t} + A_{v,k} \frac{\partial Q_v}{\partial x_k} = 0 \qquad (3)
$$

The preconditioning step consists of replacing M by another matrix Γ which multiplies each term of equation (3). Γ is further arranged by transforming back to conservative variables:

$$
\frac{\partial Q}{\partial t} + M \cdot \Gamma^{-1} \cdot \frac{\partial F_k}{\partial x_k} = 0
$$

This equation, with respect to the original equation (1) includes only the extra matrix product $M \cdot \Gamma^{-1}$ called "preconditioning operator" and the original algorithm is substantially unaltered with the only exception of the preconditioning matrix which requires the computation of a new set of eigenvalues and eigenvectors. While Λ, T in the un-preconditioned version are the eigenvalues and eigenvectors of A, it is necessary to perform the same analysis on $M \cdot \Gamma^{-1} \cdot A$. The preconditioned algorithm for the convective operator then reads:

$$M\widetilde{T}_\xi\left(I + \theta\Delta t\delta_\xi\Lambda_{\overline{A}}\right)\widetilde{N}\left(I + \theta\Delta t\delta\eta\Lambda_{\overline{B}}\right)\widetilde{P}\left(I + \theta\Delta t\delta_\xi\Lambda_{\overline{C}}\right)\widetilde{T}_\zeta^{-1}M^{-1}\Delta Q^n = -\Delta t M\Gamma^{-1}\left(\frac{\partial F_k}{\partial x_k}\right)^n$$

The term $\Lambda_{\overline{A},\overline{B},\overline{C}}$ is the eigenvalue vector of the $M\cdot\Gamma^{-1}\cdot A$ matrix, and the eigenvector matrices with a "tilda" are computed for $\Gamma^{-1}\cdot A_v$. Now, while the M matrix can be easily computed (see ref. (9)), the Γ matrix needs to be defined in order to provide the necessary eigenvalue normalization. Following Pulliam and Jespersen (10) Γ reads:

$$\Gamma = \begin{pmatrix} \dfrac{1}{(\beta'\gamma T)} & 0 & 0 & 0 & -\dfrac{\rho}{T} \\[1.5ex] \dfrac{u}{(\beta'\gamma T)} & \rho & 0 & 0 & -\dfrac{\rho u}{T} \\[1.5ex] \dfrac{v}{(\beta'\gamma T)} & 0 & \rho & 0 & -\dfrac{\rho v}{T} \\[1.5ex] \dfrac{w}{(\beta'\gamma T)} & 0 & 0 & \rho & -\dfrac{\rho w}{T} \\[1.5ex] \left(\dfrac{H}{(\beta'\gamma T)} - 1\right) & \rho u & \rho v & \rho w & \rho\left(\dfrac{\gamma}{\gamma-1} - \dfrac{H}{T}\right) \end{pmatrix}$$

in which $\beta' = \dfrac{\beta}{[1 + (\gamma - 1)\beta]}$. In the present approach we found convenient to keep β constant over the entire computational domain as $\beta = \alpha \cdot \max(M_{max}^2, \beta_{min})$. α is a user defined value which needs to be optimized as a function of β, M_{max} is the maximum Mach number, and β_{min} the minimum allowed value (to ensure that computed Mach number is below 1.).

3.2 - Turbulence modelling

The two equation model by Wilcox(11) is based on the introduction of a transport equation for the specific dissipation rate $\omega = \dfrac{\varepsilon}{\beta^* \cdot k}$ and for the turbulent kinetic energy k. The k-ω model is implemented here in its standard version with the introduction of an extra constraint on the turbulence frequency. The constraint is introduced to overcome the problem of the overproduction of turbulence in stagnation points, or in flow regions with adverse pressure gradient. Of the various possible ways to cure this known deficiency of two-equation models it was decided here to adopt what proposed by Durbin (12) who formulates a realizability constraint on the turbulent kinetic energy which results in a limitation on the possible turbulence time scales T=k/ε. This constraint, which is strictly valid only for incompressible flows and is therefore applied only in proximity to solid boundaries, reads:

$$T = \frac{k}{\varepsilon} = \min\left(\frac{k}{\varepsilon}, \frac{2k}{3v'^2 C_\mu} \cdot \sqrt{\frac{3}{8 \cdot |S^2|}}\right) \tag{4}$$

in which S is the incompressible mean strain. The term $3 \cdot \overline{v'^2}$ (v' is the velocity fluctuation orthogonal to walls) can be substituted by $2 \cdot k$ assuming isotropic turbulence. Now, recalling the definition of the specific dissipation rate ω, equation (4) can be reformulated as follows:

$$\omega = \frac{1}{T} = \frac{\varepsilon}{\beta^* k} = \max\left(\omega, \left(\frac{1}{C_\mu} \cdot \sqrt{\frac{3}{8 \cdot |S^2|}}\right)^{-1}\right) \tag{5}$$

Equation (5) represents a sort of clipping for the specific dissipation rate which cannot reach values below a certain limit, and varies depending on the local mean strain. The implementation of equation (5) in the k-ω model helped solving a large portion of the numerical problems.

4 - APPLICATIONS

Two different three-dimensional IGV have been computed under variable operating conditions. The computations have been performed using grids with a maximum number of 1.5×10^6 nodes which ensures a satisfactory discretization of the boundary layer. The second and fourth order damping weights are set equal to ½ - 1/64, and 0.0 - 1/64 for the DLR stator and for the Nuovo Pignone low subsonic case respectively. The scaling of the artificial damping in the boundary layers allowed the accuracy of the discretization close to the wall to be preserved (Michelassi and Martelli, 1998). These values have been selected on the basis of previous experience in the code validation phase.

4-1 The DLR Stator

This geometry was selected in the frame of the AGARD working group on CFD validation (13). The isentropic exit Mach number at the hub of this annular IGV is approximately 0.84. and the blade has a variable shape with height. The inlet turbulence level is approximately 1% and the blade boundary layers are assumed fully turbulent. The experimental data include the spanwise distribution, averaged in the pitchwise direction, of static and total pressure, flow angle and Mach number at a section x/Chord=1.4. The particularly interesting data allow the flow distortion induced by the stator to be evaluated. Of the various computational only those obtained on the refined grid 169×113×61 (see figure 1) results will be reported which does not show significant changes with respect to other coarser grids. Figures 2a and 2b shows the flow visualisation of the computed flow on the hub and shroud walls respectively. The plot allows the saddle point ahead of the leading edge to be clearly identified. The position of the saddle point, which indicates the formation of the horseshoe vortex, is nearly identical on the hub and on the shroud, which proves that the flow enters the IGV without any appreciable distortion.

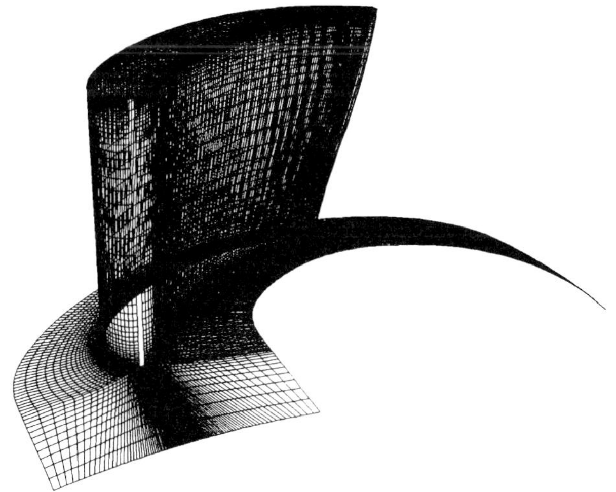

Figure 1. DLR-3D IGV grid

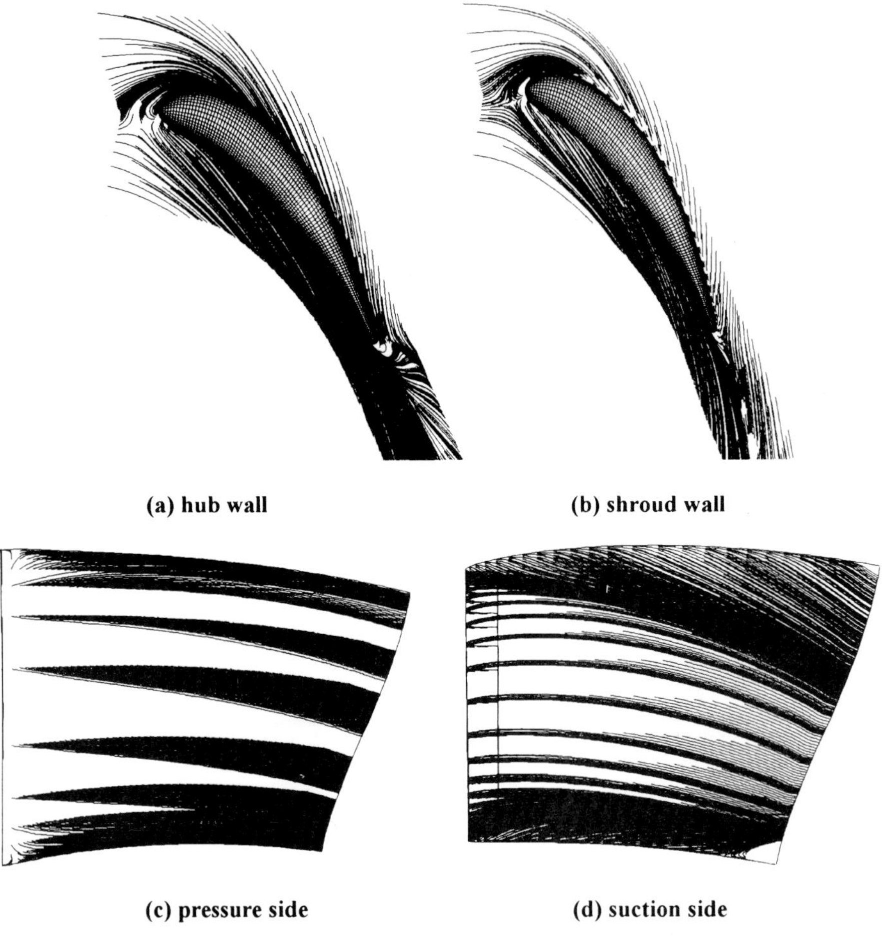

Figure 2. Flow visualisation (flow from left to right)

The strong down-wash visible in figures 2a and 2b indicates the presence of the passage vortex. While the flow structure is similar in the first 70-80% of the blade passage, some differences arise while approaching the trailing edge where on the hub wall the flow tends to shift on the suction side. The flow is more regular on the shroud wall where, downstream of the trailing edge only the passage vortex effect becomes evident. This difference between the hub and shroud wall flow pattern can be explained by looking at the flow pattern on the pressure and suction sides (figure 2c and 2d respectively). In fact, while on the pressure side the flow stays attached all the way through the trailing edge, a small separation bubble appears on the suction side hub wall in proximity to the trailing edge. Apparently, the flow distortion visible in figure 2a is caused by this small flow separation which remains a local phenomena since the flow soon reattaches.

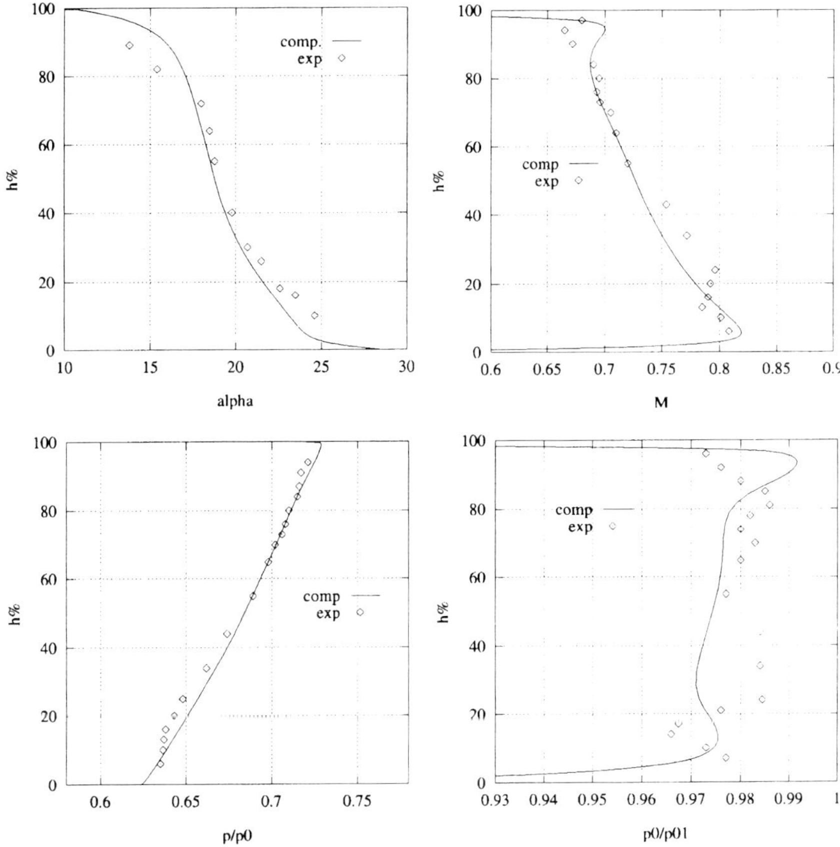

Figure 3. Comparison between measured and computed pitchwise averaged exit quantities.

Figure 3 compares computations and experiments in the section 1.4 chords downstream of the blade leading edge. Figure 3 shows that the exit flow angle is slightly well reproduced although the flow distortion in proximity to the side walls is somewhat underpredicted. This is probably due to the uncertainties in the inlet boundary layer thickness and state which cannot be accurately specified in the computation. The local Mach number is in good agreement with experiments, although the computed spikes close to the hub and shroud are not visible in the measurements which also show an up-down shape in the same region which is not captured by the computations. This up-down shape of the measured profile is likely to be caused by the secondary flows the development of which is largely controlled by the side walls inlet boundary layers. The unsatisfactory agreement of the predictions with the measurements in proximity to the side walls was common to most the predictions in the review by Dunham (13) and was possibly caused by uncertainties in the inlet flow specification. The computed static pressure distribution is in excellent agreement with the measurements, although the accuracy deteriorates in terms of total pressure profile for the same reasons indicated for the local Mach number.

4-2 The Nuovo Pignone low-subsonic IGV.

This test case was selected because of the low-subsonic nature of the flow and for the large impact of secondary flows on the blade load at different span heights. The data have been obtained during the measurements reported by Michelassi et. al (14) and Contini et al. (15).

Figure 4 shows a typical coarse grid of the three-dimensional linear cascade, together with a sketch of the axial distribution of the span height. The span height increases of approximately a factor two while moving from the inlet to the outlet of the IGV. The isentropic exit Mach number measured in reference 14 ranges from 0.11 to approximately 0.3. Since all the data are within the incompressible flow domain, it was decided to compute here only the M_{is}=0.11 case, with an inlet turbulence level of 1% and the boundary layers assumed fully turbulent. The grids adopted for the computations range from 60×17×11 for the coarse inviscid case to the 170×97×91 for the viscous refined case.

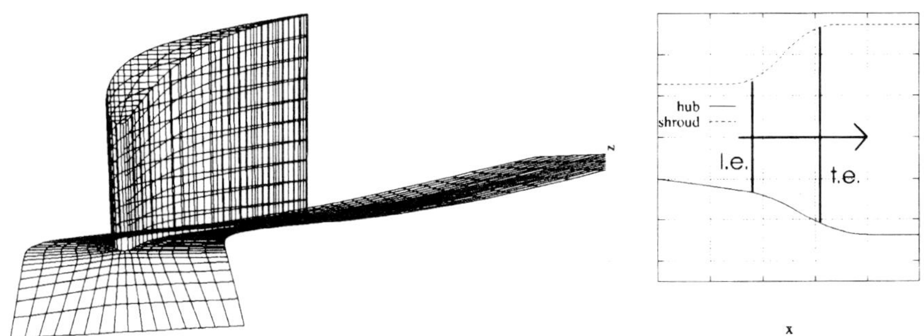

Figure 4. Coarse grid and axial distribution of the span height for the Nuovo Pignone IGV.

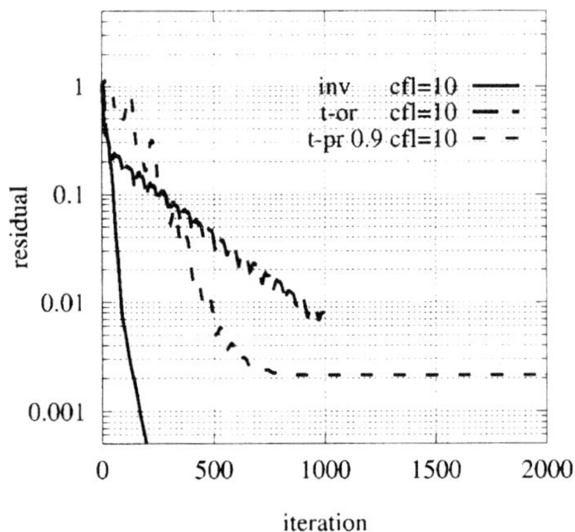

Figure 5. Convergence histories

The very low Mach number prevents from reaching a converged and accurate solution when using a solver based on conservative variables. To overcome this problem, the preconditioned version of the code ensured fast and accurate results down to exit Mach numbers of the order of 0.01. Figure 5 shows the convergence histories of the inviscid preconditioned version of the code compared with the original and preconditioned code for the turbulent flow regime. In the inviscid case the convergence is very fast, provided that the preconditioning with β=9-10 is used, otherwise the code is not able to converge. In the turbulent case, the preconditioned algorithm allows gaining approximately a factor 2.5 on the convergence rate with no significant computational overhead. Even though the code converges in the turbulent case without preconditioning, the results look bad, as proved by figure 6 which compares the measured isentropic Mach number distribution around the blade (exp) with the original (original) and preconditioned (precond) fully converged computed results. The plot shows the presence of large pressure fluctuations, accompanied by excessively high mass unbalances for the original computation. This phenomenon is present all along the three different blade heights (10%, 50%, 90%).

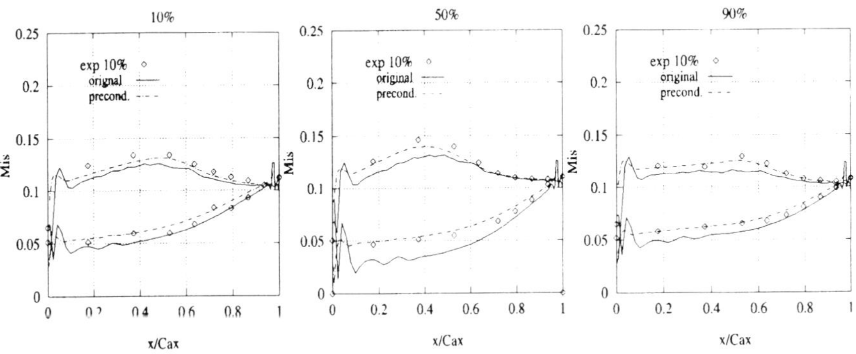

Figure 6. Effect of preconditioning on the blade isentropic Mach number.

Figure 7 shows the flow visualisation on the hub and shroud wall together with the suction and pressure sides of the IGV. The very low Mach number causes the large flow diffusion and thickening of the boundary layer upstream of the leading edge. The flow diffusion is driven by the spanwise expansion of the channel, sketched in figure 4. The large boundary layer thickness ahead of the leading edge produces a strong horseshoe vortex the size of which is proportional to the distance of the saddle point from the leading edge of the blade in figures 7a and 7b. The hub and shroud patterns are not identical on account of the asymmetry in the spanwise expansion of the channel, but still exhibits similar trends on both the suction and pressure sides. In fact the suction side branch of the horseshoe vortex interacts with the passage vortex producing the coalescence of streak lines indicated by the arrows in both figures 7a and 7b. This phenomenon was not present in the DLR IGV because of the reduced effect of the horseshoe vortex. On the pressure and suction sides, the flow visualisation show that the passage vortex is also quite strong and seems to affect the flow over the entire span height. This effect will also explain the isentropic Mach number distribution observed at various blade heights towards the trailing edge. The large flow distortion on the suction side is also due to the highly off-design conditions at which this blade, designed for transonic flow, was operated in the experiments.

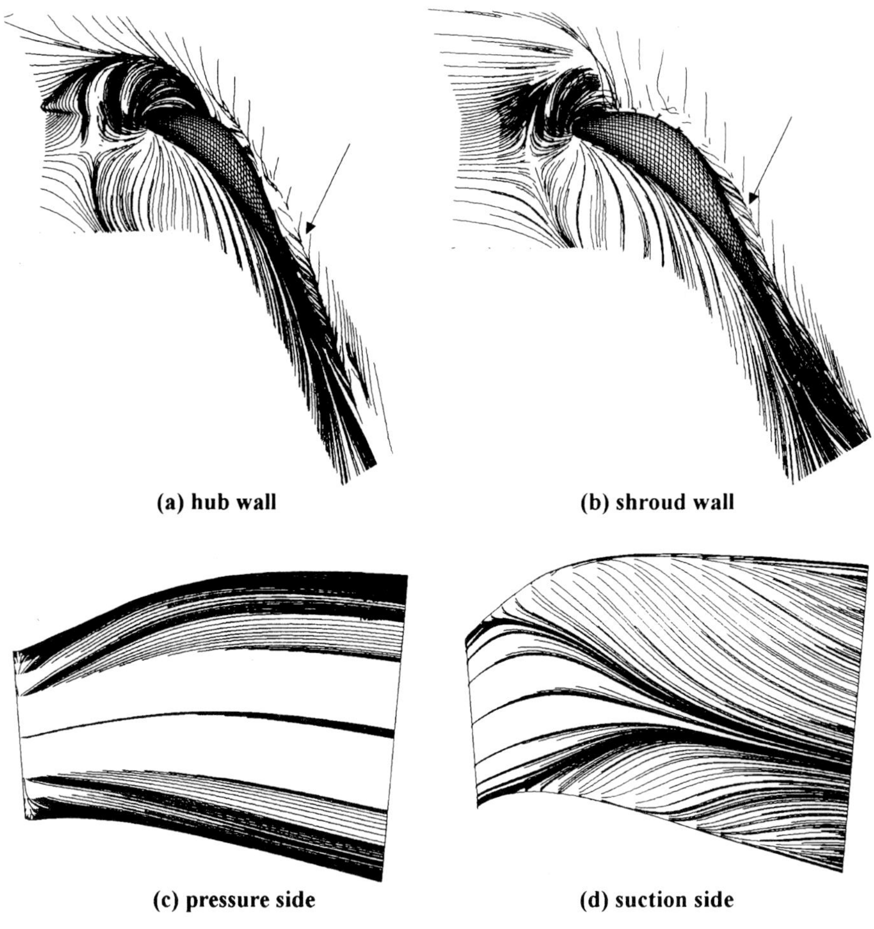

(a) hub wall **(b) shroud wall**

(c) pressure side **(d) suction side**

Figure 7. Flow visualisation (flow from left to right).

Figure 8 compares the measured (exp) isentropic Mach number around the blade for the exit Mach number of 0.11 at three different blade heights (10%, 50%, 90%) with the inviscid (inv) and turbulent calculations with the refined grid (ref). When varying the exit Mach number in the range 0.11-0.30 no substantial differences appeared on account of the absence of compressibility effects. For the highest exit Mach number only, M_{is} reaches 0.33 for x/C_{ax}=0.35. The profiles show a strong increase of M_{is} on the suction side in the first 40% of the blade, regardless of the exit Mach number. Downstream of the throat the flow decelerates and then slightly accelerates again even though the flow is subsonic and the blade height is still increasing. The differences in the computed and measured profiles at the three blade heights are not large and localised mostly near the leading edge; this effect is amplified on the suction side. On the pressure side at 50% span the admissible aerodynamic load is larger than at 10% and 90%, as can be expected because of interaction with the sidewall boundary layers (diffusing meridional geometry). The differences between the blade loads at the three blade heights tend to decrease towards the trailing edge. This is probably due to the combined effect of the flow acceleration and the increase of blade height. In fact, the discrepancies between the M_{is} profiles at the various blade heights seem to indicate the presence of the horseshoe vortex near the leading edge at 10% and 90% of the blade height. The combined effects of the horseshoe vortex dissipation and of the growth of the passage vortex might explain the coincidence of the M_{is} profiles for $x/C_{ax} > 0.6$. At 50% span the agreement between computations and experiments is good regardless of the inviscid-viscous assumption. Conversely the inviscid assumption does not allow to capture the development of the horseshoe vortex, the effect of which is visible in figures 6 at 10 and 90% span. The size and strength of the computed horseshoe vortex in proximity to the hub wall is given by the particle traces plot in figure 7a and 7b. The considerable strength of the horseshoe vortex can be inferred by the position of the saddle point, typical of the formation of this vortex, considerably upstream of the leading edge on the pressure side.

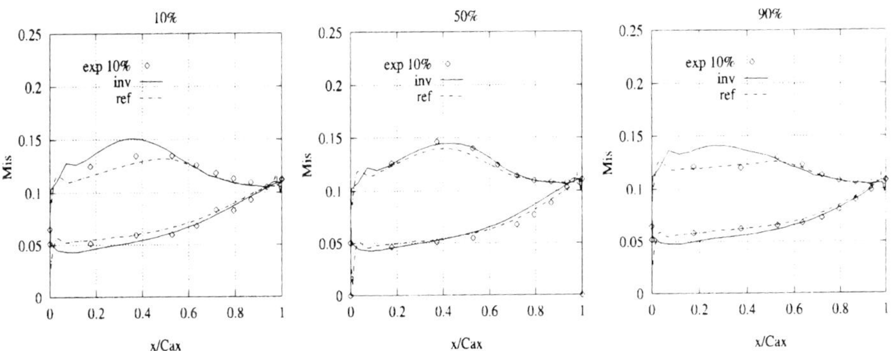

Figure 8. Inviscid and viscous blade isentropic Mach number.

5 - CONCLUSIONS

The fluid flow in three-dimensional inlet guide vanes has been analysed by means of a Navier-Stokes equations solver. The algorithm with the turbulence model and preconditioning scheme allows coping with a wide range of flows, including transonic and low subsonic conditions. The simple preconditioning scheme proved able to solve the large conservation problems encountered when solving very low Mach number flows and can be easily included in a factored scheme the structure of which is only marginally altered, thereby making this approach particularly amenable for a large number of computer codes. The two IGV selected for the investigation have a quite different behaviour because of the large difference in the exit Mach number. Nevertheless the code was able to capture the essential features of the flow field, including the large impact of the secondary flows on the blade load. The strength of the horseshoe vortex was found able to alter considerably the flow pattern in proximity to the hub and shroud walls. This effect is directly related to the boundary layer inlet state and thickness. In fact, the strong flow diffusion encountered in the low speed case upstream of the leading edge amplifies the horseshoe vortex the path of which can be traced on the suction side. The flow in the high speed case was more regular and the impact of the secondary flows reduced.

6 - ACKNOWLEDGEMENTS

The authors would like to gratefully acknowledge the discussions and help from Prof. Francesco Martelli. This work was partly supported by MURST and ASI Project grants.

7 - REFERENCES

(1) Lakshminarayana B., 1991, "An Assessment of Computational Fluid Dynamic Techniques in the Analysis and Design of Turbomachinery - The 1990 Freeman Scholar Lecture", ASME J. of Fluids Eng., Vol. 113, pp. 315-352.

(2) Mayle, R.E., 1991, "The Role of Laminar-Turbulent Transition in Gas Turbine Engines", ASME Journal of Turbomachinery, vol. 113, October, pp. 509-537.

(3) Rodi, W., Bonnin, J.-C., Buchal, T. (Ed.), 1995, Proceedings of ERCOFTAC Workshop on "Data Bases and Testing of Calculation Methods for Turbulent Flows", University of Karlsruhe, Karlsruhe, Germany, April 3-7.

(4) Michelassi, V., Rodi, W., Giess, P.-A., "Experimental and Numerical Investigation of Boundary-Layer and Wake Development in a Transonic Turbine Cascade". Aerospace Science and Technology Journal, 1998, no.3 191-204 (also IGTI Conference, 1997 Orlando, Florida).

(5) V. Michelassi, M. S. Liou, L. A. Povinelli "Implicit solution of three dimensional internal turbulent flow" NASA Technical Memorandum 103099 ICOMP-90-13. July 1990.

(6) Pulliam,T.H., Chaussee, D.S., 1981, "A Diagonal Form of an Implicit Approximate-Factorization Algorithm". Journal of Computational Physics, N. 39.

(7) Jameson, A., Schmidt, W., Turkel, E., 1981, "Numerical Solutions of the Euler Equations by Finite Volume Methods Using Runge-Kutta Time-Stepping Schemes", AIAA 81-1259.

(8) Michelassi, V., "Turbulence and Transition Modelling in Transonic Turbine Blades", Report No. 728, Institute for Hydromechanik, University of Karlsruhe, Germany, 1996.

(9) V. Michelassi, F. Martelli, (invited lecturer) "Numerical Simulation of Stator-Rotor Interaction in BRITE Turbine stage", Lecture Notes on "Blade Row Interference Effects in Axial Turbomachinery Stages", Von Karman Institute for Fluid Dynamics, 9-12 February 1998.

(10) Pulliam, T.H., Jespersen, D., Private Communication, July 18, 1996.

(11) Wilcox, D.C., 1988, "Reassessment of the Scale-Determining Equation for Advanced Turbulence Models", AIAA Journal, Vol. 26, No.11, pp. 1299-1310.

(12) Durbin, P.A., "On the k-ε Stagnation Point Anomaly", Int. J. Heat and Fluid Flow, 1996, 17, 89-90.

(13) J. Duhham, "CFD Validation for Propulsion System Components", AGARD AR-355.

(14) Michelassi, V., Martelli, F., Corradini, U., 1998, "Secondary Flow Decay Downstream of Turbine Inlet Guide Vane with End-Wall Contouring", ASME Paper 98-GT-95.

(15) Contini, D., Manfrida, G., Michelassi, V., 1998, "Secondary Flow Measurements in a Gas Turbine Cascade by 3-D Pneumatic Probes", The XIV Bi-Annual Symposium on Measuring Techniques in Transonic and Supersonic Flow in Cascades and Turbomachines", 2-4 September 1998, University of Limerick, Ireland.

An experimental investigation into the three-dimensional flow field and loss mechanisms in a two-stage axial turbine

V BREISIG, C LERNER, P PETERS, and **H PFOST**
Ruhr-University of Bochum, Germany
M DECKERS
Siemens Power Generation (KWU), Mülheim an der Ruhr, Germany

This paper presents an experimental study of the three dimensional flow field in a two-stage low speed axial flow turbine. The investigation is aimed at understanding the complex flow that occurs in such a multi-stage environment.

The air-driven turbine operates at low Mach number and exhibits the great advantage of being relatively large in dimensions. This enables very detailed five hole probe measurements to be taken behind each blade row. The turbine is equipped with prismatic (cylindrical) blading typical for reaction steam turbines.

First, a description of the test facility including instrumentation is given. Then, five hole probe measurements behind each blade row are presented and discussed with particular reference to the differences that were observed between the flow fields in the first and second stage respectively.

1 INTRODUCTION

The flow through turbomachines is of extreme complexity owing to the effects of viscosity, compressibility and unsteadiness caused by the relative motion of adjacent blade rows. But despite this, tremendous efforts have been made to improve their performance in recent years. This was achieved primarily by making use of advanced numerical methods as well as measurement techniques, which have led to considerable progress in the understanding of the flow field occuring in a real machine. Such an understanding forms the basis for the development of the so-called loss correlations which are still very widely used to predict the performance of a particular design.

Turbomachinery design is inherently associated with the identification of loss origins and an understanding of the mechanisms producing loss. An excellent overview of these various loss mechanisms was given by Denton (1). A number of more specific experimental investigations describe the influence of different parameters on the generation of loss such as flow incidence

(2), blade aspect ratio (3), inlet boundary layer profile (4) as well as the actual blade profile (5). Measurements like these are then used to develop and validate loss correlations. It must be realised however that the bulk of these studies were based on simplified models, e.g. a linear cascade blade model or a single stage turbine model, where many of the important features in a multi-stage turbine cannot be modelled adequately.

The flow through multi-stage turbines has been studied by several investigators in recent years (6), (7), (8). It was generally found that the flow field in the first stage differs greatly from that observed in the following stages. This is because the inlet flow to the embedded stages is characterised by strong non-uniformities and unsteady effects due to wake flows, secondary flows and leakage flows. Therefore, the loss sources and mechanisms producing loss differ significantly from the first stage to the following stages.

In the present investigation, a two-stage axial flow model turbine is used. The turbine is relatively large in size. Hence, a high degree of resolution from area traverse measurements can be obtained ensuring that the flow field can be resolved in great detail. The measurements were carried out with five-hole pneumatic probes. In doing so, the time-averaged flow field in the absolute frame of reference is sampled, i.e. the effects of unsteady flows was not investigated. The primary aim of the investigation is to gain a detailed physical insight into the three-dimensional flow through multi-stage turbines.

First, a description of the test facility including instrumentation is given. Then, the results obtained from design-point measurements are presented and discussed with particular reference to a) the differences between the flow fields in the first stage and the following second stage and b) the effects of the shroud leakage flow on the main flow. Finally the influence of stagger angle variation of -3° from pitch is examined.

2 TEST BED AND INSTRUMENTATION

The experimental studies were performed on an air driven low speed axial turbine at the Ruhr University of Bochum. The large scale model turbine is schematically sketched in Fig. 1.

The reaction turbine consists of two stages and has constant tip and hub diameters. The same blade profile is used for both stators and rotors. The stagger angle and the number of blades are identical. The blades are of cylindrical type, i.e. a constant profile geometry is used. The test rig is of an open circuit type. The turbine is driven by a centrifugal blower located downstream of the test section. The blower provides a pressure-difference of up to 70 mbar. Therefore a closed turbine casing is not required. Because of these small pressure differences the test rig can be simply sealed by plastic foils. The large dimensions and the design of the facility allow detailed experimental investigations of the flow within the turbine. The test-rig is characterised by a high degree of flexibility, so that different configurations can be tested (for example variations of the stage number, the axial gap between the blade rows, the stagger angle). All relevant dimensions are given in Table 1.

Before the area traverse measurements were started the performance characteristics were measured for three different stagger angles. Then area traverse measurements with five-hole probes were carried out for the stagger angle with the highest efficiency. After that the same measurements were done with the varied stagger angles (+/- 3°). The test conditions that are considered here are listed in Table 2. The Ma- and the Re-number are based on the chord length and the exit flow speed. The flow in the test rig is incompressible (Ma<0.15).

The area traverse measurements with five-hole probes were done in the four traverse planes M1-M4 (Fig. 1). The distance between the probe and the trailing edges was about 10% chord length for all four planes. Two probes, which were calibrated by the producer, were used for the investigations. They supply the static pressure, the total pressure and the three dimensional velocity vector at each measurement point. The accuracy of the pressure transmitter is +/- 0.0525 mbar and the accuracy of the absolute velocity is +/-0.19 m/s. Traverse mechanisms were used to drive the five-hole probes. The traverse mechanisms allow measurements of the flow field in radial and circumferential direction. The high number of measuring points in each traverse plane (50 height positions, 16 pitchwise positions behind one blade passage) was chosen to get a high resolution of the three dimensional flow. In order to check periodicity of the flow field, traverses at additional circumferential positions were carried out.

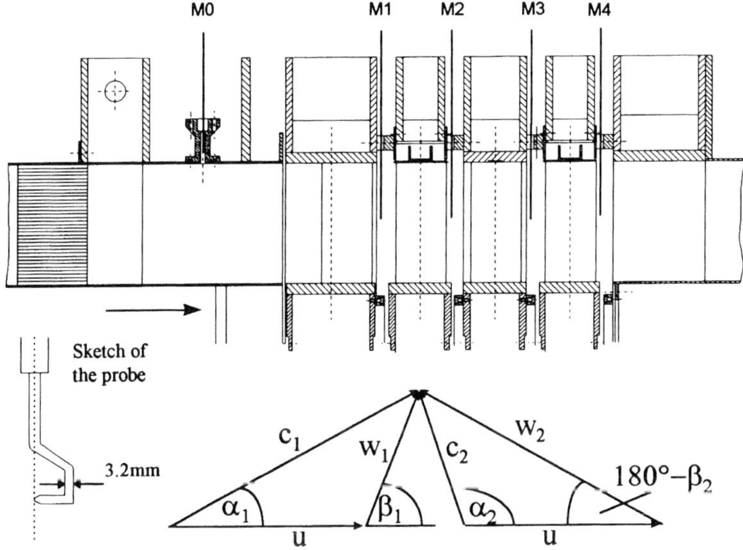

Fig.1: Two stage axial flow turbine with traverse planes and velocity triangles

Table 1: Geometric characteristics of the two stators and rotors

Blade height	Pitch at mean	Chord length	Blade number	Tip diameter
0.170m	0.069m	0.090m	68	1.660m

Table 2: Operating point data (ψ: stage loading ; $\varphi = c_1 \sin(\alpha_1)/u$; η: efficiency)

Operating point	Stagger angle	ψ/ψ_{opt}	φ/φ_{opt}	η/η_{opt}	Ma	Re	Rot.speed
1	$\beta = \beta_{DESIGN} - 3°$	1.0	0.83	0.985	0.129	2.7E5	610 rpm
2	$\beta = \beta_{DESIGN}$	1.0	1.00	1.000	0.121	2.5E5	510 rpm

The test bed is equipped with a computer-controlled data acquisition system, which stores and checks all important quantities during the tests. All quantities were reduced to standard atmosphere to allow comparisons between the test results at different ambient conditions.

———————— **Nomenclature** ————————

Ma	= Mach number	α	= absolute flow angle	φ	= flow coefficient
p	= pressure	β	= blade stagger angle	ψ	= stage loading factor
Re	= Reynolds number	η	= isentropic efficiency	ζ	= total pressure loss coefficient

3 EXPERIMENTAL RESULTS AND DISCUSSION

3.1 Data Presentation Format
In this paper all presented data were reduced to dimensionless parameters. The pressure data of the area traverse measurements were reduced to a total pressure loss coefficient. The total pressure loss coefficient is defined as follows:

$$\text{M1 and M3:}\quad \zeta = \frac{\overline{P}_{t,i-1} - P_{t,i}}{\overline{P}_{t,i} - \overline{P}_i} \qquad \text{M2 and M4:}\quad \zeta = \frac{\overline{P}_{t,r,i-1} - P_{t,r,i}}{\overline{P}_{t,r,i} - \overline{P}_i} \qquad \text{t: total; r: relative}$$

The subscripts correspond to the traverse planes in Fig.1. The circumferential averages for the spanwise distribution plots are based on the theory given by Traupel (9). According to Traupel, the pressure data should be area-averaged and the velocities should be momentum-averaged and angular-momentum averaged respectively. The flow angles in circumferential direction (α, β) and in radial direction (δ) result from the averaged velocities. The definitions of flow angles and velocities correspond to the velocity diagram in Fig.1. The secondary flow vectors were obtained by subtracting the flow vector in the mean direction from the measured local flow vector. The mean flow direction of the angle α is defined as the radial average of the measured angle α for each circumferential position. The mean flow direction of the angle δ is defined as the circumferential average of the measured angle δ for each radial position.

3.2 Inlet Conditions
In order to ensure the existence of periodic inlet flow, the velocity distribution in traverse plane M0 was taken by a five hole probe in 8 circumferential positions. The distance between the plane M0 and the leading edge of the first stator was about 200% chord length. For that reason the potential influence of the first stator was negligible. A hot wire probe was used for the measurement of the degree of inlet turbulence and to sample the inlet boundary layer profile. The degree of inlet turbulence was about 3.5%. The hub and tip absolute boundary layer thickness was 6% of blade span.

3.3 Comparison of the Two Stators for Operating Point 2
The comparison of the stator exit flows is based on measurements in plane M1 and M3. In Fig.2, the total pressure loss coefficients are presented in contour plots and the secondary flow in vector plots. The area traverse measurements were only taken over one stator pitch. The representation over two pitches was chosen to increase the clarity of the figure.

In the plots of the secondary flow velocity vectors, the length of the arrows is proportional to the secondary flow velocity. The viewing direction points into the opposite direction of the flow. For the analysis of the representation it must be taken into consideration that the inlet flow of the first stator is irrotational with an inlet flow angle of 90° and that the inlet flow of the second stator depends on the radius (Fig.3, plane M2). The inlet flow of the second stator is incident on the pressure side in the hub region between 0 and 12% of the span and on the suction side in the region between 40 and 100% blade height.

In the representation of the total pressure loss coefficient for the two stators the trailing edge wake is clearly visible. The stator blades are not stacked on the trailing edge. This explains the leaning of the trailing edge wake. The lowest total pressure loss is indicated in the free stream region between the stator blades. Around the trailing edge wake the loss increases rapidly. The total pressure loss coefficients at the midspan are of the same magnitude for both stators. Near the hub the two exit flows show a maximum local loss. The loss regions are located at the suction side of the trailing edge. The stator exit flows are clearly different in the tip region. Behind the first stator the secondary flow in the tip region causes changes in the trailing edge wake flow. Another high loss core exists at 90% of the span. The secondary flow of the second stator is so strong, that high loss material of the trailing edge wake migrates into the free stream region at 75% of the span. Further local maximum loss regions exist near the upper endwall for both stators.

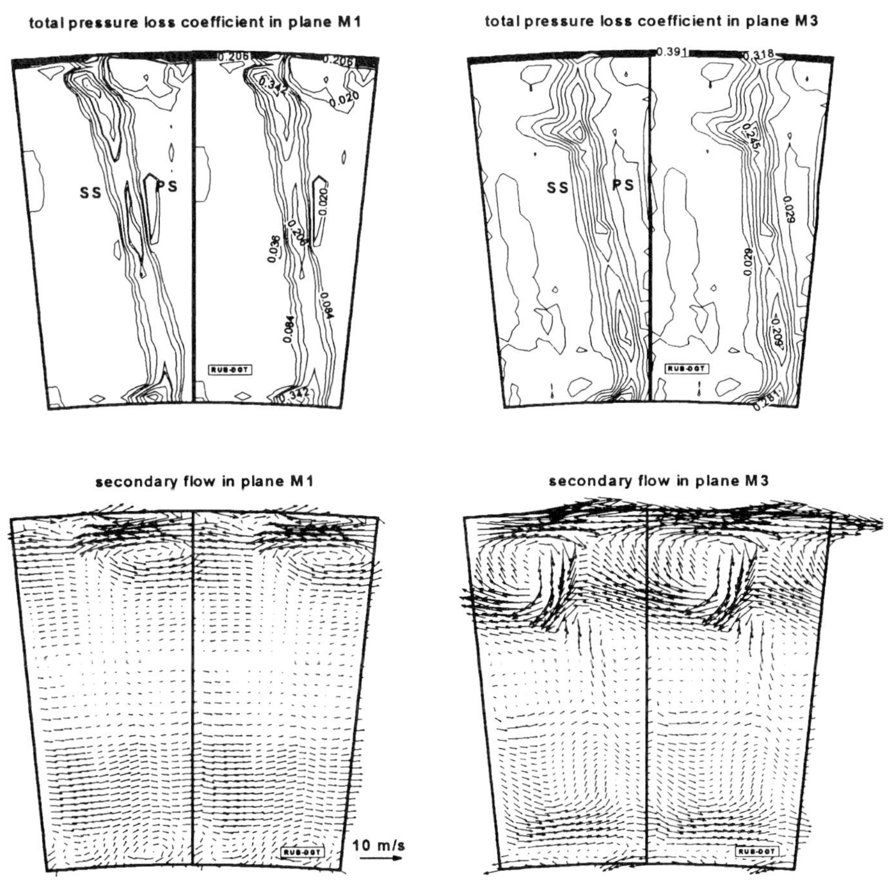

Fig.2: Total pressure loss coefficient and secondary flow in the operating point 2

The secondary flow fields of the two stators are clearly different. The secondary flow at the tip region of the second stator is stronger. The passage vortex behind the second stator covers more than 30% of the span and influences the flow over the whole pitch. For the first stator two weaker vortices can be seen in the tip region. The tip region passage vortex is characterised by a right hand rotation and half of the vortex is located in the upper axial chamber between stator and rotor casing. This vortex is clearly weaker than the passage vortex of the second stator. A left handed rotating trailing edge vortex exists in the region between 80% and 90% of the span behind the first stator, which is stronger than the one behind the second stator. At midspan there is hardly any secondary flow. Therefore, in the present turbine the midspan region is not influenced by endwall flow effects. At the vicinity of the hub the differences between the first and second stator are not significant. Both exit flows show a right hand rotating trailing edge vortex. The passage vortices of the hub regions are not visible. It is possible that the passage vortices move radially inwards towards the axial chamber between stator and rotor hub due to the radial static pressure gradient at the planes. The shapes of all vortices are elliptical, because they are cut at an angle different from the mean flow direction.

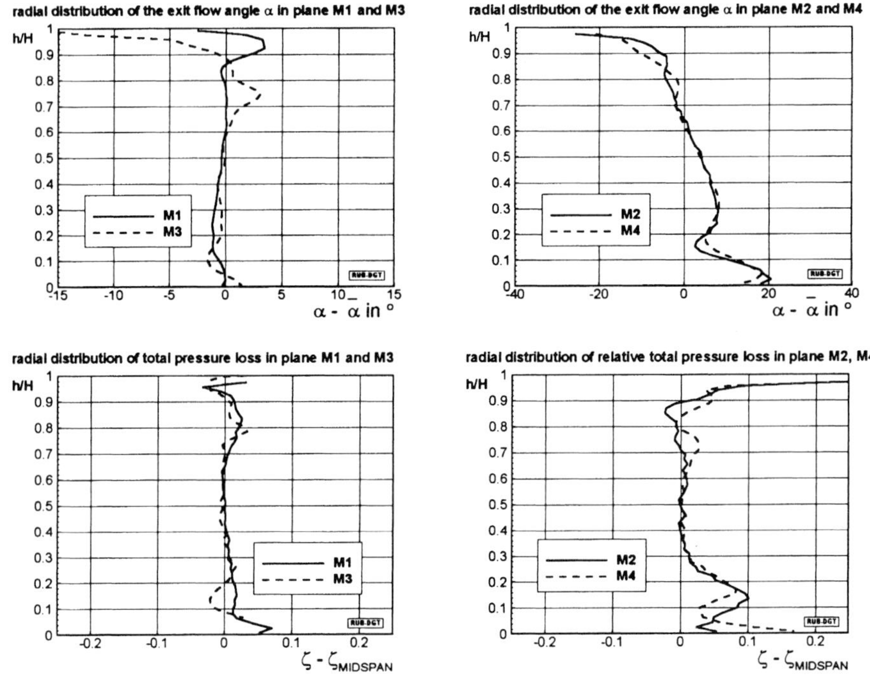

Fig.3: Radial distribution of the circumferential averaged exit flow angle α and total pressure loss coefficient ζ in the operating point 2

Fig.3 shows the distribution of the circumferential averaged exit flow angles α_1 and total pressure loss coefficients ζ behind the first and the second stator. In the region between 15% and 75% span, the exit flow angle increases from the hub to the tip due to the increasing pitch-to-chord ratio. The exit flows in the endwall regions are completely different for both stators. For the first stator the typical over- and under-turning due to secondary flows in the tip region starts at 85% of span. For the second stator the upper 30% of span are influenced by secondary flows. In addition, the over-turning at the endwall is considerably stronger. At the hub only the under-turning due to the passage vortex is visible. The over-turning disappears in the lower axial chamber. Nevertheless, it can also be seen, that the influence of the secondary flow on the exit flow is stronger for the second stator at the hub.

The non-reduced distribution of the circumferential averaged total pressure loss coefficient ζ shows, that the losses in the second stator are higher then in the first one. Noticeable is the increasing loss at 75% of span for the second stator. This maximum local loss is located at the position, where the strong secondary flow moves the high loss material of the trailing edge wake into the free stream region. The losses at midspan region are nearly constant. At the endwalls however the losses pass through a minimum followed by a loss maximum.

Overall one can say, that the secondary flow of the second stator are stronger. The reason for this effect is the different inlet flow of the first and the second stator. Therefore the losses in the second stator are higher. In addition the strong secondary flow of the second stator leads to a more non-uniform inflow into the second rotor.

3.4 Comparison of the Two Rotors for Operating Point 2

The experimental results behind the rotors are also shown in Fig.3. For the measurements behind the rotor it is not sufficient to traverse only one circumferential position with the five hole probe. Experience shows, that the flow effects of the upstream stator, like the trailing edge wake, can be seen behind the rotor. For that reason the radial traverse measurements were carried out over the whole pitch. The radial distribution of the circumferential averaged flow angle α_2 shows clearly, that the exit flow angle of the rotor depends on the radius. Especially in the tip region, the flow over the shrouded rotor leads to a significant under-turning. This under-turning weakens the over-turning of the rotor passage vortex in the endwall region. The over- and under-turning due to the passage vortex at the hub however is visible. The shapes of the radial distribution of the flow angle are not very different for the two rotors.

The distribution of the total loss coefficient is also similar for both rotors. The distributions are characterised by strong total pressure losses in the endwall regions. The total pressure loss in the vicinity of the shroud is particularly high. Between 30% and 70% of span the ζ-quantity is nearly constant. A maximum local loss exists due to the secondary flow of the rotor in the hub region between 10% and 20% of span. In the tip region a similar strong maximum is not visible. The reason for this is probably to be found in the interaction between the clearance flow and the mean flow. Finally, it can be noted, that, in spite of the different stator exit flows, the differences between the rotor exit flows of the first and the second stage are not considerable.

3.5 Comparison of the Two Stators for Operating Point 1

Operating point 1 is indicated by a stagger angle, that is 3° smaller than in operating point 2. Apart from that, the same volume flow and stage loading were chosen. The measurements were carried out in the same way as for operating point 2. The efficiency of the turbine was lower than in operating point 2.

The contour plots of the total pressure loss coefficient and the vector plots of the secondary flow for both stators are plotted in Fig.4. The 3° smaller stagger angle leads to a smaller stator exit flow angle α_1. For this reason the trailing edge wake widths must appear larger, because the angle between the flow direction and the measuring plane is smaller. Nevertheless the geometrical effect is not sufficient to explain the outgrowth of the trailing edge wakes. Therefore it could be assumed, that the reduction of the stagger angle increases the influence of the trailing edge wakes.

For both stators the endwall effects have changed. The region with high total pressure losses at the hub extends from the suction side to the pressure side. The loss maximum behind the first stator in the tip region is stronger. It is noticeable, that the migration of high loss material of the trailing edge wake behind the second stator at 75% of span is more intensive than in operating point 2.

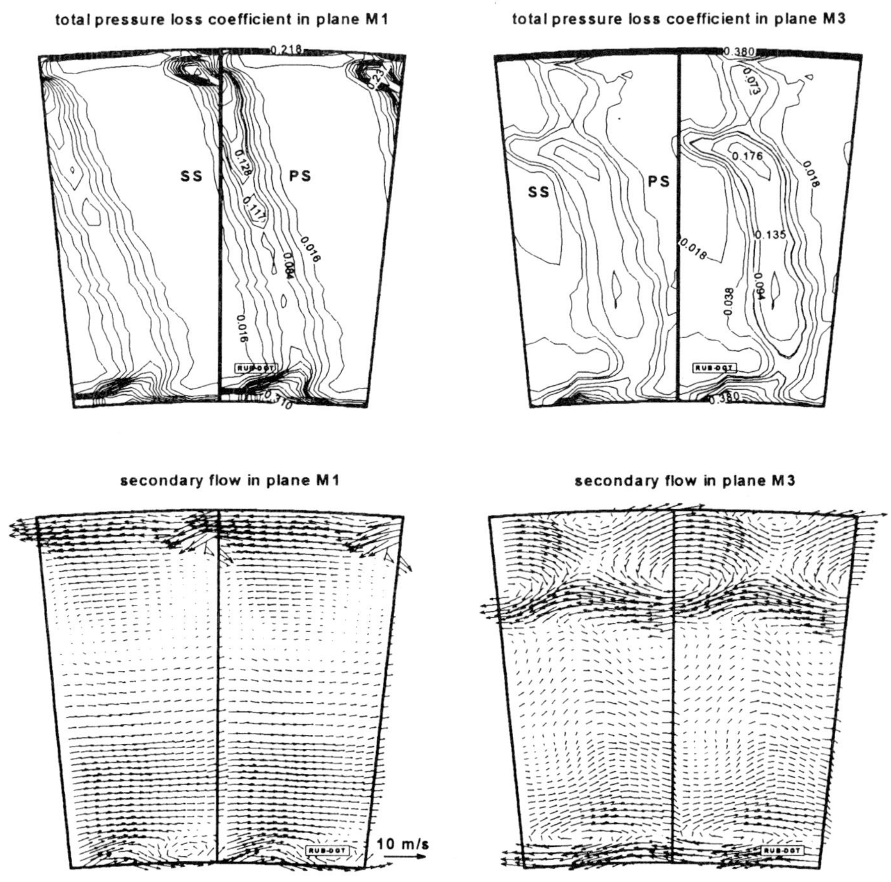

Fig.4: Total pressure loss coefficient and secondary flow in operating point 1

The secondary flow in operating point 1 is clearly different from that in operating point 2. The trailing edge vortex located behind the first stator in the tip region cannot be seen as clearly as for operating point 2. In contrast to operating point 2 a left hand rotating passage vortex is visible at the hub. A trailing edge wake cannot be seen at the hub.

The number of vortices behind the second stator increases. Above the two passage vortices in the tip region there is another left hand rotating vortex near the trailing edge. The passage vortex is less circular than in operating point 2. The trailing edge wake at the hub becomes stronger. Again, only the under-turning of the passage vortex can be seen in this region.

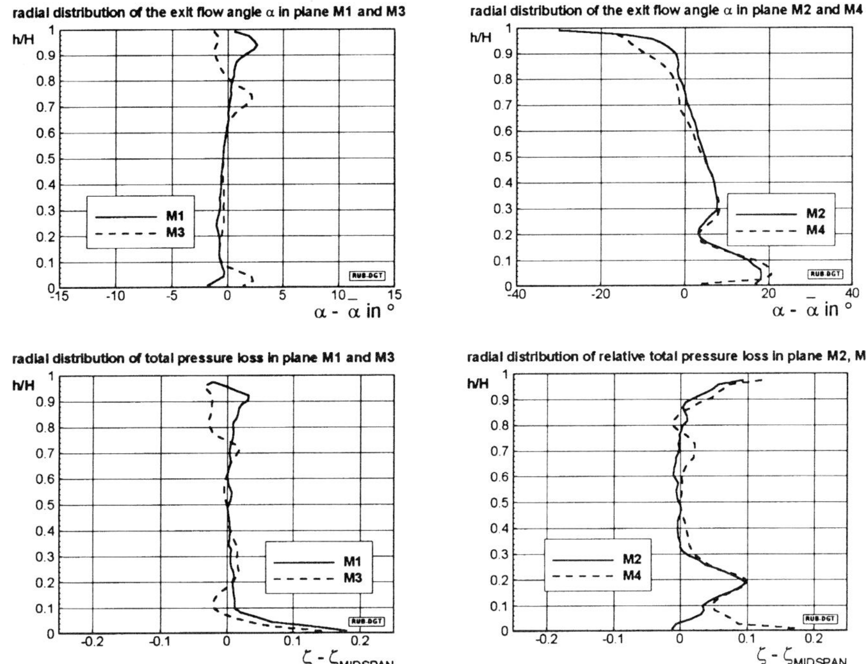

Fig.5: Radial distribution of the circumferential averaged exit flow angle α and total pressure loss coefficient ζ in operating point 1

Figure 5 presents the radial distribution of the circumferential averaged exit flow angle α_1 and the total pressure loss coefficient ζ. The distribution of the exit flow angle is more even over the span in comparison with operating point 2. The shape of the total pressure loss coefficient has not changed much. Interesting is only the rise of the total pressure losses for both stators at the hub.

3.6 Comparison of the Two Rotors for Operating Point 1

The experimental results behind the two rotors for operating point 1 are also plotted in Fig.5. The distributions of the α_2-angles are similar to the distributions in operating point 2. They also show the under- and over-turning in the hub region due to the rotor secondary flow. Compared to operating point 2 these effects are located in a greater distance from the hub. The shroud leakage flow is responsible for the intense under-turning in the tip region. The interaction of the leakage flow and the main flow modifies the flow distribution in such a way that the expected under- and over-turning due to the secondary flow disappear.

The total pressure loss coefficient distributions are nearly the same for both stages. There are only differences at the hub region. In this region the two exit flows show a local loss maximum at 20% of span followed by a local loss minimum. But the shape of the loss distribution at the hub is different. The total pressure loss increases in plane M4. On the other hand the total pressure loss decreases in the plane M2.

4 CONCLUSIONS

In this paper, results from detailed measurements of the flow field in a two-stage axial flow turbine were presented. The investigation comprised two turbine builds differing in the blade setting angle.

The results indicate that the exit flow from the second stator is by no means similar to that from the first stator. This effect can be attributed to the non-uniform inlet flow conditions that exist at inlet to the second stator. The exit flow from both rotors however were found to be very similar. Furthermore, it was found that a substantial interaction exists between the shroud leakage flow and the main flow downstream of the two rotors. As might be expected, the main effect of a reduced blade stagger angle is the generation of a different secondary flow field downstream of the blade rows but the overall flow pattern remains unchanged.

The present investigation is intended to serve as a baseline test for future experiments on the model turbine. Future testing will include different blade profiles and a more detailed study of the effects of shroud leakage flows.

ACKNOWLEDGEMENTS

The authors would like to thank Siemens Power Generation (KWU) for the financial support of the present work and the permission to publish this paper.

REFERENCES

(1)Denton, J.D., 1993,"Loss Mechanisms in Turbomachines," ASME Paper No. 93-GT-435.

(2)Yamamoto, A., and Nouse, H., 1988, "Effects of Incidence on Three-Dimensional Flows in a Linear Turbine Cascade," ASME Journal of Turbomachinery, Vol. 110, pp. 486-496.

(3)Watzlawick, R., 1991,"Untersuchung der wesentlichen Einflußfaktoren auf die Sekundärverluste in Verdichter- und Turbinengittern bei Variation des Schaufelseitenverhältnisses," Dissertation Uni BW München, Germany.

(4)Wegener, D., 1992,"Analyse der Sekundärströmungswirbel in Turbinenstatoren unter besonderer Berücksichtigung der Eintrittsgrenzschichten," Diss. Uni Bochum, Germany.

(5)Weiss, A., P., Fottner, L., 1995,"The Influence of Load Distribution on Secondary Flow in Straight Turbine Cascades," ASME Journal of Turbomachinery, Vol. 117, pp. 133-141.

(6)Lewis, K.L., 1993,"Spanwise Transport in Axial-Flow Turbines: Part 1 - The Multistage Environment," ASME Paper No. 93-GT-289.

(7)Boletis, E., Sieverding, C., H., 1991,"Experimental Study of the Three-Dimensional Flow Field in a Turbine Stator Preceded by a Full Stage," ASME Journal of Turbomachinery, Vol. 113, pp. 1-8.

(8)Walraevens, R., E., Gallus, H., E., Jung, A., R., Mayer, J., F., Stetter, H., 1998,"Experimental and Computational Study of the Unsteady Flow in a 1.5 Stage Axial Turbine With Emphasis on the Secondary Flow in the Second Stator," ASME 98-GT-254

(9)Traupel, W.,"Thermische Turbomaschinen,"Springer-Verlag, Berlin 1977.

C557/055/99

Measurements of turbine blade aerodynamic entropy generation rate

F K O'DONNELL and **M R D DAVIES**
Department of Mechanical and Aeronautical Engineering, University of Limerick, Ireland

ABSTRACT

The aim of this paper is twofold: firstly to put the case for the entropy generation rate to be used as the key parameter in assessing turbomachinery loss and to present detailed cascade boundary layer and hence entropy generation measurements. Secondly to demonstrate how surface mounted transducers may be used to measure profile loss; this opens the way for detailed loss measurements on rotating blade rows. The first aim is realised by the use of very accurate hot wire blade boundary layer traverses of the laminar section of a turbine blade suction surface in a linear cascade. The second, by using surface pressure measurements and constant temperature thin film gauges to measure both the surface shear force and the boundary layer thickness. Combined these give an accurate measurement of the profile distribution of laminar entropy generation rate.

NOMENCLATURE

Cd	Dissipation coefficient	-	β	Non-Dimensional constant	-
C_f	Skin friction coefficient	-	Δ	Portion of boundary layer within which	
L	Surface length	m		majority of entropy is generated	m
p	Pressure	N/m^2	δ	Boundary layer thickness	m
Re	Reynolds number	-	δ_2	Momentum thickness	m
$\dot{S}'''$	Entropy generation rate per unit volume	W/m^3K	δ_3	Energy thickness	m
$\dot{S}''$	Entropy generation rate per unit area	W/m^2K	μ	Dynamic viscosity	kg/ms
$\dot{S}'$	Entropy generation rate per unit length	W/mK	ρ	Density	kg/m^3
T	Temperature	K	τ_w	Aerodynamic wall shear stress	N/m^2
u	x-direction velocity	m/s			
x	Surface co-ordinate	m	**Suffices**		
y	Normal co-ordinate	m	c	Chord length	m

1.0 INTRODUCTION

The almost singular aim of turbomachinery experiments at present is the validation of predictive codes. In turbine blade rows they are primarily being developed for the prediction of work output, loss, heating load and blade unsteady forces. This paper is concerned with the second of these topics, and more specifically with the profile loss, which is thought to constitute some one third of the total.

The inability of codes to predict loss accurately has been noted by many authors including Denton[1]. Loss prediction is still, more often than not, a matter of extrapolation from past experience which, coupled with the high cost of error, can only lead to conservative designs. The long-term aim is still therefore the *a priori* prediction of loss. Achieving this requires the development of codes in parallel with measurement programs. In agreement with both Denton[1] and Denton and Cumptsy[2] the case is made in this paper for the use of entropy generation rate as the key parameter for accessing loss. It follows from this that entropy generation rate should also be the key parameter for assessing loss prediction accuracy. The accurate prediction of the whole boundary layer velocity profile is not necessarily a good test of a codes accuracy in profile loss prediction, when the entropy generation is proportional to the velocity gradient squared and most of it is generated in the inner boundary layer. Furthermore, for most turbine blade row predictions it is also necessary to accurately predict the thermal boundary layer, because the entropy generation rate is inversely proportional to absolute temperature. It is therefore proposed that the profile of static temperature and the velocity gradient squared, in a region of the boundary layer approximately equal to the energy thickness, be used to assess the profile loss prediction accuracy of numerical codes.

Although the focus here is on profile loss it is not the sole aim of the work. One of the many benefits of working with entropy generation rate as a parameter is that it may be used to measure and predict all types of loss, and importantly, heat transfer and fluid friction losses can be expressed in the same units and quantitatively compared. It also shows exactly where the loss is generated so that the possibility arises of using it as an interactive design parameter to systematically reduce the generated entropy.

The specific aim of this paper is to firstly demonstrate how the entropy generation rate may be measured in the laminar region of a linear turbine cascade suction surface boundary layer. The resulting data very clearly shows the key areas for accurate predictions to be the near wall and the leading edge regions. Secondly it is noted that the traversing method employed for these measurements is not viable in rotating transonic turbines. A method is therefore developed to measure the profile loss from surface mounted pressure and constant temperature thin film gauges. The method entails the measurement of surface shear and pressure gradients and the inference from these of the boundary layer thickness. It is shown to work accurately for the linear cascade considered.

2.0 ENTROPY GENERATION IN LAMINAR ISOTHERMAL FLOW

Assuming that the entropy is generated principally due to the shear parallel to the surface, the rate of entropy generation per unit volume due to viscous friction in an incompressible adiabatic flow is given by Bejan[3] as:

$$\dot{S}''' = \frac{\mu}{T}\left(\frac{\partial u}{\partial y}\right)^2 \tag{1}$$

From which the non-dimensional entropy generation rate per unit volume may be written as:

$$\dot{S}^{m\bullet} = \left(\frac{\partial u^{\bullet}}{\partial y^{\bullet}} \right)^{2} \qquad\qquad \text{Where: } \dot{S}^{m\bullet} = \frac{\dot{S}^{m} T \delta^{2}}{\mu u_{\infty}^{2}} \qquad\qquad (2)$$

The non-dimensional entropy generation rate per unit area is therefore simply the integral of equation (2) evaluated across the velocity boundary layer:

$$\dot{S}^{n\bullet} = \int_{0}^{1} \left(\frac{\partial u^{\bullet}}{\partial y^{\bullet}} \right)^{2} dy^{\bullet} \qquad\qquad \text{Where: } \dot{S}^{n\bullet} = \frac{\dot{S}^{n} T \delta}{\mu u_{\infty}^{2}} \qquad\qquad (3)$$

This term integrated around a blade or vane profile gives the entropy generation rate per unit surface length, or the profile entropy generation rate.

$$\dot{S}' = \int_{L_{1}}^{L_{2}} \dot{S}^{n} dx \qquad\qquad (4)$$

It is the local value, however, that is of principle interest here as it shows where on the profile the entropy is generated. Equation (3) has in the past been expressed with a dissipation coefficient, C_{d}.

$$\dot{S}^{n\bullet} = C_{d} \, \mathrm{Re}_{\delta} \qquad\qquad (5)$$

Schlichting[4] gives details of correlations of experimental work which indicate some general results in turbulent flow.

3.0 BOUNDARY LAYER TRAVERSES

As indicated by equations (1) through (3), the evaluation of local entropy generation rate requires the determination of the local boundary layer velocity gradient squared. For this to be accurately measured requires a very accurate measure of the velocity profile.

All cascade measurements were taken on the subsonic linear cascade shown in figure 1; this cascade has been closely investigated by previous authors including Fitzgerald et al[5]. The blades are scaled models of the mid-span of a rotor blade designed by Santoriello et al[6], now operational in two single-stage turbine facilities at the VKI, Brussels and DLR, Göttingen. Mid span rotor blade pressure and Nusselt number distributions have been presented by Denos et al.[7], and surface hot film measurements by Tiedermann[8] at engine Mach and Reynolds numbers. The use of such data to measure profile loss will be demonstrated in section 5.

A traverse system, driven by stepper motors, allowed the linear movement of a Dantec P11 single hot-wire probe in 10 µm increments. The traverse was mounted beneath the cascade support frame with a 4 mm diameter probe holder extended through a slot in the lower wall of the cascade; this arrangement allowed access to only the first 22.5 % of the blade suction surface. The hot-wire was positioned at 25 µm from the surface of the cascade for the first velocity measurement and then traversed normal to the surface. The traverse was performed outwards in 10 μm increments for the first 0.5mm, where it was expected that high profile definition would be required, and 100 μm for the remaining 1.5 mm. Traverses were carried out over a range of Reynolds numbers at the six test locations indicated in table 1.

The reference location of the hot-wire relative to the surface was determined by using the probe as part of an electrical circuit which is closed when the tip of the probe touches the conducting surface, as described by Azad and Burhanuddin[9]. An electron microscope was used to determine the probe geometry. The distance from the touching tip of the probe to the centre line of the 5 μm diameter wire was found to be 5 μm due to the build up of weld on the probe-support connection, shown in figure 2, this build up will be specific to each probe. The correction for this was found to have a significant effect on the determination of the profile in the critical near wall region. The degree of spatial resolution required in this region is therefore of the order of microns.

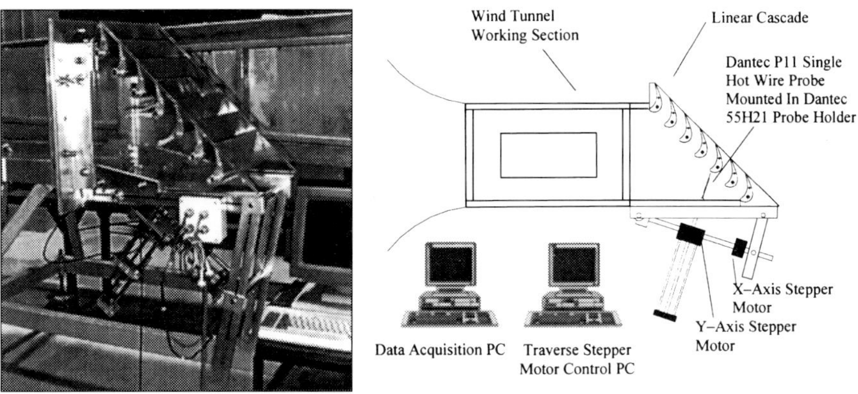

Fig.1. Linear turbine cascade boundary layer traverse experimental set-up.

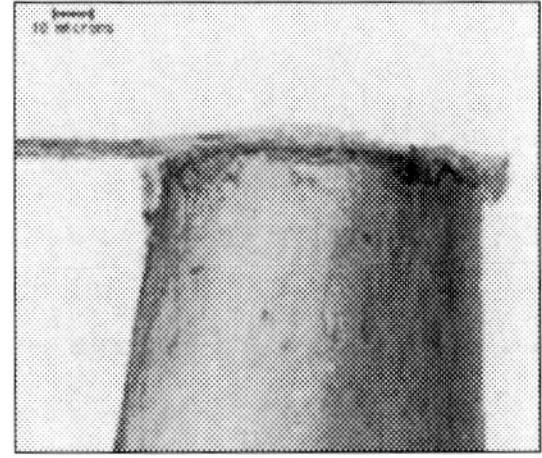

Fig.2. Electron microscope scan of hot-wire probe.

Table 1. Linear cascade velocity boundary layer traverse locations.

Test Number	1	2	3	4	5	6
% Suction Surface Length	1.8	5.3	9.3	13.6	18	22.5
Distance From Stagnation Point (m)	0.00238	0.00698	0.0129	0.01792	0.02374	0.02964

As noted by numerous authors investigating boundary layer flows with hot-wire probes, there is a marked increase in the heat transfer from the wire close to the wall. This effect leads to a notable error in the velocity indicated by the hot-wire probe in this region. Azad[10] has reviewed several theoretical and experimental investigations of this effect, the best known of which is the work of Wills[11], where a correction factor is introduced to account for the increased heat transfer. In the work presented here the correction had a significant effect on the near wall measured profiles.

4.0 Results.

Applying Wills[11] correction to the hot-wire traverses data yields the non-dimensional velocity profiles of figure 3. The free-stream velocity distribution and boundary layer thickness are shown in figure 4.

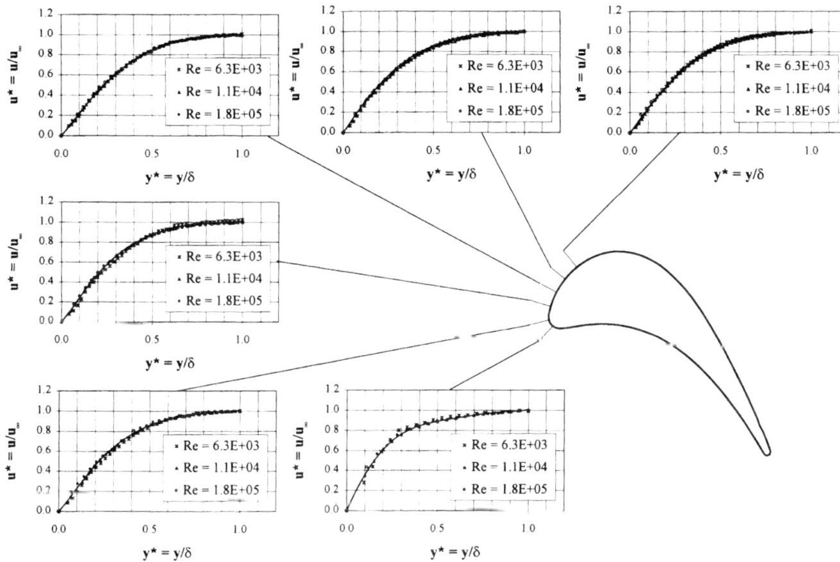

Fig.3. Non-dimensional velocity profiles measured between 1.8 % and 22.5 % suction surface length of a linear turbine cascade. $6.0 \times 10^4 < Re_C < 1.7 \times 10^5$.

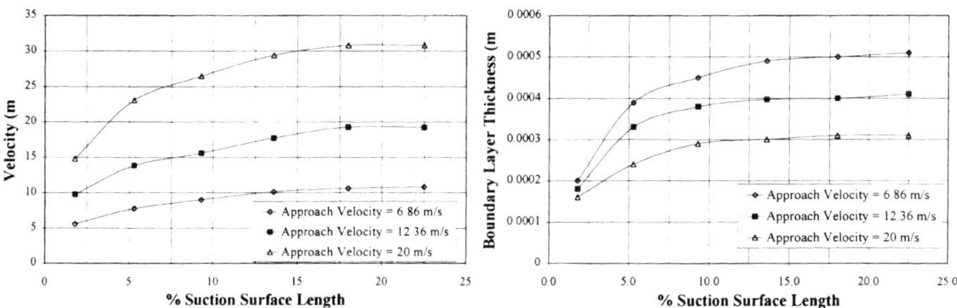

Fig.4. Velocity distribution and boundary layer growth over early suction surface of turbine cascade.

The scaled entropy generation rate per unit volume distributions in the boundary layer are shown in figure 5 at each of the six traverse locations. These distributions are independent of Reynolds number. It is clear that the majority of the entropy is generated near the wall. It is also evident that $\dot{S}'''^{*}$ is highest in the region of the blade close to the leading edge where the velocity gradients and shear work are highest.

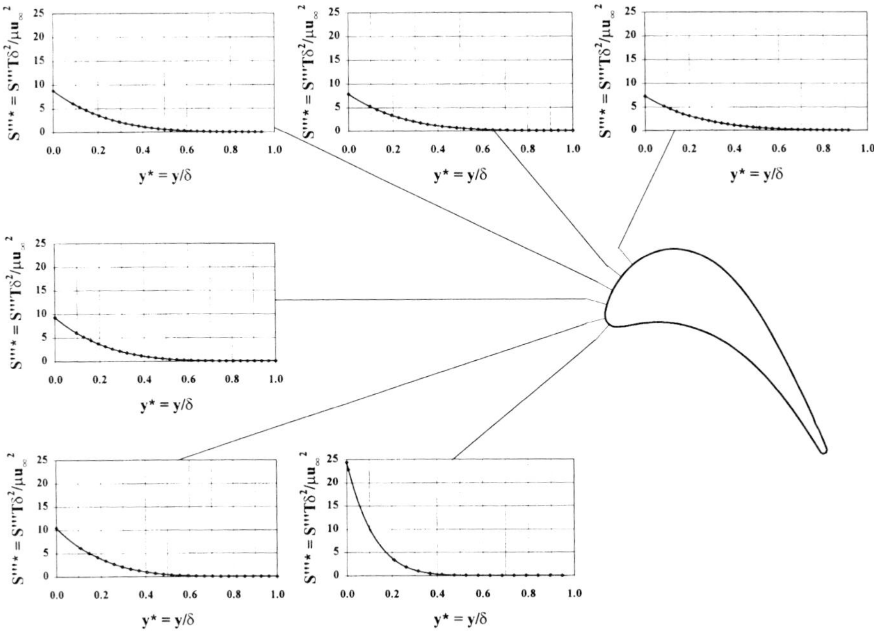

Fig.5. Distribution of Non-dimensional entropy generation rate per unit volume across the boundary layer of a linear turbine cascade, as with Fig.3. these distributions are independent of Reynolds number.

Integrating $\dot{S}'''^{*}$ across the boundary layer, in accordance with equation (3), yields the $\dot{S}''^{*}$ distribution along the blade profile as shown in figure 6. There is a steep peak in generated entropy near the leading edge due to the high shear stress in this region; this is due to the very thin boundary layer here. Such peaks in shear stress are shown by Schlichting[12] for a variety of elliptical bodies. The more slender the ellipse the greater the stress and hence loss in this area. The stream-wise resolution of these traverses was not sufficient to identify the exact location of this peak, it was therefore necessary to approximate its location and magnitude from CFD data (as will be discussed in section 5.0) in order to present the distribution of $\dot{S}''^{*}$ along the early suction surface.

The significant increase in the dimensioned entropy generation rate per unit area, $\dot{S}''$, as the Reynolds number increases, is due to the $(\partial u / \partial y)^{2}$ term in equations (1) to (3) which gives a non-linear increase in $\dot{S}''$ with Reynolds number. This emphasises the necessity for accurate velocity gradient measurement or prediction in this region. Non-dimensional plots are preferred for predictions and understanding the physics, but it is the dimensioned values that are paid for in lost turbine work, these are also shown in figure 6.

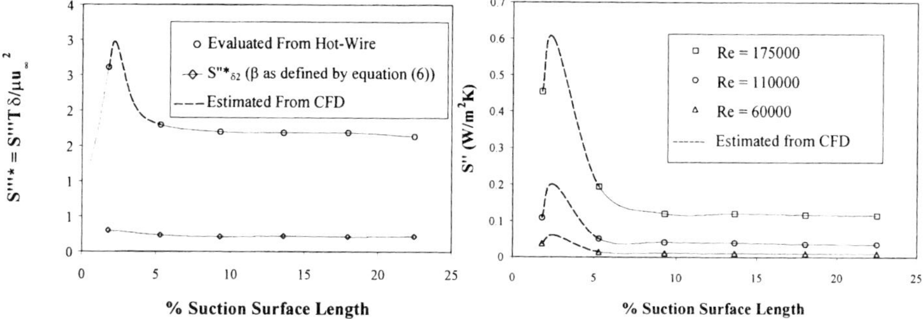

Fig.6. Non-dimensional and dimensional suction surface chord-wise entropy generation rate per unit area.

One third of the entropy generated in the considered section of the blade is generated within the first 4.5% suction surface length. This region may represent a distance of less than 1.5mm on a typical turbine blade. This emphasises the need for both high streamwise and normal spatial resolution in this region.

Previous authors have investigated entropy generation rate using the dissipation coefficient, C_d. Truckenbrodt[13] quotes results showing:

$$C_d \sim \beta \, Re_{\theta_2}^{-1} \tag{6}$$

Where the value of β is consistent at about 0.17 for laminar flows. Evaluating C_d, as defined by equation (5), yields a value of $\beta = 0.21$ for the flow considered here as shown in figure 6. This result, though higher than the value quoted by Truckenbrodt[13], is in agreement with the value of $\beta = 0.2$, suggested by Denton and Cumpsty[2] as being applicable to typical turbomachinery applications.

Presenting data in terms of C_d and β obscures the physical relevance of the parameters. C_d is nothing more than the non-dimensional entropy generation rate times a Reynolds number, and β is the non-dimensional entropy generation rate based on the momentum thickness, as shown if figure 6. The scaling presented here is very simple; the non-dimensional entropy based upon boundary layer thickness is independent of Reynolds number and the same parameter based upon momentum thickness is largely independent of surface position. As shown in figure 3 there is a notable change in the velocity profile shape in the upstream profiles in contrast to the near constant shape downstream. It is the dependence of $\dot{S}_{\delta 2}'''^*$ on the boundary layer shape, as noted by Denton[1] that leads to its dependence on stream wise position in the leading edge region. The important unanswered design question is how dependent are either of these parameters upon blade shape. Work by Benner et al.[14] has shown that the profile loss coefficient is independent of quite a large change in leading edge shape over a wide range of incidence angles. It is not known how general this result is or why it is so.

From the $\dot{S}''$ distribution the entropy generation rate per surface length, $\dot{S}'$, is evaluated from equation (4) and is presented in figure 7.

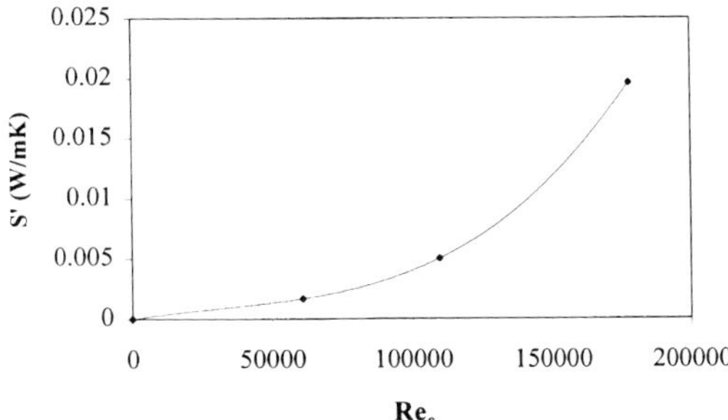

Fig.7. **Profile entropy generation rate per unit suction surface length for a linear turbine cascade between 1.8 % and 22.5 % suction surface.**

5.0 ENTROPY GNERATION RATE MEASUREMENT FROM SURFACE GAUGES

The results presented in the previous section depend upon an accurate definition of the velocity distribution within the boundary layer. However, as experimental investigations move ever closer to fully engine representative conditions, the determination of this distribution from traverses is not feasible in rotating tests. Davies and O'Donnell[15] have described in detail the development of an approximation for $\dot{S}''^*$, which may be evaluated from the non-dimensional local aerodynamic wall shear stress, τ_w^* and local boundary layer thickness, δ. The relationship, which has been shown by the authors to be valid for a wide range of well-defined laminar boundary layers, is:

$$\dot{S}''^* = \Delta^* \left[\tau_w^{*2} + \left(\frac{dp}{dx} \right)^* \Delta^* \left(\left(\frac{dp}{dx} \right)^* \frac{\Delta^*}{3} + \tau_w^* \right) \right] \quad \text{where: } \Delta^* = \frac{\Delta}{\delta} ; \; \tau_w^* = \frac{\tau_w \delta}{\mu u_\infty} ; \; \left(\frac{dp}{dx} \right)^* = \frac{dp}{dx} \frac{\delta^2}{\mu u_\infty} \quad (7)$$

By evaluating Δ^*, in a variety of laminar flows, Davies and O'Donnell[15] have determined that it may be expressed simply in terms of τ_w^* as:

$$\Delta^* = \tau_w^{*-1.5} \tag{8}$$

therefore, substitution of equation (8) into equation (7) gives:

$$\dot{S}''^* = \tau_w^{*0.5} \left[1 + \left(\frac{dp}{dx} \right)^* \tau_w^{*-2.5} \left(\frac{\tau_w^{*-2.5}}{3} \left(\frac{dp}{dx} \right)^* + 1 \right) \right] \tag{9}$$

Δ^* in equation (7) is defined as the fraction of the boundary layer within which the majority of the entropy is generated. For the Reynolds number range investigated here, this distance has been found to be approximately equal to the energy thickness, δ_3, as shown in figure 9, which is in turn equal to the entropy thickness defined by Denton[1]. Neither of these results are surprising as the entropy is generated due to the viscous generation of heat.

Equation (9) may be used to find the local and integrated profile loss if the surface shear, pressure gradient and the local boundary layer thickness are known. The pressure gradient is relatively easily measured. We will demonstrate how the other two parameters, and hence the profile loss, may be measured from surface mounted gauges.

The use of heated thin films for the measurement of aerodynamic wall shear stress has been investigated by many authors, among the first of which was Ludwieg[16] with significant contributions since then coming from Bellhouse and Schultz[17] and more recently Davies and Duffy[18].

Figure 8 shows the distributions of $C_f\,Re^{1/2}$ along the suction surface of the linear cascade. Here the value of τ_w has been evaluated both from the gradient of the measured boundary layer velocity profiles shown in figure 3 and directly using calibrated thin films. Both sets of data are in good agreement and also compare favourably with CFD predictions. A comparison is also given with the $C_f\,Re^{1/2}$ distributions around elliptical cylinders of axis ratios of 4:1, 3:1 and 2:1. A 3:1 ellipse can be fitted to a large portion of the cascade early suction surface. It is evident that the peak values of $C_f\,Re^{1/2}$ occur at the same location for both the 3:1 ellipse and the cascade, the magnitude however is notably different, and this is attributed to the cascade configuration. It is also evident that unlike the ellipse the cascade yields a near constant value of $C_f\,Re^{1/2}$ after approximately 5% suction surface length. This is due to the repressed boundary layer growth on the cascade thus maintaining the relatively high values of wall shear stress. The thin film measurements and CFD predictions are described in detail by Niven et al.[19].

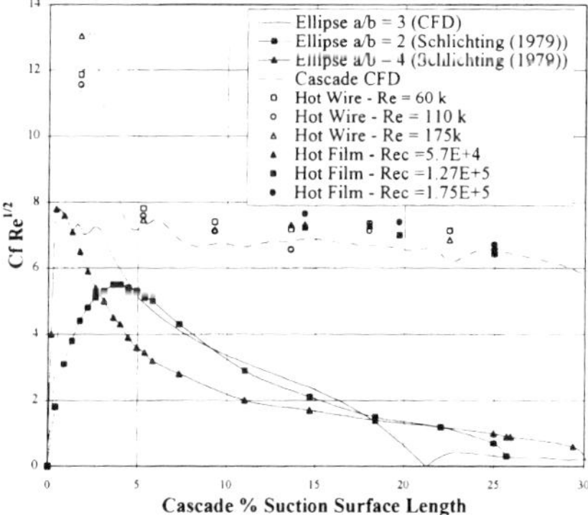

Fig.8. $C_f\,Re^{1/2}$ distribution on a linear turbine cascade blade suction surface

Schtlichting[20] gives a general relationship that may be used for predicting the boundary layer thickness from surface measurements.

$$\frac{\delta^2}{6\mu u}\frac{dp}{dx} - \frac{\delta\tau_w}{\mu u_\infty} + 2 = 0 \tag{10}$$

This is shown to work extremely well in figure 9 where the hot wire determination of the boundary layer thickness is compared with the inferred value from surface measurements. Also included in this figure are the hot wire measured energy and momentum thickness.

Using the thin film measurements and this boundary layer thickness prediction, the non-dimensional entropy generation rate per unit area, $\dot{S}''^*$, is evaluated from equation (9). At present thin film data is not available for 0% to 14% suction surface length, therefore within this region the values of τ_w^* determined from the hot-wire velocity gradients are used to illustrate the technique.

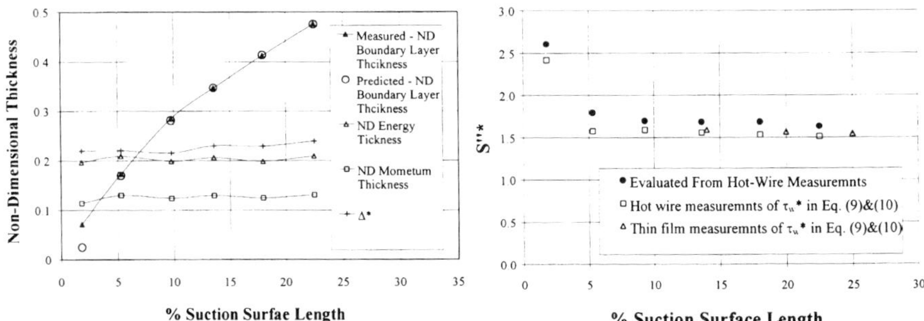

Fig.9. Predicted boundary layer growth from equation (10) compared with hot wire measurements, and energy and momentum thickness non - dimesionalised on the boundary layer thickness also profile entropy generation rate measured in three ways.

The data is very consistent and it is therefore demonstrated that it is possible to measure profile loss directly from surface thin films and pressure measurements. Before the relative significance of the entropy generated within the laminar region of the suction surface may be appreciated it will be necessary to evaluate the entropy generated along the remaining suction surface length. This is the subject of continued work being carried out by the authors.

The extension of the surface measurement technique to turbulent flow will depend on the validation of the following relationship, which may be derived from the analysis given by Davies and O'Donnell[15].

$$\dot{S}''' = \frac{\mu}{T}\left(\frac{\partial \bar{u}}{\partial y}\right)^2 + \frac{\tau_{xy}'}{T}\frac{\partial \bar{u}}{\partial y} = \frac{1}{T}\left(\frac{\partial \bar{u}}{\partial y}\right)\left(\tau_L + \tau_T\right) \qquad (11)$$

Where the turbulent and laminar shear stress are given by:

$$\tau_T = \rho\,\overline{u'v'} \quad \text{and} \quad \tau_L = \mu\frac{\partial u}{\partial y}$$

The second term in equation (11) gives the extra entropy generated due to turbulent eddies. Turbulent calibration will form the next phase of this work.

6.0 Conclusions

- Very high spatial resolution is required to accurately measure the profile loss in the critical near wall region.
- A significant fraction of the profile loss is generated around the leading edge and this fraction increases with increasing Re_c.

- Scaling the entropy generation rate on the boundary layer thickness makes it independent of Re_c and scaling it on momentum thickness makes it largely independent of Re_c and surface position.
- Surface mounted transducers may be used to measure the spatial distribution of entropy generation rate and therefore profile loss.

Acknowledgements

The authors wish to acknowledge the financial support of the University of Limerick, the support of the technical staff of department of Mechanical and Aeronautical engineering and the assistance of Michelle Starr with electron microscope imagery. The authors are also grateful for the helpful comments offered by Dr Howard Hodson at the Whittle Laboratory, Cambridge University.

References

[1] Denton, J.D., 1993, "Loss Mechanisms In Turbomachines", ASME J. Turbomachinery, Vol.115, pp.621-656.

[2] Denton,J.D and Cumpsty,N.A. 1987, "Loss Mechanisims In Turbomachines", IMechE Paper No. C260/87.

[3] Bejan, A., 1982, "Entropy Generation Through Heat And Fluid Flow", John Wiley & Sons, Inc., New York.

[4] Schlichting H., 1968, "Boundary Layer Theory", 6th Ed., Mc-Graw Hill, New York.

[5] Fitzgerald, J.E., Niven, A.J. & Davies, M.R.D., (1998), "Turbine Blade Aerodynamic Wall Shear Stress Measurements And Predictions", ASME Paper Ref. 98-GT-562

[6] Santoriello, G., Colella, A., and Colantuoni, S., 1993, "Investigation Of Aerodynamics And Cooling Of Advanced Engine Turbine Components:- Rotor Blade Aerodynamic Design" *Brite/Euram* Area 3 Aeronautics Technical Report, AER2-CT92-0044 IACA Package A.

[7] Denos, R., Sieverding, C.H., Arts, T., Brouckaert, J.F., Paniagua, G., Michelassi, V. & Martelli, F., 1998, "Unsteady Investigation Of Rotor Aerodynamics Of A Transonic Turbine Stage" 43rd ASME Gas Turbine and Aeroengine Congress, Stockholm, Sweden.

[8] Tiedemann, M., 1997, "Stator Wake Effects On The Boundary Layer Transition Of A Transonic Turbine Rotor Blade", "Eurotherm Seminar No. 55, Heat transfer in single phase flows 5" Sept 18-19, 1997 in Santorini, Greece.

[9] Azad, R.S. and Burhanuddin,S., 1983, " Measurements Of Some Features Of Turbulence In Wall Proximity", Exp. In Fluids, 1, 149-160.

[10] Azad,R.S., 1983, "Corrections To Hot Wire Measurements In Proximity Of A Wall",Report MET-7, Department of Mechanical Engineering, University of Manitiba.

[11] Wills, J.A.B., 1962, "The Correction Of Hot-Wire Readings For Proximity To A Solid Boundary" J.Fluid Mech., 55, 65-92

[12] Schlichting, H., 1979, "Boundary Layer Theory", 7th Ed., Mc-Graw Hill, New York. (pp 218)

[13] Truckenbrodt, E., 1952, "A Method Of Quadrature For The Calculation Of Laminar And Turbulent Boundary Layers In Plane And Rotational Symmetric Flow" *Ingenieur-Archiv*, Vol.20; Translated as NACA TM 1379

[14] Benner,M.W, Sjolander,S.A. and Moustapha,S.H., 1987, " Influence Of Leading Edge Geometry On Profile Loses In Turbines At Off Design Incidence; Experimental Results And An Improved Correlation", Journal of Turbomachinery, April 1987, Vol. 119.

[15] Davies,M.R.D. and O'Donnell,F.K., 1998, " Local Measurement Of Loss Using Heated Thin Film Sensors" 43rd ASME Gas Turbine and Aeroengine Congress, Stockholm, Sweden. ASME Paper Ref. 98-GT-380

[16] Ludwieg, H., 1950, "Instrument For Measuring The Wall Shearing Stress Of Turbulent Boundary Layers", NACA, TM 1285.

[17] Bellhouse, B.J. & Schultz, D.L., 1966, "Determination Of Mean And Dynamic Skin Friction, Separation And Transition In Low-Speed Flow With A Thin-Film Heated Element", J. Fluid Mech., Vol.24, Part 2, pp. 379-400.

[18] Davies, M.R.D. & Duffy, J.T., 1995, "A Semi-Empirical Theory For Surface Mounted Aerodynamic Wall Shear Stress Gauges", 40th ASME Gas Turbine and Aeroengine Congress", Paper No. 95-GT-193

[19] Niven, A., O'Donnell, F.K, Fitzgerald, J., and Davies, M., " Numerical Predictions In Support Of Thin Film Aerodynamics Shear Stress Calibration" XIV Bi-Annual Symposium On Measuring Techniques In Transonic And Supersonic Flow In Cascades And Turbomachines, University of Limerick, Ireland, 2nd – 4th September 1998.

[20] Schlichting H., 1979, "Boundary Layer Theory", 7th Ed., Mc-Graw Hill, New York. (pp 208 – 209)

Multi-stage three-dimensional Navier–Stokes computation of off-design operation of a four-stage turbine

G A GEROLYMOS and **C HANISCH**
Université Pierre-et-Marie-Curie, Paris, France

Abstract: In this paper a 3-D computational methodology for multistage turbomachinery flows is developed, validated, and applied in analyzing the reduced-massflow-rate operation of a 4-stage turbine. The flow is modelled by the 3-D Favre-Reynolds-averaged Navier-Stokes equations using the Launder-Sharma near-wall $k - \varepsilon$ turbulence closure, which are integrated using an implicit 3-order upwind solver. The computation models the 8 blade-rows, including the tip-clearance gaps of the rotors, using a 6×10^6 points grid, and computes the time-averaged flow using mixing-planes between rows. Computational results are compared with measurements for various operating points. After this validation the method is used in a first attempt to analyze the 3-D flow at reduced-massflow operation.

Nomenclature

DH	downstream H-grid
DS	deterministic-stresses
DSM	deterministic-stresses-model
dx_d	distance of DH-O boundary from the trailing-edge of the blade (Fig. 2)
dx_u	distance of UH-O boundary from the leading-edge of the blade (Fig. 2)
$h,\ h_t$	static and total enthalpy
$i,\ j,\ k$	grid indices (k = radial, i = axial for H-grids, i = around the blade for O-grids)
$k_{\text{O-OZ}}$	radial station of the O-grid at which the OZ-grid starts
ℓ_{BR}	number of grid-block
L_{BR}	total number of grid-blocks
LE	leading-edge
$M,\ M_{\text{W}}$	absolute and relative Mach-number
$\dot{m}$	massflow
$\dot{m}_{\text{D}}$	design-point massflow
MP	mixing-plane
n_w^+	nondimensional distance of the first grid-point from the wall ($n_w^+ = n\, u_\tau\, \check{\nu}_w^{-1}$ (where n is the distance-from-the-wall, u_τ the friction-velocity, and $\check{\nu}_w$ thekinematic viscosity at the wall)
n_{it}	iteration number
n_{BR}	number of blade-row
N_{BR}	total number of blade-rows
N_{B}	number of blades
n_{DOM}	number of domain within a block
$n_{\ell d}$	number of domain in global grid storage
$N_i,\ N_j,\ N_k$	number of grid-points (i-wise, j-wise, k-wise)
N_{ijk}	number of grid-points ($N_{ijk} = N_i N_j N_k$)
$p,\ p_t$	static and total pressure

$r_j,\ r_k$	geometric progression ratio for grid-points stretching (j-wise and k-wise)
RPM	revolutions per minute
O	o-grid around the blades
OZ	buffer grid between TC and O (Figs. 4)
TC	clearance-gap grid (Figs. 4)
TE	trailing-edge
T_u	turbulence intensity
$T,\ T_t$	static and total temperature
UH	upstream H-grid
(V_x, V_R, V_θ)	absolute flow velocities
(W_x, W_R, W_θ)	relative flow velocities
(x, R, θ)	cylindrical system of coordinates (x is the engine-axis)
$x_{\mathrm{in}},\ x_{\mathrm{out}}$	x-coordinates of inflow and outflow boundaries (Fig. 2)
α	absolute flow-angle ($\cos\alpha = V_m/V$; $\sin\alpha = V_\theta/V$; $V_m = \mathrm{sign}V_x\ \sqrt{(V_x^2 + V_R^2)}$)
β	relative flow-angle ($\cos\beta = V_m/W$; $\sin\beta = W_\theta/W$; $V_m = \mathrm{sign}V_x\ \sqrt{(V_x^2 + V_R^2)} \equiv W_m$)
δ_{TC}	tip-clearance height
Δs_{O}	massflow averaged entropy increase between blade-row 1 o-grid inflow and blade-row N_{BR} o-grid outflow
$\eta_{\mathrm{T\text{-}S}}$	total-to-static turbine efficiency
$\pi_{\mathrm{T\text{-}T}}$	total-to-total pressure-ratio (massflow averaged)
$\pi_{\mathrm{T\text{-}T_O}}$	total-to-total pressure-ratio between blade-row 1 o-grid inflow and blade-row N_{BR} o-grid outflow
ρ	density

Subscripts

M	pitchwise averaged quantities (meridional)
t	total
w	wall
W	relative
D	design

Superscripts

$\sim$	Favre-average
$^-$	Reynolds-average (ensemble-average)
$\smile$	function of averaged quantities that is neither a Favre-average nor a Reynolds-average
$\frown$	massflow-weighted average on a surface

Introduction

Blade-row interaction and stage matching in multistage turbomachinery off-design operation differ from the design point not only in pitchline averages [9] but also in radial distribution of flow quantities [5]. Axial flow turbines (especially nuclear plant steam turbines [6]) may operate in very low-massflow conditions resulting in severe backflow over a substantial part of the span [5]. In the multistage case these conditions appear first in the last rotor, but, as massflow diminishes, rotors further upstream are affected [34].

Multistage 3-D Navier-Stokes computations (Tab. 1), despite turbulence model deficiencies [12], offer the possibility of detailed analysis of off-design performance. There are 2 methodologies for 3-D multistage computations.

The *mixing-plane* approach (MP) introduced by Singh [35], Arts [4], Denton [11] and Dawes [10] computes the steady (time-averaged) flow in 1 representative channel per blade-row. Neighbouring blade-rows are separated by a mixing-plane, where the change of reference frame (relative to absolute and vice-versa) takes place. Each blade-row is computed in a reference frame fixed on its blades and exchanges with its neighbours pitchwise averaged quantities (meridional averages). In this way the correct radial profiles are taken into account while nonlinear unsteady effects on the time-averaged flow are neglected. This technique is the 3-D equivalent of throughflow computations. There is no overlap (or very little overlap) between the domains of neighbouring blade-rows. The variants of this approach differ in the quantities conserved across the mixing-plane at each spanwise station. Massflow, of course, is always conserved, but numerous possibilities exist for the other quantities (dynalpy, pressure, or entropy are possible choices).

The *deterministic-stresses* approach (DS) is based on an important paper by Adamczyk [1] where, by applying a suitable averaging procedure (time-average in the frame-

Table 1: Steady multistage computation methodologies (MP=mixing-plane, DS=deterministic-stresses, DSM= deterministic-stresses-model).

Reference	year	model	method	application	N_{BR}	grid/row	grid
Singh [35]	1982	3-D Euler + boundary-layer	MP	compressor-stage	2	0.003×10^6	0.007×10^6
Arts [4]	1985	3-D Euler	MP	turbine-stage	2	0.005×10^6	0.010×10^6
Celestina et al [7]	1986	3-D Euler	DS	counterrotating-propellers	2		0.057×10^6
Adamczyk et al [3]	1990	3-D Navier-Stokes (Baldwin-Lomax)	DS	turbine-stage	3	0.060×10^6	0.233×10^6
Denton [11]	1992	3-D Euler + body-forces	MP	2-stage-turbine	4	0.018×10^6	0.072×10^6
				2-stage-compressor	4	0.018×10^6	0.072×10^6
				3-stage-compressor	9	0.018×10^6	0.162×10^6
Dawes [10]	1992	3-D Navier-Stokes (Baldwin-Lomax)	MP	compressor-stage	2		0.069×10^6
				turbine-stage	2		0.047×10^6
Goyal and Dawes [24]	1993	3-D Navier-Stokes (Baldwin-Lomax)	MP	fan-bypass	3		0.164×10^6
Copenhaver et al [8]	1993	3-D Navier-Stokes (near-wall $k - \varepsilon$)	MP	compressor-stage	2	0.350×10^6	0.607×10^6
Gallus et al [18]	1995	3-D Navier-Stokes (near-wall $k - \varepsilon$)	MP	turbine-stage	2		
Turner [38]	1996	3-D Navier-Stokes (Baldwin-Lomax)	DS	2-stage-turbine	4	0.481×10^6	1.592×10^6
Jennions and Adamczyk [30]	1997	3-D Navier-Stokes (Baldwin-Lomax)	DS	turbine-stage	2	0.142×10^6	0.279×10^6
Hall [25]	1997	3-D Navier-Stokes (Baldwin-Lomax)	MP	3-stage-compressor	7	0.488×10^6	3.418×10^6
Hall [26]	1997	3-D Navier-Stokes (Baldwin-Lomax)	MP +DSM	3-stage-compressor	7	0.488×10^6	3.418×10^6
Rhie et al [33]	1998	3-D Navier-Stokes (standard $k - \varepsilon$)	DS	3-stage-compressor	7	0.200×10^6	1.420×10^6
LeJambre [31]	1998	3-D Navier-Stokes (standard $k - \varepsilon$)	DS	11-stage-compressor	22	0.300×10^6	6.600×10^6
present	1999	3-D Navier-Stokes (near-wall $k - \varepsilon$)	MP	4-stage turbine	8	0.600×10^6	4.786×10^6

of-reference of a given blade-row and average between channels to filter out channel-to-channel differences due to the multistage environment), equations describing the 3-D time-averaged flow through a representative passage of each blade-row are derived. The averaging procedure (which is in fact the 3-D equivalent of the meridional averages theory as developed by Hirsch and Dring [28] and Dring and Oates [13] [14]) introduces source terms in the equations of a given blade-row that account for the effects of neighbouring blade-rows. Adamczyk et al [2] developed a rather complicated multistage computation method, in which an N_{BR} blade-rows turbomachinery is computed using N_{BR} parallel computations. The computation corresponding to the representative passage of blade-row n_{BR} is run on a grid which discretizes the blade-row and extends over all neighbouring blade-rows [7]. The neighbouring blade-rows are not discretized, but their effect is simulated using source terms, which are evaluated at every iteration from the computations of the corresponding blade-rows. Although this approach has been used in multistage computations [3] it is both complicated and time-consuming when compared to the mixing-plane method, because computing requirements grow as N_{BR}^2 [38]. Turner [38] developed a variant to the deterministic-stresses method, where the grid of each of the N_{BR} parallel computations is truncated to include only the immediately upstream and downstream blade-rows, thus substantially reducing computing-time requirements. Rhie et al [33] further reduced the extent of overlap, to include only the immediately downstream blade-row and part of the upstream blade-row.

Recently Hall [25] developed a methodology that is essentially a mixing-plane method, but included in the computations deterministic-stresses, based on simple wake development models [26]. The originality of this approach (MP + DSM) is that it conserves the saving simplicity of the mixing-plane method, while including a correction for the deterministic-stresses. The drawback is that this correction is semi-empirical, and is not obtained as part of the computation.

The truncated unsteady methods are most certainly the best way for taking into account the multistage interaction effects. Giles [23] suggested a small-perturbation approach, based on linearization around the mixing-plane approach steady results, to compute the second-order effects of unsteadiness on turbomachinery performance [16], that could also be used to obtain models for the deterministic-stresses [17]. He [27] has used frequency filtering of unsteady signals to analyze the effect of 2 sources of unsteadiness with different chorochronic periodicities on a blade-row. Silkowski and Hall [37] used a linearized modal analysis to compute the interaction of a vibrating blade-row with its multistage environment. These methods are however at the beginning of their development.

The building block of all these methodologies is a steady and/or unsteady Reynolds-averaged Navier-Stokes solver associated with a turbulence closure. It is obvious that the quality of this solver has a paramount effect on the overall predictive capability of the method.

In the present work a mixing-plane methodology is developed and validated by comparison with measurements obtained on a 4-stage turbine (Fig. 1) tested by Evers [15] at the Institut für Strömungsmaschinen Universität Hannover (that we will abbreviate as ISUH_1 4-stage turbine). After validating the method at 3 operating points (design, part-load at design-speed, and part-load at part-speed) a computation was run at reduced-massflow condition at design speed. The results were analyzed in an attempt to understand the multistage flow structure at these conditions.

Multiblock Multidomain Strategy

The computational domain is separated in blocks and subdomains. The ISUH_1 4-stage turbine (Fig. 1) is made up of $N_{BR} = 8$ blade-rows ($n_{BR} = 1, \cdots, N_{BR}$). The computational domain extends from the user-defined inflow-boundary ($x = x_{in}$) to the user-defined outflow-boundary ($x = x_{out}$). The computational grid (Fig. 2) is made up of $L_{BR} = N_{BR} + 2 = 10$ blocks: an upstream H-grid ($\ell_{BR} = 1$), 8 O-grids corresponding to the 8 blade-rows ($\ell_{BR} = 2, \cdots, N_{BR} + 1$), and a downstream H-grid ($\ell_{BR} = L_{BR}$).

In order to define the interfaces between grid blocks a meridional grid is generated algebraically, with N_k nodes radially from hub to casing (Fig. 3). The nodes in the radial direction are streched near the hub ($\frac{1}{3}N_k$) and near the casing ($\frac{1}{3}N_k$). The stretching is geometric with ratio r_k [22]. The remaining $\frac{1}{3}N_k$ nodes are equally distributed in-between.

The block interfaces between successive blade-rows are located at mid-axial-distance between the trailing-edge of a row and the leading-edge of the next row, at every $k = \text{const}$ grid-line. This works always, even when the leading-edge and trailing-edge lines have complicated forms in the meridional plane. The interface between $\ell_{BR} = 1$ and $\ell_{BR} = 2$ (Fig. 2) is at a user-defined axial distance dx_u from the leading-edge of the $n_{BR} = 1$ blading

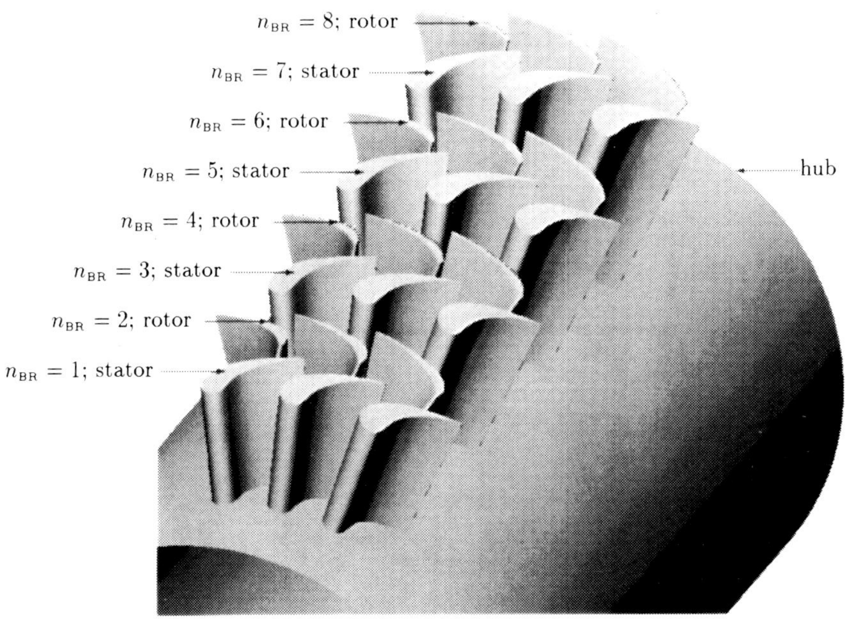

Figure 1: View of ISUH_1 4-stage turbine [15].

(this distance is the same at every $k = \mathrm{const}$ grid-line, so that the interface has the same form in the meridional plane as the leading-edge). In the same way the interface between $\ell_{\mathrm{BR}} = L_{\mathrm{BR}} - 1$ and $\ell_{\mathrm{BR}} = L_{\mathrm{BR}}$ is at a user-defined axial distance dx_d from the trailing-edge of the $n_{\mathrm{BR}} = N_{\mathrm{BR}}$ blading.

Several blocks are resolved using a single subdomain with a structured grid (UH-grid, DH-grid, O-grids around the stators). There are however blocks (Fig 4) that require more than 1 subdomains (O-grids around the rotors that have tip-clearance gaps). In this case the block (ℓ_{BR}) is meshed using 3 subdomains: the basic O-grid around the blade ($n_{\mathrm{DOM}} = 1$), the tip-clearance TC-grid ($n_{\mathrm{DOM}} = 2$), and a domain (OZ-grid) embedded into the blade O-grid ($n_{\mathrm{DOM}} = 3$) that is used to accurately resolve the flow around the tip-edges. There are therefore 4 different indices, n_{BR} denoting the blade-row, ℓ_{BR} denoting the grid block, n_{DOM} denoting the subdomain within a given block, and $n_{\ell d}$ denoting the subdomain in a global grid storage (here $n_{\ell d} = 1, \cdots, 18$). Note that the UH-grid ($\ell_{\mathrm{BR}} = 1$) is considered part of the $n_{\mathrm{BR}} = 1$ blade-row, and the DH-grid ($\ell_{\mathrm{BR}} = L_{\mathrm{BR}}$) is considered part of the $n_{\mathrm{BR}} = N_{\mathrm{BR}}$ blade-row.

The 3-D grid is generated in each block using the biharmonic generation methodology developed by Gerolymos and Tsanga [22]. This original technique introduced by Gerolymos and Tsanga [22] renders the blade O-grid independent of the exact position of the blade-tip and allows grid-clustering at the blade-tip locally (a definite advantage for multistage turbomachinery grid-generation). For the ISUH_1 4-stage turbine [15] the final grid (Fig. 5) of 4 786 730 points (without counting the O-grid points overlapped by the OZ-grid [21]) has 71 radial stations from hub to casing (Tab. 2), and 21 radial stations within

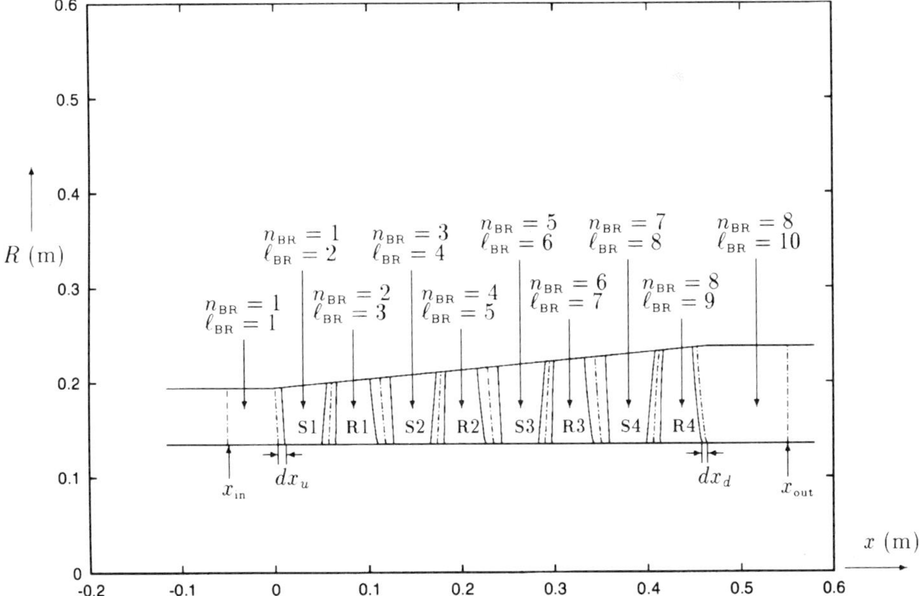

Figure 2: Meridional view of the ($\ell_{\mathrm{BR}} = 1, \cdots, 10$) grid blocks for ISUH_1 4-stage turbine [15].

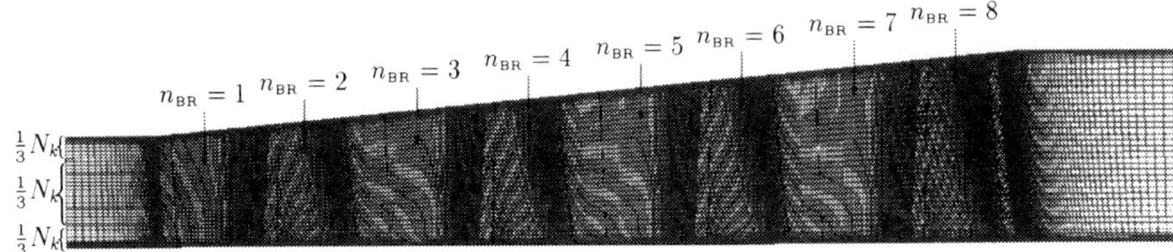

Figure 3: Meridional grid for ISUH_1 4-stage turbine [15].

the tip-clearance gaps of the rotors. This grid corresponds to $n_w^+ < \frac{3}{4}$ everywhere, except within the tip-clearances where $n_w^+ < \frac{3}{2}$. Based on grid influence studies in transonic compressors [20] the n_w^+ and the geometric stretchings used in this grid should give quite accurate results, although grid-convergence of results is not exactly achieved.

Navier-Stokes Solver

The flow is modelled by the 3-D Favre-Reynolds-averaged Navier-Stokes equations using the Launder-Sharma near-wall $k - \varepsilon$ turbulence closure [19]. The mean flow and turbulence transport equations are discretized using a finite-volume method based on MUSCL Van Leer flux-vector-splitting with Van Albada limiters. The mean flow and turbulence equations are integrated in time using a fully coupled approximately-factored implicit backward-Euler method. The single-blade-row solver is described in detail by Gerolymos et al [20],

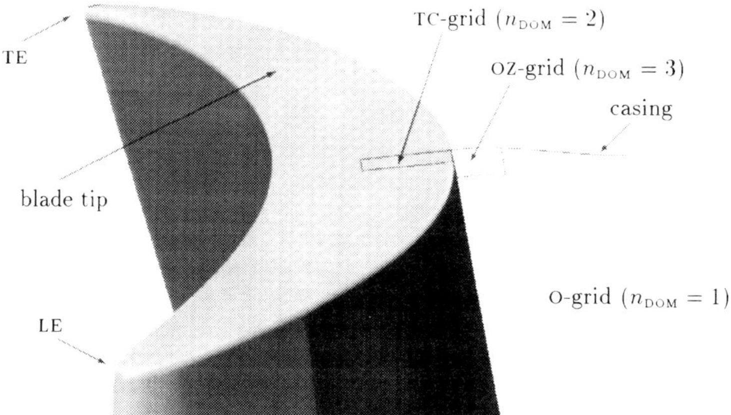

Figure 4: Subdomains for tip-clearance flow computation.

and has been validated by comparison with measurements [21].

In the computations presented in this paper an inflow reservoir condition is applied (total pressure, total temperature, flow-angles, k and ε^*). At the outflow boundary static pressure at the tip is imposed, together with a radial equilibrium condition. Pitchwise periodicity and H-O interfaces are treated using a phantom nodes technique [20]. An adiabatic no-slip condition is applied on solid walls. Care was taken to correctly model fixed and rotating parts of the hub, based on the turbine rig plans [15].

Mixing-Planes

At mixing-planes between blade-rows (O-O-interfaces) a novel phantom-nodes technique is used to exchange information. The O-grid of blade-row n_{BR} is extended upstream (downstream) into the O-grid of blade-row $n_{\text{BR}} - 1$ ($n_{\text{BR}} + 1$) using 5 phantom-nodes (Fig. 6). The upstream-phantom-nodes correspond to $j \in [N_j(n_{\text{BR}}) + 1, N_j(n_{\text{BR}}) + 5]$ for $i \in [i_{u_1}(n_{\text{BR}}) + 1, i_{u_2}(n_{\text{BR}}) - 1]$ at every radial station k. The downstream-phantom-nodes correspond to $j \in [N_j(n_{\text{BR}}) + 1, N_j(n_{\text{BR}}) + 5]$ for $i \in [i_{d_2}(n_{\text{BR}}) + 1, N_i(n_{\text{BR}})] \cup [1, i_{d_1}(n_{\text{BR}}) - 1]$ at every radial station k. The axial and radial locations of these phantom-nodes are taken from the axial and radial locations of the $i = $ const grid-line of the neighbouring domain that passes through the middle point of the interface. The phantom-nodes are equidistant pitchwise. Thus, the upstream-phantom-nodes of blade-row n_{BR} are located at

$$x_{\text{O}}(n_{\text{BR}}, i, j, k) = x_{\text{O}}(n_{\text{BR}} - 1, 1, N_j(n_{\text{BR}} - 1) - (j - N_j(n_{\text{BR}})), k)$$
$$R_{\text{O}}(n_{\text{BR}}, i, j, k) = R_{\text{O}}(n_{\text{BR}} - 1, 1, N_j(n_{\text{BR}} - 1) - (j - N_j(n_{\text{BR}})), k)$$
$$\forall\, i \in [i_{u_1}(n_{\text{BR}}) + 1, i_{u_2}(n_{\text{BR}}) - 1] \quad \forall\, j \in [N_j(n_{\text{BR}}) + 1, N_j(n_{\text{BR}}) + 5] \quad \forall\, k \tag{1}$$

and the downstream-phantom-nodes of blade-row n_{BR} are located at

$$x_{\text{O}}(n_{\text{BR}}, i, j, k) = x_{\text{O}}(n_{\text{BR}} + 1, N_i(n_{\text{BR}} + 1)/2 + 1, N_j(n_{\text{BR}} + 1) - (j - N_j(n_{\text{BR}})), k)$$
$$R_{\text{O}}(n_{\text{BR}}, i, j, k) = R_{\text{O}}(n_{\text{BR}} + 1, N_i(n_{\text{BR}} + 1)/2 + 1, N_j(n_{\text{BR}} + 1) - (j - N_j(n_{\text{BR}})), k)$$
$$\forall\, i \in [i_{d_2}(n_{\text{BR}}) + 1, N_i(n_{\text{BR}})] \cup [1, i_{d_1}(n_{\text{BR}}) - 1] \quad \forall\, j \in [N_j(n_{\text{BR}}) + 1, N_j(n_{\text{BR}}) + 5] \quad \forall\, k \tag{2}$$

The downstream phantom-nodes are also distributed equidistantly pitchwise (Fig. 6). At every iteration these phantom-nodes are updated using the meridional averages of the neighbouring blade-row at the corresponding (x, R) location. The computational grids

Table 2: Grid domains for ISUH_1 4-stage turbine[15] grid.

ℓ_{BR}	n_{BR}	n_{ld}	N_{B}	N_i	N_j	N_k	r_j	r_k	N_{ijk}	$k_{\text{O-OZ}}$
1	1	1 (UH)	29	31	31	71		1.47	68231	
2	1	2 (O)	29	181	41	71	1.44	1.47	526891	
3	2	3 (O)	30	181	41	71	1.44	1.47	526891	54
		4 (TC)		181	17	21		1.50	64617	
		5 (OZ)		181	21	31			117831	
4	3	6 (O)	29	181	41	71	1.44	1.47	526891	
5	4	7 (O)	30	181	41	71	1.44	1.47	526891	54
		8 (TC)		181	17	21		1.50	64617	
		9 (OZ)		181	21	31			117831	
6	5	10 (O)	29	181	41	71	1.44	1.47	526891	
7	6	11 (O)	30	181	41	71	1.44	1.47	526891	54
		12 (TC)		181	17	21		1.50	64617	
		13 (OZ)		181	21	31			117831	
8	7	14 (O)	29	181	41	71	1.44	1.47	526891	
9	8	15 (O)	30	181	41	71	1.44	1.47	526891	54
		16 (TC)		181	17	21		1.50	64617	
		17 (OZ)		181	21	31			117831	
10	8	18 (DH)	30	61	41	71		1.47	177571	

used are O-grids. To compute the meridional averages for updating the upstream-phantom-nodes (downstream-phantom-nodes) of blade-row n_{BR} an auxiliary H-grid is created in the neighbouring blade-row $n_{\text{BR}}-1$ ($n_{\text{BR}}+1$), at the (x, R) locations of the phantom-nodes. The flowfield of the neighbouring blade-row is interpolated in the auxiliary H-grid. The radial grid-surfaces being continuous between blade-rows, a surface interpolation procedure is applied. The O-grid of the neighbouring blade-row at a given $k = $ const surface is divided in triangles using the (i, j) cells diagonal, and after determining the triangle that contains a point of the auxiliary grid, a linear area-coordinates interpolation is applied [40]. The interpolation coefficients are computed only once at the beginning of the iterations, so that the computational cost of the interpolation is negligible. The interpolation is done on the variables $[\bar{\rho}, \bar{\rho}\tilde{V}_x, \bar{\rho}\tilde{V}_R, \bar{\rho}\tilde{V}_\theta, \check{h}_t, \text{k}, \varepsilon^*]^{\text{T}}$. Then meridional averages are computed

$$\rho_{\text{M}}(n_{\text{BR}} \pm 1, x, R) = \frac{N_{\text{B}}(n_{\text{BR}} \pm 1)}{2\pi} \int_{\theta_-(n_{\text{BR}}\pm1,x,R)}^{\theta_+(n_{\text{BR}}\pm1,x,R)} \bar{\rho}\,d\theta$$

$$\rho_{\text{M}}(n_{\text{BR}} \pm 1, x, R)\, V_{x_{\text{M}}}(n_{\text{BR}} \pm 1, x, R) = \frac{N_{\text{B}}(n_{\text{BR}} \pm 1)}{2\pi} \int_{\theta_-(n_{\text{BR}}\pm1,x,R)}^{\theta_+(n_{\text{BR}}\pm1,x,R)} \bar{\rho}\tilde{V}_x\,d\theta$$

$$\rho_{\text{M}}(n_{\text{BR}} \pm 1, x, R)\, V_{R_{\text{M}}}(n_{\text{BR}} \pm 1, x, R) = \frac{N_{\text{B}}(n_{\text{BR}} \pm 1)}{2\pi} \int_{\theta_-(n_{\text{BR}}\pm1,x,R)}^{\theta_+(n_{\text{BR}}\pm1,x,R)} \bar{\rho}\tilde{V}_R\,d\theta$$

$$\rho_{\text{M}}(n_{\text{BR}} \pm 1, x, R)\, V_{\theta_{\text{M}}}(n_{\text{BR}} \pm 1, x, R) = \frac{N_{\text{B}}(n_{\text{BR}} \pm 1)}{2\pi} \int_{\theta_-(n_{\text{BR}}\pm1,x,R)}^{\theta_+(n_{\text{BR}}\pm1,x,R)} \bar{\rho}\tilde{V}_\theta\,d\theta$$

$$\text{k}_{\text{M}}(n_{\text{BR}} \pm 1, x, R) = \frac{N_{\text{B}}(n_{\text{BR}} \pm 1)}{2\pi} \int_{\theta_-(n_{\text{BR}}\pm1,x,R)}^{\theta_+(n_{\text{BR}}\pm1,x,R)} \text{k}\,d\theta$$

$$\varepsilon^*_{\text{M}}(n_{\text{BR}} \pm 1, x, R) = \frac{N_{\text{B}}(n_{\text{BR}} \pm 1)}{2\pi} \int_{\theta_-(n_{\text{BR}}\pm1,x,R)}^{\theta_+(n_{\text{BR}}\pm1,x,R)} \varepsilon^*\,d\theta \qquad (3)$$

where $\theta_\pm$ are the pitchwise periodicity boundaries (Fig. 6).

These averages are used to update the phantom nodes at the beginning of each iteration. It is important to note that a particular treatment is applied at the O-O-interface ($j = N_j$) at the end of every iteration. After extensive testing on various compressor and turbine configurations it was found that forcing continuity at the interface by taking the average value from each side leads to space-wise perturbation of the solution and is

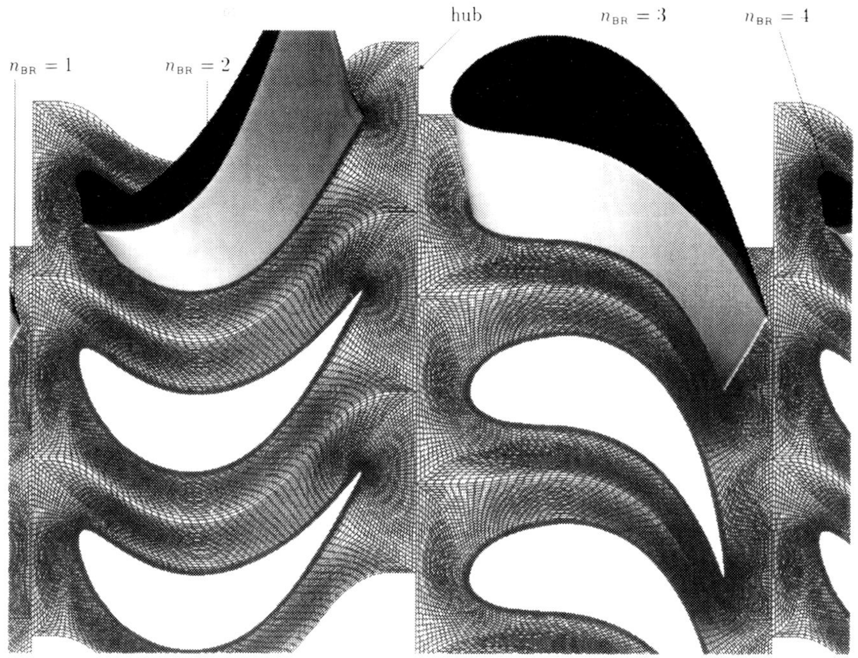

Figure 5: Partial view of the grid at the hub of ISUH_1 4-stage turbine [15].

detrimental for global massflow conservation. It was found that a relaxation technique that exchanges the meridional averages from each side is the best way to ensure continuity at the interfaces (in the sense of meridional averages) and global massflow conservation. At the end of each iteration the meridional averages on each side of the interface $x_{\text{o-o}}(n_{\mathrm{BR}}, n_{\mathrm{BR}} \pm 1, R)$ are computed. The pitchwise gradients on each side of the interfaces are taken from the solution (denoted SCH), but at each radial station the meridional averages are exchanged between blade-rows

$$
\begin{aligned}
{}^{n+1}\bar{\rho}\ (n_{\mathrm{BR}}, x, R, \theta) &= {}^{\mathrm{SCH}}\bar{\rho}\ (n_{\mathrm{BR}}, x, R, \theta) + [\ \rho_{\mathrm{M}}\ (n_{\mathrm{BR}} \pm 1, x, R) - \rho_{\mathrm{M}}\ (n_{\mathrm{BR}}, x, R)\ ] \\
{}^{n+1}\tilde{V}_x\ (n_{\mathrm{BR}}, x, R, \theta) &= {}^{\mathrm{SCH}}\tilde{V}_x\ (n_{\mathrm{BR}}, x, R, \theta) + [\ V_{x_{\mathrm{M}}}\ (n_{\mathrm{BR}} \pm 1, x, R) - V_{x_{\mathrm{M}}}\ (n_{\mathrm{BR}}, x, R)\ ] \\
{}^{n+1}\tilde{V}_R\ (n_{\mathrm{BR}}, x, R, \theta) &= {}^{\mathrm{SCH}}\tilde{V}_R\ (n_{\mathrm{BR}}, x, R, \theta) + [\ V_{R_{\mathrm{M}}}\ (n_{\mathrm{BR}} \pm 1, x, R) - V_{R_{\mathrm{M}}}\ (n_{\mathrm{BR}}, x, R)\ ] \\
{}^{n+1}\tilde{V}_\theta\ (n_{\mathrm{BR}}, x, R, \theta) &= {}^{\mathrm{SCH}}\tilde{V}_\theta\ (n_{\mathrm{BR}}, x, R, \theta) + [\ V_{\theta_{\mathrm{M}}}\ (n_{\mathrm{BR}} \pm 1, x, R) - V_{\theta_{\mathrm{M}}}\ (n_{\mathrm{BR}}, x, R)\ ] \\
{}^{n+1}k\ (n_{\mathrm{BR}}, x, R, \theta) &= {}^{\mathrm{SCH}}k\ (n_{\mathrm{BR}}, x, R, \theta) + [\ k_{\mathrm{M}}\ (n_{\mathrm{BR}} \pm 1, x, R) - k_{\mathrm{M}}\ (n_{\mathrm{BR}}, x, R)\ ] \\
{}^{n+1}\varepsilon^*\ (n_{\mathrm{BR}}, x, R, \theta) &= {}^{\mathrm{SCH}}\varepsilon^*\ (n_{\mathrm{BR}}, x, R, \theta) + [\ \varepsilon_{\mathrm{M}}^*\ (n_{\mathrm{BR}} \pm 1, x, R) - \varepsilon_{\mathrm{M}}^*\ (n_{\mathrm{BR}}, x, R)\ ]
\end{aligned}
\tag{4}
$$

The continuity of meridional averages at O-O-interfaces that is obtained using this relaxation technique is illustrated by considering the interface between rotor R3 ($n_{\mathrm{BR}} = 6$) and stator S4 ($n_{\mathrm{BR}} = 7$) and the interface between stator S4 ($n_{\mathrm{BR}} = 7$) and rotor R4 ($n_{\mathrm{BR}} = 8$) and plotting pitchwise averages of density ρ_{M}, axial velocity $V_{x_{\mathrm{M}}}$, pressure p_{M}, total pressure $p_{t_{\mathrm{M}}}$, and total temperature $T_{t_{\mathrm{M}}}$ (Fig. 7). The results presented are for the design-point of the turbine (they are typical of results obtained at the other interfaces and/or other operating points ; in fact they illustrate the worst continuity conditions obtained). It is seen

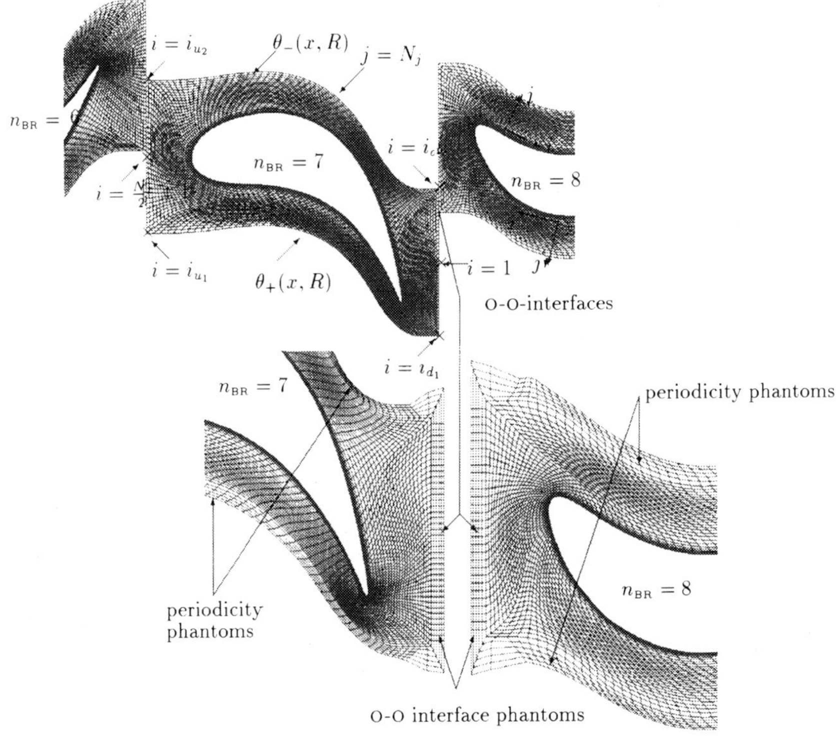

Figure 6: Phantom-nodes strategy for O–O interfaces.

that the mixing-plane methodology used ensures excellent conservation at the interfaces.

These results (Fig. 7) are in complete disagreement with the work of Rhie et al [33] who reported important discontinuities and even erroneous radial gradients at the interfaces of a 3-stage compressor using the mixing-plane method. In fact the continuity obtained in this work is better than what was obtained by Rhie et al [33] using the deterministic-stresses method. This is attributed to the relaxation procedure used in the implementation of the mixing-plane method, and also to the fact that using 5 phantom nodes ensures continuity not only of values but also of their streamwise derivatives at the interfaces.

Comparison with Measurements

Data were available for the ISUH_1 4-stage turbine [15] at design-speed and at 75% of design-speed (Fig. 8). Computations were run at 4 operating points (Tab. 3): the design-point (point 1), a part-load point at design-speed (point 2), a part-load point at part-speed (point 3) and a reduced-massflow point at design-speed (point 4). Detailed rake measurements at the exit planes of the rotors were available for points 1, 2, and 3 [15]. These data were used for the validation of the methodology developed in the present paper. Point 4 was computed because it is just after the limit between normal part-load operation and reduced-massflow operation, where the efficiency-massflow curve of the turbine drops

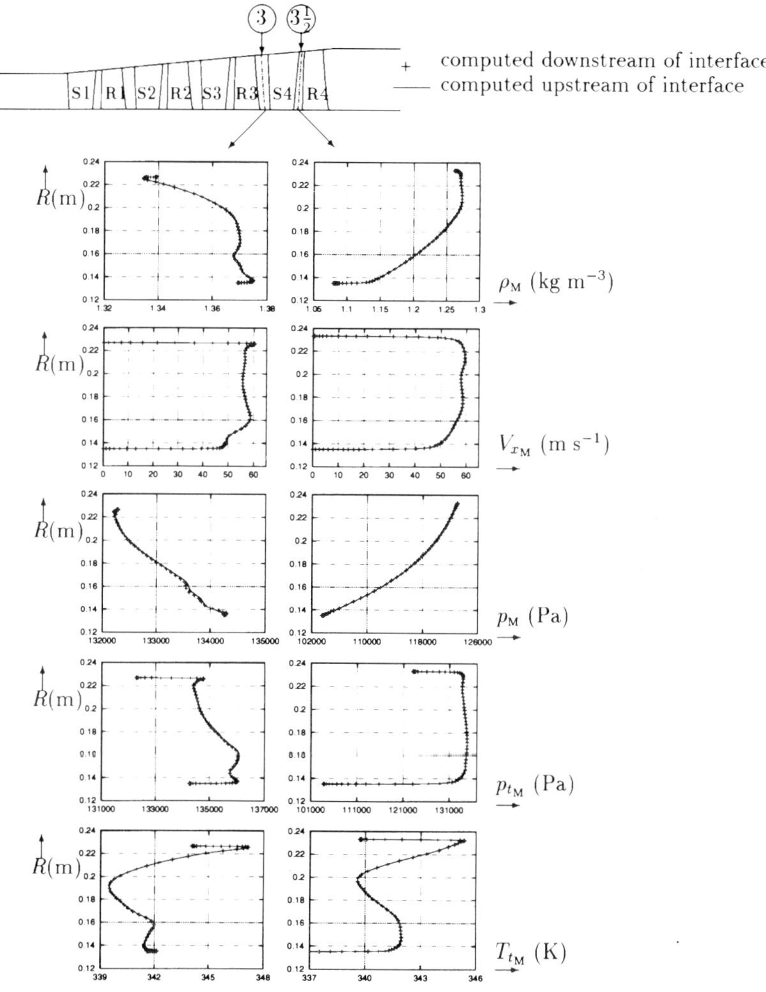

Figure 7: Continuity of meridional averages at O-O interfaces at design-point of ISUH_1 4-stage turbine [15].

rapidly (Fig. 8). Although detailed data were not available for operating point 4, it was interesting to use the method in studying the particular conditions of high negative incidence and important mismatch between the stages that characterize this point.

Table 3: Computed operating points for ISUH_1 4-stage turbine [15].

point		$\dot{m}$	RPM
1	design	$\dot{m}_D = 7.8$ kg s^{-1}	RPM$_D$ = 7500
2	part-load	$0.53\,\dot{m}_D$	RPM$_D$
3	part-speed	$0.45\,\dot{m}_D$	0.75 RPM$_D$
4	reduced-massflow	$0.41\,\dot{m}_D$	RPM$_D$

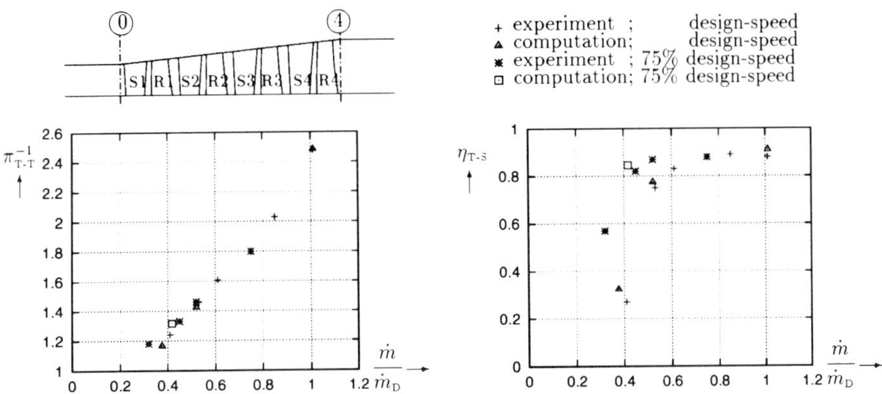

Figure 8: Computed and measured operating map for ISUH_1 4-stage turbine [15].

The performance characteristics of the turbine are defined [15] by the total-to-total pressure-ratio $\pi_{\text{T-T}}^{-1}$ and the total-to-static efficiency $\eta_{\text{T-s}}$ between stations 0 and 4 (Fig. 8)

$$\pi_{\text{T-T}}^{-1} = \frac{\widehat{p}_{t_0}}{\widehat{p}_{t_4}} \qquad \eta_{\text{T-s}} = \frac{1 - \dfrac{\widehat{h}_{t_4}}{\widehat{h}_{t_0}}}{1 - \left(\dfrac{\widehat{p}_4}{\widehat{p}_{t_0}}\right)^{\frac{\gamma-1}{\gamma}}} \tag{5}$$

In the computations $\widehat{p}_{t_0}$ and $\widehat{p}_{t_4}$ are massflow-weighted averages of total pressure at stations 0 and 4, $\widehat{h}_{t_0}$ and $\widehat{h}_{t_4}$ are massflow-weighted averages of total enthalpy at stations 0 and 4, and $\widehat{p}_4$ is the massflow-weighted average of static pressure at station 4 (the detailed averaging procedure used for processing the measurements is not given in the data description [15]). The pressure-ratio curve is quite accurately predicted by the method, for all operating points (Fig. 8). Comparison of total-to-static efficiency of the turbine is less satisfactory. The computations systematically overpredict efficiency. The overestimation of efficiency is however roughly the same ($\sim 2\%$) for all operating points, so that the efficiency-massflow curves are slightly shifted between computation and experiment. The experimental accuracy in determining the efficiency is given as $\pm 0.8\%$ [15]. There are many possible explanations for this discrepancy: 1) the computations model neither the small gaps between the rotating and stationary parts of the hub nor the labyrinth clearances below stators [15], and as a consequence neglect the associated losses; 2) the computational grid used, although rather fine by comparison with published work in the field, is not sufficient for grid convergence of the method [20], both in radial and in blade-to-blade resolution and also in the tip-clearance gap (Tab. 2), and this may partly explain the discrepancies in efficiency; 3) the turbulence model used (Launder-Sharma near-wall $k - \varepsilon$) has deficiencies, both in predicting flow detachment [19] and flow transition [21]; 4) unsteady effects are expected to have an influence on performance [18]; 5) efficiency is a very sensitive quantity and recent precise measurements using a fast aspirating probe in a multistage compressor [36] indicate substantial differences in efficiency ($\pm 2\%$) between blades of the same blade-row (passage-to-passage differences). Note that Turner [38] in his DS multistage computations of a 2-stage turbine also overpredicted efficiency by $\sim 2\%$, so that the accuracy obtained by the method may be considered state-of-the-art. What is important is that the computations correctly predict the drop in efficiency with decreasing massflow (Fig. 8), but for a shift of the curve towards slightly higher efficiency ($+2\%$).

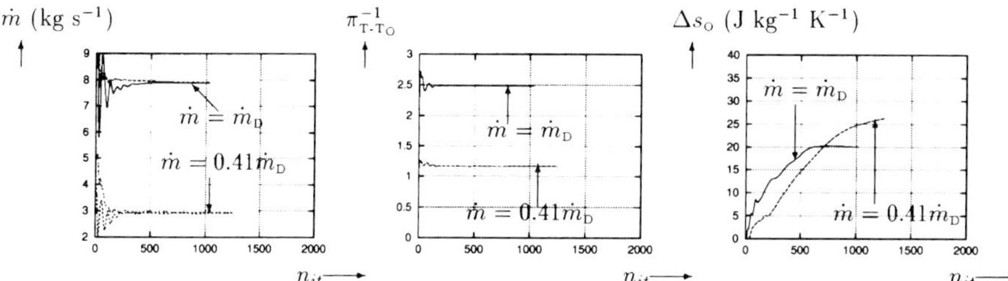

Figure 9: Convergence of computations at design speed for ISUH_1 4-stage turbine [15].

All the computations were run until full convergence (Fig. 9) of massflow, pressure-ratio and entropy-rise between inflow (station 0) and outflow (station 4). More iterations are needed for the computation of off-design points (Fig. 9) because of substantial flow detachment, in agreement with similar observations in compressor rotor computations.

Measurements were available at rotor exit planes [15]. Computations were compared with measurements for the design-point ($\dot{m} = \dot{m}_D$) and the part-load point at design-speed ($\dot{m} = 0.53\dot{m}_D$). Computed results are pitchwise massflow-weighted averages of total pressure p_{t_M} and total temperature, and absolute flow-angle α_M (computed from mass-weighted pitchwise-averaged velocities and measured from the meridional direction). The radial distributions of these quantities were compared with rake measurements (Fig. 10).

At inflow measured profiles of p_{t_M}, T_{t_M}, and α_M were applied as boundary conditions. A turbulence intensity of $T_u = 3\%$ was imposed at inflow for all the computations, while the boundary-layer thicknesses applied at inflow (estimated from measured p_{t_M} distributions) were $\delta_{HUB} = 5$ mm at the hub, and $\delta_{CASING} = 10$ mm at the casing (these quantities are important in estimating the k and ε^* radial distributions at inflow, which are applied as boundary conditions [20]).

Comparison of computational and experimental results is globally satisfactory (Fig. 10). At turbine exit (station 4) the agreement between computation and experiment is quite satisfactory, both at design-point ($\dot{m} = \dot{m}_D$) and at part-load operation ($\dot{m} = 0.53\dot{m}_D$). Total pressures p_{t_M} are quite accurately predicted at turbine exit, as well as flow-angles α_M. Total temperatures T_{t_M} lower by 2–3 K than the measured ones are predicted at turbine exit (station 4), for both operating points: this is consistent with the overestimation of efficiency by 2%. For the design-point, the α_M peak near the hub that is associated with the rotor secondary flows is predicted at a lower radial location than the measured one, but Gallus et al [18] have shown that this may be due to unsteady flow effects, that are of course neglected in this steady calculation.

Comparison of computations with measurements is less satisfactory in the intermediate planes, especially at rotors R2 and R3 exit. It is possible that these discrepancies are due to unsteady effects, the hub and labyrinth clearances, or in the case of α_M distributions to potential effects due to the leading-edge of the stators (it is not known whether the experimental data were obtained as averages at different pitchwise locations with respect

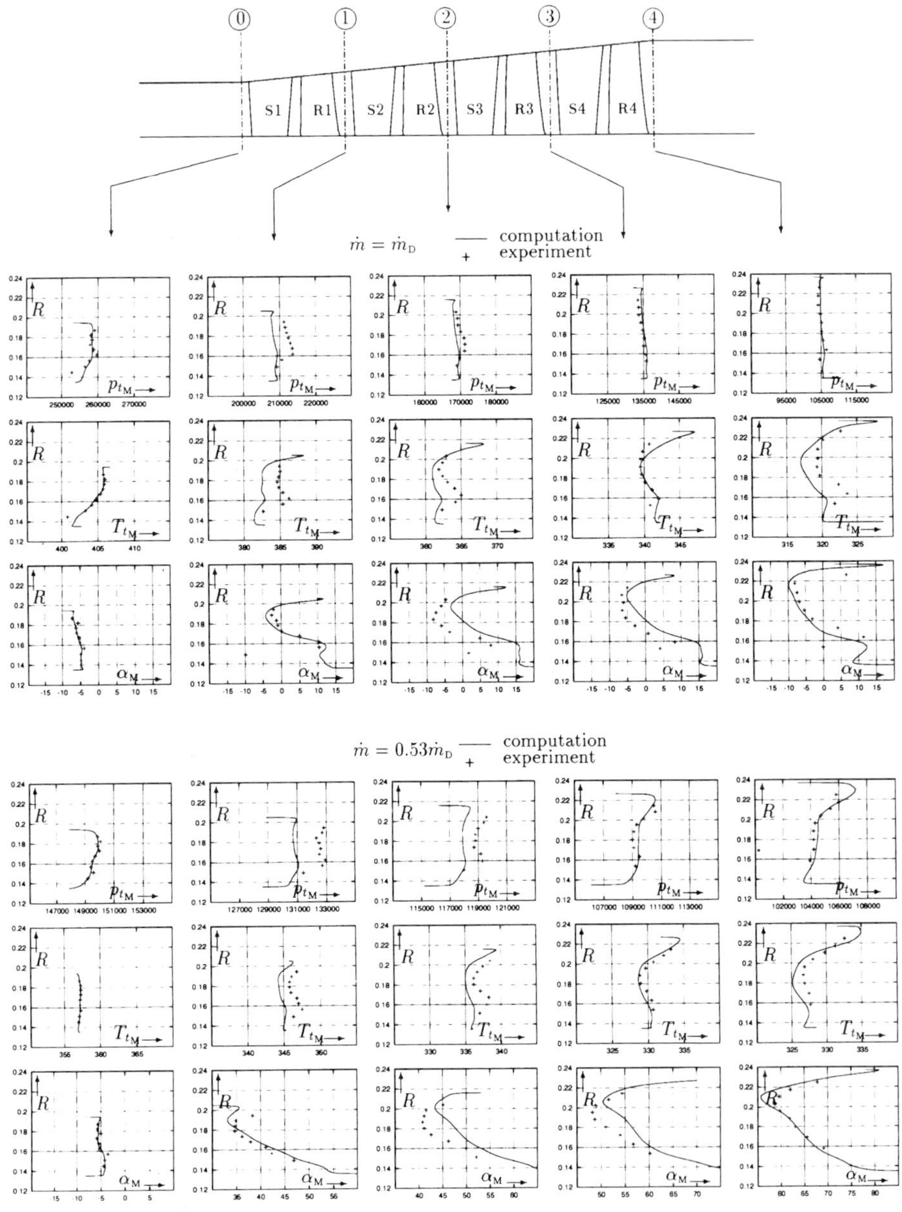

Figure 10: Comparison of computation and experiment at 2 operating points ($\dot{m} = 0.53\dot{m}_\mathrm{D}$ and $\dot{m} = 0.53\dot{m}_\mathrm{D}$) at design speed for ISUH_1 4-stage turbine [15] (R[m], p_{t_M}[Pa], T_{t_M}[K], α_M[deg]).

to the stators blades, or if they were just rake measurements at stator mid-passage).
Another important reason for these discrepancies may be the effect of the upstream struts
present in the experimental set-up [15], and presumably responsible for the slight inlet
swirl (Fig. 10), which were not modelled in the computations: this might explain why the
agreement between computations and experiment is improved as one moves downstream
toward the exit.

Globally the comparison with measurements is quite satisfactory, despite the discrepancies discussed above. In particular the variations in α_M distributions between the design-
point ($\dot{m} = \dot{m}_D$) and the part-load operation ($\dot{m} = 0.53\dot{m}_D$) is correctly predicted. The
difference between design and part-load α_M values increases from the rotor R1 exit (~ 35
deg) toward the rotor R4 exit (~ 65 deg): this difference corresponds to off-design negative
incidence for the stators. The results obtained at part-load demonstrate that the method
is capable of correctly predicting the off-design operation in multistage turbomachinery.

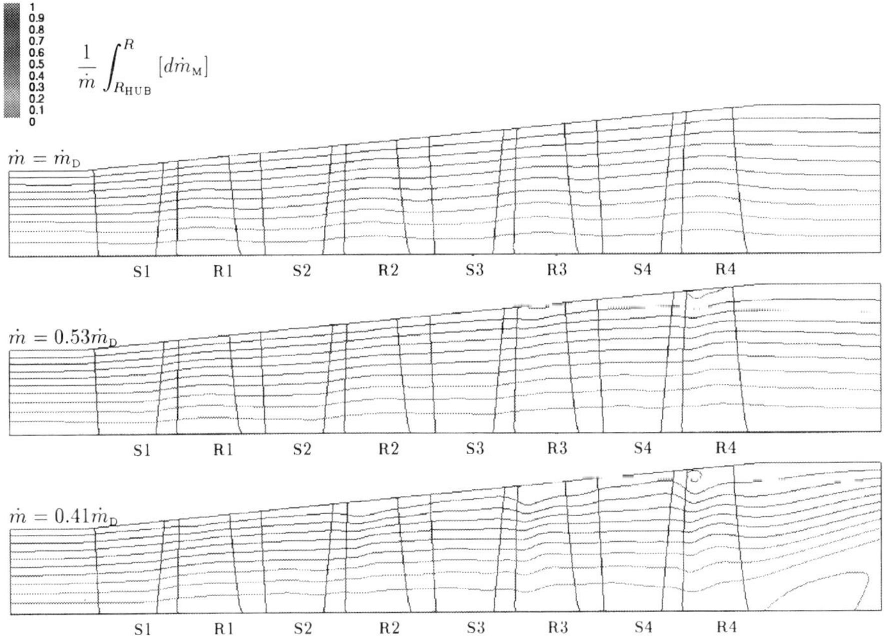

Figure 11: Meridional streamlines for the 3 operating points at design-speed for ISUH_1
4-stage turbine [15].

Reduced-Massflow Operation

It has been observed experimentally both for isolated rotor configurations [5] and for
multistage turbines [34] that at very low massflow operation substantial separation of

meridional streamlines appears at the hub. Petrovic and Rieß [32] developed a throughflow computational method for analyzing such low-load reduced-massflow operation, in a single-stage turbine, whose geometry is the same as the repetitive stages of the 4-stage turbine computed in the present work. Using adequate loss and deviation models Petrovic and Rieß [32] obtained very good agreement with experiment. Throughflow computations however do not offer detailed 3-D flow information to help in understanding the complex physical phenomena dominating the operation of multistage turbines at reduced massflow.

Although only global performance experimental data for operating point 4 (Tab. 3) were available in the database [15], computations were used for analyzing the corresponding multistage flowfield. The inflow conditions were obtained by linear extrapolation of the inflow conditions given for the other operating points at design-speed (the outflow pressure was approximately the same for all operating points, while inflow conditions varied [15]). Comparison of computed global performance results with measurements (Fig. 8) for this reduced-massflow point ($\dot{m} = 0.41\dot{m}_{\mathrm{D}}$) was as satisfactory as for the design-point and for the part-load point examined in the last section on the validation of the method. Based on the previous validation, the good agreement of global performance with measurements, and the generality of the flow-model used, it is believed that the computational results for the reduced-massflow operating point ($\dot{m} = 0.41\dot{m}_{\mathrm{D}}$) are sufficiently accurate to give a realistic picture of the 3-D flow through the turbine.

The 3-D flowfield is computed using O-grids (Fig. 5). A postprocessing H-grid is created, based on the mid-passage meridional grid (Fig. 3), and the flowfield is interpolated into this grid (using an interpolation procedure analogous to the one used for the mixing-planes phantom-nodes) to compute the pitchwise averaged quantities. The resulting throughflow streamlines for the 3 computed operating points at design-speed (Fig. 11) are in agreement with the experimental observations of Rieß and Evers [34]. At design-point ($\dot{m} = \dot{m}_{\mathrm{D}}$) the meridional streamlines are well behaved (Fig. 11), with the usual wavy shape often observed in turbine stages [39], and characteristic of free-vortex design [29] (the present turbine is of the free-vortex type with $\sim$50% reaction at mid-span [15]). At part-load ($\dot{m} = 0.53\dot{m}_{\mathrm{D}}$) the streamlines change both near the hub and near the casing (Fig. 11): the massflow fraction passing through the hub region is substantially reduced, expecially in the last 2 stages, and a recirculation region appears near the tip of the last 2 rotors. At reduced-massflow ($\dot{m} = 0.41\dot{m}_{\mathrm{D}}$) the features at the tip and the hub are much more pronounced (Fig. 11): quite large recirculating regions appear near the tip of rotors R4 and R3, and to a lesser extent of rotor R2; the massflow fraction passing through the hub region diminishes, with massive separation at the exit of rotor R4 causing the meridional streamlines to move swiftly upward. In this last case the outflow boundary condition is ill-posed, and induces premature reattachment before the outflow boundary. It is believed that this does not influence the flowfield structure in the rotor, but future computations will discretize the outlet diffuser (the flowpath has constant height but increasing radius [15]).

These differences in meridional streamlines pattern are associated with high negative incidence on the rotor blades (the stagnation-point moves away from the leading-edge to a point on the convex-side of the blade). Plots of pitchwise-averaged relative flow-angles β_{M} at the inlet planes of the rotors (Fig. 12) illustrate the increase in incidence as the massflow is reduced. At the design-point the flow-angle is the same for all the rotors (the turbine is designed to have the same blade sections in all stages at a given radius [15]). As the massflow diminishes the incidence, which can be represented by the difference of the flow-angle from its design value at the same radius (Fig. 12), increases, especially at

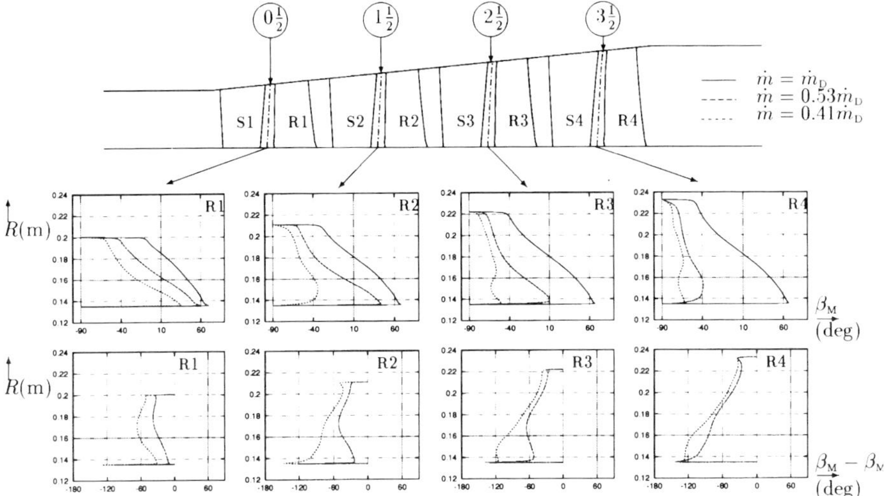

Figure 12: Relative flow-angle (β_M) and difference from the design value ($\beta_\mathrm{M} - \beta_{\mathrm{M_D}}$) at the inlet of the rotors for the 3 operating points at design-speed for ISUH_1 4-stage turbine [15].

the hub, where it is as high as $\beta_\mathrm{M} - \beta_{\mathrm{M_D}} = -120$ deg at the hub of rotor R4 at reduced-massflow ($\dot{m} = 0.41\dot{m}_\mathrm{D}$). At the rotor tip the incidence does not exceed -40 deg (Fig. 12). Petrovic and Rieß [32] performed throughflow computations of the low load operation of a single-stage version of this turbine (only the first of the 4 repetitive stages) and obtained satisfactory comparison with experiment, using loss and deviation correlations that included adequate spanwise distributions, which were considered essential for the success of the throughflow method. This clearly means that intense radial mixing occurs inside the rotor. The present 3-D computations offer the possibility of detailed examination of the flow structure in the blade passages. Examination of the Mach-number relative to the blades of the last stage ($\check{M}$ in the stator S4, and $\check{M}_\mathrm{w}$ in the rotor R4), at 25%-span and at 75%-span, illustrates (Fig. 13) the difference between the design-point ($\dot{m} = \dot{m}_\mathrm{D}$) and the reduced-massflow point ($\dot{m} = 0.41\dot{m}_\mathrm{D}$). At design-point the flow is very well-behaved, with the stagnation-points located correctly at the leading-edge of the blades (Fig. 13): as a consequence the blade-boundary-layers are very thin and the flow remains attached. The situation changes completely at reduced-massflow ($\dot{m} = 0.41\dot{m}_\mathrm{D}$) where both the stator S4 and the rotor R4 operate at high negative incidence, with the stagnation-points located far away from the leading-edge, on the convex-side of the blade (Fig. 13).

There is however a marked difference in flow structure between the stator and the rotor. There is massive separation on the concave-side of the blade at both spanwise stations within the stator S4 (Fig. 13). The flow structure appears much more complicated in the rotor R4 at reduced-massflow ($\dot{m} = 0.41\dot{m}_\mathrm{D}$). At 25%-span there is a region of low velocity located at a distance from the concave-side of the blade. Between this region and the blade-surface there is a region of high velocity. The same structure, although less pronounced is observed at 75%-span.

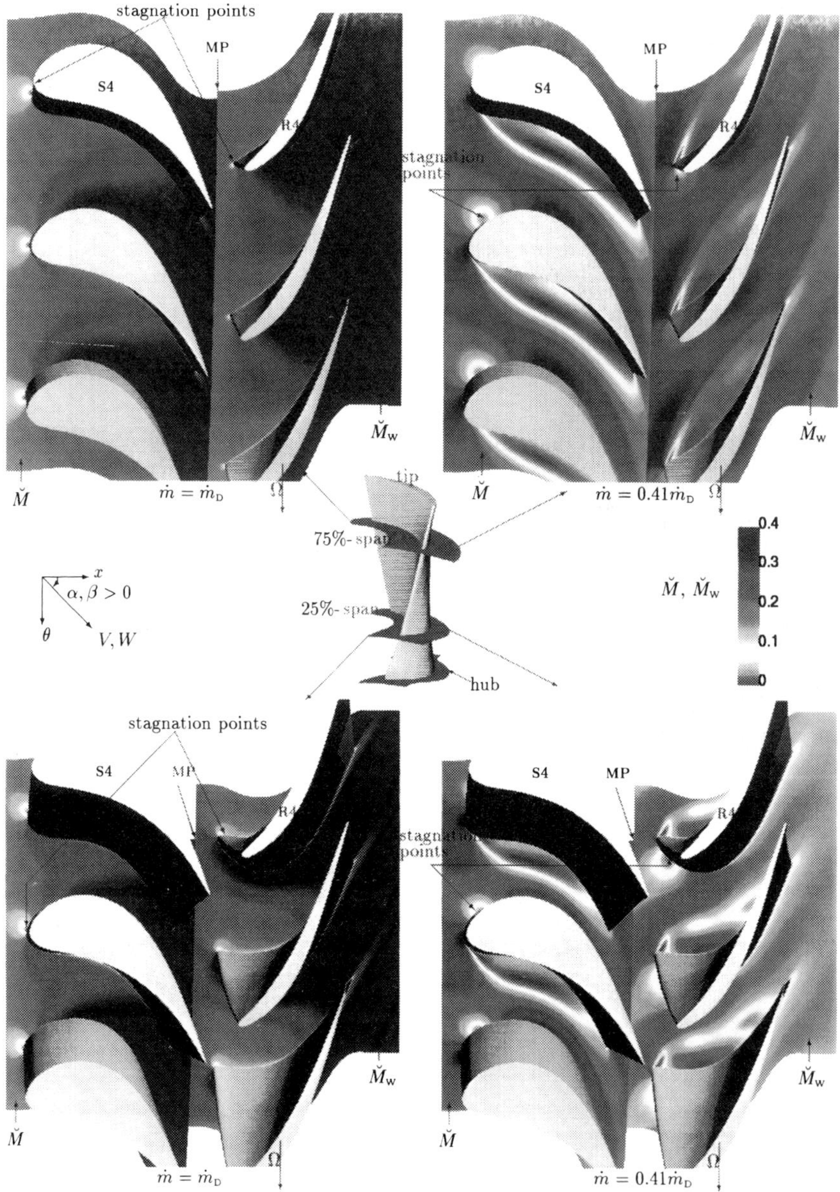

Figure 13: Plots of $\check{M}_{\mathrm{w}}$ for rotor R4 and $\check{M}$ for stator S4 at 25%-span and at 75%-span, at design-point ($\dot{m} = \dot{m}_{\mathrm{D}}$) and at reduced-massflow operation ($\dot{m} = 0.41\dot{m}_{\mathrm{D}}$), at design-speed for stage-4 of ISUH_1 4-stage turbine [15].

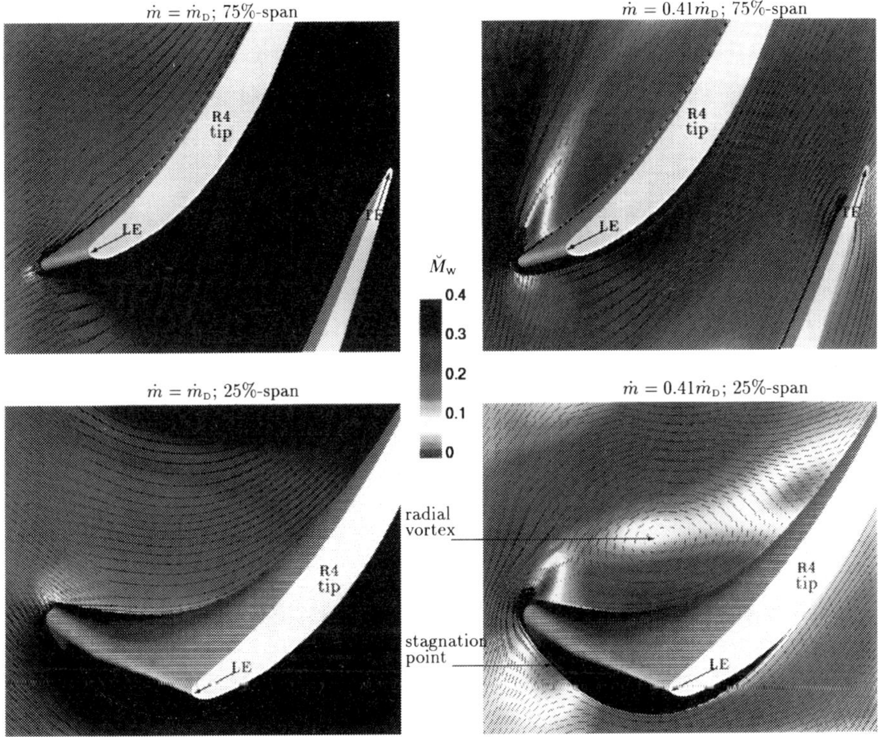

Figure 14: Plots of $\breve{M}_\mathrm{w}$ and $\breve{M}_\mathrm{w}$-vectors at 25%-span and at 75%-span, at design-point ($\dot{m} = \dot{m}_\mathrm{D}$) and at reduced-massflow operation ($\dot{m} = 0.41\dot{m}_\mathrm{D}$), at design-speed, for rotor R4 of ISUH_1 4-stage turbine [15].

A close-up view of the leading-edge region of rotor R4, where the $\breve{M}_\mathrm{w}$-vectors are plotted over the iso-$\breve{M}_\mathrm{w}$ (Fig. 14) shows that, at reduced massflow ($\dot{m} = 0.41\dot{m}_\mathrm{D}$), the relative flow impinges on the convex-side of the blade, is then strongly accelerated around the leading-edge toward the concave-side of the blade, where a vortex core is clearly visible at 25%-span, but is not seen at 75%-span. In fact the flow is highly 3-D (Fig. 15). A front-view of $\breve{M}_\mathrm{w}$-vectors and selected streamlines (Fig. 15), at reduced-massflow ($\dot{m} = 0.41\dot{m}_\mathrm{D}$), shows that there are very strong radial velocities on the concave-side of the blade near the leading-edge region. A strong radial-vortex is formed, that convects the air upward, creating an important blockage of the passage. This vortex is deflected by the casing at ∼75%-span, and flows downstream with reduced strength. The vortex-core seen at 25%-span (Fig. 14) is the core of this radial-vortex, and the difference in flow structure at

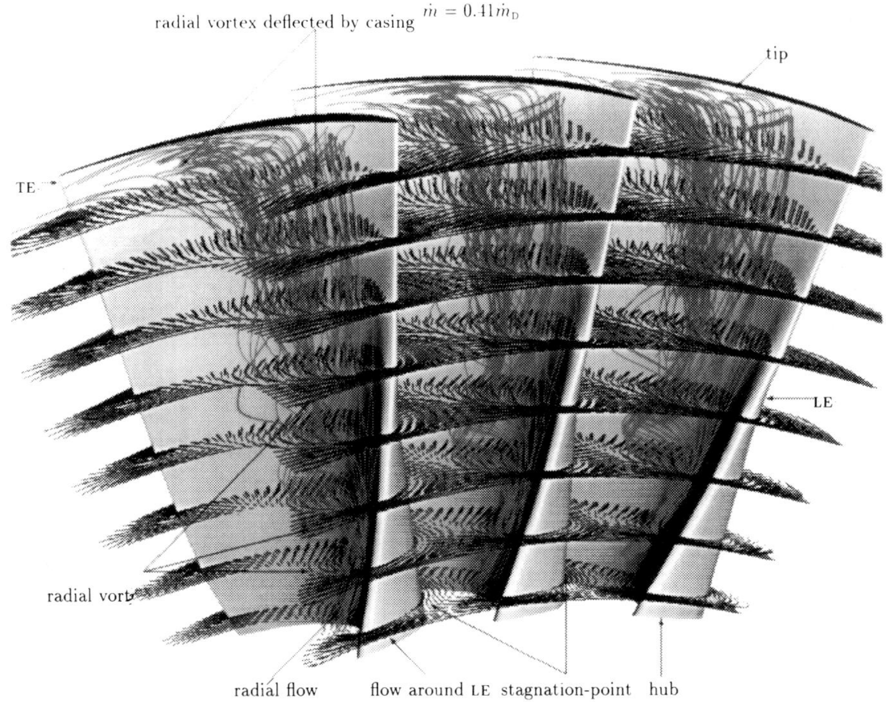

Figure 15: Front-view of $\check{M}_{\mathrm{w}}$-vectors and selected relative-flow streamlines for reduced-massflow operation ($\dot{m} = 0.41\dot{m}_{\mathrm{D}}$) at design-speed for rotor R4 of ISUH_1 4-stage turbine [15].

75%-span (Fig. 14) is due to the fact that the vortex is deflected by the casing (Fig. 15). Between the radial-vortex and the concave-side of the blade fluid flows radially upward (Fig. 15), escaping the vortex, corresponding to the high velocity region between the vortex-core and blade (Fig. 14): when this fluid impinges on the casing, near the leading-edge of the tip sections of the blade (Fig. 15) it creates the recirculating region seen in the meridional streamlines near the tip (Fig. 11). It is clear from this analysis that it is not possible to define blade-to-blade streamsurfaces, and throughflow models based on blade-sections on these streamsurfaces are not representative of the physics of the flow at reduced-massflow.

In the tip-clearance region substantial differences appear between the design-point and the reduced-massflow point (Fig. 16). The tip-clearance was 0.4 mm [15]. $\check{M}_{\mathrm{w}}$-vectors at mid-distance between the blade-tip and the casing, for the 2 operating points, shows that at reduced-massflow ($\dot{m} = 0.41\dot{m}_{\mathrm{D}}$), in the leading-edge neighbourhood, the leakage flow is from the convex-side of the blade toward the concave-side of the blade, suggesting that

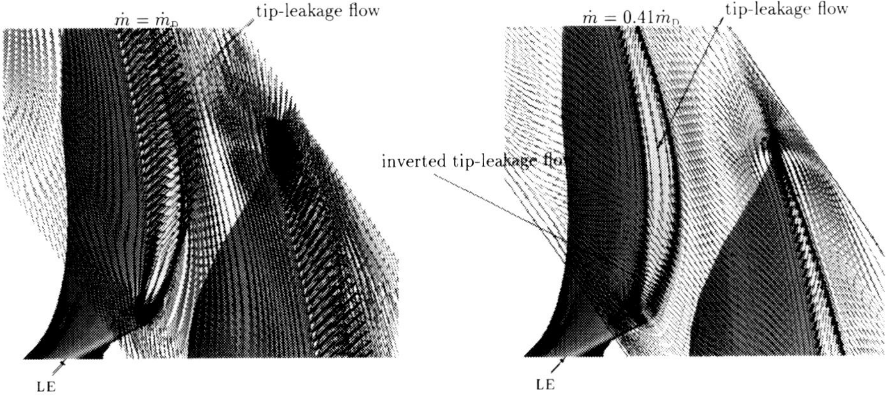

Figure 16: Plot of $\check{M}_\mathrm{w}$-vectors at 25%-span and at 75%-span, at design-point ($\dot{m} = \dot{m}_\mathrm{D}$) and at reduced-massflow operation ($\dot{m} = 0.41\dot{m}_\mathrm{D}$), at design-speed, for rotor R4 of ISUH_1 4-stage turbine [15].

pressure is lower on the concave-side of the blade (negative lift). That is why the terms convex-side and concave-side were systematically used in the text instead of suction-side and pressure-side.

Conclusions

In the present work a multistage steady 3-D Navier-Stokes methodology, using near-wall $k - \varepsilon$ turbulence closure, and a mixing-plane approach between blade-rows, was developed and validated by comparison with available measurements in a 4-stage turbine. The methodology includes an automatic multistage biharmonic grid-generation procedure for the blades and eventual tip- or hub-clearances. The particular implementation of the mixing-plane interfaces, using 5 phantom nodes, applies conservation of both meridional averages and their streamwise gradients at the interface. Excellent continuity of meridional averages was obtained at the mixing-plane interfaces, negating the conclusions of some authors concerning insufficient continuity at the mixing-plane interfaces (as compared with the deterministic stresses approach). Comparison with measurements is state-of-the-art, despite a slight overestimation of efficiency and some discrepancies observed at planes between the rows (at the turbine exit very good agreement is obtained). The same level of agreement with experiment was obtained at design and at off-design or part-speed conditions.

The methodology was then used in a first attempt to analyze the complex 3-D flow observed at reduced-massflow operation ($\sim$40% of design massflow). The prediction of turbine performance at these severe off-design conditions is quite satisfactory. It is important to note the generality of the method, based on the 3-D Navier-Stokes equations, with a turbulence closure developed and validated independently of particular turbomachinery configurations. At reduced-massflow operation high negative incidence (stagnation-point on the convex-side of the blades) is observed for both the stators and the rotors, with values as high as -120 deg at the hub of the last rotor. The convergence of the computations at these difficult conditions is very satisfactory, demonstrating the robustness of the method. This robustness is attributed to the use of an implicit 3-order upwind scheme and a careful implementation of a near-wall $k - \varepsilon$ model that is independent of wall-distance or wall-orientation (and offers as a consequence great computational simplicity).

Detailed analysis of the flowfield structure at reduced-massflow operation ($\sim$40% of design massflow) reveals massive separation on the concave-side of the last stator. In the last rotor highly 3-D flow is observed. Because of the high negative incidence (stagnation-point on the convex-side of the blades) a strong radial-vortex is created near the hub, on the concave-side of the blades, in the leading-edge neighbourhood. This radial-vortex flows upward and is deflected by the casing. It creates high blockage in the interblade channel. These results suggest that the physics of the flow at reduced-massflow operation are very different from the usual spanwise-mixing theories: radial-mixing is not due to secondary flows, but to a massive radial flow over a substantial part of the interblade channel. It is indeed difficult, if at all possible, to define blade-to-blade streamsurfaces, in the classical throughflow theory sense.

It is believed that such 3-D Navier-Stokes computations will enhance current understanding of the low-load operation of multistage axial turbines, and will help in determining flow instabilities and associated severe blade vibration observed at these conditions.

References

[1] Adamczyk, J.J., "Model Equation for Simulating Flows in Multistage Turbomachinery," ASME Paper 85-GT-226, 1985.

[2] Adamczyk, J.J., Mulac, R.A., and Celestina, M.L., "A Model for Closing the Inviscid Form of the Average-Passage Equation System," *J. Turbom.*, Vol. 108, 1986, pp. 180-186.

[3] Adamczyk, J.J., Celestina, M.L., Beach, T.A., and Barnett, M., "Simulation of 3-D Viscous Flow within a Multistage Turbine," *J. Turbom.*, Vol. 112, 1990, pp. 370-376.

[4] Arts, T., "Calculation of the 3-D, Steady, Inviscid Flow in a Transonic Axial Turbine Stage," *J. Eng. Gas Turb. Power*, Vol. 107, 1985, pp. 286-292.

[5] Bessay, M., and Pluviose, M., "Etude Aérodynamique sur Maquette d'un Etage Terminal de Turbine à Vapeur, dans les régimes partiels de Fonctionnement," *Houille Blanche*, Vol. 1982-2, 1982, pp. 219-226.

[6] Blanchet, J.M., Bourcier, P.B., and Malherbe, C., "Surveillance des Turbines à Vapeur de Grande Puissance, Vue du Constructeur et de l'Exploitant," *Rev. Franç. Méc.*, Vol. 1986-4, 1986, pp. 199-207. .

[7] Celestina, M.L., Mulac, R.A., and Adamczyk, J.J., "A Numerical Simulation of the Inviscid Flow Through a Counterrotating Propeller," *J. Turbom.*, Vol. 108, 1986, pp. 187-193.

[8] Copenhaver, W.W., Hah, C., and Puterbaugh, S.L., "3-D Flow Phenomena in a Transonic, High-Throughflow, Axial-Flow Compressor Stage," *J. Turbom.*, Vol. 115, 1993, pp. 240-248.

[9] Cumpsty, N.A., *Compressor Aerodynamics*, 1989, Longman, pp. 80-91.

[10] Dawes, W.N., "Toward Improved Throughflow Capability: The Use of 3-D Viscous Flow Solvers in a Multistage Environment," *J. Turbom.*, Vol. 114, 1992, pp. 8-17.

[11] Denton, J.D., "The Calculation of 3-D Viscous Flow through Multistage Turbomachines," *J. Turbom.*, Vol. 114, 1992, pp. 18-26.

[12] Denton, J.D., "Loss Mechanisms in Turbomachines," *J. Turbom.*, Vol. 115, No. 4, 1993, pp. 621-656.

[13] Dring, R.P., and Oates, G.C., "Throughflow Theory for Nonaxisymetric Turbomachinery Flow: Part I — Formulation," *J. Turbom.*, Vol. 112, 1990, pp. 320-327.

[14] Dring, R.P., and Oates, G.C., "Throughflow Theory for Nonaxisymetric Turbomachinery Flow: Part II — Assessment," *J. Turbom.*, Vol. 112, 1990, pp. 328-337.

[15] Fottner, L., ed., "Test Cases for Computation of Internal Flows in Aero Engine Components," AGARD Advisory Report 275, 1990, pp. 365-375.

[16] Fritsch, G., and Giles, M.B., "Second-Order Effects of Unsteadiness on the Performance of Turbo-machines," ASME Paper 92-GT-389, 1992.

[17] Fritsch, G., and Giles, M.B., "An Asymptotic Analysis of Mixing Loss," *J. Turbom.*, Vol. 117, 1995, pp. 367-374.

[18] Gallus, H.E., Zeschky, J., and Hah, C., "Endwall and Unsteady Flow Phenomena in an Axial Turbine Stage," *J. Turbom.*, Vol. 117, 1995, pp. 562-570.

[19] Gerolymos, G.A., and Vallet, I., "Implicit Computation of the 3-D Compressible Navier-Stokes Equations using $k - \varepsilon$ Turbulence Closure," *AIAA J.*, Vol. 34, 1996, pp. 1320-1321.

[20] Gerolymos, G.A., Tsanga, G., and Vallet, I., "Near-Wall $k - \varepsilon$ Computation of Transonic Turbo-machinery Flows with Tip-Clearance," *AIAA J.*, Vol. 36, No. 10, October 1998 (in print).

[21] Gerolymos, G.A., and Vallet, I., "Tip-Clearance and Secondary Flows in a Transonic Compressor Rotor," ASME Paper GT-98-366, 1998 (to appear *J. Turbom.*).

[22] Gerolymos, G.A., and Tsanga, G., "Biharmonic 3-D Grid Generation for Axial Turbomachinery with Tip-Clearance," *J. Prop. Power*, Vol. 15, 1999, in print

[23] Giles, M.B., "An Approach for Multistage Calculations Incorporating Unsteadiness," ASME Paper 92-GT-282, 1992.

[24] Goyal, R.K., and Dawes, W.N., "A Comparison of the Measured and Predicted Flow Field in a Modern Fan-Bypass Configuration," *J. Turbom.*, Vol. 115, 1993, pp. 273-282.

[25] Hall, E.J., "Aerodynamic Modeling of Multistage Compressor Flowfields — Part 1: Analysis of Rotor/Stator/Rotor Aerodynamic Interaction," ASME Paper 97-GT-344, 1997.

[26] Hall, E.J., "Aerodynamic Modeling of Multistage Compressor Flowfields — Part 2: Modeling Deterministic Stresses," ASME Paper 97-GT-344, 1997.

[27] He, L., "Method of Simulating Unsteady Turbomachinery Flows with Multiple Perturbations," *AIAA J.*, Vol. 30, 1992, pp. 2730-2735.

[28] Hirsch, C., and Dring, R.P., "Throughflow Models for Mass- and Momentum-Averaged Variables," *J. Turbom.*, Vol. 109, 1987, pp. 362-370.

[29] Horlock, J.H., *Axial Flow Turbines*, 1966, Butterworth, pp. 149-151.

[30] Jennions, I.K., and Adamczyk, J.J., "Evaluation of the Interaction Losses in a Transonic Turbine HP-Rotor/LP-Vane Configuration," *J. Turbom.*, Vol. 119, 1997, pp. 68-76.

[31] LeJambre, C.R., Zacharias, R.M., Biederman, B.P., Gleixner, A.J., Yetka, C.J., "Development and Application of a Multistage Navier-Stokes Solver: Part II — Application to a High-Pressure Compressor Design," *J. Turbom.*, Vol. 120, 1998, pp. 215-223.

[32] Petrovic, M., and Rieß, W., "Throughflow Calculation in Axial Flow Turbines at Part Load and Low Load," *VDI Berichte*, Vol. 1185, 1995, pp. 309-326.

[33] Rhie, C.M., Gleixner, A.J., Spear, D.A., Fischberg, C.J., Zacharias, R.M., "Development and Application of a Multistage Navier-Stokes Solver: Part I — Multistage Modeling using Bodyforces and Deterministic Stresses," *J. Turbom.*, Vol. 120, 1998, pp. 205-214.

[34] Rieß, W., and Evers, B., "Die Strömung in mehrstufigen Turbinen mit langen Schaufeln bei Schwachlast- und Leerlaufbetrieb," *VGB Kraftwerkstechnik*, Vol. 65, 1985, pp. 1020-1026.

[35] Singh, U.K., "A Computation and Comparison with Measurements of Transonic Flow in an Axial Compressor Stage with Shock and Boundary-Layer Interaction," *J. Eng. Gas Turb. Power*, Vol. 104, 1982, pp. 510-515.

[36] Suryavamshi, N., Lakshminarayana, B., and Prato, J., "Aspirating Probe Measurements of the Unsteady Total Temperature Field Downstream of an Embedded Stator in a Multistage Axial Flow Compressor," *J. Turbom.*, Vol. 120, 1998, pp. 156-169.

[37] Silkowski, P.D., and Hall, K.C., "A Coupled Mode Analysis of Unsteady Multistage Flows in Turbomachinery," *J. Turbom.*, Vol. 120, 1998, pp. 410-421.

[38] Turner, M.G., "Multistage Turbine Simulations with Vortex-Blade Interaction," *J. Turbom.*, Vol. 118, 1986, pp. 643-653.

[39] Vavra, M.H., *Aerothermodynamics and Flow in Turbomachines*, 1960, Wiley, pp. 452-454.

[40] Zienkiewicz, O.C., *The Finite Element Method*, 1977, McGraw-Hill, pp. 164-168.

Throughflow analysis for cooled turbines

S GEHRING and **W RIESS**
Institute for Turbomachinery, University of Hannover, Germany

The feeding of additional fluid into the main flow and the interaction of main flow and coolant flow affect the aerodynamic performance of the blading and cause changes of the overall flow capacity and efficiency of the turbine. A modified distribution of the pressure drop over the stages and different thermodynamic properties of the fluid during the expansion are results of the mixing of main and coolant flow. Computing systems which take into account all effects of coolant are necessary for the flow field calculation and performance prediction of cooled multi-stage turbines.

Because of the very limited effort necessary for preparation, the low consumption of computing time and the high quality of results attainable with suitable programs, 2D-throughflow calculation methods are of high practical value for design studies, parameter variation and optimization. The code presented here is based on the finite element method and includes models for the radial distribution of losses and for spanwise mixing. The calculation comprises models for different cooling configurations, e.g. convection cooling with trailing edge outlet and film cooling. The changes in mass flow, momentum, energy and entropy are derived from the conservation laws in main flow direction. The code is applied to a model turbine. Calculations are executed for different cooling systems and different coolant mass flows to show the influence on performance and efficiency.

INTRODUCTION

The development of modern gas turbines tends to an increasing maximum power output in combination with high overall efficiency. These design goals can be reached by higher inlet temperatures coupled with optimized pressure ratios. However higher turbine inlet temperatures require complex cooling configurations for blading, discs and endwalls. Cooling systems of high temperature gas turbines consist of a combination of convection cooling optimized by impingement cooling and turbulators in the coolant channel, trailing edge ejection and film cooling.

For the optimization of performance and efficiency the complete machine has to be analysed including flow pattern analysis and heat transfer calculation.

During the design process throughflow calculations are useful tools for flow field analysis and performance prediction of multistorey turbines. The well known method attained an advanced state of development and allows routine calculation of many operating conditions. The advantage of this method is a short calculation time connected with sufficient accuracy, thus variations of many design parameters and complex optimization are possible.

The purpose of this paper is to present a new code for throughflow calculations of cooled axial turbines. The new extensions consider the influence of complex cooling configurations in modern gas turbines on their aerodynamical behaviour.

CALCULATION METHOD

Basic Theory

The calculation method is based on the classical throughflow theory which reduces the Navier Stokes equations to a greatly simplified formulation of equations [1,2]. The basic computer code for adiabatic flow in axial turbines includes improvements which take into account radial distribution of secondary and clearance losses, radial mixing and off-design losses. This theory is published in several articles [3,4,5] and will only be repeated here as far as necessary for the understanding of extensions which consider cooling effects.

For the 2D calculation the momentum equation is projected on the meridional hub-to-shroud surface. With the approximation of viscous effects by friction forces, introducing body forces to replace the turbine blades and using the stream function ψ the equation can be written in the following form:

$$\frac{\partial \psi^2}{\partial r^2} + \frac{\partial \psi^2}{\partial z^2} = \left[\frac{\partial \psi}{\partial r} \frac{\partial}{\partial r}\left(\frac{1}{\rho r b}\right) - \frac{\partial \psi}{\partial z} \frac{\partial}{\partial z}\left(\frac{1}{\rho r b}\right) + \frac{2\pi}{\dot{m}} q \right] \rho r b \tag{1}$$

$$q = f(geometry, \vec{W}, i, rc_u, s)$$

____ **Nomenclature** ____

b	blockage factor	Re	Reynolds number		Superscripts
c	absolute velocity	r	radius		0 total
c_p	specific heat	s	entropy		
h	enthalpy	T	temperature		Subscripts
i	rothalpy	Tu	turbulence intensity	c	coolant
k	mass flow coefficient	$\vec{W}$	relative velocity vector	h	hot gas
		w	relative velocity	in	turbine inlet
l	streamline length	α	heat transfer coefficient	mix	mixed
$\dot{m}$	mass flow	Δh^0_{net}	netto total enthalpy change	out	rotor outlet
$\dot{m}_0$	inlet mass flow			r	radial direction
Nu	Nusselt number	η^{TT}	total to total efficiency	S	blade surface
n_s	number of blades	η^{TS}	total to static efficiency	u	tangential direction
p	pressure	μ	mass flow ratio	z	axial direction
$\dot{Q}$	heat flow	κ	ratio of specific heats		
q	$\dot{Q}/\dot{m}$	ρ	density		
R	gas constant	ψ	stream function		

The right side depends on the geometry, the velocity, the streamfunction itself, the total enthalpy or in the case of a rotor the rothalpy, the tangential momentum, the entropy and the thermodynamic fluid properties. The geometry terms in equ. (1) include also parameters to consider the lean of the blading.

Equ. (2) defines the streamfunction:

$$\frac{\partial \psi}{\partial r} = w_z \frac{2\pi}{\dot{m}_0} \rho rb \; ; \qquad \frac{\partial \psi}{\partial z} = -w_r \frac{2\pi}{\dot{m}_0} \rho rb \tag{2}$$

where b is the tangential blockage factor ($b = 1 - d/t$, d is the tangential blade thickness and t is the blade pitch). In order to achieve an easier definition of boundary conditions the mass flow is introduced into the streamfunction equations. Calculating a turbine without extraction or ejection the boundary conditions are defined by: $\psi = 1.0$ along the casing and $\psi = 0.0$ along the hub.

The main independent variables of equ. (1) are the total enthalpy respectively rothalpy, the tangential momentum and the change in entropy. The geometry and the other terms are fixed or a result of the iterative calculation, e.g. the velocities and the thermodynamic fluid properties. Solving equ. (1) the energy equation is used to calculate the rothalpy and empirical models are introduced to estimate the entropy increase and the deviation angle. For the loss coefficient prediction at design conditions Traupel´s loss models are applied [6]. Petrovic and Riess [3] introduced a new model for a more realistic distribution of the secondary and clearance losses based on extensive experimental investigations by Groschup [7]. In order to avoid an unrealistic entropy accumulation near the endwalls, because the viscous effects between the streamlines are neglected in the basic equations, a new radial mixing model was developed. The model realizes a mathematical redistribution of total enthalpy, entropy and angular momentum. The cascade exit flow angle is calculated by a modified sine rule which also depends on the local values of the secondary and clearance loss coefficient. Using a new off-design loss model [8] calculations over a wide range of operating conditions are possible.

The governing equations are solved using the finite element method with eight-node elements. The finite element equations are formed by applying the Galerkin procedure of weighted residuals, while the resulting system of linear equations is solved by the frontal solving method. Because the circumferential velocity is imposed to the meridian velocity considering the distribution of the moment of momentum $r \cdot c_u$, the code is stable if the meridian velocity is subsonic ($M_{merid} < 1$).

Cooling flow addition
High temperature gas turbines require cooling air for blading, discs and casing. The mixing of coolant and main flow causes changes in mass flow, energy and momentum which influences the aerodynamic performance of the blading. Figure 1 shows schematically the different cooling techniques in a stator blade row.

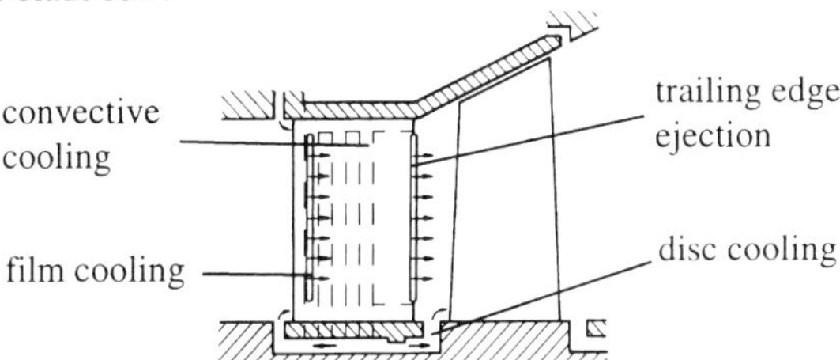

Fig. 1: Cooling techniques in a stator blade row

In the code presented here convection cooling, coolant ejection at the trailing edge outlet and film cooling was implemented. The influence of the cooling is considered by modified equations for the variables of the governing equ. (1), which are the total enthalpy respectively rothalpy, the entropy and the momentum in tangential direction. The solution of the equ. (1) with the modified variables gives the new distribution of the streamfunction and with equ. (2) the distribution of the meridional velocity.

In case of convection cooling changes of energy and entropy are considered in the blading. Therefore the energy equation of the basic code is extended by additional terms to describe the heat transfer across the blading. The rothalpy change can be calculated by integrating equ. (3) along the streamlines.

$$-\frac{n_s \bar{\alpha}}{\pi r \rho w_z}(T_h - \bar{T}_S)dl = dq = di \tag{3}$$

The mean values of the heat transfer coefficient $\bar{\alpha}$ and the surface temperature of the blading $\bar{T}_S$ cannot be calculated in the throughflow calculation. Therefore the specific heat flow rate or the heat transfer coefficient and the surface temperature has to be introduced from external heat transfer calculations. In this paper empirical correlations are used to calculate the heat transfer coefficient, because no heat transfer calculations are done. The heat transfer coefficient at the leading edge is approximated by a correlation for cylinders [9].

$$Nu = \left[0.945 + 4.48 \cdot \left(Tu \cdot \frac{\sqrt{Re}}{100} - 3.99 \cdot \left(Tu \cdot \frac{\sqrt{Re}}{100}\right)^2\right)\right] \cdot \sqrt{Re} \tag{4}$$

The Reynolds number is calculated using the leading edge diameter and the turbulence intensity Tu is assumed to 5%. Along the blade surface correlations for the convection over a flat plate are used. The transition from laminar to turbulent flow is assumed to be situated at 10% of the curved surface.

The entropy is affected by the losses and the heat flow across the surfaces. It is assumed that both can be calculated separately so that the entropy change along a streamline in a turbine with convective cooling can be written in the following form:

$$\Delta s = \Delta s_{loss} - \frac{q}{T_h} \tag{5}$$

In case of a trailing edge ejection or film cooling further extensions are necessary to consider the mixing of coolant and main flow. For the purpose of modelling the mixing, it is assumed that the axial extension of the mixing zone is small and that the mixing process can be represented by expressions derived from the conservation laws.

A general mixing equation is given with equ. (6), where Ω represents the three main variables of the equ. (1) rothalpy, moment of momentum and the entropy. The subscripts in equ. (6) h, c and mix denote the hot gas, the coolant and the mixed flow.

$$\Omega_{mix} = \frac{\dot{m}_h}{\dot{m}_{mix}} \cdot \Omega_h + \sum_i \left(\frac{\dot{m}_{c,i}}{\dot{m}_{mix}} \cdot \Omega_{c,i}\right) + \Delta\Omega \tag{6}$$

If Ω describes the total enthalpy respectively the rothalpy, equ. (6) changes into the energy equation and the term $\Delta\Omega$ is equal to zero. If Ω is the momentum in tangential direction some assumptions are necessary, because frequently the momentum of the coolant and the dissipation of kinetic energy described by the term $\Delta\Omega$ are not exactly known. For the coolant passing the trailing edge outlet it will be assumed, that the coolant is ejected in direction of the blade angle. For the film cooling it is assumed that the coolant flow is ejected in main flow direction and does not affect the outlet flow angle. In both cases it is supposed that the momentum change by the dissipation of kinetic energy due to viscosity is small and can be neglected.

The third mixing equation determines the entropy of the mixed flow. In this equation $\Delta\Omega = \Delta s_{mix}$ describes the entropy production caused by the irreversibility of the mixing process, which depends on the static thermodynamical properties of the mixed flow. For the calculation of those properties the conservation law of momentum in axial direction is applied. In the open literature several simple 1D models can be found.

The two implemented methods based on models of Traupel [6] and Hartsel [10] use simplified formulations of the momentum equation. Traupel has assumed an instantaneous mixing without pressure and area change whereby the pressure terms in the momentum equation vanish:

$$\dot{m}_{mix} \cdot c_{z,\,mix} = \dot{m}_h \cdot c_{z,\,h} + \dot{m}_c \cdot c_{z,\,c} \tag{7}$$

The second model considers pressure changes, but uses a mixing plane which is fixed during the mixing process, equ. (8).

$$\dot{m}_{mix} \cdot c_{z,\,mix} + p_{mix} \cdot A_{mix} = \dot{m}_h \cdot c_{z,\,h} + p_h \cdot A_h + \dot{m}_c \cdot c_{z,\,c} \tag{8}$$

Using these models the thermodynamic properties of the mixed flow including the entropy change can be calculated:

$$\Delta s = c_p \cdot \ln\frac{T_{mix}}{T_h} - R \cdot \ln\frac{p_{mix}}{p_h} \tag{9}$$

The pressure term vanishes in case of Traupels model. The entropy change in turbines with convective cooling and coolant ejection can be estimated with a combination of equ. (5) and equ (9).

The increasing mass flow caused by coolant ejection is considered introducing the scaling coefficient k in the stream function equations, equ. (10), which influences the streamfunction similar to the blockage factor b.

$$\frac{\partial\psi}{\partial r} = w_z\frac{2\pi}{k \cdot \dot{m}_0}\rho r b \; ; \qquad \frac{\partial\psi}{\partial z} = -w_r\frac{2\pi}{k \cdot \dot{m}_0}\rho r b \tag{10}$$

The mean value of the scaling coefficient is defined by equ. (11):

$$k = \frac{\dot{m}_0 + \dot{m}_c}{\dot{m}_0} \tag{11}$$

Without cooling, k is equal to unity and with every coolant ejection k increases. The scaling coefficient can vary along the blade height and the local value depends on the local ratio between coolant and hot gas flow.

NUMERICAL ALGORITHM

For calculating turbines with convective cooling the given distribution of the heat transfer per area or mean values of the blade temperature and the heat transfer coefficient are redistributed to the mesh points. The first step during the calculation is the estimation of the total enthalpy respectively the rothalpy change by integrating equ. (3) along the streamlines according to the heat transfer at every grid point. Equ. (5) gives the entropy change.

The coolant ejection at the trailing edge is considered to be situated in the inlet plane of an axial gap. The mixing equ. (6) delivers the properties of the mixed flow at the grid points in this plane. Furthermore the coefficient k will be increased at these points according to the continuity equation, fig. 2. In the case of film cooling the ejection holes are positioned in a finite element row and the variables of the mixed flow are calculated at the downstream limit of the element row, fig. 2. The new values of the variables are distributed along the streamlines downstream of the mixing plane.

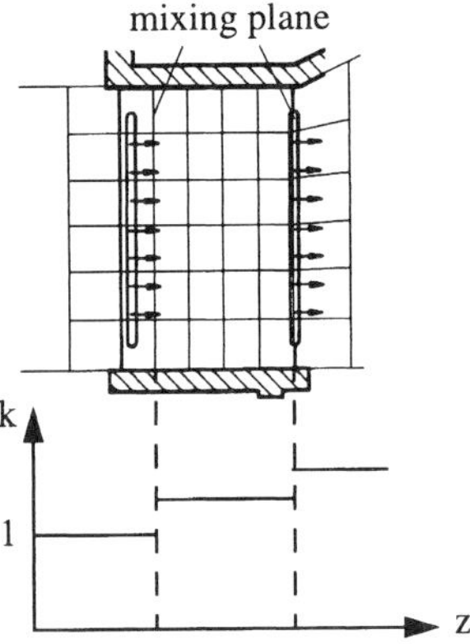

Fig. 2: Increasing coefficient k for a row of film cooling holes and a trailing edge outlet

TEST CASE

In order to show the effects of different cooling systems without other complexities the new code has been applied to a model turbine based on the institute´s single stage air turbine. The blading has been designed according to the free-vortex-law with a degree of reaction of about 50% at the mean height. Detailed data of the turbine has been published in [11]. The calculations of the cooled model turbine are done for the nominal rotation speed n_0 = 7500 rpm with modified thermodynamical properties at the inlet T_{in}^0 = 1623.15 K, p_{in}^o = 1.36 bar and a reduced

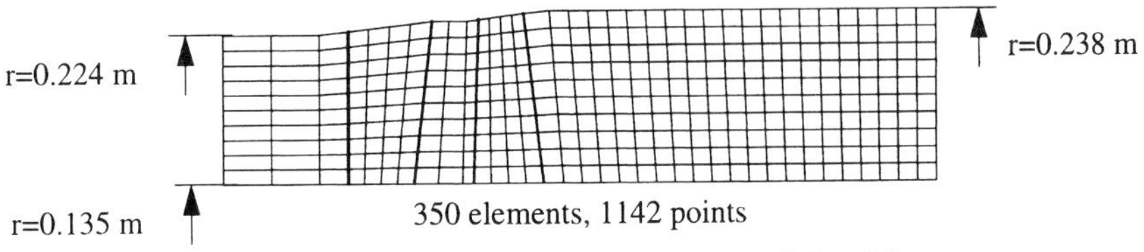

Fig. 3: Finite element grid of the single stage model turbine

mass flow compared to the design mass flow of the air turbine. Figure 3 shows the finite element grid. The thermodynamic values of the coolant are set for all calculations to $T_C^0 = 623.15$ K and $p_C^o = 1.36$ bar. The blade surface temperature is set to a uniform value $T_S = 1187.15$ K.

The next figures show the results along the streamline $\psi = 0.5$ of the non-adiabatic model turbine. The distribution of the total enthalpy and the heat transfer coefficients for a convective cooling system are presented in fig. 4. Particularly in the stator the effect of the convection cooling can be seen. At the leading edge the correlation gives high values for the heat transfer coefficient which causes the total enthalpy to decrease rapidly. Downstream of the stagnation point the heat transfer coefficients are estimated by using the model for laminar flow along a flat plate. At about 10% of the curved blade surface the transition point is assumed to be situated. Downstream of this point the correlation for turbulent flow is used and the heat transfer coefficient rises. The distribution of the heat transfer coefficient corresponds qualitatively with the values in the literature [9,12], but it must be emphasized here that simplifying assumptions are used to estimate the mean value of the heat transfer coefficient between pressure and suction side.

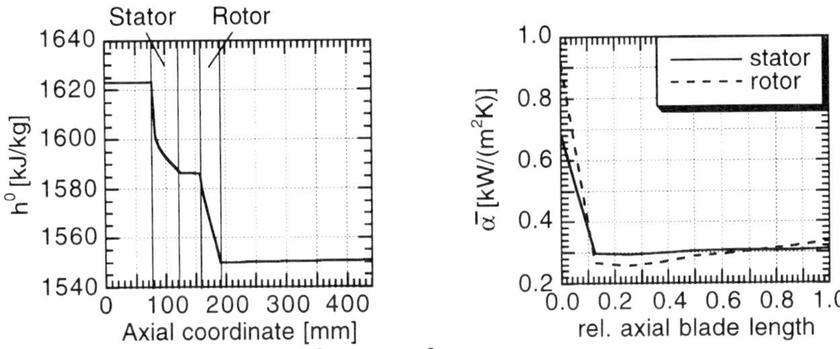

Fig. 4: Convective cooling of stator and rotor:
    Distribution of total enthalpy and of heat transfer coefficient along the streamline ψ=0.5

Fig. 5 shows the result for a film cooled blading. The calculation was done with 1% coolant addition in the first and the third vertical element row of both cascades and 1% coolant ejection at the trailing edge of the stator. Since the mixing models give nearly the same results only Traupel's [6] model is used here. The increasing mass flow coefficient k, shown in the right diagram, depends on the given ratio between the local coolant flow and the inlet mass flow. The mixing

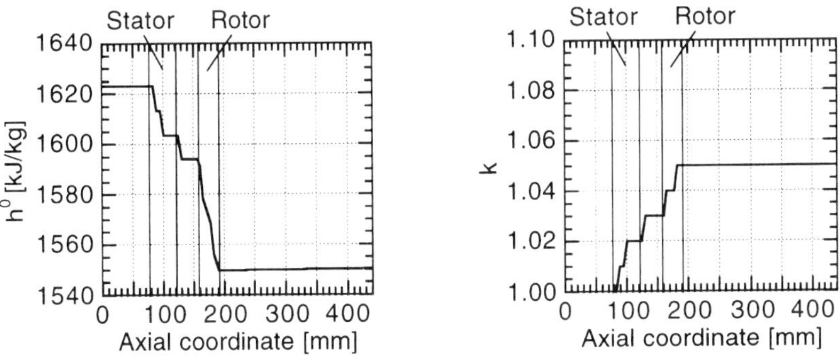

Fig. 5: Film cooling in stator and rotor, trailing edge ejection in the stator:
    Distribution of total enthalpy and of mass flow coefficient along the streamline ψ=0.5

calculations are done at the limits of the finite elements and result in a decreasing total enthalpy. Fig. 6 shows the total enthalpy and the mass flow coefficient for variable coolant ejection along the blade height. Due to an ejection from the disc cooling upstream the stator the mass flow coefficient at the stator exit increases near the hub. A second disc cooling flow and the trailing edge ejection give the higher values at the rotor inlet. Therefore the gradient of the coefficient k is stronger near the hub. Caused by the film cooling the coefficient increases in the rotor. Near the casing the gradient of k takes into account an ejection at the blade tips. The diagram at the left shows the distribution of the total enthalpy.

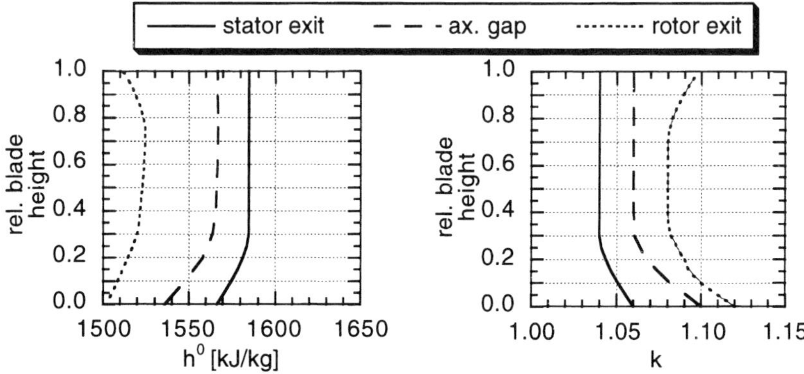

Fig. 6: Test turbine with rotor/stator film cooling and stator trailing edge ejection. Variable coolant ejection along the blade height.

Fig. 7 shows the total to total efficiency, the total to static efficiency and the aerodynamical work normalized by the inlet flow. The basic definitions of the parameters are given in equ. (12) and (13), which are based on the conservation law of energy between the turbine inlet and rotor exit plane.

$$\Delta h^0_{net} = h^0_{in} - h^0_{out} + \sum_i \mu_{c,i}(h^0_{c,i} - h^0_{out}) + \frac{\dot{Q}}{\dot{m}_0} \qquad \text{with} \quad \mu_{c,i} = \frac{\dot{m}_{c,i}}{\dot{m}_0} \tag{12}$$

$$\eta^{TT} = \frac{\Delta h^0_{net}}{h^0_{in} \cdot \left[1 - \left(\frac{p^0_{in}}{p^0_{out}}\right)^{\frac{\kappa-1}{\kappa}}\right] + \sum_i \mu_{c,i} h^0_{s,c,i} \left[1 - \left(\frac{p^0_{c,i}}{p^0_{out}}\right)^{\frac{\kappa-1}{\kappa}}\right]}$$

$$\eta^{TS} = \frac{\Delta h^0_{net}}{h^0_{in} \cdot \left[1 - \left(\frac{p^0_{in}}{p_{out}}\right)^{\frac{\kappa-1}{\kappa}}\right] + \sum_i \mu_{c,i} h^0_{s,c,i} \left[1 - \left(\frac{p^0_{c,i}}{p_{out}}\right)^{\frac{\kappa-1}{\kappa}}\right]} \tag{13}$$

As this definition does not consider the energy flow connected with trailing edge ejection in the rotor, this coolant addition is an exception. Therefore the useful total enthalpy difference Δh^0_{net} has to be diminished by the pumping work of the rotor which is necessary to transport the trailing edge coolant within the blades to its radius of ejection. Calculations were done for different mass flows with different cooling configurations and in case of coolant ejection with constant mass flow coefficients *k* along the blade height. The values of k have been kept constant for variable mass flows.

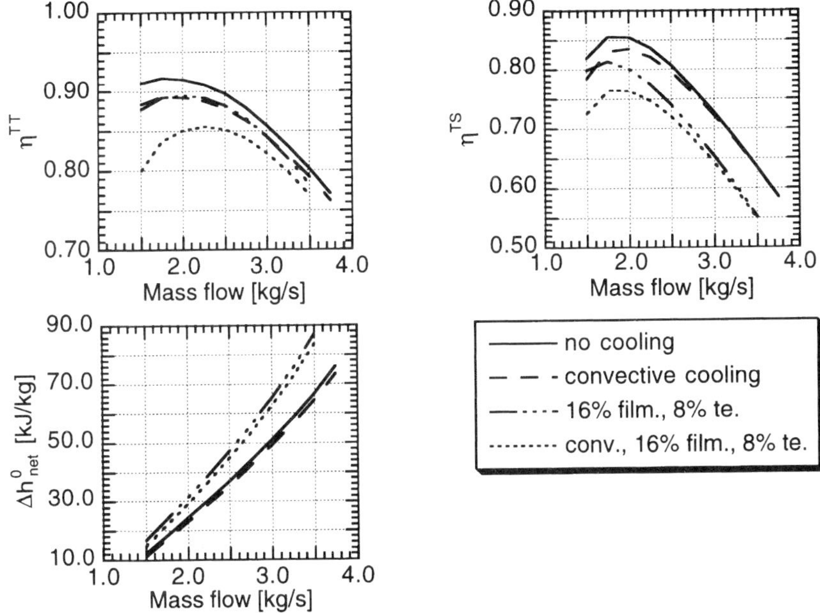

Fig. 7: Characteristics of the model turbine with different cooling systems

The closed convection cooling decreases the mean temperature of the expansion process whereby the aerodynamic work and the efficiencis are reduced. The lower temperatures caused by the heat flow result in lower volume flow and therefore the characteristics have an offset to higher mass flows.

In case of the open cooling system 4% film cooling ejection in the first and the third vertical element row of both cascades are considered. Furthermore 4% coolant ejection at the trailing edges is assumed. At constant inlet mass flow the coolant ejection increases the aerodynamic work, because the coolant contributes to the expansion process, but the efficiencies are decreased by the lower mean temperature of the cooled process. The total to static efficiency is reduced more by the cooling than the total to total. This means that the increasing mass flow due to the coolant ejection causes higher velocities in this low performance turbine. The characteristics Δh^0_{net} and η^{TS} are shifted to lower inlet mass flows by the coolant addition.

The lowest efficiencies are calculated for the combination of coolant ejection and convection. In this case the effects of both cooling systems superpose: The coolant ejection causes an offset of the characteristic Δh^0_{net} to lower mass flows, but compared with the open cooling system without convection the characteristic is shifted again to higher mass flows. The effects of the film cooling on the heat transfer coefficient are neglected.

CONCLUSIONS

A throughflow method for cooled axial turbines based on a finite element procedure has been described. The new code considers convection cooling, coolant ejection at the trailing edge and film cooling. For the convection cooling the code allows to use given distributions of the heat flux or heat transfer coefficients in combination of blade surface temperatures. If only the surface temperature is known, correlations for the heat transfer coefficient are used. Coolant ejection can be considered at every downstream limit of the finite elements. The increasing mass flow in a turbine with coolant ejection is taken into account by a new mass flow coefficient in the definition of the stream function. Variable coolant ejections along the blade height are possible.

The new method is applied to a single stage model turbine to show the capabilities of the code. Calculations were done for different inlet mass flows and different cooling configurations. The low computing time allows fast calculations of the overall characteristics in a wide range of load considering the influence of complex cooling configurations on the aerodynamic behaviour of the blading.

ACKNOWLEDGMENT

This work is part of the AG Turbo research program, a cooperative effort between Industry, Universities and National Research Centres, financially supported by the German Ministry of Education, Science, Research and Technology under contract No. 0327041I.

REFERENCES

[1]   Wu, C.H.: A General Theory of Three-Dimensional Flow in Subsonic and Supersonic Turbomachines of Axial-, Radial- and Mixed-Flow Types, NACA TN 2604, 1952.

[2]   Hirsch, C.; Warzee, G.: A Finite Element Method for Through-Flow Calculations in Turbomachines, ASME Paper 76-FE-12, 1976.

[3]   Petrovic, M.; Riess, W.: Through-Flow Calculation in Axial Turbines at Part Load and Low Load. 1st European Conference " Turbomachinery - Fluid Dynamic and Thermodynamic Aspects", 1995.

[4]   Petrovic, M.; Riess, W.: Off-Design Flow Analysis of LP Steam Turbines. 2nd European Conference " Turbomachinery - Fluid Dynamic and Thermodynamic Aspects", 1997.

[5]   Petrovic, M.; Riess, W.: Off-Design Flow Analysis and Performance Prediction of Axial Turbines. ASME Paper 97-GT-55, 1997.

[6]   Traupel, W.: Thermische Turbomaschinen, Vol. 1. Springer-Verlag, Berlin, 1988.

[7]   Groschup, G.: Strömungstechnische Untersuchungen einer Turbinenstufe im Vergleich zum Verhalten der ebenen Gitter ihrer Beschaufelung. Diss. Univ. Hannover, 1977.

[8]   Zehner, P.: Calculation of Four-Quadrant Characteristics of Turbines. ASME Paper 80-GT-2, 1980.

[9]   Henrich, E.: Wärmeübergang an Turbinenschaufeln verschiedener aerodynamischer Auslegung. Technischer-Bericht MTUM-B92ET-0056, MTU München, 1992.

[10] Hartsel, J.E.: Prediction of Effects of Mass.Transfer Cooling on the Blade Row Efficiency of Turbine Airfoils. AIAA Paper 72-11, 1972.

[11] Groschup, G.:Through Flow Calculations in Axial Turbomachinery. AGARD AR 175, ed. C. Hirsch, J.D. Denton, 1981.

[12] Lakshmirayana, B.: Fluid Dynamics and Heat Transfer of Turbomachinery. John Wiley & Sons, Inc., 1996.

C557/073/99

Measurements of complex air flow phenomena inside the rotor of an operating industrial gas turbine

D REGNERY, U HOEPPNER, N VORTMEYER, and **K NITSCHE**
Siemens AG, Power Generation Group, Mülheim, Germany

SYNOPSIS

A specific feature of the Siemens heavy duty gas turbines is the rotor design consisting of discs being held together by a centre tie rod. This enables the transport of the cooling air needed for the turbine blades through the rotor. Due to the highly complex fluid dynamics inside the rotor (rotating side walls and centrifugal forces) appropriate numerical calculation methods for the cooling air flow have only recently been developed.

To get a better understanding of the complex flow phenomena and for validating the numerical procedures temperatures and pressures were measured inside the rotor under all operating conditions. New methods had to be developed for continous measurement of static pressures inside the rotor of an operating gas turbine.

The paper gives a detailed description of the instrumentation, the procedures for adaptation to the harsh environment and a method for correction for the influence of temperature gradients and centrifugal forces on the measured values. The experimental results provided a database for validating the CFD calculations. Experimental results are shown and discussed referring to calculated values, giving an impression of the complex cooling air flow.

NOMENCLATURE

c	[m/s]	velocity
p	[N/m^2]	static pressure
n	[min^{-1}]	rotational speed
A	[m^2]	area
D	[m]	diameter
L	[m]	$= R_o - R_u$, length of the tube segment
R	[Nm/kgK]	gas constant

R_o [m]   Radius at upper end of tube section
R_u [m]   Radius at lower end of tube section
T   [K]   temperature
V   [m³]   volume
ρ [kg/m³]   density
ω [s⁻¹] $= 2\pi n$, angular speed

Indices

o   upper, at the larger radial position
u   lower, at the smaller radial position

1 INTRODUCTION

Currently, there are three designs of rotors for heavy duty gas turbines, namely monoblock rotors, disc-type rotors with a central tie bolt and disc-type rotors with a number of circumferentally arranged bolts. The disc-type rotor with a central tie rod has been used in all Siemens heavy duty gas turbines since 1960 /1/. This rotor design allows the rotor mass to be kept relatively low and, at the same time, allows for an intensive transport of cooling air from the cooling air extraction ports in the compressor to the rotating, film cooled blades of the various turbine stages.

The flow path of secondary air inside the rotor is geometrically characterised as sequence of drillings, holes, cavities and concentric cylinders, all rotating at high speed and subject to large temperature gradients. An efficient design of the gas turbine secondary air systems requires precise calculation methods of the cooling air flow from the extraction ports to the blade roots. The methods of calculation have been steadily improved for decades and calibrated using tests performed on prototype full scale gas turbines in the turbine test bed. In order to achieve major advances in the calculation methods, the air flow in the rotor of the new model V84.3A gas turbine prototype was comprehensively tested experimentally. Using these data a new generation of mathematical modelling was established, which was communicated in an earlier paper /2/.

The required high level of accuracy of the validation test data for the improved computer codes and at the same time the high standards concerning reliability and durability of the test installation turned out to be a real challenge for the Siemens team of test engineers. New concepts concerning the instrumentation and the evaluation methods of the raw data had to be developed. The purpose of this paper is to give an account of the test team's solution as an example of current state of the art testing technology in the sector of large turbomachinery.

2 BASIC CONSIDERATIONS

Figure 1 gives a section through the prototype gas turbine indicating the path of the cooling air through the rotor. There are three supplies for turbine stages 1, 2 and 3 respectively, each being extracted from the compressor at an appropriate pressure level.

To receive reliable pressure data at the secondary air extraction and outlet ports and the other positions indicated in figure 1 the pressure transducers must necessarily rotate with the shaft. Pressure transducers at the indicated positions would have to bear the high centrifugal forces resulting from about 8500 g near the blade roots and temperatures between 250°C and 400°C without damage and problems with stability, sensitivity shift and linearity. Even if appropriate masterpieces of modern sensor technology were available, the heavily stressed rotor would not allow for relatively large insertion bores near the inlet and outlet ports or other modifications. Mounting the sensors without machining would have caused an inadmissible disturbance of the flow field. Since static pressure transducers also need a reliable power supply, a measurement with pressure transducers mounted directly at the measurement positions turned out to be impossible.

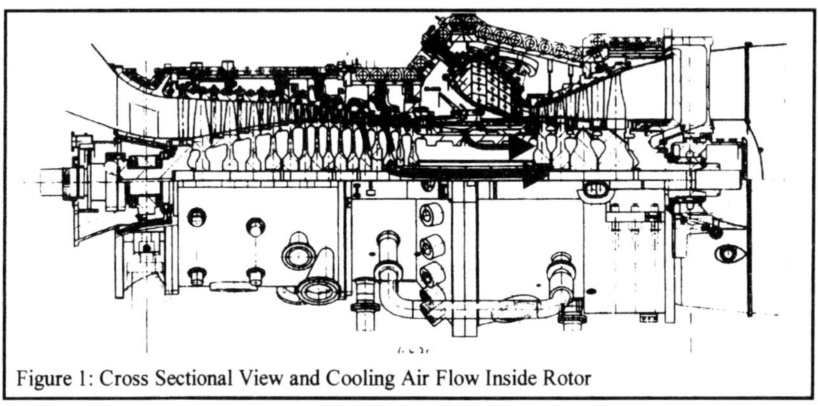

Figure 1: Cross Sectional View and Cooling Air Flow Inside Rotor

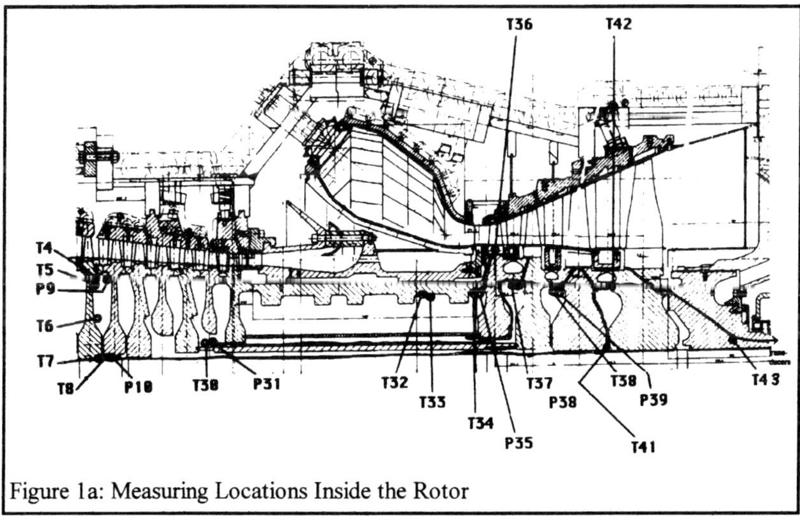

Figure 1a: Measuring Locations Inside the Rotor

The solution of the problem was achieved with an instrumentation method as has been frequently used for strain gauge measurements of blade vibrations. Power supplies, signal conditoning and other rotating electronic equipment are mounted in a bolt with a multitude of insertion bores at the turbine end of the rotor where temperatures are below 100°C and centrifugal loads are at approx. 2500 g. This bolt (Figure 5) is large enough to carry individual pressure transducers for each measuring position thus eliminating the need for a scanivalve which is known to be troublesome. The pressure transducers and the measuring positions have to be linked with tubes which form the impulse lines. The hardware was installed by skilled fitters who are experts in the installation of wires and tubes that are resistant to high centrifugal forces. Measuring temperatures is comparatively easy by simply introducing thermocouples at the measurement locations and using a telemetry system.

Along with finding an appropriate instrumentation technique, the basic question was the interpretation and correction of the pressure transducer signals since the air column in the impulse lines is influenced by a field of varying g-forces and temperatures. A correction scheme had to be established in order to evaluate the correct pressures inside the rotor from the measured values.

3 CORRECTION TECHNIQUE AND ALGORITHMS

The air inside the pressure tubes does not flow once the rotor has reached its nominal speed. One end of a tube is sealed by the pressure transducer, the other receives the pressure from the cooling air system. The following assumptions can be made
- constant rotational speed (ω=const)
- no airflow (c=0, stationary column of air)
- slow temperature changes (dT/dt << 1)

- ideal gas (air, equation of state)

$$\frac{p}{\rho} = RT \qquad (1\text{-}1)$$

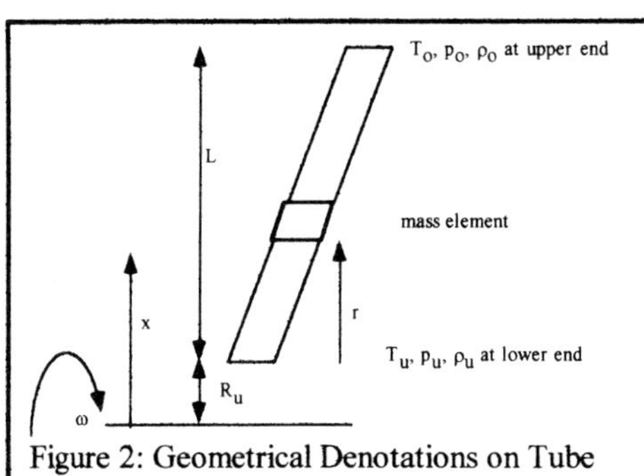

Figure 2: Geometrical Denotations on Tube Segment

- Regarding a limited section of a tube, temperature changes along this section are assumed to be linear (denotations explained in figure 2, mass element marked red):

$$\Delta T = T_o - T_u$$
$$T(r) = T_u + \frac{r - R_u}{L}(T_o - T_u) = T_u + (r - R_u)\frac{\Delta T}{L} \qquad (1\text{-}2)$$

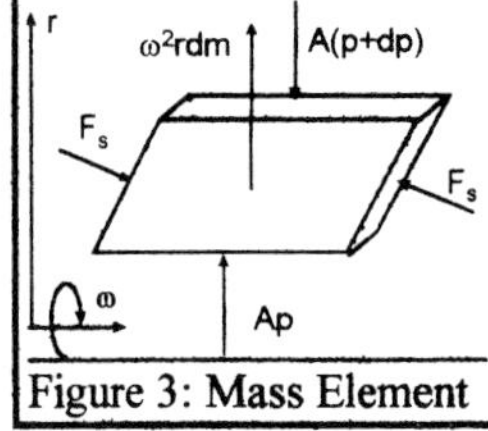

Figure 3: Mass Element

The equilibrium in an inclined column of air with centrifugal forces can be written as in figure 3.

$$A(p + dp) = Ap + \omega^2 r dm$$
$$Adp = \omega^2 r dm = \omega^2 r \rho dV = \omega^2 r \rho A dr \qquad (1\text{-}3)$$

Inserting the equation of state (1-1) and radial temperature distribution (1-2) leads to

$$\frac{dp}{p} = \frac{\omega^2}{R} \frac{r}{T_u + (r - R_u)\frac{\Delta T}{L}} \, dr \tag{1-4}$$

Transformation of coordinates gives:
$$x = r - R_u \Rightarrow dx = dr \tag{1-5}$$

$$\frac{dp}{p} = \frac{\omega^2 L}{R\Delta T}\left(\frac{x}{T_u \frac{L}{\Delta T} + x} + \frac{R_u}{T_u \frac{L}{\Delta T} + x} \right) dx \tag{1-6}$$

$$\Rightarrow \int_{p_u}^{p_o} \frac{dp}{p} = K \int_0^L (I_1 + I_2)\,dx \tag{1-7}$$

Where integrals I_1 and I_2 can be expressed as

$$\int_0^L (I_1)\,dx = \int_0^L \left(1 - \frac{T_u \frac{L}{\Delta T}}{T_u \frac{L}{\Delta T} + x} \right) dx = L - \frac{T_u L}{\Delta T} \ln \frac{\frac{T_u L}{\Delta T} + L}{\frac{T_u L}{\Delta T}} = L - \frac{T_u L}{\Delta T} \ln\left(\frac{T_o}{T_u} \right) \tag{1-8}$$

$$\int_0^L I_2\,dx = \int_0^L \left(\frac{R_u}{T_u \frac{L}{\Delta T} + x} \right) dx = R_u \ln\left(\frac{T_u \frac{L}{\Delta T} + L}{T_u \frac{L}{\Delta T}} \right) = R_u \ln\left(\frac{T_o}{T_u} \right) \tag{1-9}$$

$$\Rightarrow \ln \frac{p_o}{p_u} = \frac{\omega^2 L}{R\Delta T}\left(L - \frac{T_u L}{\Delta T} \ln\left(\frac{T_o}{T_u} \right) + R_u \ln\left(\frac{T_o}{T_u} \right) \right) \tag{1-10}$$

This equation allows for correction of the pressure in a tube segment with any inclination between vertical and horizontal which is subject to centrifugal loads and a temperature field.

If the tube segment lies parallel to the rotor axis just experiencing a temperature gradient ($R_o = R_u \Rightarrow L = 0$) there is no pressure change as can be seen from the equilibrium of forces. Only the density varies. Any case with the tube not lying perpendicular to the rotor axis can be reduced to just considering the changes in radial direction.

If there is no temperature change ($T_u = T_o$) the expression can be simplified

$$dp = \omega^2 r \frac{p}{RT_u} \, dr \Leftrightarrow \int_{p_u}^{p_o} \frac{dp}{p} = \frac{\omega^2}{RT_u} \int_{R_u}^{R_o} r\,dr \tag{1-11}$$

$$\Rightarrow \ln \frac{p_o}{p_u} = \frac{\omega^2}{2RT_u}\left(R_o^2 - R_u^2 \right) \tag{1-12}$$

This can also be derived by inserting an infinitely small ΔT in equation (1-10) and applying de l'Hopitals rule.

For calculation of the whole impulse line the tube is divided into convenient segments with assumed linear changes in temperature and a single direction of the tube. This method requires temperature measurements at both ends of each segment.

4 EXPERIMENTAL SETUP AND INSTRUMENTATION

4.1 Test Bed and Gas Turbine

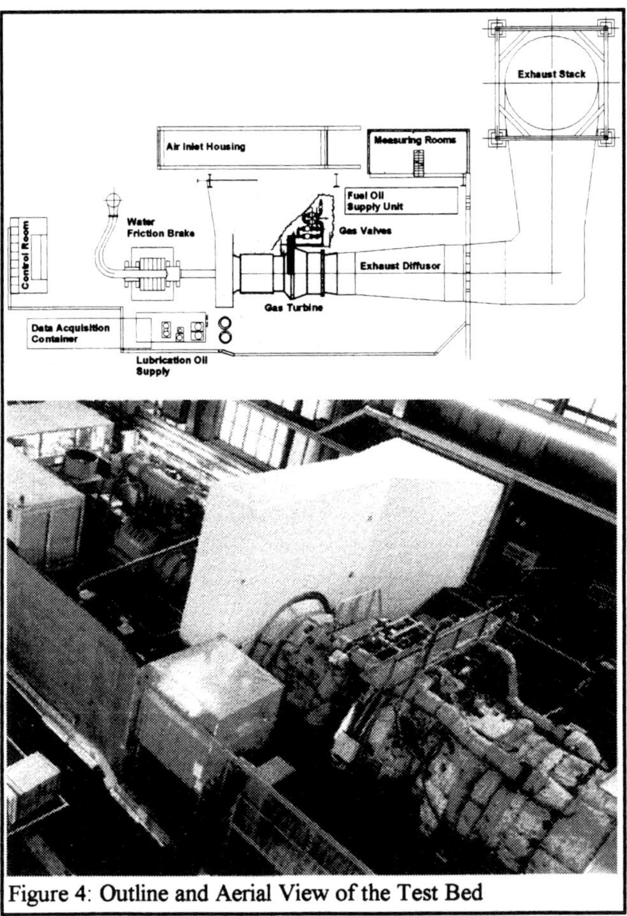

Figure 4: Outline and Aerial View of the Test Bed

The gas turbine test bed is equipped with a water friction brake enabling full load operation of heavy duty gas turbines even at off-design speed up to 200 MW /3/. Typically, prototype testing is performed covering and exceeding the complete range of operating conditions. Figure 4 gives an outline of the test bed and an aerial view of the water friction brake and the test bed with a gas turbine.

The gas turbine chosen was the 60Hz V84.3A Siemens heavy duty gas turbine. It features the standard single shaft splined disc type rotor, a 15-stage compressor and a 4-stage turbine. The gas turbine is cold end drive and delivers roughly 170 MW under full load /4/.

Figure 1a indicates the measurement locations inside the rotor in a cross section, where T refers to a thermocouple and p to a pressure measurement. One thermocouple was installed beside each pressure tap and several more were installed to collect information on the temperature distribution in special parts of the rotor.

All pressure transmitting tubes and all thermocouples were led through specially drilled holes to the turbine end of the rotor. As required for the correction scheme, some additional thermocouples were placed alongside the tubes to gain the temperature distribution.

In order to limit the number of transmission channels, temperatures inside the drillings for the wires and tubes in turbine stages one through three were assumed to be the same as at the thermocouples T34, T37, and T38, respectively. As the locations were less than 20 mm apart, the temperature error would be very small.

4. 2 Pressure Correction Scheme

As discussed earlier, each pressure tube has to be divided into several sections with an assumed linear temperature profile. The length of each segment and the number of thermocouples is tuned with respect to the routing of the entire tube and to best fit a linear temperature profile in each segment. Introducing the abbreviation

$$\Rightarrow \ln\frac{p_o}{p_u} = \frac{\omega^2 L}{R\Delta T}\left(L - \frac{T_u L}{\Delta T}\ln\left(\frac{T_o}{T_u}\right) + R_u \ln\left(\frac{T_o}{T_u}\right)\right) = K_{u-o} \tag{1-15}$$

the segments can be recombined.

For example, the impulse line for the pressure p_9 consists of the radial segments 5-7 and 41-42, the axial segments 7-41 and 43-transducer p_9, and the inclined segment 42-43 where the numbers correspond to figure 1a. In order to get the true pressure p_{9corr} at the measuring location from the measured pressure p_9 the individual correction factors of the segments have to be combined as follows:

$$
\begin{aligned}
K_9 &= \ln\left(\frac{p_{9corr}}{p_9}\right) = \ln\left(\frac{p_{9corr}}{p_{9-7}}\frac{p_{9-7}}{p_{9-41}}\frac{p_{9-41}}{p_{9-42}}\frac{p_{9-42}}{p_{9-43}}\frac{p_{9-43}}{p_9}\right) \\
&= \ln\left(\frac{p_{9corr}}{p_{9-7}}\right) + \ln\left(\frac{p_{9-7}}{p_{9-41}}\right) + \ln\left(\frac{p_{9-41}}{p_{9-42}}\right) + \ln\left(\frac{p_{9-42}}{p_{9-43}}\right) + \ln\left(\frac{p_{9-43}}{p_9}\right) \\
&= K_{7\to5} + 0 + K_{42\to41} + K_{43\to42} + 0
\end{aligned}
\tag{1-14}
$$

The second and the last values are equal to zero since the corresponding sections are parallel to the rotor axis (cf. figure 1)

The overall correcting factors can be calculated in the same way for all measuring locations. The true pressures are then expressed as

$$p_{xkorr} = p_x e^{K_x} \tag{1-16}$$

4.3 Instrumentation

Installation of wires and tubes inside a rotor being exposed to centrifugal accelerations of up to 8500 g´s (blade root) or even 16500 g´s (near the blade tip, for strain gauges) needs an experienced crew of fitters and engineers.

As mentioned above, all tubes and thermocouple wires together with the wiring for some 50 strain gauges was led to the turbine end of the rotor through drilled holes. There connections were made to pickups and power supplies. Figure 5 displays a measuring location on a disc during rotor assembly and a view of the rotor end after final assembly of all instrumentation wires.

The pressures and temperatures measured were quasi static values. So the thermocouples were directly connected to a 16 channel multiplexing telemetry transmitter. The pressure tubes were connected to miniature pressure transducers of piezoresistive type. A second multi channel transmitter with an integrated power supply and bridge amplifier was linked to the pickups.

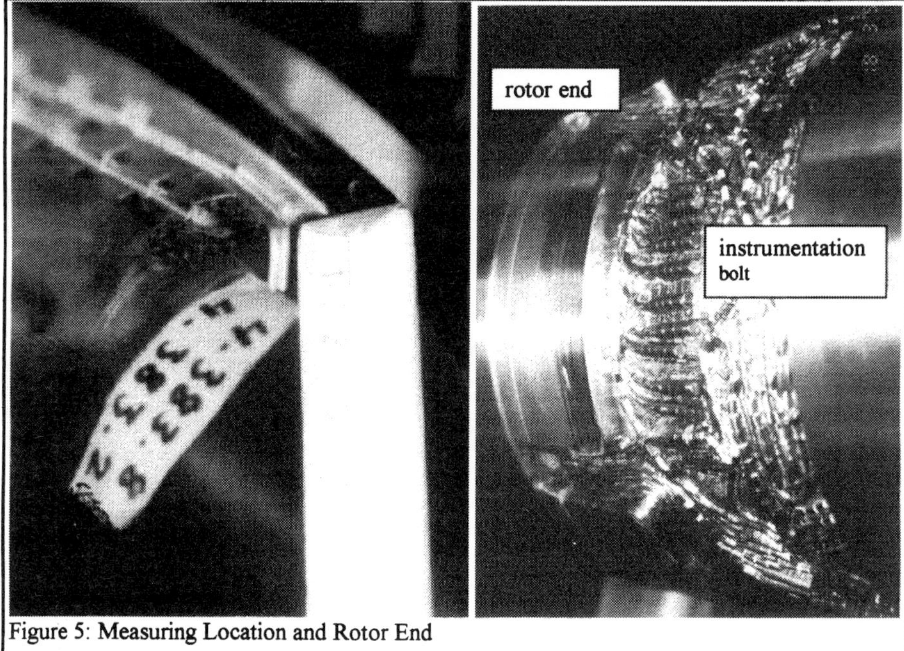

Figure 5: Measuring Location and Rotor End

Prior to installation one transducer was tested for zero shift under centrifugal loads during the balancing of a rotor which was equipped with some instrumentation anyway. As balancing is performed under vacuum, the transducer output gives an immediate picture of the acceleration sensitivity. The transducer proved to be very well within its specification. All transducers were checked for their temperature shift in an oven. The complete signal transmission line was calibrated over its entire operating range by applying pressure to the transducers.

All signals were decomposed in the receiver unit inside the data acquisition room. They were displayed for integrity checks (figure 6) and monitoring purposes and were stored on hard disc for further analysis.

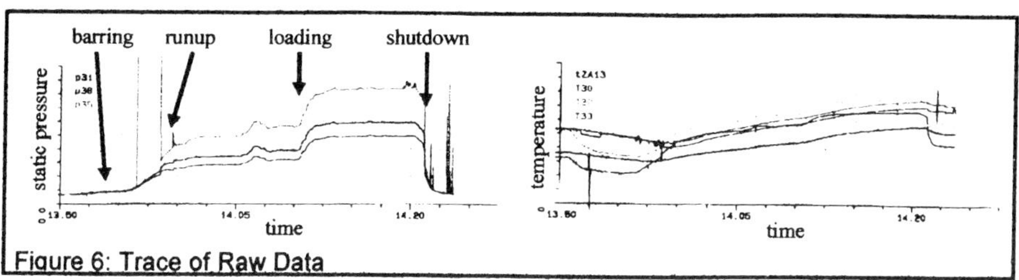

Figure 6: Trace of Raw Data

6 RESULTS

Figure 7 shows the raw data and the corrected values plotted against time for a typical test run. After runup the gas turbine was rapidly loaded, held at constant load for a short while and then loaded to more than 90% of rated power. Shutdown occured from high load due to a sudden shortage in fuel oil supply making pressures drop rapidly on the right hand side of the graph. Depending on the position of the measuring locations inside the rotor and thus the way of the tubes, the corrections lie between 2 and 18% of the measured values. Repeatability of the measurements proved to be very good. Only a slight sensitivity shift of the sensors was detected over the number of tests and was taken into account by recalibration.

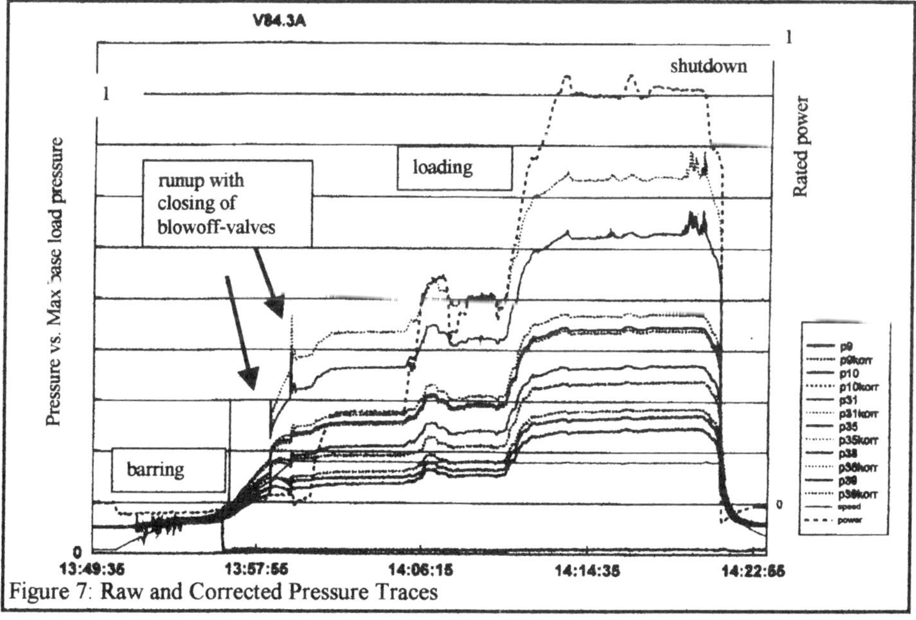

Figure 7: Raw and Corrected Pressure Traces

All changes in the operating conditions of the gas turbine can be clearly identified and analysed in terms of effect on the cooling air supply. Examples are closing of blow off valves during startup, loading/deloading the gas turbine or adjusting the IGV. According to the measurements the readings show sufficient cooling air pressures for all operating conditions.

From correlating pressure traces at the beginning and the end of individual cooling air ducts losses can be calculated. The temperature measurements (figure 6 right hand side) not only give information for correction of the pressure measurements but deliver information on heating of the cooling air inside the rotor (and thus on the temperature rise available for cooling) as well as on thermal stresses and on the cooling down behaviour after shutdown /1/.

7 SUMMARY

A technique was developed and applied to enable static pressure measurements inside the rotor of an operating gas turbine. This technique allows for use of 'over the counter' pressure transducers as they can be installed at one rotor end in an area with a comparatively low load in terms of centrifugal forces and temperatures, it moreover allows accessibility of the transducers. The transducers are connected to the measurement location with a very thin tube serving as impulse line. An algorithm was found to correct for the influences of the varying temperatures and centrifugal forces on the air column inside the impulse line.

The results were helpful in getting for the first time a thorough in depth view into the airflow characteristics inside a disc-type rotor with a central tie bolt /1/. They serve as a database for the perfection of already highly sophisticated CFD tools needed for optimized secondary air paths of future gas turbine designs.

REFERENCES
/1/   Janssen, M.; Joyce, J.S.: 35-Year Old Splined-Disc Rotor Design für Large Gas
        Turbines, ASME paper 96-GT-523
/2/   Brillert, D., Reichert, A. W., Simon, H., Cooling Air Flow in a Multi Disc Industrial
        Gas Turbine Rotor, ASME paper 98-GT-136
/3/   Boehm, W., Raake, D., Regnery, D., Seume, J., Terjung, K., Testing the Model
        V84.3A Gas Turbine - Experimental Techniques and Results, PowerGen´96, Orlando,
        FL, December 4-6 1996
/4/   Siemens AG, The Siemens .3A Gas Turbine Family, Siemens AG, Power Generation
        Group, 1995

Design Methods

C557/103/99

The aerodynamic design and testing of a supersonic turbine for rocket engine application

U WÅHLÉN
Space Propulsion Division, Volvo Aero Corporation, Trollhättan, Sweden

ABSTRACT

This paper describes the development at Volvo Aero Corporation in Sweden of a one-stage supersonic impulse turbine typical of rocket engine applications for high pressure ratios. Aerodynamic design, design variants and performance testing are discussed. Results from flow analysis of the complete turbine stage are compared with test results. The results derived from this and similar turbine designs were later used in the development of the full-scale oxygen (LOX) turbine used in the Vulcain engine in the European launcher Ariane 5.

NOTATION

		Subscripts	
c	chord length	0	stagnation value
c	absolute flow velocity	0	position at turbine inlet
c_0	isentropic velocity	1	position before stator nozzles
M	Mach number	2	position after stator
p	static pressure	3	position after rotor
s	pitch	N	nozzle
U	blade speed at mean radius	rel	value in the relative frame
w	relative flow velocity	R	rotor
Y	stagnation pressure loss coefficient	ts	total to static value over turbine
α	absolute flow angle	tt	total to total value over turbine
β	relative flow angle		
γ	ratio of specific heats		
$\eta_{N,R}$	nozzle and rotor efficiency		
π	pressure ratio		

1 INTRODUCTION

In recent years there has been an increasing interest in supersonic turbines, as they make high specific power per stage possible, which minimises the turbine's weight, complexity and cost as well as the amount of driving fluid. The challenge is to achieve all this with a minimum drop in efficiency.

Supersonic turbines can be found in for example turbopump turbines for rocket engines using the gas generator cycle. Common characteristics of this type of turbine are high pressure ratio and low rotational speed, which in combination leads to low isentropic velocity ratios (U/c_0). Therefore the most suitable design is a one or two stage supersonic turbine of the impulse type, which has better efficiency at low U/c_0. Frequently, the rotational speed is not limited by the mechanical strength of the turbine but instead by the margin required for pump cavitation and also, in some cases, by the rotor dynamics. The pump cavitation margin is dependent on the pump medium, which can be seen in, for example, the difference in rotational speed in cryogenic pumps using liquid oxygen or hydrogen.

The information found in the literature on the design and testing of supersonic turbines is, however, relatively limited, examples are (1), (2), (3). Furthermore, there are few comparisons between test results and stage flow analysis, which would contribute to the understanding of the complex flow in the supersonic turbine.

This paper describes the development at Volvo Aero Corporation in Sweden of a one-stage supersonic impulse turbine, typical of rocket engine application for high pressure ratios. This turbine evolved from early designs in sub-scale and then later to the full-scale oxygen (LOX) turbine used in the Vulcain engine in the European launcher Ariane 5. The LOX turbine development work was performed under Fiat Avio responsibility in the frame of the development of the Vulcain engine, SEP as prime contractor under contracts of ESA/CNES. The paper concentrates mainly on one of the successful sub-scale designs. Aerodynamic design, design variants and performance testing are discussed. Results from viscous and non-viscous CFD flow analysis of the complete stage are compared with turbine test results. The aim of this work is a better understanding of the development potential of this type of turbine design.

2 ONE-STAGE SUPERSONIC TURBINE

2.1 Design Concept

The gas generator cycle is used in the main engine Vulcain for the European launcher Ariane 5. Hydrogen and oxygen are burned in the gas generator driving both the hydrogen (LH2) and oxygen (LOX) turbine that are arranged in parallel. The gas leaving the turbines is exhausted to the atmosphere, and therefore, the available pressure ratio is high - about 14 for the LOX turbine, see table 1. The rotational speed is low leading to a very low isentropic velocity ratio ($U/c_0 = 0.08$).

The LOX turbine evolved from early designs in sub-scale and later to the full-scale LOX turbine used in the Vulcain engine (4), (5).

The turbine described in this paper was a sub-scale design aimed at development work: the sub-scale design was chosen for practical reasons. It was one of a number used for testing different types of turbine component designs. To achieve good comparison between the early sub-scale designs, this one was also manufactured in the sub-scale.

The turbine was designed as a one-stage supersonic turbine with high pressure ratio and very low rotational speed. The turbine was also designed for a medium with high specific heat which leads to high isentropic velocities and, thereby, in combination with these parameters, to very low U/c_0. Table 2 gives the main parameters of the design.

Table 1   Vulcain LOX turbine specification

Inlet total temperature	K	875
Inlet total pressure	kPa	6350
Outlet total pressure	kPa	450
Rotational speed	rpm	13290
Shaft power	kW	3076

Table 2   Design parameters

Number of stator 1 nozzles	18
Degree of admission	100 %
M_2	2.4
α_2	74
Number of rotor blades	79
$M_{2\,rel}$	2.2
$M_{3\,rel}$	1.9
$\Delta\beta$	133
Aspect ratio h/c	0.85
Solidity s/c	0.39
Rotor mean radius (mm)	93.75
Stage pressure ratio, π_{ts}	26.6
Blade speed/isentropic velocity U/c_0	0.08

2.2 Nozzle

Stator 1 nozzle design can be made with low losses up to high Mach numbers. The important part is to reduce the trailing edge losses and create good inlet flow conditions for the rotor.

The stator was designed as a two dimensional asymmetrical nozzle using the symmetry line in the centre of a symmetrical nozzle as a wall. The throat section chosen was sharp with an angle corresponding to half the Prantdl-Meyer angle of the intended Mach number. As this type of design gives very short axial chord of the nozzles, the axial length of the turbine can be reduced. The supersonic part was designed using the method of characteristics with boundary layer correction. The pre-swirl from the inlet manifold reduced the flow turning needed.

In addition, designs with two dimensional symmetrical nozzles and wrapped around Laval nozzles were manufactured. These designs were also tested in the turbine rig, but did not show any performance advantage.

2.3 Supersonic Rotor

The rotor blade was designed with supersonic relative inlet and outlet Mach numbers and a relatively large flow deflection. The rotor was designed to have started supersonic flow and since the axial Mach number is subsonic the unique incidence has to be considered (6). The method of characteristics with boundary layer displacement correction was applied for the profile design. In addition, the rotor was designed with a shroud to reduce the tip leakage loss and increase the blade damping.

To reach good efficiency in these types of turbines, the matching of the different components must also be considered. In addition, different alternative rotor designs were tested and the sensitivity of the leading edge design was investigated.

2.4 Continued Development

The information from the testing of the sub-scale designs was later used in the design of a full-scale turbine, i.e. the Vulcain LOX turbine. The by testing verified design methods were then applied. For the full-scale turbine a tandem supersonic/subsonic outlet guide vane was designed. Due to the large flow turning and low rotational speed the outlet swirl from the rotor is large. Therefore, the outlet guide vanes have to turn the flow by nearly 60 degrees to achieve axial flow and reduce the supersonic inlet Mach number to subsonic velocities.

3 CALCULATION METHOD

The CFD code used for the analysis of the turbine was the general viscous three-dimensional flow solver VOLSOL (7) developed at Volvo. The code solves the Euler equation or Reynolds-averaged Navier-Stokes equations. The numerical method is an explicit three-stage Runge-Kutta time-marching method with optimized local time-step. The equations are discretized in finite volume form with a third order accurate upwind biased scheme for the convective (Euler) fluxes and second order centered scheme for the diffusive (viscous) fluxes. The solver uses structured multi-block grid. The tangential averaging algorithm calculates the fluxes through the stator/rotor interface with absolute flux conservation. The multi-grid technique was used for convergence acceleration.

Chien´s (7) low Reynolds k-ε turbulence model was used in the three-dimensional stage calculation for the viscous computation. However, near the hub and tip walls Kauto & Launder´s (9) low Reynolds k-ε turbulence model was used.

The three dimensional computational model consisted of 11 blocks and 251 885 nodes. The mesh was refined to achieve a Y^+ less than one at the first cell on the blade profiles. Between the stator and rotor components a patch block was used to allow different number of cells in the radial direction. The mesh of the model at the mean-line radius is shown in figure 1. Perfect gas was assumed in the calculations.

4 TURBINE TESTING

Turbine stage performance was verified in the turbine test rig at Volvo, see figure 2. The test rig uses compressed air from a 130 ton air magazine 85 m below the ground level. Water is used to give the pressure head. The air is preheated and an ejector can be used to increase the pressure ratio over the turbine. The rotational speed is controlled by a water brake, which is also used to measure the turbine shaft torque. The large air magazine enables testing under stabilised conditions over a long period and at different operation points.

The mass flow was measured by an orifice meter. The turbine inlet and outlet total pressure as well as the total temperature were measured using pitot rakes. Static pressure was measured at several stations: at the hub and tip before and after each component.

Downstream of the rotor a straight diffuser is used to give dynamic pressure recovery and to avoid choking.

The test is run under simulated performance conditions and the performance is scaled by similarity rules to the rocket engine conditions. The test conditions are given in table 3. The power levels chosen can be much lower than under hot running conditions in the rocket engine, mainly because of the large difference in the gas constant between the hydrogen/oxygen combustion gas and air. The turbine inlet pressure level is given by the air magazine pressure and the inlet total temperature chosen is high enough to avoid condensation. Therefore, simpler material can be used in the special development test hardware, thereby decreasing the cost and lead time for manufacturing the test hardware.

Table 3 Turbine test conditions in air

Inlet total pressure	800 kPa
Inlet total temperature	500 K
Range of pressure ratio tested π_{ts}	4.9-35.5
Range of rotational speed tested, rpm	2150-8480
Range of velocity ratio tested, U/c_0	0.035-0.136

5 RESULTS AND DISCUSSION

5.1 Test results

The performance results are presented in form of the total and static efficiency curves versus the U/c_0 in figure 3. The large difference between the levels of the static and total efficiencies are due to the high outlet Mach number. Therefore, it is possible to reach a high overall turbine efficiency if the dynamic pressure can be recovered with low losses in the outlet guide vane and diffusor.

The dependency of the pressure ratio can also be seen from the measurement points at constant pressure ratio when changing the rotational speed. Peak efficiency is achieved for a pressure ratio, π_{ts}, near 20 and slightly decreasing with increased pressure ratio, see figure 3. Therefore the values at the higher pressure ratio are relatively closely spaced. This value was lower than expected from the design and no further increase of the turbine power above the peak value could be reached. The same effect can be seen in figure 4 where only the outlet condition is influenced when further decreasing the outlet pressure.

The curves, in figure 3, show that the turbine runs far below its optimum efficiency due to the design limitation of the rotational speed. Thus, the possible maximum performance level can be expected at a velocity ratio between 0.4-0.5.

The pressure ratio over the rotor can be seen in figure 4, which shows that the turbine has positive reaction at a pressure ratio slightly above 15.

5.2 Numerical results and comparison with test

To have a good comparison between the test results and the computer model, the CFD computation uses also air. Thereby will there be relatively small difference, between the test and analysis, in the ratio of specific (γ) heats and the Reynolds number. The difference of γ is also small between the gas generator gas and the air test conditions, at the nozzle throat about 1.3 %. For the Reynolds number however the difference will be larger, about 2.5 times larger Reynolds number for the rotor in the gas generator gas, and the influence from this has therefore to be regarded.

Both Euler and Navier-Stokes solutions predict a higher mass flow compared to the test. This is expected since no corner radii are used in the model and constant inlet boundary conditions are applied. The true geometrical nozzle area is therefore smaller and the blockage can be expected to be somewhat larger. This is however estimated to have little effect when comparing non-dimensional variables. Furthermore, the CFD models do not include any model for the tip leakage, which mainly influences the rotor losses.

To get an estimate of the relation between the viscous losses and the shock losses the Euler and Navier-Stokes (N-S) calculations are compared, see table 4. These data are presented in the form of nozzle and rotor efficiency and stagnation pressure loss coefficient, which are defined by:

$$\eta_N = \left(\frac{c_2}{c_{2is}} \right)^2 \qquad \eta_R = \left(\frac{w_3}{w_{3is}} \right)^2 \qquad (1)$$

$$Y_N = \frac{p_{01} - p_{02}}{p_{02} - p_2} \qquad Y_R = \frac{p_{02rel} - p_{03rel}}{p_{03rel} - p_3} \qquad (2)$$

Values are also given from the test results, however an assumption of the effective flow area between the nozzle and rotor must be made since no total pressure measurement was performed between the components.

The viscous losses for the inlet part of the nozzle are very small: almost all losses are shock losses in the expansion part and at the trailing edge of the nozzle. In the CFD results an oblique shock can be seen at the curved surface near the nozzle trailing edge, see figure 6. The cause of this oblique shock is over expansion due to higher outlet pressure than ideally. The main reason for the higher outlet pressure are the matching to the rotor and the radial gradient of the pressure, which in computation will be exaggerated due to the axi-symmetric CFD model compared to the plane nozzles in the design. The radial variation in pressure will also influence the position of this oblique shock and near the hub it is seen at the nozzle trailing edge. The expansion at the trailing edge and the wake can also be seen in figure 6. In figure 5 the entropy is plotted at a station downstream of the nozzle and the losses caused by the above described flow phenomena can bee seen. The losses in the nozzle are somewhat larger than expected from the original design but good comparison of the component losses can be seen in table 4.

The rotor flow is started and an flow acceleration can be seen at the suction surface followed by a deceleration with an increase of the boundary-layer and a separation bubble which reattach, leading to entropy generation. Although, the viscous losses for the rotor are significant, the shock losses dominate. In the entropy plot downstream of the rotor shown in figure 7, the effect of the nozzles low entropy region can be seen at the hub region. The in test predicted component loss has a lower value than from the computation. Simplification of the CFD model probably contributes to this difference e.g. the leakage over the tip shroud is not simulated.

Table 4 Component losses

Component loss	Euler	N-S	Test
η_N	0.91	0.91	0.92
η_R	0.81	0.78	0.74
Y_N	0.54	0.56	0.45
Y_R	1.09	1.34	1.67

CONCLUSION

The development of the one-stage supersonic sub-scale turbine led to the testing of a number of different aerodynamic designs. From the experience gained, good design methods could be derived and later used in the design of the LOX turbine for the Vulcain engine. The development of the LOX turbine has been very successful and all the performance goals have been reached. The design methods derived where thereby successfully applied and thus showing their validity.

Combining turbine testing with viscous three-dimensional stage calculation gives supplementary information that helps to increase the overall knowledge concerning the design results and helps explaining the observed flow phenomena. This combination can lead to further improvements in supersonic turbine design and to the development of different design methods.

ACKNOWLEDGEMENTS

This work was supported by the Swedish National Space Board.

REFERENCES

1. Stratford B.S., Sansome G. E., "Theory and Tunnel Tests of Rotor Blades for Supersonic Turbines", A.R.C., Reports and Memoranda No. 3275, December 1960
2. Binsley R. L., Boynton J. L., "Aerodynamic Design and Verification of a Two-Stage Turbine With a Supersonic First Stage", Journal of Engineering for Power, April 1978
3. Kurzrock J., W., "Experimental Investigation of Supersonic Turbine Performance", ASME 89-GT-238, 1989
4. Håll U., "Aerodynamic Design and Test of a Supersonic Turbine for a Rocket Engine", Proceeding of symposium: 'Fluid dynamics and space', VKI, Rhode-Saint-Genèse, 25 & 26 June 1986.
5. Wåhlen U., "The Aerodynamic Design of the Turbines for the Vulcain Rocket Engine", AIAA 95-2536
6. Billonnet G., "Flow Computation and Blade Cascade Design in Turbopump Turbines", ASME 88-GT-248, 1988
7. Groth P., Mårtensson H., Eriksson L-E., "Validation of a 4D Finite Volume Method for Blade Flutter", ASME 96-GT-429
8. Chien K., Y., "Predictions of channel and boundary-layer flows with a low Reynolds number turbulence model", AIAA Journal, 20(1): pp 33-38, 1982
9. Kato M., Launder B., E., "The modelling of turbulent flow around stationary and vibrating square cylinders", ninth symposium on turbulent shear flow, August 16-18 1993
10. Andersson S., Lindeblad M., Wåhlén U., "Performance Test Results for the Vulcain 2 Supersonic/Transonic Turbine", AIAA-98-399, 1998

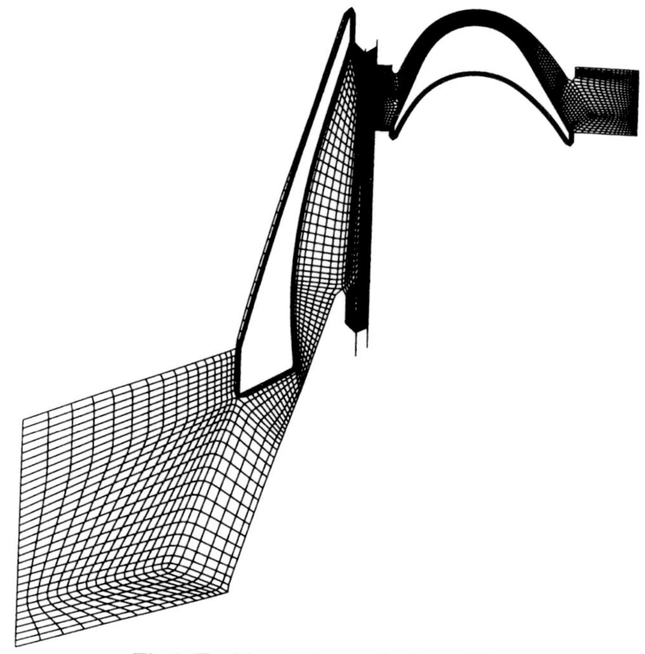

Fig 1 Turbine stator and rotor grid

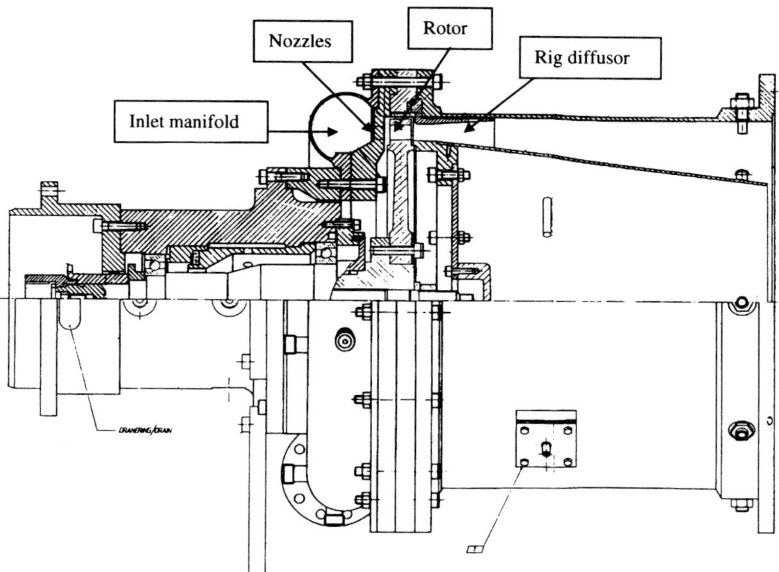

Fig 2 Turbine test rig

C557/103 © IMechE 1999

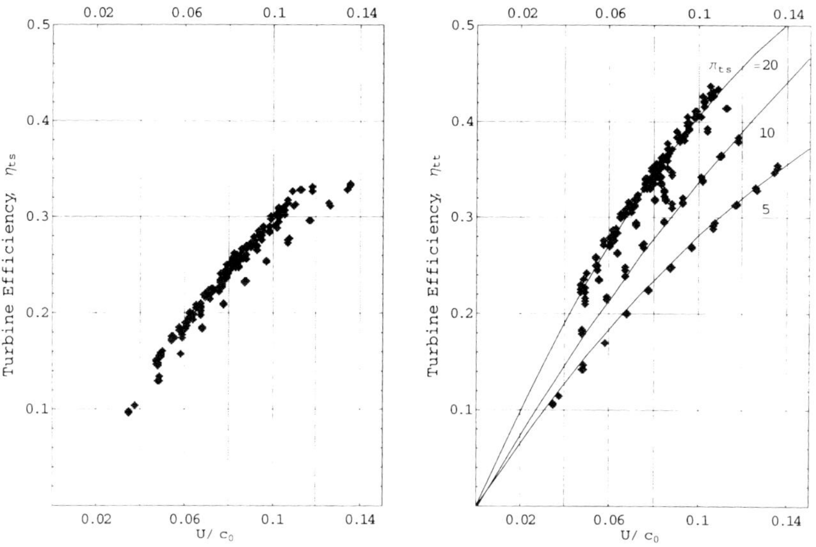

Fig 3   Turbine performance

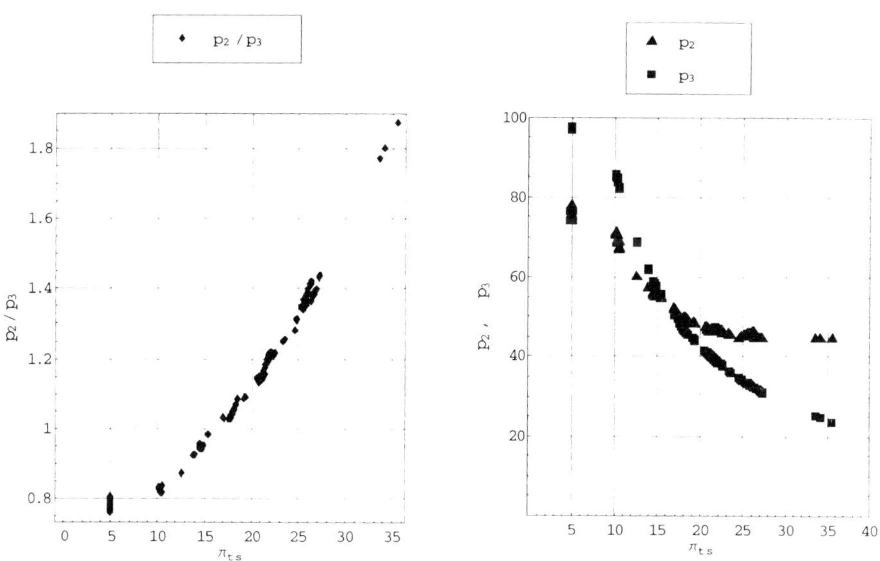

Fig 4   a)   Measured mean pressure ratio over rotor b) Measured mean static pressure before and after rotor.

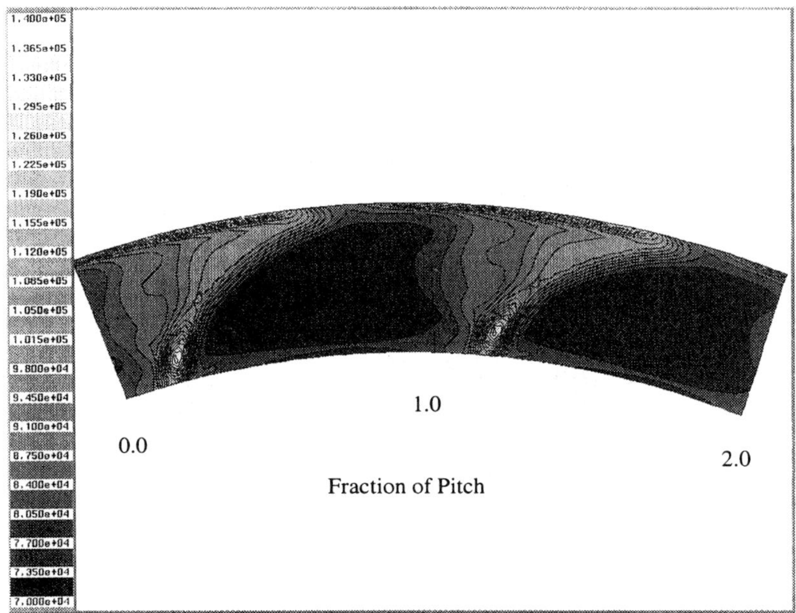

Fig 5   Entropy contours downstream of nozzle.

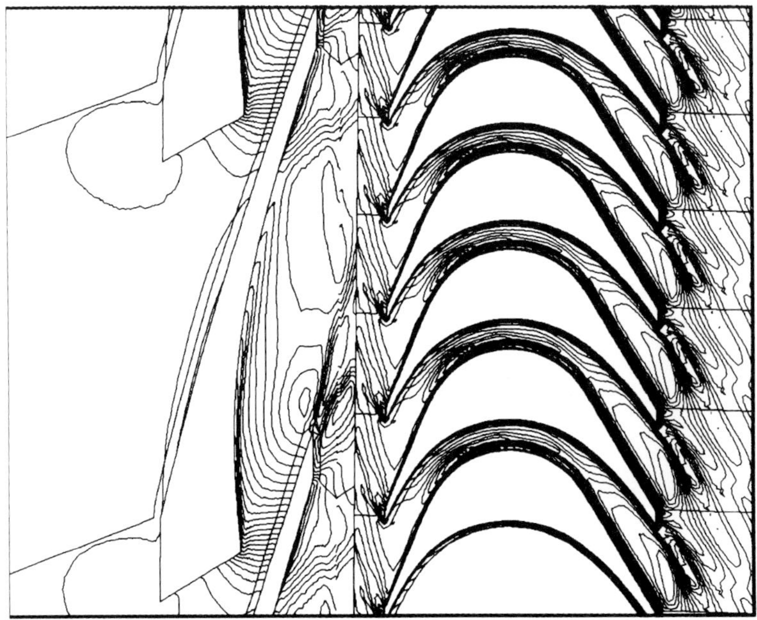

Fig 6   ISO-Mach number contours at mean radius.

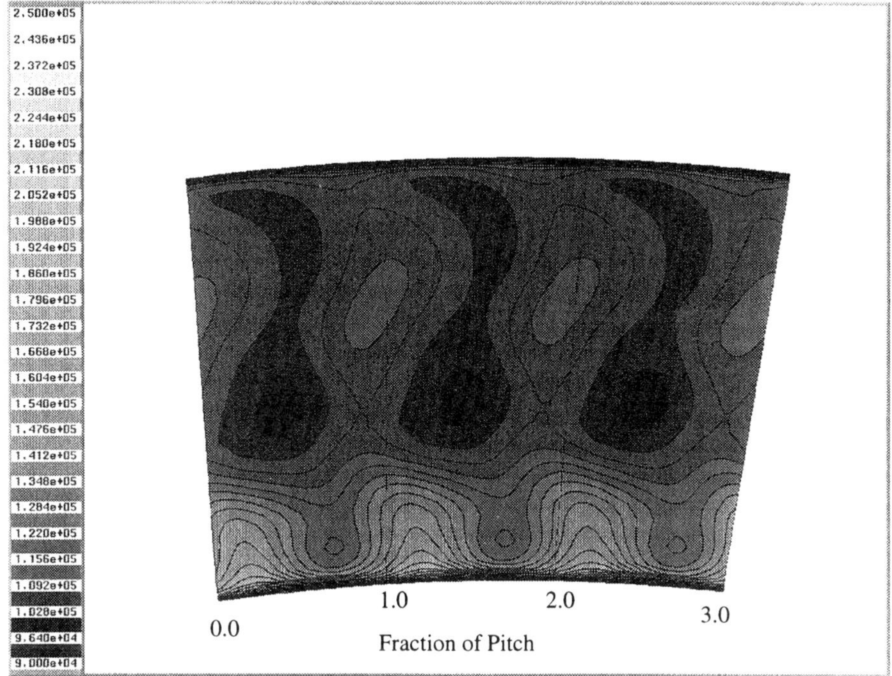

Fig 7 Entropy contours downstream of rotor

C557/154/99

Three-dimensional turbine blade design using a Navier–Stokes solver and Artificial Neural Network

S PIERRET and **R A VAN DEN BRAEMBUSSCHE**
Von Karman Institute, Rhode-Saint-Gènese, Belgium

ABSTRACT

Improving turbine efficiency by applying non-radial stacking and three-dimensional design techniques has received considerable attention in the recent years. A big source of losses is the spanwise non-uniformity of the next stage inlet flow angle resulting from the non-uniformity of the outlet flow angle of the preceding blade row. This non-uniformity can be reduced by adjusting the 2D sections along the span and/or by leaning the blades.

The present method describes the design of a 3D blade geometry built by a radial stacking of several 2D blade sections which are provided by a 2D design system. A 3D Navier-Stokes solver is used to check the blade performance and to update the requirements imposed for the next design of the 2D blade sections.

The 2D sections are designed using an Artificial Neural Network (ANN). The latter one constructs an approximate model (response surface) using a database containing the 2D Navier-Stokes solutions obtained from previous designs. It is used for the optimization of the 2D blade geometry by means of Simulated Annealing (SA). The optimum 2D geometry is then verified by a 2D Navier-Stokes solver.

This procedure results in a considerable speed-up of the design process by reducing both the interventions of the operator and the computational effort. It also allows the design of more efficient blades, satisfying both the aerodynamic and mechanical constraints. The method has been used to design different types of turbine blades of which one example will be presented.

1. INTRODUCTION

The flow around a 3D blade is very often highly 3 dimensional due to secondary flows, non-radial stacking, endwall contouring and variation of blade shape in the spanwise direction. Beside the redistribution of losses, this 3 dimensional flow also modifies the blade exit flow angles. In particular, the secondary flows are known to create an overturning very close to the endwalls and a region of underturning some distance away from them. The effects of this deviation on the exit velocity profile and on the inlet incidence angles to the next blade row can be very large and can substantially increase the incidence losses on the next blade row (1).

In recent years there has been considerable effort to reduce secondary flows and therefore improve turbine efficiency by applying non-radial stacking and three dimensional design techniques. However, there is no consensus as to what is the optimum stacking law. It seems that the main benefits of three dimensional design is to compensate for secondary flows and to make the flow more uniform as it enters the next blade row (2). Experience has shown that the reduction of the non-uniformity of the outlet flow angle is more easily achieved by using spanwise variations of blade exit angle than by non-radial stacking. Changing the blade exit angle, to make the outlet flow more uniform, will anyway result in some kind of lean toward the endwalls as the largest adaptation of the exit blade angle is required in this region.

This paper describes an automatic procedure for 3D blade design by adapting the 2D sections in order to make the outlet flow angle more uniform. The 2D sections are designed by the method described in (3) and (4).

2. DESCRIPTION OF THE 3D DESIGN PROCEDURE

The required inlet and outlet flow angles as well as an estimation of the total and static pressure at several spanwise positions, for all blade rows of a multistage machine are normally provided by a throughflow calculation (Fig. 1). The 3D requirements, upstream and downstream the blade row, are then split into N 2D requirements and used for the design of the N 2D blade sections by means of the 2D design procedure.

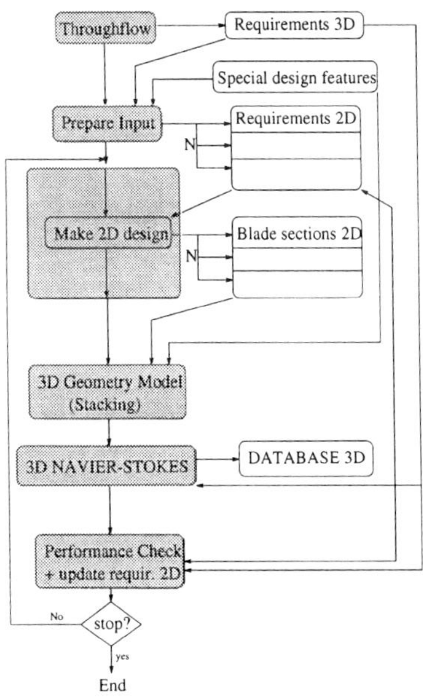

Figure 1: General algorithm of the method

They are then used to build the 3D geometry by placing them along a prescribed stacking line. The 3D flow in this new geometry, computed by means of a 3D Navier-Stokes solver, is stored into database 3D which can be used later for the optimization of the stacking line.

Due to 3D effects, the aerodynamic performances will be different from the imposed ones and from the ones obtained with the 2D Navier-Stokes solver. Those performances are verified by checking the objective function and a new design iteration is started if its value is too big. Before starting a new iteration one will change the requirements used to design the 2D sections. Those requirements are shifted in the direction opposite to the difference between the 3D solution and the 2D solutions. As an example, suppose that the outlet flow angle at mid span imposed by the throughflow is -65 degrees, the 2D design system will then produce a blade which in 2D has an outlet flow angle of -65 degrees. However, due to 3D effects, the outlet flow angle predicted by the 3D flow solver might be -67 degrees. Then the outlet flow angle imposed at the next iteration for the 2D design system is decreased to -63 degrees such that in 3D the outlet flow angle should come closer to the imposed value (-65 degrees).

The main components of this procedure are discussed in more detail in the next 3 sections, namely : the 2D design method, the 3D Navier-Stokes solver and the 3D geometry model. The other components such as the neural network, the optimization algorithm (simulated annealing) and objective function are very similar to the ones developed for the 2D design system (3, 4).

3. THE 2D DESIGN METHOD

The design procedure for each 2D section is very similar to the one presented in (3) and the objective function is described in (4). It is based on the following principle. A Navier-Stokes solver, referred to as a model of high physical fidelity but high computational cost, is used to compute the 2D flow around the geometry. This solver uses as input : the geometry ($\vec{G}$) and boundary conditions ($\vec{BC}$). The output is : the performances ($\vec{P}$) characterized by the efficiency, the turning and the Mach number distribution on the suction and pressure side. The results of all the computations performed with this solver (called computer experiments) are stored into a database under a parametric format ($\vec{G}$, $\vec{BC}$, $\vec{P}$) and used to construct an approximate model of lower physical fidelity but lower computational cost. One then starts a number of optimization iterations using this simpler, cheaper approximate model. At the end of this optimization phase, one has recourse to the high-fidelity model to verify the lower-fidelity model. The optimization then continues using again the simplified model after it has been recalibrated (improved) by means of the results of the latest calculation.

The approximate model is built by means of an Artificial Neural Network (ANN) which is an interpolator defining the response surface by learning the relation between performance, boundary conditions and geometry :

$$\vec{P} = \mathcal{N}(\vec{BC}, \vec{G}) \tag{1}$$

After mapping the database samples, the ANN is able to generalize, meaning that it can predict the performance of a new geometry, not present in the database. As the time for ANN learning is proportional to the number of training samples, it is sometimes of interest to use a sub-database (training database) containing only blade samples that are similar to the blade being designed. As we have N_{PS} sections which might be very different from hub to tip, one has to use N_{PS} training databases and N_{PS} approximate models.

The optimization algorithm is either an heuristic search procedure, called "simulated annealing" or a genetic algorithm. The optimization is driven by minimizing an objective function which is the sum of several penalty terms (constraints are transformed into penalties). They measure the quality of each blade section independently. There are mechanical constraints such as 2D cross section area, minimum and maximum momentum of inertia, angular position of the axis of minimum momentum of inertia, minimum blade thickness, leading edge and trailing edge radius of curvature. Geometrical constraints are also imposed to control the curvature close to the leading edge and on the rear part of the suction side. Blades need to perform also well at off-design conditions. One must therefore avoid to design blade geometries which have very good performance but are operating very close to separation and are likely to separate at slightly off-design conditions. Some constraints are therefore imposed on the Mach number distribution such as : minimum acceleration on the front part near the leading edge, absence of inflection point on some part of the Mach distribution and a no-separation safety margin controlled by the Pohlhausen and Buri parameters (4). It is also possible to add some penalties which measure the quality of the whole 3D blade shape, e.g. the smoothness of the spanwise variation of some geometrical parameters or some aerodynamic characteristics.

4. THE 2D/3D NAVIER-STOKES SOLVER (TRAF2D/3D)

The TRAF2D/3D (TRAnsonic Flow 2D/3D) Navier-Stokes solver is used to predict the performance of the blades. It has been developed (5) in a research project on viscous cascade flow simulation by the University of Florence and NASA (ICASE and ICOMP).

The TRAF3D code is able to calculate the flow in linear cascades as well as annular ones and rotating blade passages with tip clearance flow. The Reynolds-averaged Navier-Stokes equations are solved using a Runge-Kutta scheme in conjunction with accelerating techniques (local time stepping, residual smoothing and Full-Approximation-Storage (FAS) multigrid). These equations are discretized using finite volumes and a cell-centered discretization with artificial dissipation. A very low level of artificial viscosity is ensured by eigenvalue scaling.

The two-layer mixing length model of Baldwin and Lomax is used to compute the turbulent quantities. The transitional criteria of Baldwin and Lomax is adopted on the blade surface while on the end walls, the shear layer is assumed to be fully turbulent from the inlet boundary. The transition can also be imposed at a given point on both suction and pressure surfaces.

A three-dimensional computation using 300 000 points requires a memory of 90 MB and a CPU time of about 2 hours on the ALPHA station (type 500/500). Two-dimensional computations are usually performed with 12 000 points with means a memory of 3 MB and a computational time of 5 minutes.

5. THE 3D GEOMETRY MODEL

A critical element in the success of any shape optimization method is its capability to generate a great variety of physically realistic shapes. Ideally, the geometry model should allow as much geometry flexibility as possible with as few design variables as possible.

A 3D geometry model has been developed which builds the 3D shape by stacking a number of 2D sections. It is done in 5 distinct steps :

1. Definition of N_{PS} primary sections on the cylindrical surfaces of radius R_{PS}, using the 2D geometry model (see Fig. 2 a and 2 b).

   Each blade section is defined by two Bézier curves, one for the suction side and one for the pressure side. At the leading edge they are connected in a way to ensure continuity up to the second derivatives. At the trailing edge they are connected by a circle of given diameter with continuity up to the first derivative.

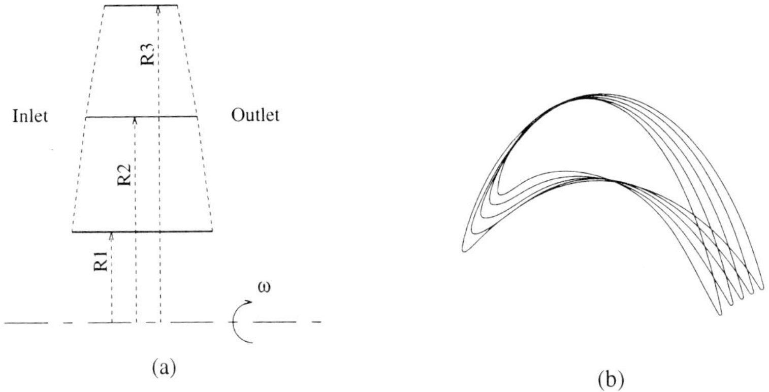

Figure 2: Geometry model : primary sections : (a) location, (b) definition

2. Definition of the stacking point : the trailing edge, the leading edge or the gravity center.

3. The stacking line is defined by two Bézier curves at both endwalls and a straight line in between for both the tangential lean and the meridional sweep (Fig. 3). The first control point of the lower Bézier curve is the starting point. The second control point is defined by the angle α_1 and the length P1. The third control point is located at the radial distance C1 and imposing the continuity of the slope at the junction point between this Bézier curve and the straight part at an angle α_2 with the radial direction. The straight part extends until a radial position defined by C2. The upper Bézier curve is defined similarly to the lower one using the parameters P2 and α_3, assuring continuity of the slope at the junction point with the straight part. The straight part length will be zero if the sum of C1 and C2 is equal to 1.0.

   This method defines both the tangential lean and meridional sweep using only 7 parameters.

4. Generation of additional 2D sections by spanwise interpolation of the blade parameters or of the blade point coordinates.

5. Definition of the endwall shape using Bézier curves and computation of the intersection with the 3D blade.

6. RESULTS

This design procedure has been tested with the redesign of a 3D blade defined by five 2D sections radially stacked around their gravity center between cylindrical endwalls. The mechanical constraints imposed at each section are : prescribed value of the 2D blade section area, minimum and maximum momentum of inertia and angle between the axis of minimum momentum of inertia and the axial direction, the radius of the trailing edge circle, the axial chord and the number of blades. The tip to hub radius ratio is 1.19 and the blade height to hub axial chord ratio is 2.85.

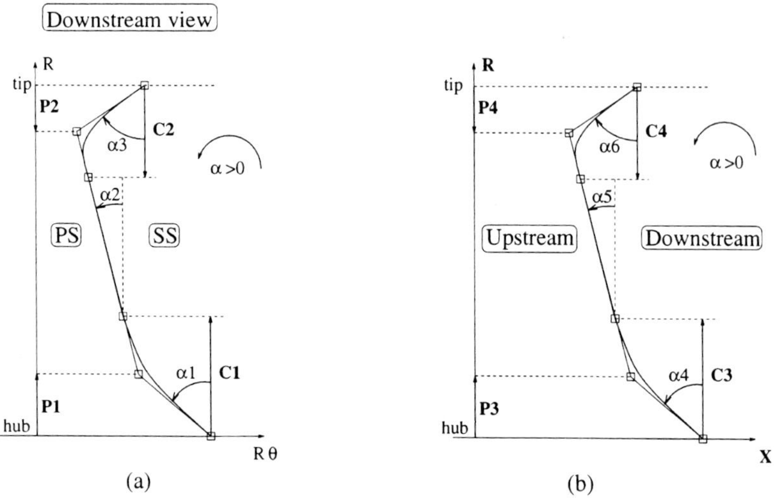

Figure 3: Geometry model : tangential lean and meridional sweep laws

The aerodynamic conditions imposed are : the spanwise distribution of inlet flow angle in blade-to-blade plane (Fig. 4), zero inlet flow angle in the meridional plane, constant inlet total pressure, constant inlet total temperature, outlet static pressure at hub, 1.5 C_{ax} downstream the trailing edge.

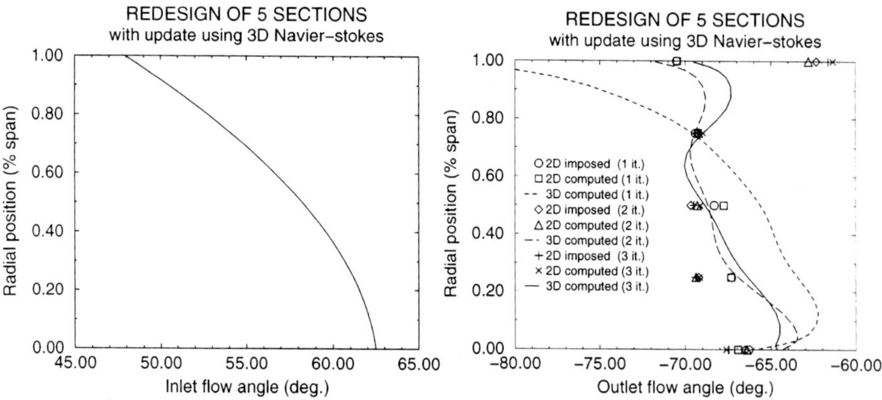

Figure 4: Spanwise distribution of the blade-to-blade inlet flow angle

Figure 5: Spanwise distribution of the blade-to-blade outlet flow angle

The outlet flow angle is imposed at an axial distance 0.2 C_{ax} behind the trailing edge in five radial positions equally distributed from hub to tip and represented by the circles on figure 5.

The initial 2D database contains 15 samples and the computational time needed to train the ANN and to run the optimization algorithm are respectively 20% and 40% of the time needed by the 2D Navier-Stokes solver.

Three design cycles have been made. Each of them consists of the generation of 5 blade sections by the 2D design system and the analysis of the resulting blade with the 3D Navier-Stokes solver. Only 3 iterations composed of a 2D blade optimization and 2D Navier-Stokes analysis are made for the design of each of the 5 blade sections. The blade geometrical parameters, boundary conditions and blade performance are added to the database after each 2D Navier-Stokes solution.

Despite of the small number of 2D design iterations the blade sections obtained in the first design cycles show an outlet flow angle (in 2D) which is less than one degree from the imposed value.

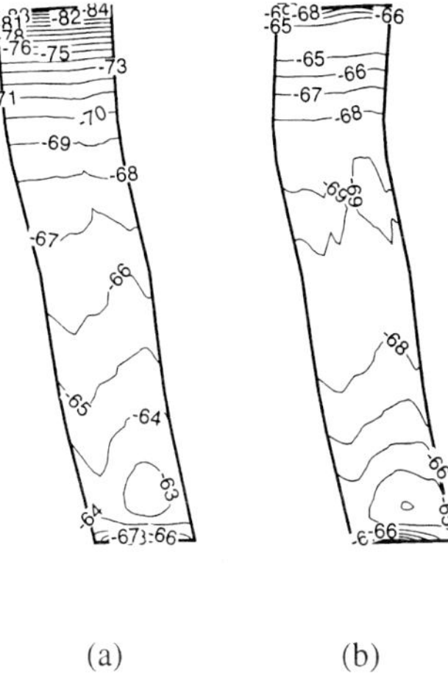

(a)   (b)

Figure 6: Contour lines of constant blade-to-blade outlet flow angle 0.2 Cax % downstream the trailing edge after the first (a) and the third (b) iterations

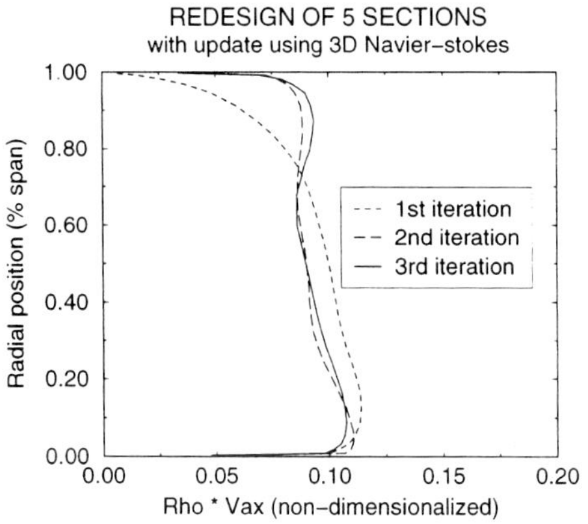

Figure 7: Spanwise distribution of the mass flow

These five sections are used to build the 3D blade geometry by means of the 3D geometry model. The outlet flow angle computed by the 3D Navier-Stokes solver is represented by the dashed line and shows a discrepancy of + 1 degree at the hub, + 3 degrees over 3/4 of the blade height and - 16 degrees at the tip.

At this point the 2D requirements are updated to compensate for the 3D flow effects on the outlet flow angle. A relaxation factor of 0.5 is used for stability reason. The 2D requirements are changed by -0.5 degree at the hub, - 1.5 degrees at the mid and + 8 degrees at the tip (represented by the full squares) after which a second design cycle is performed.

After 3 iterations using the extended 2D database (which now contains 3 * 5 additional samples) and the updated 2D approximate model, the new outlet flow angle (represented by the squares) is close to the newly imposed 2D outlet flow angle (Fig. 5).

After stacking of the blade sections, a second 3D Navier-Stokes analysis is performed which predicts an outlet flow angle represented by the long-dashed line. This outlet flow angle is much closer to the imposed one. The discrepancy is 2 degrees at the hub, - 0.5 degree at mean and - 1 degree at the tip.

Finally a third design cycle is performed which results in the outlet flow angle represented by the solid line. The error is now of + 1.5 degree at the hub, - 1 degree at mean and + 1 degree at the tip.

The final results show a remarkable improvement with respect to the first 3D Navier-Stokes computation. The more uniform outlet flow angle (Fig. 5) and massflow (Fig. 7) provide a more uniform flow (Fig. 6) and hence a better incidence angle on the next blade row. However the spanwise variation of the outlet flow angle still shows some oscillations indicating that more than 5 sections should be used for the design of such a 3D blade or that an other radial position of the 5 sections should be selected.

Figure 8 (a) shows how the spanwise static pressure distribution has changed due to a higher outlet flow angle between the hub and 70 % of the span and a lower flow angle near the tip.

Figure 8 (b) shows the evolution of the enthalpy loss coefficient in function of the spanwise position. The main improvement occurs in the tip region. The overall loss coefficient for the 3 blades is : 7.6 %, 8.0 % and 7.8 % respectively indicating that the blade sections of all designs are optimized and that a more uniform outlet flow is not at the expense of the efficiency. The main improvement is a more uniform outlet flow angle after the third iteration and therefore substantial efficiency gain should come from the better incidence angle on the next blade row.

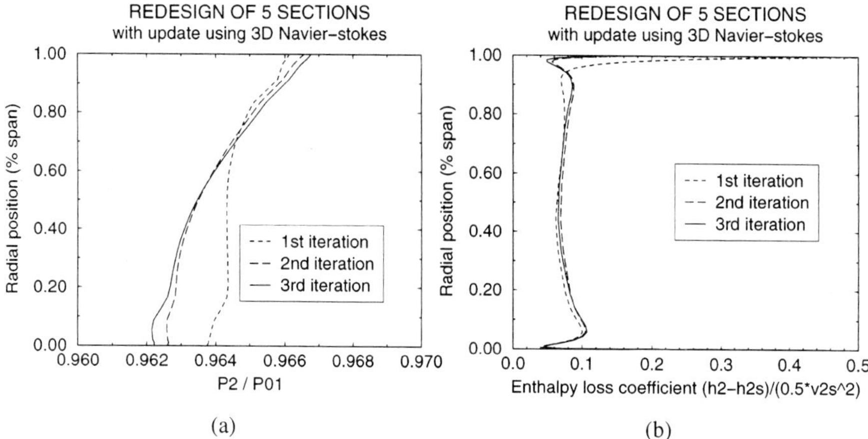

Figure 8: Spanwise distribution of the outlet static pressure (a) and of the enthalpy loss coefficient (b)

Figure 9 shows the 2D blade sections after the 1st and the 3rd design cycle. As expected, the largest changes occurred at the tip section, where the increased throat resulted in a decrease of the outlet flow angle.

Figure 10 shows the 3D blade shape used for the 1st and 3rd 3D Navier-Stokes computations and a side view of the blade after the third iteration.

7. CONCLUSIONS

We have shown that, using the method presented in this paper, only a few Navier-Stokes computations are needed to define an optimized blade satisfying both the aerodynamic and the mechanical requirements.

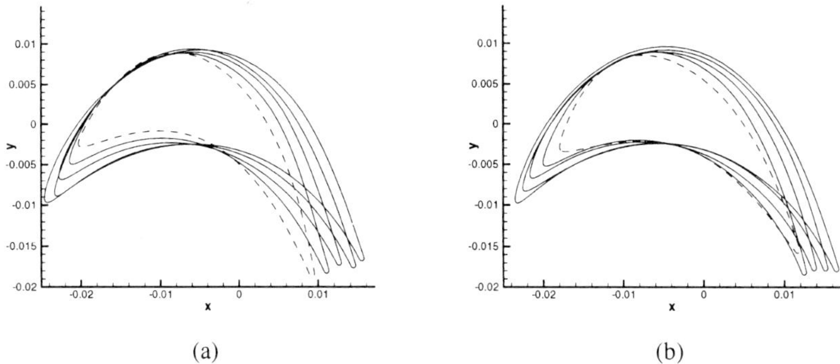

(a)   (b)

Figure 9: Stacking of the 5 primary sections after the first and the third 3D Navier-Stokes computations

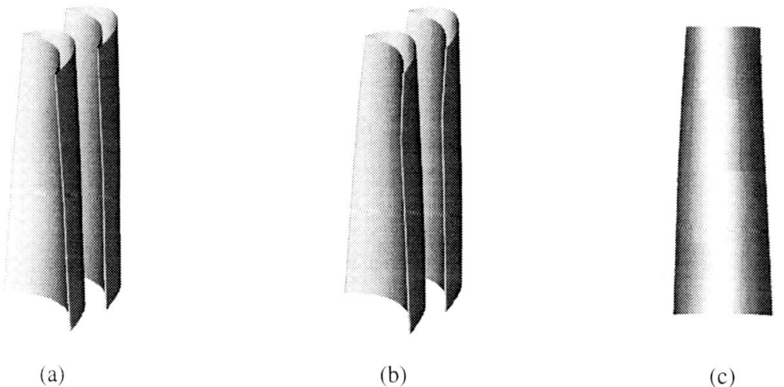

(a)   (b)   (c)

Figure 10: 3D Blade geometry in the first (a) and third (b) 3D Navier-Stokes computations and the side view after the third iteration(c)

The efficiency of the method results from :

- a fully automated design procedure without operator intervention. This was possible mainly due to the development of an objective function which translates the judgement of an experienced designer concerning the quality and the reliability of the solution into a single number which can be handled by a computer.

- the use of an objective function which includes some penalties accounting for the qualities of the Mach number distribution at off-design conditions in order to ensure good performance of the blade over a wide range of operating conditions without the cost of Navier-Stokes computations at several off-design conditions.

- the use of an artificial neural network and a database able to acquire experience from previous designs and efficiently use it for subsequent designs.

- the use of a robust geometry model which avoids the generation of unrealistic blades but which nevertheless still has enough flexibility to cover a wide range of different blade types.

- the optimization algorithm, based on simulated annealing or genetic algorithm, guaranteeing that the system will not be trapped in a local minimum and that from time to time new innovative designs are found and stored in the database.

This design method allows the exploration of more design options in a given period of time than with traditional methods and can be used with any Navier-Stokes solver.

ACKNOWLEDGMENTS

The first author gratefully acknowledges the financial support by the "Fonds pour la formation à la Recherche dans l'Industrie et l'Agriculture" (FRIA). The von Karman Institute acknowledges the computer support provided by Ansaldo Energia s.p.a. Special thanks also to Dr. A. Arnone for making his solver available.

References

[1] C.H. Sieverding. Recent progress in the understanding of basic aspects of secondary flows in turbine blade passages. *Transactions of the ASME*, pages 248–257, April 1985.

[2] J.D. Denton, A.M. Wallis, D. Borthwick, J. Grant, and I. Ritchey. The three-dimensional design of low aspect ratio 50% reaction turbines. *In aerodynamics of Turbomachinery, IMechE Seminar publication 1996-21, Latest advances in the aerodynamics of Turbomachinery with special emphasis upon unsteady flows, UK*, pages 109–120, December 9-10 1996.

[3] S. Pierret and R.A. Van den Braembussche. Turbomachinery blade design using a Navier-Stokes solver and artificial neural network. *ASME paper 98-GT-4*, 1998.

[4] S. Pierret and R.A. Van den Braembussche. Turbomachinery blade deign using a navier-stokes solver and artificial neural network. *Proc. of RTA/AVT Symposium on Design Principles and Methods for Aircraft Gas Turbine Engines, Toulouse*, 1998.

[5] A. Arnone and E. Benvenuti. Three-dimensional Navier-Stokes analysis of a two-stage gas turbine. *ASME 94-GT-98*, 1994.

The role of research in the aerodynamic design of an advanced low-pressure turbine

N W HARVEY and **J C COX**
Rolls-Royce plc, Derby, UK
V SCHULTE
BMW Rolls-Royce GmbH, Dahlewitz, Germany
R HOWELL and **H P HODSON**
Whittle Laboratory, Cambridge University, UK

SYNOPSIS

The new BR715 engine features an advanced Low Pressure (LP) turbine aerodynamic design which includes "high lift" aerofoils. This paper shows how every part of the design methodology has been derived from extensive research instigated by both Rolls-Royce (RR) and BMW Rolls-Royce (BRR). Of particular importance has been the investigation of unsteady wake-boundary layer interaction on low Reynolds number blading and the generation of a database for the validation of the design methods.

Preliminary results from cold flow rig testing have verified the design aims. In particular the aerodynamic efficiency is approximately 1% above the prior technology base.

1 INTRODUCTION

The BR715 21k lb. thrust engine powers the new B717-200 regional aircraft and follows the BR710, already in service with Gulfstream and Canadair executive jets. As part of achieving the BR715 programme an advanced technology LP turbine (figure 1) has been designed by RR, which draws on technology derived from the BR710 and RR Trent engine programmes, as well as University research sponsored by the two companies.

1.1 Overall aerodynamic philosophy

In modern high bypass ratio fans the LP turbine can be one third of engine weight. For small engines, maintaining the same levels of stage work as in larger ones would result in excessively heavy LP turbines with too many stages. Thus stage pressure ratios and Mach number (Mn) levels are higher for the BR700 series of engines compared to the Trent.

There is a strong trade-off between LP turbine efficiency and engine fuel burn. Losses must be minimised at the given operating point. The aspect ratios of the LP blading for the

BR715 are 4-5, and so the losses are largely 2-D in origin.

Achieving an acceptable design is therefore largely an exercise in optimising the 2-D aerodynamics within the additional constraints of cost and weight. To help do this the research programme described in this paper was undertaken. This has included work on high speed and low speed cascades (with and without upstream rotating bars), on a large scale low speed rotating rig and on engine scale cold flow LP turbine rig tests.

The result has been a new high lift design with aerofoils at lift coefficients approximately 20% higher than those of the Trent and the original BR710, see figure 2. This has provided not only the required weight and cost reductions - but also improvements in aerodynamic efficiency. Key to this approach is the control of boundary layer separation and transition on the aerofoil surfaces in the unsteady engine environment.

2 2-D AERODYNAMIC DESIGN

The final aerodynamic style for the BR715 LP turbine blading is based on the research cascade profile "H", see Curtis et al. (1996). Figure 3 shows the velocity distribution (in steady flow) compared with a conventional "datum" Trent profile. The key features are:

1. Continuous acceleration on the suction surface (A) from the leading edge to the peak Mn at (B). The lift is maximised, especially the nose loading, within the constraints of maintaining an adequate positive incidence margin. This depends on a number factors: the incidence variation over the engine working line, the margin required for errors in engine matching and also future thrust growth, and the allowance that must be made locally for the effects of secondary flow deviations from the upstream blade row.

2. The profile is strongly aft-loaded, with the peak Mn typically downstream of the geometric throat (C). The benefits of this in reducing profile loss were identified by Hoheisel et al. (1987). The value of back surface diffusion, and thus the peak Mn level, is directly related to the target lift coefficient. As well as being a feature of the 2-D aerodynamics, such strong aft-loading also has benefits in controlling 3-D secondary flows.

3. The inviscid design of the velocity profile has just a straightforward diffusion on the back surface from the peak Mn to the trailing edge. In steady, viscous flow, the suction surface (laminar) boundary layer separates shortly after the start of the diffusion, forming a separation bubble which reattaches before the trailing edge (D).

4. In steady flow the presence of the suction surface separation bubble appears on the lift plot as a characteristic plateau up to the transition point, followed by a sharp pressure recovery. The design must ensure that the bubble reattaches before the trailing edge - otherwise there is an open separation bubble with very high losses. However, if reattachment is too early it results in unnecessary (turbulent) boundary layer growth after the bubble on the back surface and thus also adds to the loss. A means of accurately predicting bubble size and behaviour is therefore crucial to successfully implementing this design style.

5. On the pressure surface the lowest possible Mns are be achieved (E), thus maximising lift while avoiding pressure surface separation bubbles. These can cause additional loss, if large enough, and once they form no more lift is achievable on the pressure surface. Ideally pressure side diffusion should be avoided completely. However, this also related to the suction side nose loading (and positive incidence margin) and the blade thickness. For the BR715 LP turbine, which has "thin solid" blading to further reduce cost, the compromise between these means that some pressure surface diffusion is unavoidable.

3 HIGH LIFT PHILOSOPHY

Hourmouziadis (1989) was the first to describe in detail the issues around the design of "high lift" blading and the behaviour of laminar separation bubbles on the suction surface - inevitable for blading with high levels of back surface diffusion at low Reynolds numbers.

However, during the design of the Trent 700 and 800 LP Turbines, it was understood that there was a trade-off between aerodynamic performance and the cost and weight benefits of blading at lift coefficients above the conventional. This was based on an early high lift design "L4" (velocity distributions for which are shown in figure 8), tested in steady flow cascade. Figure 4 compares the steady flow cascade losses of the datum profile and profiles "H" and "L4". The higher loss of "L4" at engine Reynolds number levels (10^5 to 1.3×10^5) was attributed to the higher suction side velocities outweighing the reduction in aerofoil numbers.

With the imperative of achieving maximum aerodynamic efficiency, therefore, the Trent series of LP Turbines and the original BR710 all featured conventional levels of lift and low levels of back surface diffusion. Only in subsequent research was the need to compromise between efficiency, cost and weight successfully eliminated, in the form of profile "H". There have been two key elements in achieving this:

1. "Flap" testing, which allowed the equivalent of over 40 different profiles to be tested - the cost of which would otherwise have been prohibitive, see Curtis et al. (1996).
2. The development of the cascade testing to include the simulation of wake passing, see Banieghbal et al. (1995).

Figure 6 shows the arrangement of the "flap" cascade tests. By compromising the cascade periodicity the suction surface velocity distributions on the test profile could be widely varied by moving the flap (also by adding inserts to the adjacent blade and applying suction to the flap). This is described in more detail in Curtis et al. (1996).

The effective profile losses were determined by measuring the loss of the suction side boundary layer, and assuming the pressure side and trailing edge losses per aerofoil were unchanged. Similarly an effective pitch was calculated from the increased lift of the measured velocity distributions. The results, in terms of loss and pitch relative to the datum are plotted in figure 7. This showed that the conventional design was optimum, but a 20% lift increase was possible with perhaps only a 10% increase in loss.

The velocity distribution of profile "H" , described in section 2, was derived from these results. Figure 4 shows that in steady flow this has a slightly better efficiency than the datum profile, for Reynolds numbers above 10^5. This was better than expected from the flap tests and indicates that there was also some reduction in pressure side or trailing edge losses.

The cascade tests with bar passing (see figure 5) gave the further, very beneficial, result that in the wake passing environment profile "H" is much more efficient than conventional for Reynolds number above 7.5×10^4. Comparing the losses with and without wake passing shows that the datum aerofoil is more sensitive to unsteadiness than the high lift one.

4 SUCTION SURFACE BOUNDARY LAYER BEHAVIOUR

The suction side boundary layer is responsible for most of the 2-D loss of an LP turbine aerofoil, see Curtis et al. (1996). As described in section 2, a high lift aerofoil optimised in steady flow requires a closed separation bubble, but with a minimum length of turbulent boundary layer after reattachment. The current design philosophy requires that the predicted boundary layer behaviour be acceptable in steady (and thus unsteady) flow.

Steady flow modelling is achieved using a quasi 3-D blade-to-blade Euler code with a strongly coupled integral boundary layer. Key to using such a method is its validation, and here the early high lift profile "L4", tested in steady flow cascade at engine representative Reynolds and Mach numbers has been used. Figures 8a and b compare the measured and calculated surface velocity distributions for "L4" at Reynolds numbers of 10^5 and 4×10^5. The bubble behaviour is very well calculated including the location of transition.

4.1 Wake passing

The linear cascade tests with simulated wake passing show that in the unsteady environment profile "H" has significantly lower loss than the conventional one. Although only an idealised representation of the real multi-stage engine situation, confidence in this approach had been established by Banieghbal et al. (1995) who showed that the boundary layer behaviour in such a cascade was a good approximation to that in the Trent 700 LP turbine.

The interaction of wakes with laminar separation bubbles is complex, as shown in the investigations of Schulte & Hodson (1994) and Halstead et al. (1995). A comprehensive review of the whole field is given in the VKI Lecture Series on Blade Row Interference effects in Turbomachinery, see Hodson (1998). However, a brief description is as follows:

1. Figure 9 shows a model space-time diagram for the state of the suction surface of a 2-D aerofoil section that, in steady flow, would have a closed separation bubble on the back surface. The time ordinate (y axis) is normalised by the period of the upstream wake passing.

2. On the first part of the surface the boundary layer is laminar and the momentum thickness Reynolds number sufficiently low for it to remain unaffected by the wakes.

3. In the diffusing region downstream of peak velocity the boundary layer in steady flow would separate. However, due to the turbulence in the wake it is now susceptible to create a time dependent transitional flow regime (which extends across the span). This initiates the development of turbulent spots in the boundary layer (A in figure 9) at each wake passing.

4. By analogy with the behaviour of turbulent spots, the front of the transitional flow travels at about 88% of the local freestream velocity and when it reaches the aerofoil trailing edge (point B in figure 9) the boundary layer is turbulent.

5. The rear of the transitional flow only travels at about 50% of the local free stream velocity so that the chordwise extent and duration of the transitional flow increases as the trailing edge is approached. Furthermore, the transitional flow becomes fully turbulent so that each wake passing is associated with turbulent rather than transitional flow.

6. Behind the turbulent spot is a becalmed region of, effectively, laminar flow. The rear of this region travels at about 30% of the freestream velocity. Even if the aerofoil has a separation bubble present in steady flow, the rear of becalmed region continues unaffected to the aerofoil trailing edge (point C in figure 9).

7. At the nominal separation location, once the becalmed region has gone by the laminar boundary layer separates under the effect of the diffusion (point E) and the bubble gradually grows, but never has time to develop its full characterisation. It only continues in time until the next turbulent spot initiated by the next wake (at point D) suppresses it again (point F).

There should be an optimum wake passing frequency such that the rear of the becalmed region from one wake reaches the aerofoil trailing edge at just the same moment as the front of the turbulent spot initiated by the next wake. This is the situation shown in figure 9, and results in minimum loss generation on the suction surface. This is verified by cascade results.

Schulte and Hodson (1996 and 1997) have shown how to calculate the time resolved boundary layer, provided the turbulence distribution in the wake is known.

5 PRESSURE SURFACE BOUNDARY LAYER BEHAVIOUR

The boundary layers on the pressure surfaces of LP blading have not typically received the same consideration as those on the suction surfaces. This is because, based on attached flow observations, the loss is less than 20% of the total profile 2-D value. In addition, even if a pressure side separation bubble was present it was understood that the local entropy generation could only be low since the local freestream velocity was low. The presence of a strong acceleration after the bubble also led to the expectation that even with the boundary layer reattaching as turbulent it should remain thin to the trailing edge and keep losses low.

However, research has now shown a number of significant loss sources that can arise from the presence of pressure side laminar separation bubbles:

1. The blockage of the bubble raises the suction side velocities and if extreme enough changes the velocity distribution such that transition occurs much earlier, increasing the loss. This can, however, be easily checked by using a standard steady flow CFD code. Figure 10 shows measured and calculated velocity distributions for a conventional thin solid profile, at -10° incidence, which has a pressure side bubble even at design flow conditions. An inviscid 2-D blade-to-blade calculation has also been included to illustrate how the bubble blockage elevates the Mach numbers on both aerofoil surfaces. As can be seen the CFD code calculates the time-averaged effect of the bubble very well.

2. The bubble is unsteady itself, even without additional unsteadiness in the form of an incoming wake. The effect of this is seen in figure 11 which shows a typical wake traverse downstream of the conventional, datum profile at -10° incidence in steady flow cascade. The time-averaged loss profile clearly shows a 'tail' extending from the pressure side to beyond mid passage. This can effectively double the pressure side loss. As yet the only means of mitigating it is to keep the size of any pressure side bubbles to a minimum.

3. Radial migration of fluid in the separation bubble, feeding additional fluid into one or both of the end wall secondary flow vortices. This was identified by Scrivener et al. (1991), but the magnitude of the effect is still to be determined. It is difficult to separate from the effect that as the early pressure side diffusion increases then, at the end walls, the cross-passage pressure gradient also increases and thus exacerbates the secondary flows - even if no separation bubble is present.

6 RESEARCH LP TURBINE VALIDATION

In order to mitigate the risk of going straight from 2-D cascade, even with simulated wake passing, to cold flow rig or engine the new high lift philosophy was first verified in a 1½ stage low speed rotating rig. This was tested in the 'Peregrine' facility at the Whittle Laboratory, Cambridge - described by Hodson et al. (1994b).

Two designs of the second Nozzle Guide Vane (NGV2) were tested, one based on the conventional datum cascade profile and one on profile "H". The first vane (the rig inlet guide vane) and the rotor were of conventional design, for both tests. All the data was taken at the design Reynolds number of NGV2, 1.3×10^5, while the incidence onto NGV2 was varied by changing the rotor speed (with the rig mass flow and axial velocity fixed).

From figure 12, which shows the measured mid-height downstream loss traverses for the two second vanes, it can be seen that:

1. Over the whole incidence range the 2-D loss for the high lift design is less than that of the conventional profile - verifying the cascade result.

2. The high lift profile shows considerably better performance than the conventional design at negative incidence, the loss of the latter increasing considerably with the growth of the pressure side bubble - possibly due to 2-D unsteady eddy shedding from the bubble.

3. The high lift profile retains its beneficial performance to at least +10° incidence.

Although not shown here, area traverses behind the two second vanes demonstrated that the secondary flows were only slightly increased for the high lift design - due both to the high aspect ratio (about 5) and the strong aft-loading of the profile velocity distributions. Overall this work gave confidence in applying the high lift philosophy to the BR715 LP Turbine.

7 COLD FLOW RIG TEST VALIDATION

Figure 13 presents the variation of measured turbine efficiency against average blade row Reynolds number. The original target efficiency, derived from previous Trent and BR710 LP Turbine rig test results, is included for comparison and shows that the BR715 LP Turbine has achieved an aerodynamic efficiency approximately 1% above this prior technology base.

The variation in efficiency between cruise and take-off conditions is quite large, at about 1%, but is not of concern. Since the BR715 exceeded its bid efficiency at cruise, this means it is simply better than expected at take-off. Also, this variation is still significantly less than that seen in other large civil turbofans which can be as much as 2%, see Ashpis (1997). Detailed analysis of the blading boundary layer behaviour, measured using hot film gauges, is underway to compare it with that observed in cascade and the Peregrine facility.

Considerable data has been taken from aerofoil pressure tappings at a number of conditions. Results taken at the off-design condition of 120% speed and design work, serve to illustrate features of the design and provide further validation of the design tools. Figures 14a and b present measured static pressures at the mid height and tip section (90% height) for the second vane compared with calculations from steady flow CFD, and show:

1. The mid-height pressure distribution is as designed. There is a pressure side separation bubble, but its blockage is small.

2. On the suction surface, there is no visible evidence in the time-averaged pressure distributions of the presence of the back surface separation bubble. It seems to have been completely suppressed by wake passing. This was also observed in the Peregrine tests.

3. At the tip section there is a large pressure side separation bubble, causing significant blockage which is well calculated by the CFD. The pressures on the suction side are over estimated by the CFD which appears to have under estimated the secondary flow blockage, and possibly the secondary flow itself, near the casing.

4. Although it cannot be measured directly the CFD shows strong migration in the pressure side bubble towards the casing, illustrated in the flow visualisation from the CFD in figure 15. This was generated by injecting particles into the pressure side boundary layer at the leading edge, at selected heights up the span. Most of this flow is caught up in the highly vortical pressure surface separation bubble - which is calculated to migrate largely out to the casing. The excellent match between calculated and measured pressures on the pressure surface is circumstantial evidence that this is a real effect and is captured by the CFD.

8 CONCLUSIONS

A new high lift LP Turbine blading design philosophy has been developed from an extensive research programme and applied to the design of the BR715 engine.

Key to the approach is the control of boundary layer separation and transition on both suction and pressure surfaces of the blading, in particular in the unsteady environment.

Preliminary results from cold flow rig test indicate that the turbine has behaved better than expected, and that the efficiency has exceeded the previous technology base by about 1%.

9 ACKNOWLEDGEMENTS

Work reported here has been carried out with the support of Rolls Royce plc, BMW Rolls-Royce and the Defence Research Agency (MoD and DTI), Pyestock. The authors would like to thank them for funding it and their permission to publish this paper.

10 REFERENCES

1. Ashpis, D, 1997, "Low Pressure Turbine Flow Physics Program", Minnowbrook II, Workshop on Boundary Layer Transition in Turbomachines, Syracuse University.
2. Banieghbal MR, Curtis EM, Denton JD, Hodson HP, Huntsman I, Schulte V, Harvey NW, Steele AB, 1995, "Wake Passing in LP Turbine Blades", AGARD CP-571.
3. Cobley K, Coleman N, Siden G, Arndt N, 1997, "Design of New Three Stage Low Pressure Turbine for the BMW Rolls-Royce BR715 Engine", ASME 97-GT-419.
4. Curtis EM, Hodson HP, Banieghbal MR, Denton JD, Howell RJ, Harvey NW, 1996, "Development of Blade Profiles for LP Turbine Applications", ASME 96-GT-358.
5. Halstead DE, Wisler DC, Okiishi TH, Walker GJ, Hodson HP, Shin HW, 1995, "Boundary Layer Development in Axial Compressors and Turbines. Parts 1-4", ASME 95-GT-461/2/3/4.
6. Hoheisel H, Kiock R, Lichtfuss HJ, Fottner L, 1987, "Influence of Free-Stream Turbulence and Blade Pressure Gradient on Boundary Layer and Loss Behaviour of Turbine Cascades", Journal of Turbomachinery, April 1987, Vol. 109, pp 210-219.
7. Hodson HP, Huntsman I, Steele A, 1994, "An Investigation of Boundary Layer Development in a Multistage LP Turbine", Journal of Turbomachinery, July 1994, Vol. 116, pp 375-383.
8. Hodson HP, Banieghbal MR, Dailey GM, 1994, "The Analysis and Prediction of the Effects of Bladerow Interactions in Axial Flow Turbines", IMechE Conference on Turbomachinery, October 1994.
9. Hodson HP, 1998, "Blade Row Interactions in Low pressure Turbines", VKI Lecture Series 1998-02, Blade Row Interference Effects in Axial Flow Turbomachinery Stages.
10. Hourmouziadis J, 1989, "Aerodynamic Design of Low Pressure Turbines", AGARD Lecture Series, 167.
11. Schulte V, Hodson HP, 1994, "Wake Separation Bubble Interaction in Low Pressure Turbines", AIAA-94-2931.
12. Schulte V, Hodson HP, 1996, "Unsteady Wake-induced Boundary Layer Transition in High Lift LP Turbines", ASME 96-GT-486.
13. Schulte V, Hodson HP, 1997, "Prediction of the Becalmed Region for LP Turbine Profile Design", ASME 97-GT-398.
14. Scrivener CTJ, Connolly CF, Cox JC, Dailey GM, 1991, "Use of CFD in the Design of a Modern Multistage Aero Engine LP Turbine Design", IMechE C423/056.

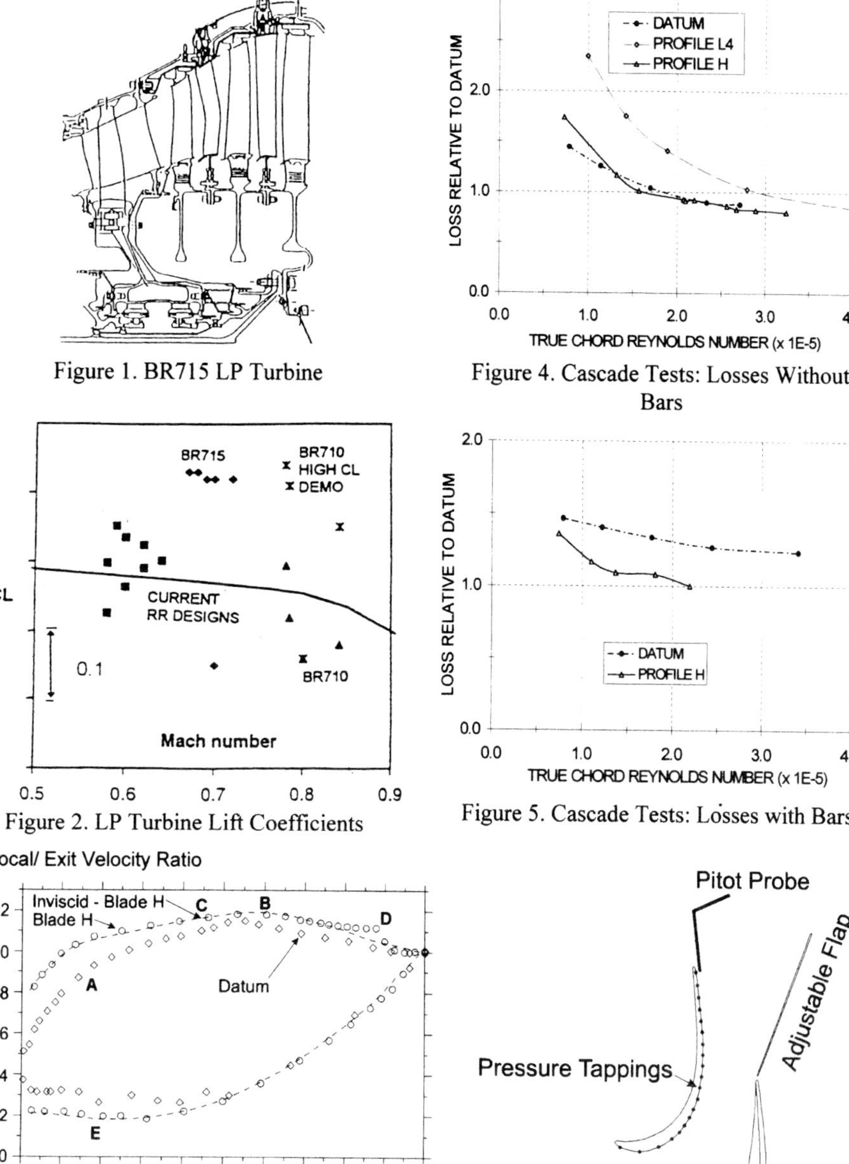

Figure 1. BR715 LP Turbine

Figure 4. Cascade Tests: Losses Without Bars

Figure 2. LP Turbine Lift Coefficients

Figure 5. Cascade Tests: Losses with Bars

Figure 3. Velocity Distributions for Research Profiles

Figure 6. Flap Test Cascade

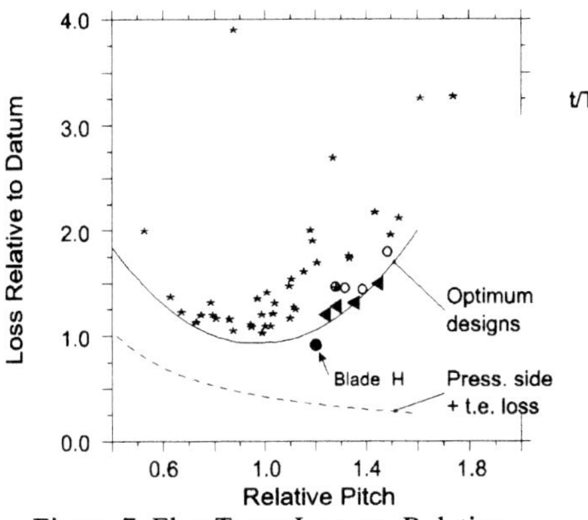

Figure 7. Flap Tests: Loss vs. Relative Pitch

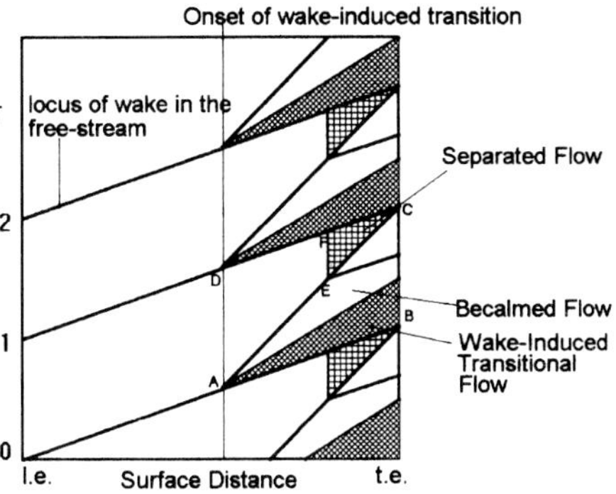

Figure 9. Schematic distance-time diagram showing effect of becalmed region on a separated boundary layer and illustrating optimum wake-passing frequency

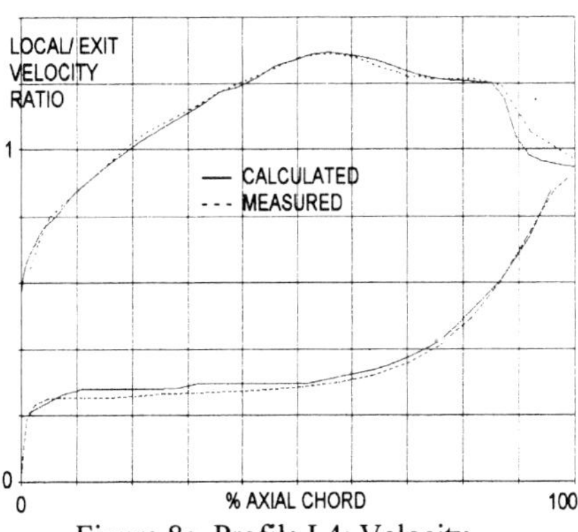

Figure 8a. Profile L4: Velocity Distribution at 10^5 Reynolds number

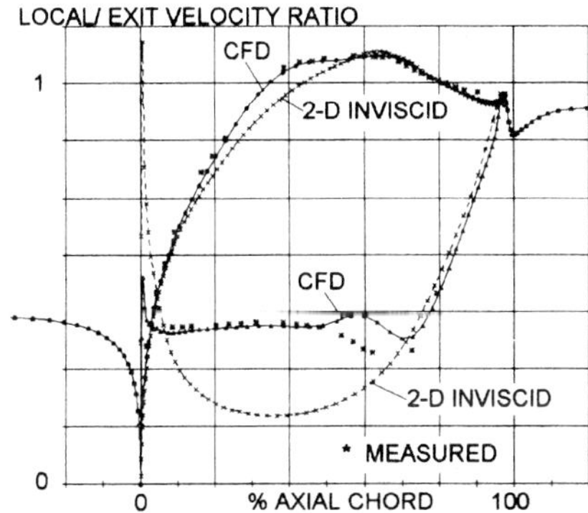

Figure 10. Conventional Profile: Velocity Distributions at -10° Incidence

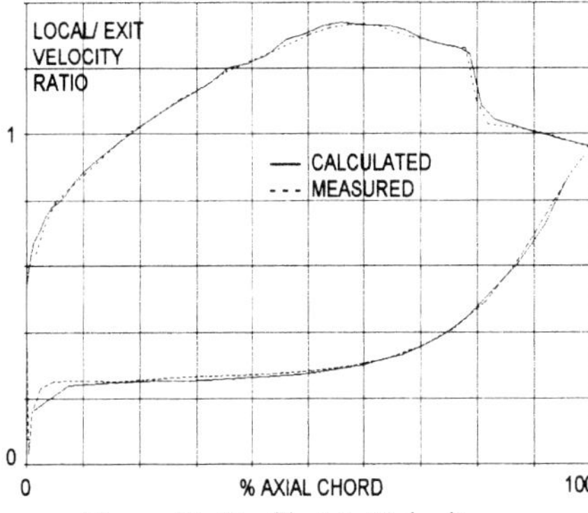

Figure 8b. Profile L4: Velocity Distribution at 4×10^5 Reynolds number

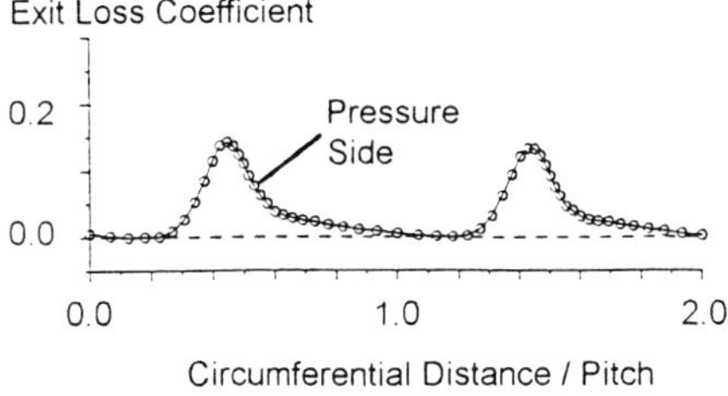

Figure 11. Conventional Profile. Downstream Loss Traverse at -10° Incidence

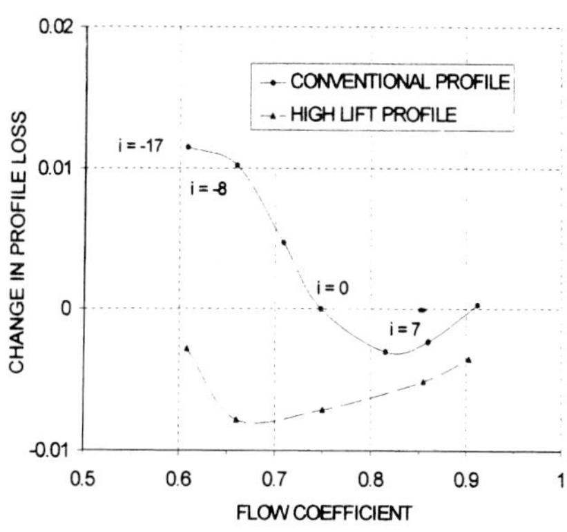

Figure 12. Peregrine LP Turbine. Mid-height Losses for NGV2

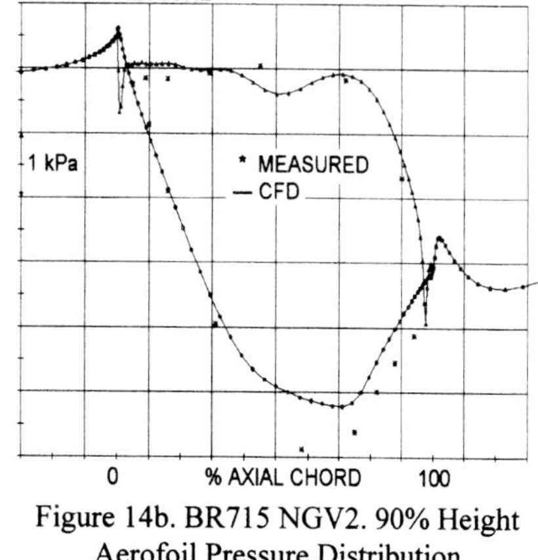

Figure 14b. BR715 NGV2. 90% Height Aerofoil Pressure Distribution

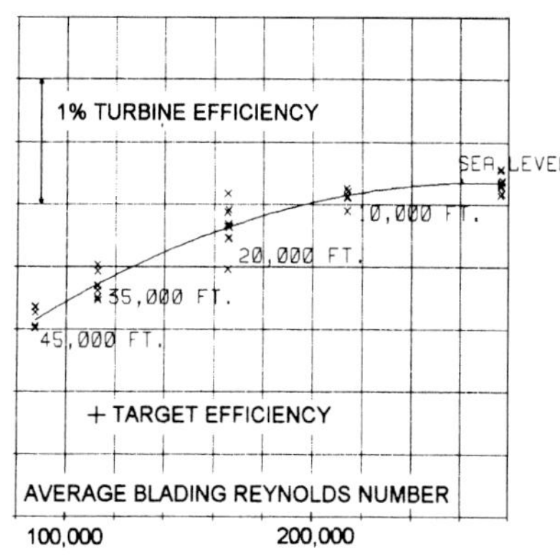

Figure 13. BR715 LP Turbine. Efficiency against Reynolds number

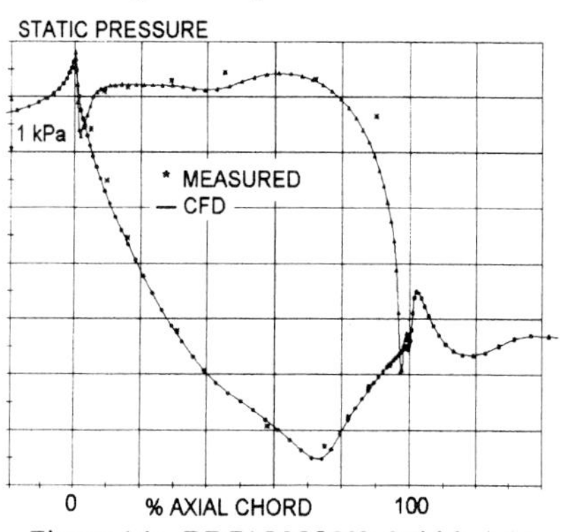

Figure 14a. BR715 NGV2. Mid-height Aerofoil Pressure Distribution

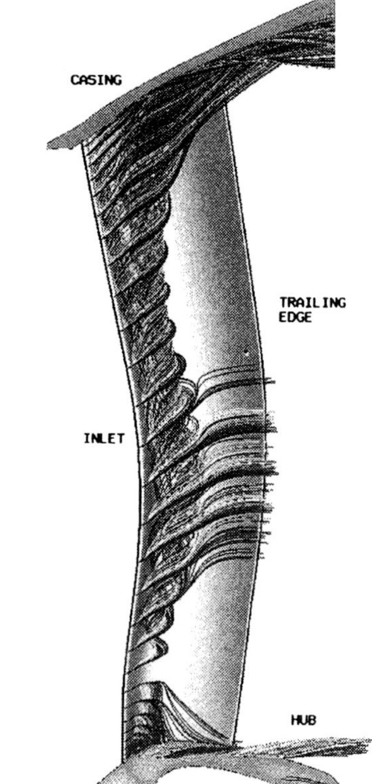

Figure 15. BR715 NGV2. CFD Calculated Pressure Side Flow Visualisation

Transition and Turbulence

C557/142/99

The prediction of flow on a flat plate with a circular leading edge under zero and non-zero pressure gradient

C YAKINTHOS and **A GOULAS**
Laboratory of Fluid Mechanics and Turbomachinery, Aristotle University of Thessaloniki, Greece

SYNOPSIS

The prediction of flow on a flat plate with a circular leading edge for two turbulence levels of low intensity under zero and non-zero pressure gradient is presented. In the experiment, recirculation of flow starts after the curved leading edge and a separation bubble is formed. The method of prediction is based on the control volume approach where the discretized Navier-Stokes equations are solved using the pressure correction algorithm. The low Reynolds k-ε turbulence model is used in combination with the Kato-Launder modification for the turbulence generating term. The computational results are compared with the measurements of Rolls-Royce Plc. The use of the low Reynolds k-ε model with the Kato-Launder modification gave good predictions for flows with relatively high free stream turbulence and the velocity distributions were in general in good agreement with the experimental data. For the low turbulence level case, although the results are very good, there was a need to adjust the relaxation parameters in the iterative solution procedure because of the appearance of instabilities under certain values.

1. INTRODUCTION

The flow around a flat plate with a circular leading edge is of general interest because it combines stagnation points with a region of flow separation and laminar to turbulent boundary layer transition. A boundary layer developing on a turbomachinery blade starts as a laminar layer and becomes later turbulent, a case which is well understood. When though a separation zone occurs, the flow becomes more complicated. The size of the separation zone, and so the transition from laminar to turbulent flow is influenced by the free-stream turbulence and by any external pressure gradient. Small values of turbulence in the free stream region produce larger separation bubbles compared to those with increased turbulence levels. Correspondingly negative pressure gradients give also smaller bubbles. The transition region is affected in a similar way. In the past there have been numerical predictions of the flow around a flat plate using CFD techniques. A summary of the experimental and computational results for some cases can be found in ref (1). Most of the methods presented give good prediction of the flow compared to the experimental data. The cases where problems are indicated, are in those where pressure gradients are present. The combination of flow separation in the presence of pressure gradient and their effects on boundary layer transition creates a very difficult problem although it is well known that situations like this are common place in the turbomachinery field. The amount of published experimental data for this combination is minimal and the only source is the work carried out by Rolls-Royce Plc. in the context of a Brite-Euram project.

The present paper has as an aim to examine and evaluate the methods, which have been developed for the prediction of transition over a flat plate, if they are capable, to predict accurately the flow conditions over a plate with a circular leading edge and in combination of flow separation and pressure gradient similar to those appearing in turbine flows.

2. THE SIMULATION OF THE FLOW

2.1 The numerical procedure.

A 2-D CFD code based on the control volume approach and developed at the Laboratory of Fluid Mechanics & Turbomachinery was used to carry out the predictions. The code is based on the pressure correction technique and solves the discretized momentum equations for the primary variables of the flow, u, v, p, and the k and ε turbulent parameters for turbulent flows. It uses the HLPA convection scheme after Zhu (3) and a co-located arrangement where all the primary variables are stored at the center of each cell of a curvilinear grid. The momentum interpolation is introduced, Rhie & Chow (4), taking also into account the remarks of Majumdar (5). Turbulence is modelled using the low Reynolds version of k-ε model of Launder & Sharma (6). The reason for chosing this model, despite the well known limitations, has to do with the fact that is robust and it is also widely used by industry. The two equations for k and ε are given below:

$$U_j \frac{\partial k}{\partial x_j} = \frac{\partial}{\partial x_j}\left\{\left(v + \frac{v_t}{\sigma_k}\right)\frac{\partial k}{\partial x_j}\right\} - \overline{u_i u_j}\frac{\partial U_i}{\partial x_j} - \varepsilon + D$$

$$U_j \frac{\partial \varepsilon}{\partial x_j} = \frac{\partial}{\partial x_j}\left\{\left(v + \frac{v_t}{\sigma_\varepsilon}\right)\frac{\partial \varepsilon}{\partial x_j}\right\} - C_{\varepsilon 1}f_1 \frac{\varepsilon}{k}\overline{u_i u_j}\frac{\partial U_i}{\partial x_j} - C_{\varepsilon 2}f_2 \frac{\varepsilon^2}{k} + E$$

(1)

Where, v_t is the eddy viscosity defined as: $\quad -\overline{u_i u_j} = v_t\left(\frac{\partial U_i}{\partial x_j} + \frac{\partial U_j}{\partial x_i}\right) - \frac{2}{3}\delta_{ij}\kappa$ (2)

and is related to k and ε through the Kolmogorov-Prandtl relation as $v_t = C_\mu f_\mu k^2/\varepsilon$. The expressions and values of the extra terms and parameters appearing in these equations are defined as:

$$D = \left(\frac{\partial \sqrt{k}}{\partial y}\right)^2 , \quad f_\mu = \exp\left[\frac{-3.4}{(1+R_T/50)^2}\right] , \quad E = 2\nu\nu_t\left(\frac{\partial^2 U}{\partial y^2}\right)^2 , \quad R_T = k^2/\nu\varepsilon$$

$$f_1=1, f_2=1-0.3\exp(-R^2{}_T), \ C\mu=0.09, \ C_{\varepsilon 1}=1.44, \ C_{\varepsilon 2}=1.92, \ \sigma_k=1.0, \ \sigma_\varepsilon=1.3$$

The term $\nu_t\left(\dfrac{\partial U_i}{\partial x_j} + \dfrac{\partial U_j}{\partial x_i}\right)$ in equation (2) represents the generation of turbulence energy. By defining the dimensionless strain S and dimensionless vorticity Ω after Kato & Launder (7) as:

$$S = \frac{k}{\varepsilon}\sqrt{\frac{1}{2}\left(\frac{\partial U_i}{\partial x_j} + \frac{\partial U_j}{\partial x_i}\right)^2} , \quad \Omega = \frac{k}{\varepsilon}\sqrt{\frac{1}{2}\left(\frac{\partial U_i}{\partial x_j} - \frac{\partial U_j}{\partial x_i}\right)^2} \qquad (3)$$

one can have the turbulence generation term as follows

$$P_k = f_\mu C_\mu \varepsilon S\Omega \qquad (4)$$

Preliminary computations showed that this modification leads to a remarkable reduction in energy generation near the stagnation point while it has no effect in a simple shear flow. This affects the flow in the regions after the front stagnation point and in the vicinity of walls where the predicted turbulence has lower values than those obtained when the classic expression of P_k is used. Initial runs showed that this substitution led to computational results with a very good agreement with the experimental ones. However, other researchers reported this modification for the value of P_k produces larger recirculation regions and does not constitute a general modelling framework, Chen, Lien & Leschziner (8). The authors found that indeed, this formulation gives larger bubbles from the size calculated when using the classic expression for P_k but the final results are close enough to the experimental data, Yakinthos & Goulas (9).

2.2 Initial and boundary conditions

The initial conditions for k and ε were based on the turbulence level at inlet. Depending on the desired turbulence level that had to be simulated, appropriate values of these two parameters in the entire flow field were calculated. The basic idea was taken from the Frenkiel's formula of turbulence decay:

$$Tu = c\left(\frac{x}{d}\right)^{-5/7} \qquad (5)$$

where, x is the distance from the point where is fixed a desired value of turbulence level and d is the diameter of the rods of the turbulence generating grid. From another point of view, d is a length scale of turbulence. In fact a "virtual" turbulence generating grid is used in the simulation by fixing the d according to the x distance at the inlet of the computational grid. In equation (5), c takes a value from 0.71 to 0.85 depending on the experimental data. For this study it was found that a value for $c=0.83$ was the best choice to fit the experimental data for the decay of turbulence. The initial value of k was taken form the inlet turbulence, while for the ε, the initial value was calculated from the formula $\varepsilon=k^{3/2}/L$, where $L=d$. This was chosen after Rodi & Scheuerer (10). They suggested to use for the length scale the same value as the diameter of rods.

The boundary conditions consisted of extrapolated values from inner points for the pressure on walls and outflow regions and an overall continuity condition at the outflow of the flow field. No boundary condition for ε at the wall was used as the low Reynolds k-ε model was employed.

Regarding the exit conditions, a special treatment in the case of non-zero pressure gradient has to be applied. This has to do with the different height of the channel and the fact that the flow on either side of the plate is not equally distributed (fig. 1).

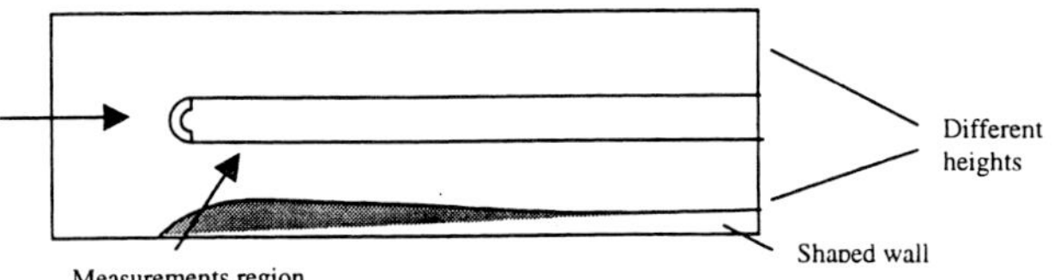

Figure 1. The flat plate setup and the two different exit heights

In the case of zero pressure gradient the fraction $\frac{flowout\ at\ south}{flowout\ at\ north}$ was equal to 1 to ensure that the front stagnation point will be on the axis of symmetry. In cases where the flow is asymmetric because of the position of the plate in the wind-tunnel it is necessary either to know the exact ratio or to ensure that the stagnation point lies on the axis of symmetry. So in the case of non-zero pressure gradient as this ratio was unknown it was first fixed to a number less than unity. During the iterations the position of stagnation point was changing. When the errors of the primary variables were small and close to the convergence criterion the position of stagnation point was then fixed with the corresponding value for the above ration and the solution was left finally to converge to the residual limits prescribed. Because of the asymmetry in the set-up the "bottom" surface only was under the influence of the pressure gradient, while the "top" surface of the plate will face an almost zero pressure gradient. Therefore the velocity profiles and flow behavior will be different on each side of the plate.

2.3 The grid simulating the flow.
A C-type grid was used to simulate the flow around the flat plate where the east boundary at the computational grid was for flat plate's wall. Fig. 2 shows the grid used for the simulation for the case of non-zero pressure gradient. When a coarse grid (30x300 nodes) was used there was a failure to predict the flow. The reasons were the small number of nodes inside the boundary layer where the steep gradients of ε cannot be simulated (Rodi (11) suggested that at least 50-60 nodes inside the boundary layer are needed when low Reynolds k-ε model is used) and the coarse resolution in stations downstream of the reattachment point. When a very fine grid was used (100x900), the computational time was very large, especially when sometimes instabilities in the iterative solution, as in the case of low free stream turbulence, occurred. The final grid used has 81x380 nodes with 50 nodes inside the boundary layer in the direction normal to the plate surface and a local refinement near the reattachment point. The first node (cell-center) near the wall was at a distance of 0.03mm giving values for y^+ around 0.1 for all cases. The time needed to obtain results for each case of turbulence level and pressure distribution using different FORTRAN compilers is presented in the following table.

	Pentium Pro 200 MHz	PPC 604 133 MHz	PPC G3 250MHz	HP PA 50MHz
30x300	~30min	~1hr	~50min	~2hrs
80x380	~2hrs	~3hrs	~2hrs40min	~5hrs
100x900	~8hrs	~15hrs	~~10hrs	~20hrs

2.4 Iterative procedure.

For the solution of the tri-diagonal system of the discretized equations for all the primary variables, the Stone's (12) algorithm was used. The code was regarded as having converged when the normalized square root of the sum of square of errors at each cell for u, v, p separately where less than 5.E-06. The numerical scheme described above was used to predict the flow around a flat plate with a circular leading edge and for cases where experimental conditions as given in the next section were available.

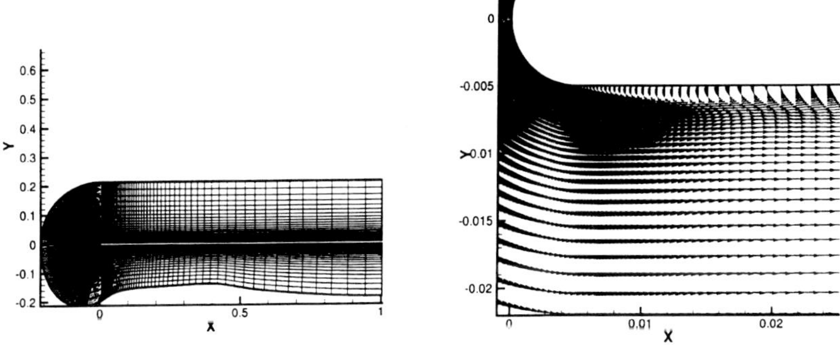

Figure 2. The C-Type grid used in the computations for non-zero pressure gradient and a vector plot at the recirculation zone

3. THE EXPERIMENTAL INFORMATION.

The experiments used here to evaluate the method were carried-out at Rolls-Royce Plc. by Coupland & Brierley (2), using a hot-wire anemometry technique under both zero and non-zero pressure gradient in the flow. A flat plate with a circular leading edge with a radius R=5mm has been inserted in the measuring section of a wind tunnel. The inlet velocity of the air was 5m/s under two turbulence levels, one with a value at the free stream flow of Tu=0.20% and a second with a value of Tu=2.3%. The existence of the circular leading edge caused separation of the boundary layer at the end of the curved surface followed by a recirculation region and a reattachment point on the flat plate. Two sets of measurements were carried out: one with zero pressure gradient which is achieved be having the wind-tunnel walls parallel and a second one with a pressure gradient in the direction of main flow. The pressure distribution was developed by introducing a specially designed contoured wall opposite to the flat plate on the wind tunnel wall. Measurements were taken on the flat plate's side opposite to the shaped wall, "bottom" surface, see fig.1. The shape of the wall gave a negative pressure gradient dP/dx<0 in the region of the leading edge, resulting in an acceleration of the flow. In general, measurements showed that with the use of a contoured wall, the size of the recirculation zone was smaller than the one measured with the zero pressure gradient distribution.

In the next section, a detailed comparison between the predictions obtained using the technique developed in the previous section and the experimental results will be given.

4. RESULTS

4.1 Tu=0.20%, U=5m/s, zero pressure gradient
The computational results for this case indicated a recirculation region. The velocity profiles normalized with the maximum velocity of air at stations downstream of the leading edge are shown in the next figures together with a comparison with the experimental data.

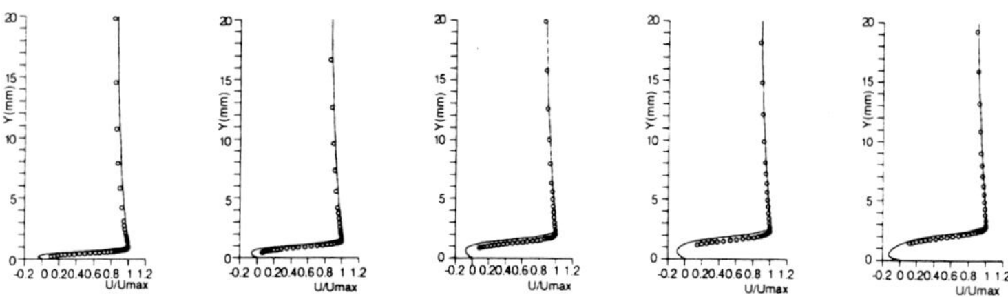

Figure 3. Velocity profiles at x=6, 8, 11, 13, 17 mm from the leading edge.
Symbols are for experimental data

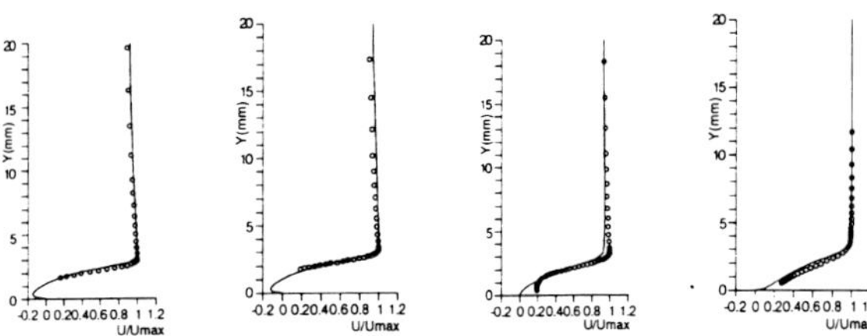

Figure 4. Velocity profiles at x=20, 23, 30, 35 mm from the leading edge
Symbols are for experimental data

From these figures a good agreement between the prediction and the experimental data at all the stations is shown. The predicted thickness of the recirculation bubble in the earlier stations is a little larger than this one from the experiment-this may arise from the Kato-Launder modification-but in general with a medium-sized grid one can have good predictions. The size of the recirculation zone for this turbulence level is calculated to be about 32mm (*reattachment position minus starting position of 5mm*). In the computational procedure instabilities appeared until convergence and therefore the relaxation factor had to be adjusted. The reason for this behavior can be the following: it is well known that the k-ε turbulence model is not very good for cases where the free stream turbulence level is below 1%. The modification by Kato and Launder which reduced significantly the production term of turbulence's kinetic energy made the model capable to work at lower turbulence levels. Nevertheless the model it is not capable to predict the large flow structures detached from the recirculation zone at low frequencies. The instabilities observed had to do with the latter. The adjustment to the relaxation factor required damped these fluctuations by "averaging" the flow over a "final" period, in number of iterations, longer that the shedding frequency of the large structures allowing converge of the solution. Figure 5. shows an example of the convergence history before and after adjusting the relaxation factors.

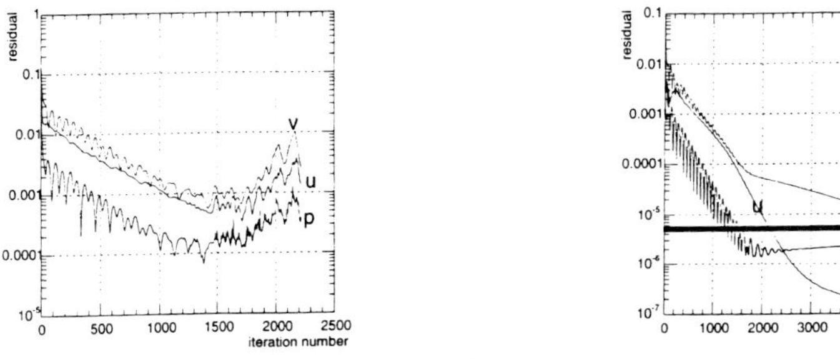

Figure 5. Convergence history before and after adjusting the relaxation factors

To obtain an overall view of the flow field, the integral boundary layer parameters such as displacement thickness, momentum thickness and shape factor are plotted and compared with the experimental data in the following figures.

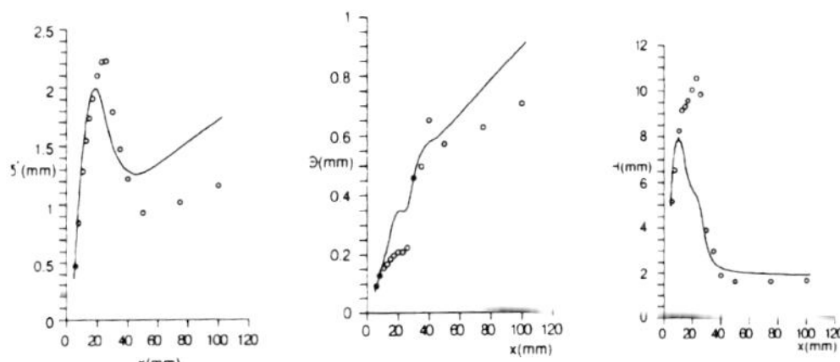

Figure 6. Displacement – momentum thickness and shape factor for Tu=0.20%
Zero pressure gradient. Symbols are for experimental data

From these figures it is clear that there is in general a good agreement with the experimental data except the momentum thickness.

4.2 Tu=2.3%, U=5m/s, zero pressure gradient
The predicted velocity profiles for this case and their comparisons with the experimental data are given in the next figures.

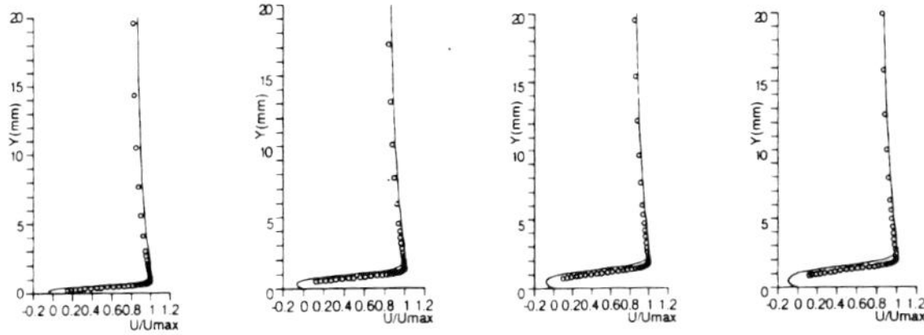

Figure 7. Velocity profiles at x=6, 8, 11, 13 mm from the leading edge.
Symbols are for experimental data

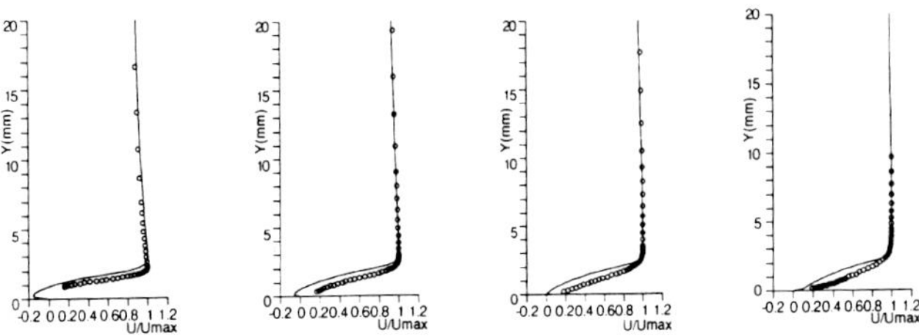

Figure 8. Velocity profiles at x=15, 19, 21, 23 mm from the leading edge.
Symbols are for experimental data

The computational results are in good agreement with the experimental data. It is clear also here that the Kato-Launder modification of the turbulence generation term predicts still a slightly larger recirculation region than the one observed in the experiment. The size of the predicted bubble is ~16mm. No major problems were reported at the iteration procedure and the code reached convergence with no significant instabilities. This enforces the point made in the previous paragraph regarding the turbulence model.

The boundary layer information compared with the experimental data are shown in the next figures.

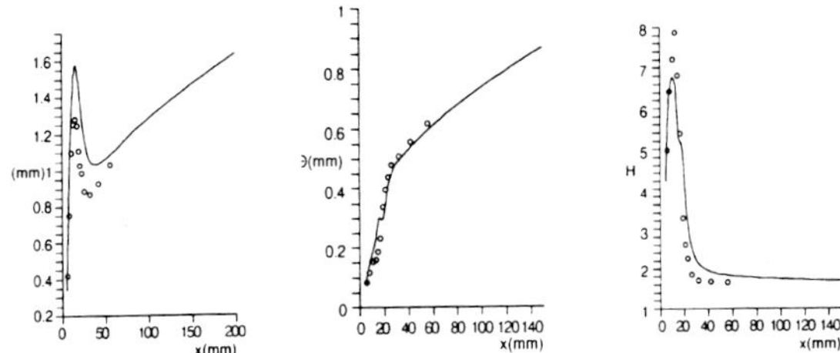

Figure 9. Displacement – momentum thickness and shape factor for Tu=2.3%
Zero pressure gradient. Symbols are for experimental data

The predictions of the boundary layer parameters are quite good and in general are in a agreement with the experimental data. Problems occur in the region of separation but the differences are not significant. From this point of view, the modification of Kato-Launder in the low Reynolds k-ε model behaves better in larger turbulence levels.

4.3 Tu=0.20%, U=5m/s, non-zero pressure gradient.

The pressure gradient in the region of the leading edge is negative. As a result there is an acceleration of the flow and therefore a smaller separation bubble for both test cases of Tu=0.20 and 2.3% and a smaller transition region. The velocity distributions and the comparisons between predictions and experiments are presented in the following figures.

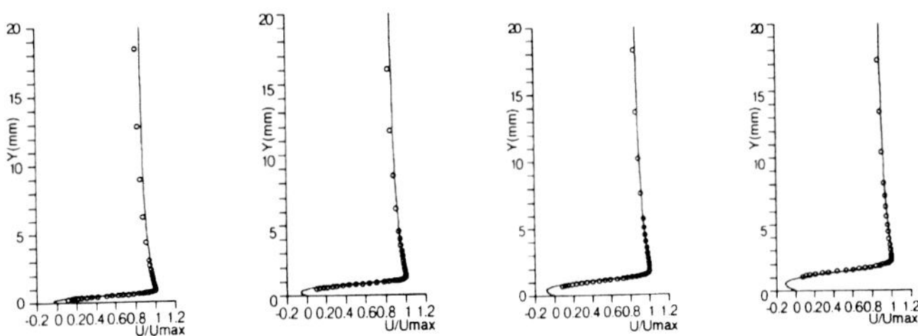

Figure 10. Velocity profiles at x=6, 8, 11, 17 mm from the leading edge.
Symbols are for experimental data

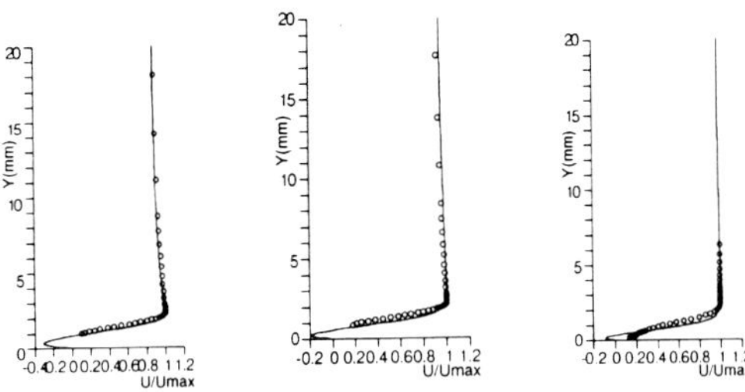

Figure 11. Velocity profiles at x=20, 23, 26 mm from the leading edge.
Symbols are for experimental data

The first remark is that the predicted length of the recirculation region is about 21mm and is smaller than the one formed with the same turbulence level and with zero pressure gradient. The velocity distributions are in very good agreement with the experimental ones. During the calculations the problem of instability discussed in section 4.1 also appeared. The significant difference though was that the "shedding" of the large flow structures was occurring on the "top" surface while the solution for the "bottom" surface was converging steadily. The reason for this is that the velocities occurring in the free stream above the "bottom" surface are higher than those of the "top" surface and therefore the "shedding" frequency will be higher in the "bottom" surface and damped by the numerical solution as discussed earlier, while the "top" one needs different relaxation factor to achieve convergence in the "average" solution. The boundary layer integral parameters are shown in the next figure and compared with the experimental ones.

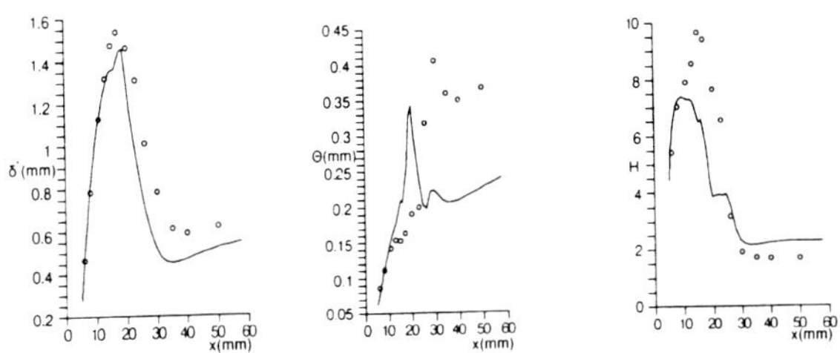

Figure 12. Displacement – momentum thickness and shape factor for Tu=0.20%
Non-zero pressure gradient. Symbols are for experimental data

From these plots it is clearly shown that the prediction of boundary layer information fails for the momentum thickness and shape factor especially in the region of recirculation of the flow while the displacement thickness is predicted satisfactorily. The results for the low turbulence level for both pressure gradients indicate that for the recirculated flows a further modification must be done in the solution procedure. The Kato-Launder modification is not enough to compensate for the physical problem of flow structure shedding. A solution to this modelling problem of low turbulence and the way that turbulence models behave under these conditions is required.

4.4 Tu=2.3%, U=5m/s, non-zero pressure gradient.

For this case, the results are shown in the following figure where the velocity distributions are presented.

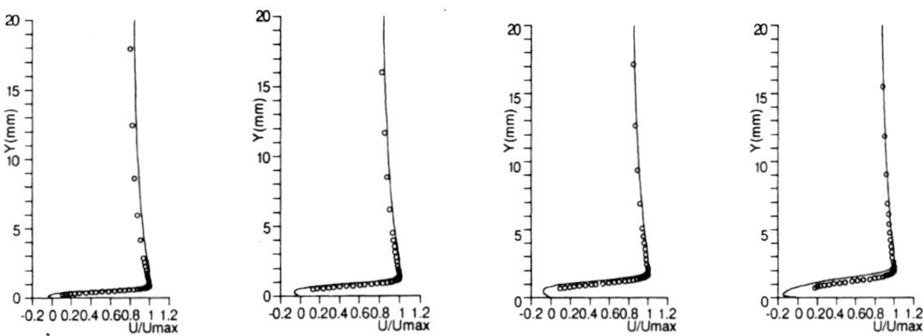

Figure 13. Velocity profiles at x=6, 8, 11, 15 mm from the leading edge.
Symbols are for experimental data

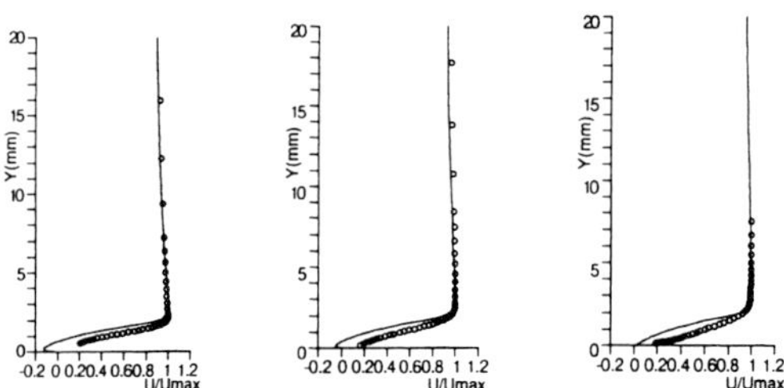

Figure 14. Velocity profiles at x=17, 19, 21 mm from the leading edge.
Symbols are for experimental data

The predictions of the flow at each measurement station downstream the flow show a slightly larger separation bubble but in general the velocity profiles are in a good agreement with the experimental ones. The size of the bubble is about 15mm (smaller than this one predicted for zero pressure gradient). In general the low Reynolds number k-ε model with the Kato Launder modification behaves better under a larger turbulence level.

The integral boundary layer parameters are shown in the following figures.

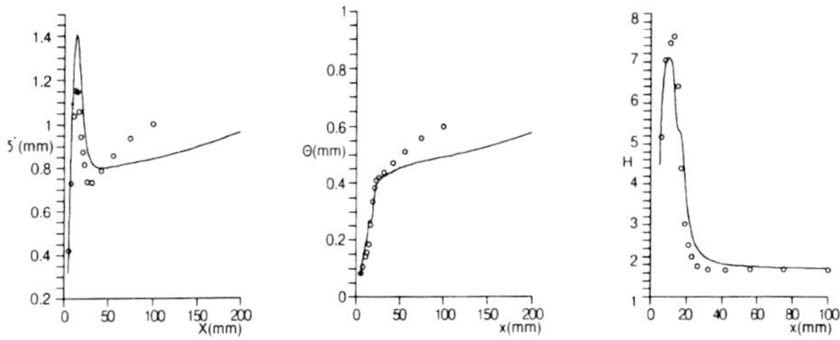

Figure 15. Displacement – momentum thickness and shape factor for Tu=2.3%
Non-zero pressure gradient. Symbols are for experimental data

Very good agreement is observed between experimental data and computational results for the shape factor and the initial past of the boundary layer. There is a discrepancy in the fully turbulent region beyond the recirculation bubble which again has to do with the vortical structures shed downstream of the recirculation bubble.

5. CONCLUSION

In this work, the prediction of flow on a flat plate with a circular leading edge under zero and non-zero pressure gradient for two turbulence levels has been presented and compared with experimental results. Such flows, especially when separation occurs under low turbulence levels, are difficult to be simulated by CFD codes. The difficulties arise from the way that turbulence will be modelled and the size of the grid that will be used. The classical turbulence models in some cases fail to predict well the fluid flow. The use of the low Reynolds k-ε model with the Kato-Launder modification in the turbulence generating term gave good predictions for flows with relatively high free stream turbulence and the velocity distributions were in general in good agreement with the experimental data. For the low turbulence level case although the results are very good there was a need to adjust relaxation parameters because of the appearance of instabilities under certain values. It has been numerically observed that such phenomena are linked to low frequency "shedding" of flow structures from the back region of the recirculation bubble. Further work on the turbulence modelling is required to improve the prediction and take into account the low frequency effects.

6. REFERENCES

1.   ERCOFTAC Bulletin 24, March 1995
2.   COUPLAND J., BRIERLEY D. H., Transition in Turbomachinery flows, Final report BRITE/EURAM Project AERO-CT92-0050, January 1996
3.   ZHU J., A Low Diffusive and Oscillation Free Convcetion Scheme. Communications in Applied Numerical Methods, Vol. 7, 1991, Willey & Sons, Ltd
4.   RHIE C. M. and CHOW W. L., A Numerical Study of the Turbulent Flow Past an Isolated Airfoil with Trailing Edge Separation, AIAA Journal, vol. 21, pp. 1525-1532, 1983
5.   MAJUMDAR S., Role of Underrelaxation in Momentum Interpolation for Calculation of Flow with Nonstaggered Grids, Numerical Heat Transfer, vol. 13, pp. 125-132, 1988

6.   LAUNDER B. E. and SHARMA, B. I., Application of the Energy-Dissipation Model of Turbulence to the Calculation of Flow Near a Spinning Disc, Letters in Heat and Mass Transfer, vol. 1, 1974, pp. 131-138

7.   KATO M. and LAUNDER B. E., The Modelling of Turbulent Flow Around Stationary and Vibrating Square Cylinders, Proc. 9nth Symposium on Turbulent Shear Flows, Kyoto, 1993.

8.   CHEN W-L, LIEN F-S and LESCHZINER M. A, The Prediction of transition with an Elliptic Solver using Linear and Non-Linear Eddy-Viscosity Models, ERCOFTAC Bulletin 24, March 1995, pp. 31-33

9.   YAKINTHOS C. and GOULAS A., The Prediction of Flow on a Flat Plate with a Circular Leading Edge, Sixth International Symposium on Computational Fluid Dynamics, Vol III, 1995, Nevada

10.   RODI W. and SHEUERER G., Calculation of Heat Transfer to Convection-Cooled Gas Turbine Blades, Journal of Engineering of Gas Turbine and Power, transactions of the ASME, vol. 107, July 1985, pp. 620-627.

11.   RODI W., Experience With Two-Layer Models Combining the k-ε Model with a One-Equation Model Near The Wall, 29th Aerospace Sciences Meeting, 1991, Nevada

C557/025/99

Unsteady transition measurements using a Ludwieg tube

R SCHOOK, H C DE LANGE and **A A VAN STEENHOVEN**
Department of Mechanical Engineering, Eindhoven University of Technology, The Netherlands

ABSTRACT

A new set-up based on a Ludwieg tube is developed for studying wake induced transition. A special wake generator is used which makes it possible to translate cylinders with high velocities in front of a test plate. The maximum Mach number is approximately 0.5 while the background turbulence level can reach values from 0.5% up to 4.5%.

The principle and design of the test section are presented. The flow conditions can be deduced from pressure measurements in the tube. These conditions are slightly different from those in a conventional Ludwieg tube due to the change in geometry which is needed to incorporate the wake generator.

A number of heat transfer experiments are presented. Both, steady and unsteady transition phenomena are investigated. It is shown that the new set-up is a useful tool for wake induced transition research.

1 INTRODUCTION

The boundary layer heat transfer changes significantly when laminar to turbulent transition occurs (1). Especially for turbomachine design this transition is of major importance for an accurate description of the heat transfer. Many factors like turbulence level, turbulence length scale, pressure gradient and compressibility affect transition. An important factor is unsteadiness due to the process of rotor stator interaction. One of the reasons for this is the need for reliable transition models which describe this interaction. The rotor blade heat transfer is influenced by the upstream stator wakes. These wakes initiate earlier transition which results in a higher mean heat flux than would be expected from bypass transition phenomena.

Unsteady transition experiments can be found in literature (2,3,4). However, most of these experiments are performed using profiles initiating pressure gradients. Only few flat plate data are available (5,6).

A number of models containing the unsteady effects can be found (5,6,7,8,9,10). The Narasimha and Johnson transition models are extended to describe wake induced transition. To incorporate the unsteady effect, sometimes the wake is assumed to cause a fully turbulent region which initiates transition instantaneously. More sophisticated models also should take into account wake spreading and calmed region effects. However, no general theory exists which takes all variables into account. To extend existing boundary layer transition models with the unsteady effects more transition experiments are needed.

In the past it has been shown that the Ludwieg- or expansion tube is suitable to study bypass boundary layer transition (11,12). This paper describes a new set-up to perform wake induced transition experiments. It is based on the same Ludwieg tube used in earlier experiments (13). To generate the wakes a special wake generator is developed. Basically, the set-up consists of a mechanism which translates cylinders in front of a sensor plate.

In this paper the development of this set-up is described. Also preliminary test results are discussed including pressure- and heat transfer measurements.

2 EXPERIMENTAL SET UP: THE LUDWIEG TUBE

With the Ludwieg tube (figure 1) well defined expansion waves can be generated. The tube, with a length of $10m$ and a cross section of $0.1 \times 0.1m^2$, is connected to a dump tank (volume of $0.43m^3$) via a test section. A diaphragm is situated in between the test section and the dump tank. The pressure in the tank is brought to nearly vacuum ($\sim 300 \ N/m^2$) while the pressure in the test section and the tube can be adjusted between vacuum and atmospheric. When the diaphragm is ruptured an expansion wave moves through the test section and the tube. During this expansion a steady flow is obtained for about $50ms$.

A flat sensor plate with heat flux gauges is mounted in the test section between the side walls. Just upstream the leading edge two sites are available for turbulence generating grids. These grids are constructed from rods with diameters ranging from 1 to $4mm$. The resulting turbulence level reached values from 0.5% up to 4.5% depending on the Mach number. The level is determined by a hot-wire positioned $10mm$ above the leading edge of the plate. Also a pressure gauge is available at one of the side walls.

In the conventional tube the Mach number ranged from 0.15 to 0.56 while the unit Reynolds number ($Re_u = U/v$) could reach values up to $4 \cdot 10^6 \ m^{-1}$.

2.1 Translating cylinder mechanism

To study wake induced phenomena a wake generator had to be incorporated in the test section. It is important to design the set-up such that it can operate at sub-atmospheric pressure. This implies that the complete wake generator must be situated in a tank.

A schematic view of the tube, the test section and the wake generator is presented in figure 1. The wake generator consists of two parallel moving belts with a length of $1.5m$. These belts are driven by four discs (diameter $0.15m$). Each belt contains 60 positions for cylinders. One disc is mounted on a shaft which is led outside the vacuum tank by a special sealing. With this construction it is possible to translate cylinders in front of the sensor plate while the gear and the engine are outside the vacuum tank. The distance from the cylinders to the leading edge is adjustable and set at 20*mm*.

The difference between the earlier and the new test section is the translating bar mechanism. It has been mounted in a rectangular box that has to have an open connection with the Ludwieg tube. This connection implies the presence of two slits in upper and bottom plate of the tube and a buffer volume. The buffer volume influence is minimised by filling the free space in the tank with foam. After reducing it an extra volume of approximately 16 liter remained.

The wake generator is constructed for a maximum translating velocity of 50*m/s*. This means that high velocity ratios, i.e. translating cylinder velocity divided by the mean stream velocity, can be reached. Due to the high velocities the cylinders must weight as little as possible. To that end cylinders of carbon fibre are used with which high velocities are obtained without vibrations or belt problems.

An overview of set-up conditions is given by:
- Mach number: up to 0.5 (mean stream velocity: 150 *m/s*)
- Unit Reynolds number: up to $4.10^6 \, m^{-1}$
- Turbulence level: 0.5% up to 4.5%
- Translating cylinder velocity: up to 50 *m/s*
- Minimum mesh size cylinders: 25*mm*
- Cylinder diameter: up to 6*mm*

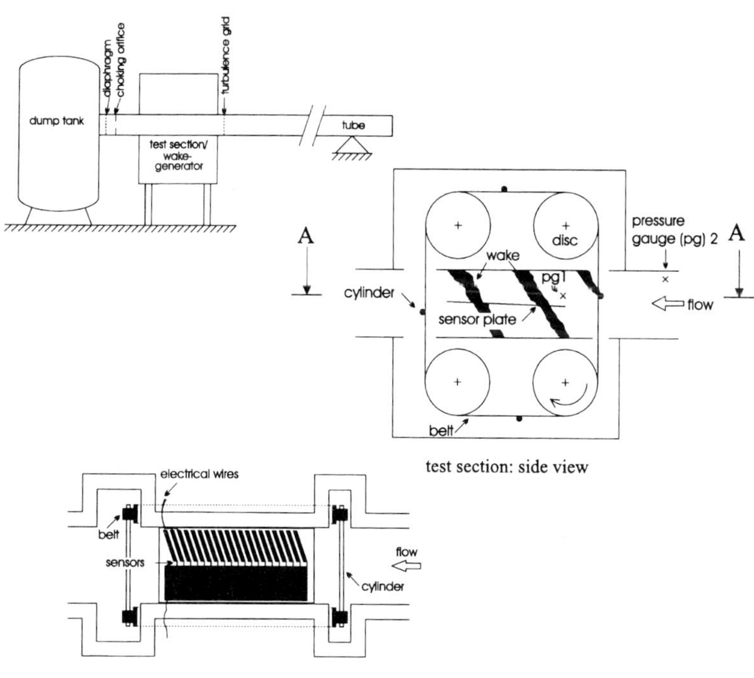

Figure 1: Experimental set-up, the Ludwieg tube.

2.2 Thin film gauges

The heat transfer is measured by a thin film technique. The test plate consists of a $1mm$ stainless steel plate and a $3mm$ glass layer with a sharpened leading edge. This glass (B-270) serves as a substrate for 20 titanium chemical vapour deposited resistors. These sensors (thickness $70nm$) do not disturb the boundary layer developing on the plate. In figure 1 view AA the sensor geometry is presented. The sensors have a length of $3mm$ and a width of $10\mu m$. The sensor resistance is about $1k\Omega$ while the resistance of the connected electrical wires is approximately 50Ω.

During the experiment the temperature of the gas which flows along the plate decreases. This results in a change of the sensor resistance depending on the temperature according to the relation:

$$R = R_0(1 + \alpha_0(T - T_0))$$

In this equation T_0 is the initial sensor temperature (reference condition at $T_0 = 20°C$), R_0 the initial sensor resistance and α_0 the resistance temperature coefficient which is approximately: $3.0 \cdot 10^{-3}\ K^{-1}$. This coefficient is determined for each sensor.

Each sensor is connected to a Wheatstone bridge which converts the resistance variation in a voltage variation. A schematic view is given in figure 2. Due to the expansion the sensor voltage decreases. The resulting signal voltage is amplified 250 times by a two-step operational LF357 amplifier and recorded, at a frequency of $50kHz$ per sensor, in a computer using an AT-MIO-64 interface board.

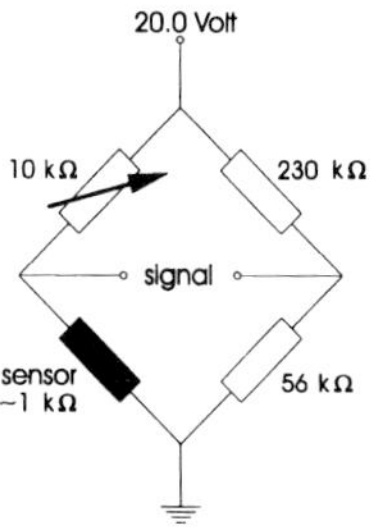

Figure 2: Wheatstone bridge.

The temperature profile perpendicular to the glass is reconstructed by solving the one-dimensional heat conduction equation:

$$\rho c \frac{\partial T}{\partial t} = \frac{\partial}{\partial z}\left(k \frac{\partial T}{\partial z} \right)$$

with ρ, c, and k the substrate properties which are known accurately ($\rho = 2550\ kg/m^3$, $c = 860$ J/kgK and $k = 0.92\ W/mK$). Because of the short test time, the thermal front does not reach the bottom of the substrate during the experiment. So, it can be assumed that the bottom temperature remains at the initial temperature. This boundary condition together with the sensor temperature measured are used to solve the heat conduction equation and to deduce the surface heat flux.

3 PRESSURE MEASUREMENTS

The flow conditions in the test section can be calculated by assuming one dimensional gas dynamics. The Mach number is adjusted by setting the ratio of the tube cross section and the opening in the choking orifice. It follows (14) that the ratio of the pressure during the experiment (p_{tube}) and the initial pressure (p_{ini}) equals:

$$\frac{p_{tube}}{p_{ini}} = \left[1+\left(\frac{\gamma-1}{2}M\right)\right]^{\frac{-2\gamma}{\gamma-1}} .$$

In a conventional Ludwieg tube the pressure remains nearly unchanged during the experiment. In the new set-up however, this is not the case. Due to the wake generator the test section is connected to the buffer volume as explained before. This implies that the geometry in the section is not ideal. To test flow conditions, the pressure signal is recorded at two different positions. Position 1 in figure 3 shows the signal taken by a Sensit pressure gauge just above the leading edge of the sensor plate. The second gauge is situated $0.3m$ upstream in the tube (Kistler pressure gauge; the signal is sampled at a frequency of $50kHz$). At this position 1D flow conditions are assumed. Both signals have a similar shape and only a time shift difference when compared with each other. From this it is concluded that the flow along the plate indeed can be described by one dimensional gas dynamics.

The pressure signal is not constant during the expansion. Transition will be studied in the time interval between $20ms$ and $50ms$ after diaphragm rupture. The corresponding Mach number changes 20%. The mean Mach number in all the measurements described is 0.33 and can be reproduced within 3% accuracy.

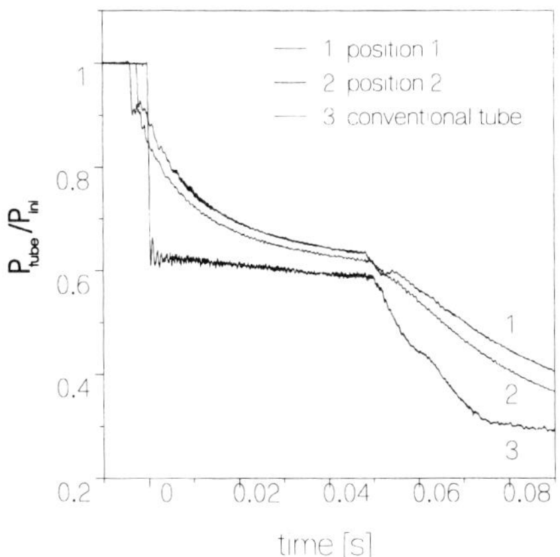

Figure 3: pressure signals for the conventional and the new set-up.

4 HEAT FLUX MEASUREMENTS

All the flux experiments are performed at a unit Reynolds number of $2.4 \cdot 10^6 \, m^{-1}$. This value is adjusted by setting the initial pressure, i.e. the kinematic viscosity, and is based on a Mach number of 0.33 corresponding to a mean stream velocity of 106m/s.

For a laminar boundary layer the theoretical heat flux is given by (15):

$$q''_{lam}(x) = 0.322 \frac{k}{x} \left(T_w - T_r\right) \mathrm{Re}_x^{1/2} \mathrm{Pr}^{1/3} \quad ,$$

while for a turbulent boundary layer the flux equals:

$$q''_{tur} = 0.0288 \frac{k}{x} \left(T_w - T_r\right) \mathrm{Re}_x^{4/5} \mathrm{Pr}^{1/3} \quad .$$

In these equations T_w is the wall temperature of the plate and k is the thermal conductivity. T_r is the recovery temperature which is a function of the Mach number and the temperature at infinity:

$$T_r = T_\infty \left(1 + \tfrac{1}{2}\left(\gamma - 1\right)M^2 \sqrt[n]{\mathrm{Pr}}\right)$$

For a laminar flow n equals 2 while for a turbulent flow n is 3. The above mentioned expressions are used to compare the data obtained with in the steady as well as the unsteady transition experiments.

4.1 Steady transition measurements

To obtain transitional flow in the new set-up some experiments are performed. First a measurement is done without a turbulence generating grid (figure 4). The data together with the theoretical fluxes for the mean Mach number and the maximum Mach number obtained during the expansion are shown. It can be seen that the heat flux, especially for the first sensor positions, agrees well. The increase of trailing edge sensor heat fluxes might be due to flow acceleration or transition start. To check this trailing edge pressure measurements have to be performed.

Bypass transition experiments are done for three different turbulence generating grids positioned 0.15m in front of the leading edge. These grids have rod diameters of 1, 2 and 3 mm, with corresponding turbulence levels 1.8, 2.2 and 3.4% respectively. In figure 5 the mean heat fluxes are given together with the laminar and turbulent values. The absolute value of the turbulent flux remains lower than the expected value. This also is observed in earlier experiments (13).

The time dependent flux signals for 6 successive sensors during one experiment are shown in figure 6. Areas of local flux maxima seem to grow in width and height, with increasing streamwise positions. It is assumed that these are the turbulent spots which indicate the transition process (16). So, it is concluded that also in the new test section boundary layer transition can be initiated and measured.

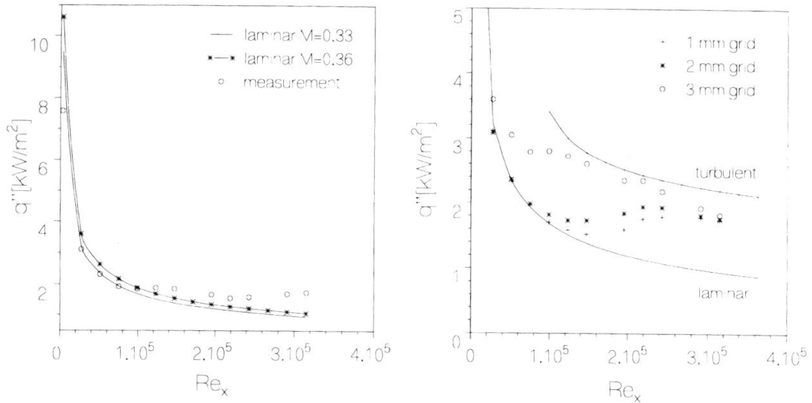

Figure 4 (left): laminar heat flux measurement.
Figure 5 (right): transitional heat flux measurements.

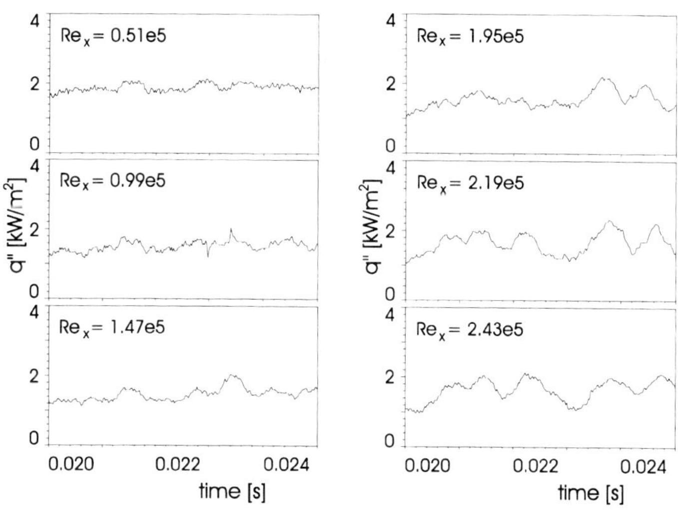

Figure 6: Heat flux for different gauges.

4.2 Unsteady transition measurements

The result of a wake induced transition experiment is presented in figures 7 and 8. In this experiment the belt moves at a velocity of 21.7*m/s* i.e. a velocity ratio of 0.2. No turbulence generating grids are used. The number of cylinders (diameter *6mm*) mounted on the belt is 12. This implies that within the test time interval, six cylinders are passing the test plate and consequently six times a new wake is generated. These wakes result in an increase of heat flux as can be seen in figure 7 (flux changes are visible; not the absolute values). It shows the flux signal for 7 streamwise sensors. While the flow develops along the plate the wake width increases.

Figure 8 presents the mean heat fluxes with moving and non-moving cylinders (laminar). As expected, it follows that in general the heat flux increases when wake induced transition occurs. This does not hold for the first sensor ($Re_x = 3120$). At this position the flux is less than the laminar value. It might be explained by the fact that when the wake passes, no boundary layer develops but separation takes place instantaneously.

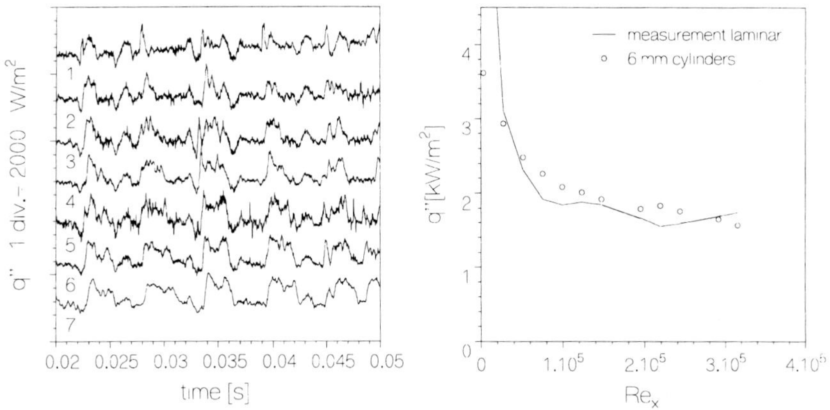

Figure 7 (left): Heat flux for different gauges while cylinders are moving.
Figure 8 (right): Heat flux measurement with moving cylinders (time mean flux).

CONCLUSIONS AND PROGRESS

From the experiments described in this paper it is concluded that the Ludwieg tube is suitable for studying wake induced transition. The new set-up makes it possible to do experiments in a wide range of parameters. The translating cylinder velocity combined with the high mean stream velocity even makes it possible to study the influence of compressibility on unsteady transition. Also measurements with both a turbulence generating grid and translating cylinders can be performed to test existing models.

Future experiments will focus on the effect of changing the cylinder speed, diameter and mesh size i.e. wake-wake interaction. In fact the wake structure is changed by altering above mentioned parameters. Both heat flux and hot-wire measurements will be performed. Finally, the data obtained must contribute to a general model for unsteady transition.

ACKNOWLEDGEMENTS

The authors gratefully acknowledge Professor M.E.H. van Dongen from the section Gas Dynamics and Aeroacoustics, Department of Applied Physics at the Eindhoven University of

Technology. His knowledge and support made it possible to perform the experiments described.

REFERENCES

[1] Consigny, H., and Richards, B.E., 1982, 'Short duration measurements of heat transfer rate to a gas turbine rotor blade,' *Journal of Engineering for Power*, vol. 104, pp. 542-551.

[2] Hodson, H.P., 1984, 'Boundary layer and loss measurements on the rotor of an axial flow turbine,' *Journal of Engineering for Gas Turbines and Power*, vol. 106, pp. 391-399.

[3] Doorly, D.J., 1983, 'Modeling the unsteady flow in a turbine rotor passage,' *Journal of Turbomachinery*, vol. 110, pp.27-37.

[4] Schobeiri, M.T., and Pappu, K., 1997, 'Experimental study on the effect of unsteadiness on boundary layer development on a linear turbine cascade,' *Experiments in Fluids*, vol. 23, pp. 306-316.

[5] Funazaki, K., 1996, 'Unsteady boundary layers on a flat plate disturbed by periodic wakes: part 1 - Measurements of wake affected heat transfer and wake induced transition model,' *Journal of Turbomachinery*, vol. 118, pp. 327-336.

[6] Pfeil, H., Herbst, R., and Schröder, T, 1983, 'Investigation of the laminar-turbulent transition of boundary layers disturbed by wakes,' *Journal of Engineering for Power*, vol. 105, pp.130-137.

[7] Addison, J.S., and Hodson, H.P., 1992, 'Modeling of unsteady transitional boundary layers,' *Journal of Turbomachinery*, vol. 114, pp. 580-589.

[8] Hodson, H.P., Addison, J.S., and Shepherdson, C.A., 1992, 'Models for unsteady wake induced transition in axial turbomachines,' *J. Phys. III France*, vol. 2, pp. 545-574.

[9] Mayle, R.E., and Dullenkopf, K., 1990, 'A theory for wake induced transition,' *Journal of Turbomachinery*, vol. 112, pp. 188-195.

[10] Mayle, R.E., and Dullenkopf, K., 1991, 'More on the turbulent strip theory for wake induced transition,'' *Journal of Turbomachinery*, vol. 113, pp. 428-432.

[11] Hogendoorn, C.J., 1997, '*Heat transfer measurements in subsonic transitional boundary layers*,' PhD-thesis, Eindhoven University of Technology.

[12] Hogendoorn, C.J., de Lange, H.C., and van Steenhoven, A.A., 1996, 'Heat transfer measurements in transitional boundary layers,' *Proceedings of 2nd European Thermal Sciences and 14th UIT National Heat Transfer Conference*, Edizioni ETS, pp. 593-600.

[13] Schook, R., Lange, H.C., de, and Steenhoven, A.A., van, 1998, 'Effects of compressibility and turbulence on bypass transition,' ASME Paper No. 98-GT-286.

[14] Owczarek, J.A., 1964, '*Fundamentals of Gas Dynamics*,' International Textbook Company.

[15] Schlichting, H., 1979, '*Boundary Layer Theory*,' McGrawHill Book Company, seventh edition.

[16] Emmons, H., 1951, 'The laminar turbulent transition in a boundary layer: part 1,' *J. Aero. Sci.*, vol. 18, pp. 490-498.

Calculation of transition in turbine cascades by conditioned Navier–Stokes equations

J STEELANT
Section of Aerothermodynamics (TOS-MPA), ESTEC/ESA, Noordwijk, The Netherlands
E DICK
Department of Flow, Heat, and Combustion, University of Gent, Belgium

To simulate transitional heat transfer in turbine cascades, the conditionally averaged Navier-Stokes equations are used. To cover both the physics of freestream turbulence and the intermittent flow behaviour during transition, a turbulence weighting factor τ is introduced. A transport equation is presented for this τ-factor including convection, diffusion, production and sink terms. In combination with the conditioned Navier-Stokes equations, this leads to improvements in the calculation of flow characteristics in both the transitional layer and the freestream. The method is validated on transitional heat transfer measurements in a linear turbine cascade at typical operational conditions.

INTRODUCTION

By-pass transition emanates mainly from high free-stream turbulence and leads to a transition far further upstream than what would be expected for natural transition. In contrast with natural transition which emanates from the breakdown of amplified disturbances within the boundary layer, by-pass transition is mainly determined by the free-stream turbulence which affects the pre-transitional (pseudo-laminar) layer directly by diffusion and indirectly by pressure fluctuations. This diffusion of turbulent eddies has an intermittent character and is first localized in the outer part of the laminar boundary layer.

A similar but different kind of intermittent behaviour is also seen during the transition where the flow in the boundary layer is characterized by distinct turbulent and laminar phases alternating in function of time. In the latter, the intermittent behaviour is quantified by the intermittency factor γ. This factor is the relative fraction of time during which the flow is turbulent at a certain position. It evolves from 0% at the transition point up to 100% at the end of transition.

The relative fraction of time can also be taken to quantify the intermittent behaviour of the turbulent eddies on the underlying pseudo-laminar boundary layer. This parameter, in the sequel referred to as freestream factor ω, is 0% near the wall and tends to 100% in the freestream.

NOTATION

c	chord length [m].	ρ	density.
$\tilde{c}$	local velocity amplitude, $\sqrt{\tilde{u}^2 + \tilde{v}^2}$.	τ	turbulence weighting factor.
d	rod diameter of turbulence grid [m].	ω	freestream factor.
k	turbulence kinetic energy.	2	outlet value.
K_M	compressible acceleration parameter.	i	inlet
M	local isentropic Mach number.	is	isentropic
s	distance along the blade surface starting at the leading edge stagnation point [m].	l	laminar state.
		t	turbulent state.
		∞	local freestream
Tu	turbulence intensity, $100\sqrt{k}/U(\%)$.	le,∞	entrance plane of the cascade.
γ	intermittency factor.	$\sim$	conditioned Favre-average.
μ	dynamic viscosity.	$-$	conditioned Reynolds-average.

TRANSITION MODEL

When modelling transition, it is essential to take intermittency into account. Global time averaging used for classical turbulence modelling is not appropriate in intermittently changing flows. To describe the transitional zone and the outer layer zone, it is necessary to use conditional time averaging. These averages are taken during the fraction of time the flow is laminar or turbulent respectively. As we are only interested in the state of the flow, i.e. laminar or turbulent, at a certain position, it is sufficient to make use of a turbulence weighting factor τ, which is the sum of the intermittency factor γ and the freestream factor ω: $\tau = \gamma + \omega$. As a consequence, this factor τ incorporates two effects: firstly the diffusion of freestream turbulent eddies into the boundary layer and secondly the transport and growth of the turbulent spots during transition. Hence, this factor is 0% in the vicinity of the wall within the pre-transitional boundary layer, and 100 % in the freestream and inside a fully turbulent boundary layer. Near the wall, the τ-factor is identical to the intermittency factor γ as $\omega = 0$ close to the wall.

Applying the conditionally averaging technique onto the Navier-Stokes equations leads to a set of laminar and turbulent equations for mass, momentum and energy [1, 2]. These conditioned equations differ from the original Navier-Stokes equations by the presence of source terms which are function of the turbulence weighting factor τ, the laminar and turbulent state:

$$\frac{\partial \overline{U}_l}{\partial t} + \frac{\partial \overline{F}_l}{\partial x} + \frac{\partial \overline{G}_l}{\partial y} = \frac{\partial \overline{F}_{vl}}{\partial x} + \frac{\partial \overline{G}_{vl}}{\partial y} + \frac{1}{2(1-\tau)}\overline{S}^{\tau}(\tau, \overline{U}_l, \overline{U}_t)$$

$$\frac{\partial \overline{U}_t}{\partial t} + \frac{\partial \overline{F}_t}{\partial x} + \frac{\partial \overline{G}_t}{\partial y} = \frac{\partial \overline{F}_{vt}}{\partial x} + \frac{\partial \overline{G}_{vt}}{\partial y} + \frac{1}{2\tau}\overline{S}^{\tau}(\tau, \overline{U}_l, \overline{U}_t).$$

Writing out the mass equations results in:

$$\frac{\partial \overline{\rho}_l}{\partial t} + \frac{\partial \overline{\rho}_l \tilde{u}_l}{\partial x} + \frac{\partial \overline{\rho}_l \tilde{v}_l}{\partial y} = \frac{1}{2(1-\tau)}S_{\rho}^{\tau}, \qquad \frac{\partial \overline{\rho}_t}{\partial t} + \frac{\partial \overline{\rho}_t \tilde{u}_t}{\partial x} + \frac{\partial \overline{\rho}_t \tilde{v}_t}{\partial y} = \frac{1}{2\tau}S_{\rho}^{\tau},$$

$$\text{with} \quad S_{\rho}^{\tau} = (\overline{\rho}_l - \overline{\rho}_t)\frac{\partial \tau}{\partial t} + (\overline{\rho}_l \tilde{u}_l - \overline{\rho}_t \tilde{u}_t)\frac{\partial \tau}{\partial x} + (\overline{\rho}_l \tilde{v}_l - \overline{\rho}_t \tilde{v}_t)\frac{\partial \tau}{\partial y}.$$

Similar expressions can be written for the momentum and energy equations. In order to close the equation, τ has to be determined. The factor τ has, both in streamwise and in normal direction, a spatial distribution which largely depends on the conditions of the mean flow and the boundary layer. For simple cases, one can set up an algebraic expression. For more general flows, a transport equation is needed. A transport equation has been constructed for the intermittency factor γ in previous work [2]. To evaluate the turbulence weighting factor τ, the transport equation is extended with a diffusion and a dissipation term and is given by

$$\frac{\partial \overline{\rho}\,\tilde{u}\tau}{\partial x} + \frac{\partial \overline{\rho}\,\tilde{v}\tau}{\partial y} \;=\; D_\tau + P_\tau - E_\tau. \tag{1}$$

The value of τ is set at 1 at the inlet. Prior to the transition point, determined by a correlation (see [3]), the value of τ is set to zero on the wall. The derivation of equation (1) is based on flat plates and can be found in [3]. The terms D_τ, P_τ and E_τ represent respectively diffusion, production and dissipation of τ:

$$D_\tau = \frac{\partial}{\partial x_i}[\mu_\tau \frac{\partial \tau}{\partial x_i}], \quad P_\tau = 2f_\tau(1-\tau)\sqrt{-\ln(1-\tau)}\beta\overline{\rho}\,\tilde{c}, \quad E_\tau = 2.5\mu_\tau \frac{\tilde{c}}{\tilde{c}_\infty^2}\frac{\partial \tilde{c}}{\partial n}\frac{\partial \tau}{\partial n}.$$

The coefficient μ_τ quantifies the diffusion of the turbulent freestream eddies into the boundary layer with f_{μ_τ} being a damping function.

$$\mu_\tau \;=\; 33\mu f_{\mu_\tau} T u_{le,\infty}^{-0.69}[-ln(1-\tau)]^{-\frac{5}{6}(1-\tau)}, \quad f_{\mu_\tau} = 1 - \exp\left[-256\left(y\rho_\infty c_\infty/\mu_\infty\right)^2\right].$$

The function f_τ models the distributed breakdown near the onset with $f_\tau = 1$ for $\tau > 0.45$ and $f_\tau = 1 - \exp\left[-1.735\mathrm{tg}(5.45\tau - .95375) - 2.2\right]$ for $\tau < 0.45$. The β-function is related to the growth of the turbulent spots in combination with Mach number effects (f_M) and pressure gradient effects (f_K):

$$\beta^2 \;=\; 1.25 \times 10^{-11} T u_\infty^{7/4} f_K f_M \left(\frac{\rho_\infty U_\infty}{\mu_\infty}\right)^2,$$

$$f_M \;=\; \left[1 + 0.58 M^{0.6}\right]^{-2}, \quad f_K = 1 + f_\tau(PRC - 1),$$

$$PRC \;=\; \begin{cases} \left(474 T u_{le,\infty}^{-2.9}\right)^{\left[1-\exp(2\times10^6 K)\right]} & K < 0, \\ 10^{-3227 K^{0.5985}} & K > 0, \end{cases}$$

As the evaluation of Tu_∞ is not evident, it is determined algebraically. Throughout the cascade, the local freestream turbulence level is set at $Tu_\infty = Tu_{le,\infty}(U_{le,\infty}/U_\infty)^{3/2}$. In the presence of a shock, detected in the calculation through the presence of a sudden pressure rise, the level is set at $Tu_\infty = 15\%$ downstream of the shock [4]. In the wake-region the Tu_∞-level is set at a physical value of 20%. The local compressible pressure gradient parameter K_M is given by: $K_M = -\frac{\mu_\infty}{\rho_\infty^2 U_\infty^3}\mid 1 - M^2 \mid \frac{dp}{dx}$.

TEST CASES

The capability of the present transition model is demonstrated on a 2D-turbine guide vane operating at typical design conditions. This guide vane has been experimentally investigated by Arts *et al.* [5] in order to understand the influence of Mach number, turbulence intensity and Reynolds number on the transitional heat transfer distribution. The blade profile has a chord of $c = 67.647mm$ and a pitch to chord ratio of $g/c = 0.85$. The total inlet temperature is set at approximately T=420K. The wall temperature is considered to be at an almost constant level of 300K during the measurement. To validate the present transition model, three different test cases are considered which are described in table 1. The incoming turbulence level Tu_i is measured 55mm upstream of the leading edge plane ($x/c_{ax} = -1.487$). As no length scale or dissipation were measured, the turbulence level at the leading edge Tu_{le} is estimated by the general correlation for turbulence decay proposed by Roach [6]: $Tu = 0.80 \, (x/d)^{-\frac{5}{7}}$. Hence, the rod diameter d is considered here as representative for the turbulence length scale. Mayle showed in [7] that within a realistic spectrum of turbulence length scales the transition length is effected up to a factor of two.

Table 1: Description of the different test cases.

Case	$M_{2,is}$	$Tu_i(\%)$	$Tu_{le}\,(\%)$	$Re_{c,2}$
MUR239	0.922	6	4.52	2.10^6
MUR245	0.924	4	3.34	2.10^6
MUR241	1.089	6	4.52	2.10^6

The computational domain consists of two blocks. The first is a O-grid with 433 nodes along the blade and 73 nodes normal to the blade. The first point is situated at a location $y^+ < 1$. The second block is a H-grid of 217×49 nodes and is placed in line with the outlet angle. Figure 1 shows the geometry and the location of the different blocks.

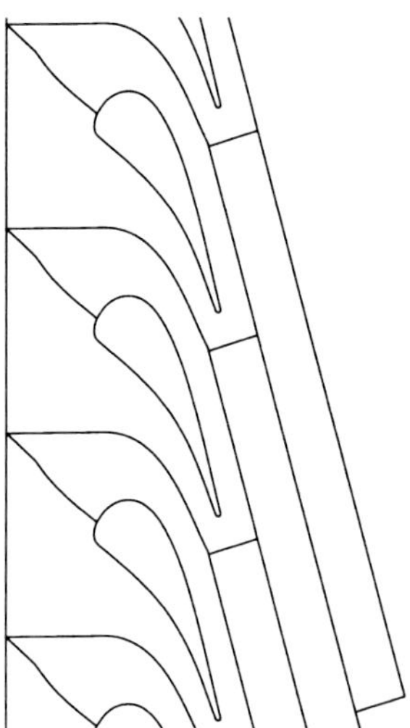

Figure 1: Geometry and block location of the 2D turbine cascade

C557/036 © IMechE 1999

TURBULENT RESULTS

Before moving to the results of the present transition model, it is important to recognize how the used turbulence model behaves in a turbine cascade flow. It is generally known that, in the vicinity of stagnation regions, the classical turbulence models generate a too high turbulence kinetic energy. As a consequence the corresponding heat transfer rate is far too high in comparison to experimentally measured data.

A remedy exists in reformulating the production term according to Kato and Launder [8]. Classically this term is written as $P_k = \mu_t S^2$ where S is the magnitude of the shear tensor: $S = \sqrt{S_{ij}S_{ij}}$. Using $P_k = \mu_t S\Omega$ instead reduces the excessive raise near stagnation regions but falls back to the same analytical expression for boundary layer flows. The value Ω is the magnitude of the vorticity tensor $\Omega = \sqrt{\Omega_{ij}\Omega_{ij}}$. Applying this modification to the k-ϵ model of Yang and Shih [9] results in a much better heat transfer distribution with a lower stagnation peak (see fig. 2). The heat transfer coefficients on the pressure side are placed on the left hand side, while those on the suction side are placed on the right hand side. The parameter s is the distance along the blade surface in meter.

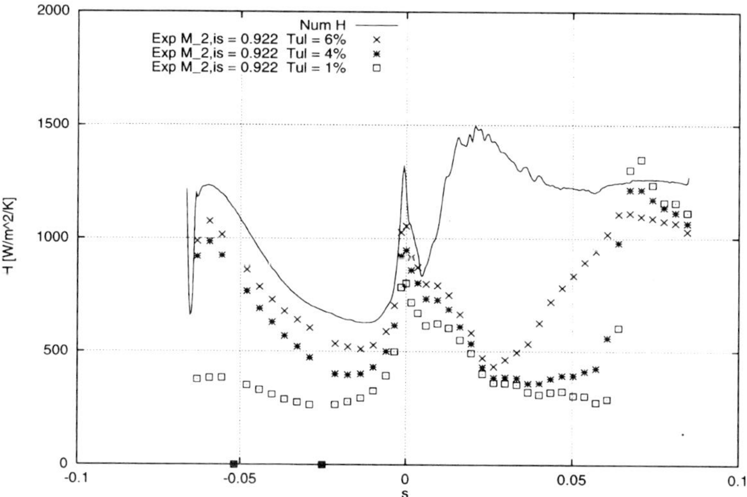

Figure 2: Fully turbulent heat transfer distribution for MUR239 ($Tu_i = 6\,\%$, $M_{2,is} = 0.922$) obtained with modified k-ϵ model; full line: calculated, symbols: experiments (other Tu_i also shown).

Although the obtained values are much lower, they still remain too high in comparison to the experiments both in the stagnation region and at the end of the suction surface. Similar overpredictions have been regularly reported by other researchers, e.g. see [10] and [11]. In the sequel, only the k-ϵ turbulence model of Yang and Shih with the Kato-Launder modification is used. To interpret the transitional results, one should keep in mind the possible overprediction of the turbulent heat transfer due to the underlying turbulence model.

TRANSITION RESULTS

From the turbulent results, one can clearly notice that the obtained transition happens too early and too rapid. This defect is generally seen by all kinds of two-equation turbulence models. Applying the present transition model on test case MUR239 results in the heat transfer distribution given in figure 3. The overall distribution agrees very well with the experimental results which is mainly due to the correct prediction of the evolution of the turbulence weighting factor near the wall. The predicted heat transfer rate in the

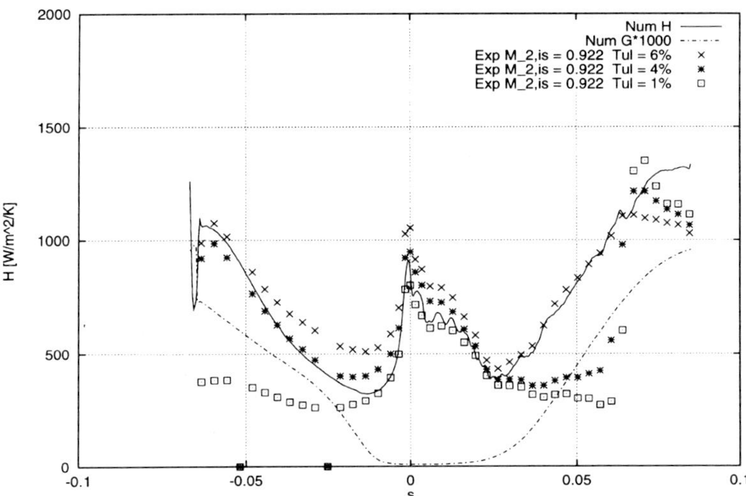

Figure 3: Heat transfer distribution for MUR239 (Tu_i =6 % $M_{2,is}$ = 0.922); full line: calculated heat transfer, dashed line: intermittency ($\times$ 1000), symbols: experiments (other Tu_i also shown).

pretransitional boundary layer is generally lower than the experimental values both on the suction and pressure side. At high freestream turbulence, velocity fluctuations are induced in the boundary layer prior to transition. This inherently results in higher heat transfer rates [12]. This mechanism is however not incorporated in the present model. A possible method dealing with freestream turbulence effects is the model proposed by Volino [13]. Here, laminar values are obtained which are almost identical to the low Tu-level date ($Tu = 1\%$). As mentioned above, the fully turbulent level is still higher than the experimental level. This can be attributed to the underlying turbulence model. The near-wall distribution of the turbulence weighting factor τ, which corresponds with the intermittency factor γ, indicates that the transition is almost completed on the suction side (dashed line in fig. 3). At the pressure side, the transition is not completed as the intermittency reaches the 70% value near the trailing edge. The Mach number distribution is given in figure 4 together with experimental results taken at slightly different Mach numbers $M_{2,is}$ = 0.875 and 1.02 and at the lowest turbulence level.

Lowering the turbulence level to 4% (MUR245) reduces the spot production parameter such that the thinner laminar layer allows the existence of a shock wave (fig. 5). The sudden raise in heat transfer is very well predicted but it ends with a too large overshoot.

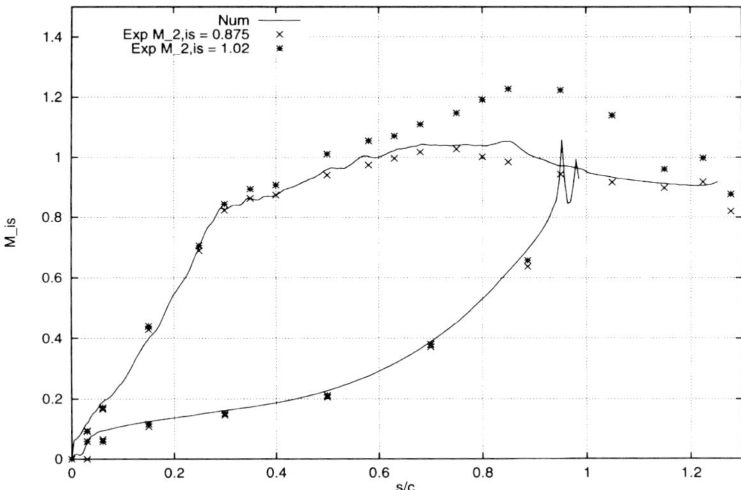

Figure 4: Isentropic Mach number distribution for MUR239 at $M_{2,is} = 0.922$; full line: calculated, symbols: experiments at $M_{2,is} = 0.875$ and 1.02 with $Tu_i = 1\%$

On the pressure side, the intermittency level is lower than in the previous case due to the lower turbulence level. On the suction side, however, the intermittency undergoes a steep increase downstream of the shock. This leads to a shorter transition zone ending before the trailing edge. The corresponding Mach number distribution is given in fig. 6 and differs slightly on the suction side from the result in fig. 4. The thinner laminar layer allows a longer persistence of the low pressure region than in the MUR239 case. This region is then abruptly ended by the shock wave.

Increasing the outlet isentropic Mach number from $M_{2,is} = 0.922$ in MUR239 to $M_{2,is} = 1.089$ in MUR241, results in a lower pressure on the suction side but has little or no effect on the pressure side (fig. 7). The higher acceleration in the region $s/c = 0.4$ to 0.6 compared to MUR239 decreases the spot growth resulting in $\gamma = 15\%$ at position $s = 4$cm instead of the 26% value in the MUR239 case (fig. 8). The deceleration from $s/c = 0.6$ to 0.8 increases the spot growth again with a sharp raise in heat transfer as a consequence. The slightly accelerated region further downstream attenuates this raise. Near the aft, the heat transfer rate ends with a peak due to the presence of a shock at the trailing edge. On the pressure side the intermittency distribution is almost identical to MUR239 due to the identical pressure distribution.

CONCLUSION

The use of the conditioned Navier-Stokes equations in combination with a transport equation for the turbulence weighting factor allowed the calculation of transitional heat transfer distributions in a turbine cascade operating at typical design conditions. The numerically obtained results are in good correspondance with the experimental data. Besided pressure gradient and turbulence level, the effect of compressibility needs to be taken into account to assure the correct prediction of both the transition onset and length. The

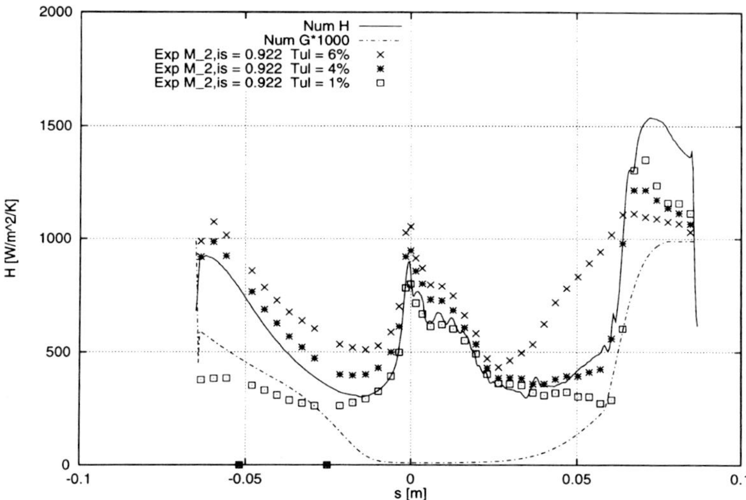

Figure 5: Heat transfer distribution for MUR245 (Tu_i =4 % and $M_{2,is} = 0.924$); full line: calculated heat transfer, dashed line: intermittency ($\times$ 1000), symbols: experiments at $M_{2,is} = 0.924$ (other Tu_i also shown).

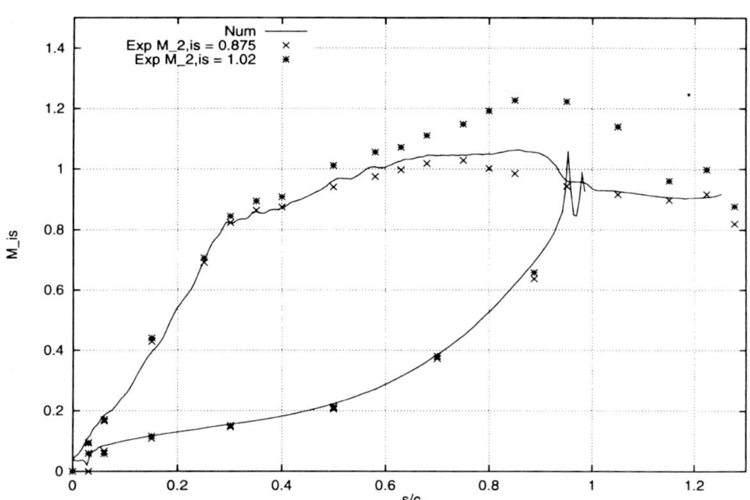

Figure 6: Isentropic Mach number distribution for MUR245 at $M_{2,is} = 0.924$; full line: calculated, symbols: experiments at $M_{2,is} = 0.875$ and 1.02 with $Tu_i = 1\%$.

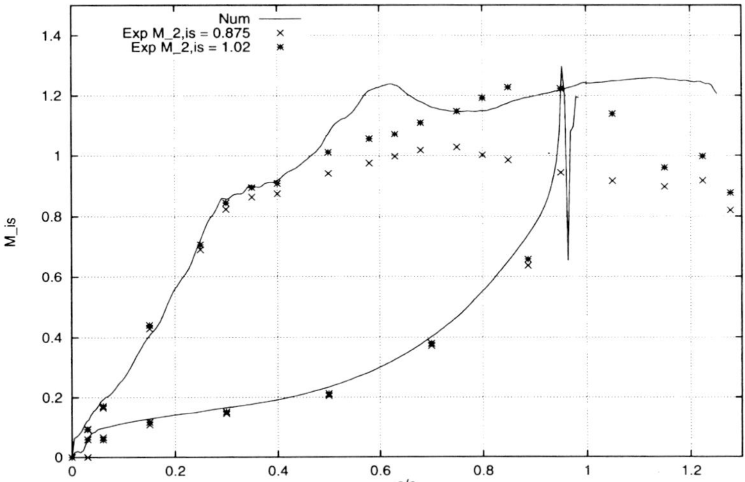

Figure 7: Isentropic Mach number distribution for MUR241 (Tu_i=4% and $M_{2,is} = 1.089$); full line: calculated, symbols: experiments at $M_{2,is} = 0.875$ and 1.02 with $Tu_i = 1\%$.

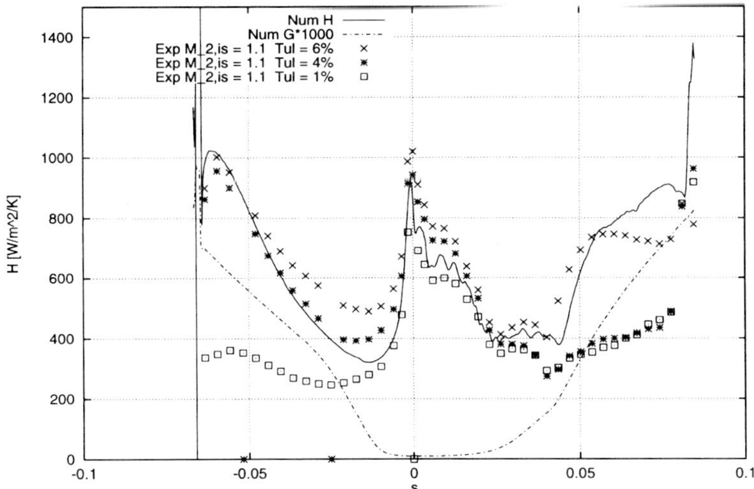

Figure 8: Heat transfer distribution for MUR241 (Tu_i=4% and $M_{2,is} = 1.089$); full line: calculated heat transfer, dashed line: intermittency ($\times$ 1000), symbols: experiments (other Tu_i also shown).

present model, however, does not allow the prediction of the higher heat transfer rates in the pretransitional boundary layer. This pleads for a further extension of the model to incorporate the physics of the freestream pressure fluctuations affecting the underlying boundary layer.

ACKNOWLEDGEMENTS

The research reported here was granted under contract G.0283.96 by the Flemish Science Foundation (F.W.O.). Part of the work is related to research activities, undertaken in the European Space Agency's Facilities, in the framework of a fellowship granted through the Training and Mobility of Researchers Programme financed by the European Community. The authors wish to thank Prof. R. Mayle and Prof. T. Arts for their helpful discussions during the course of this work.

REFERENCES

[1] Steelant J. and Dick E. , 'Modelling of Bypass Transition with Conditioned Navier-Stokes Equations coupled to an Intermittency Equation'. *Int. J. for Numerical Methods in Fluids*, 23:193–220, 1996.

[2] Steelant J. and Dick E. , 'Calculation of Transition in Adverse Pressure Gradient Flow by Conditioned Equations'. *ASME 96-GT-160*, 1996.

[3] Steelant J. and Dick E. , 'A Transport Equation of a Turbulence Weighting Factor for Modelling By-Pass Transition'. In K.D. Papailiou *et al.*, editor, *'Proc. of 4th European CFD Conference, Athens'*, pages 535–540. J. Wiley, 1998. ISBN 0-471-98579-1.

[4] Lee S., Lele S.K. and Moin P. , 'Isotropic Turbulence Interacting with a Weak Shock Wave'. *J. of Fluid Mechanics*, 251:533–562, 1993 and corrigendum 264:373-374, 1994.

[5] Arts T., Lambert de Rouvroit M. and Rutherford A.W. , 'Aero-Thermal Investigation of a Highly Loaded Transonic Linear Turbine Guide Vane Cascade'. Technical Report TN-174, Von Karman Institute, 1990.

[6] Roach P.E. , 'The Generation of Nearly Isotropic Turbulence by means of Grids'. *Heat and Fluid Flow*, 8:2:82–92, 1989.

[7] Mayle R.E. , 'A Theory for Predicting the Turbulent Spot Production Rate'. *ASME 98-GT-256*, 1998.

[8] Kato M. and Launder B.E. , 'Modified Form of Turbulent Flow around Stationary and Vibrating Square Cylinders'. In *Proc. Turbulent Shear Flows, Kyoto*, pages 10-4, 1993.

[9] Yang Z. and Shih T.H. , 'New Time Scale Based $k - \epsilon$ Model for Near-Wall Turbulence'. *AIAA J.*, 31(7), 1993.

[10] Zhang L.J. and Glezer B. , 'Turbine Airfoil External Heat Transfer Measurement in a Hot-Cascade'. *ASME-Paper 97-GT-327*, 1997.

[11] Biswas D. and Fukuyama Y. , 'Calculation of Transitional Boundary Layers with an Improved Low-Reynolds-Number Version of k-ϵ Turbulence Model'. *J. of Turbomachinery*, 116:765–773, 1994.

[12] Mayle R.E., Dullenkopf K. and Schulz A. , 'The Turbulence that Matters'. *J. of Turbomachinery*, 120:402–409, 1998.

[13] Volino R.J. , 'A New Model for Free-Stream Turbulence Effects on Boundary Layers'. *J. of Turbomachinery*, 120:613–620, 1998.

Transition studies on a two-dimensional NACA63 isolated airfoil at high Mach numbers

S SVENSDOTTER, J HU, and **T FRANSSON**
Heat and Power Technology, Royal Institute of Technology, Stockholm, Sweden

Experimental and numerical studies have been performed on the boundary layer transition process on an isolated NACA63 airfoil for varying inlet Mach numbers, incidence angles and free-stream turbulence levels. The experimental results show that for a free-stream turbulence of 0.3% and an inlet Mach number of 0.5, transition starts at about 60% chord for zero incidence and ends at about 85% chord. For increasing incidence, the suction side transition onset rapidly moves towards the leading edge and at 3° incidence, the boundary layer is entirely turbulent. The pressure side transition onset moves towards the trailing edge for increasing incidence and the transition zone is prolonged. Increasing Mach number gives an earlier and shorter transition zone and increased turbulence intensity gives an earlier transition onset. The numerical investigation captures the same trend as the experiments, but fails to detect the same transition positions.

1 INTRODUCTION

Studies on isolated airfoils in incompressible flows and low free stream turbulence levels have been performed for a number of years with the main emphasis on studying aircraft wings (1), (2), (3). These early tests were performed with low inlet velocity and large models. The Reynolds numbers were thus large, but the flows were incompressible. A few tests have been performed later with somewhat higher inlet velocities (4), (5) but no investigations seem to have been performed for high subsonic Mach numbers. The aims of the presented study are to provide experimental data at combinations of higher Mach numbers and increased free-stream turbulence level than found in previous open literature and to also compare these with a few results from numerical predictions.

2 TEST FACILITY

The VM100 wind tunnel at HPT is run continuously with air from a 1MW compressor. The free stream gas conditions can be varied from 290 to 400K and 1 to 4 bars respectively; the maximum air flow is 4.7kg/s. The inlet Mach number can be varied continuously from 0.3 to 2.0. An outlet fan and an outlet valve provide the opportunity to adjust the outlet pressure level, thus the Mach number and the Reynold's number can to some extent be varied independently. This was however not done for the investigation presented below. The measurement section is 100X105 mm^2 and the cross section of the stagnation chamber is 0.0625m^2 which gives a contraction ratio of 5.95. A turbulence grid can be positioned 1.9 chord upstream of the airfoil. The inlet static pressure is measured over an area between 0.5 and 1.6 chord upstream of the leading edge, in order to verify the periodicity of the inlet flow, and the outlet static pressure is measured at 1 and 1.25 chord downstream of the trailing edge. A sketch of the wind tunnel can be seen in Figure 1. The blade is mounted in a turnable window in order to vary the flow incidence angle.

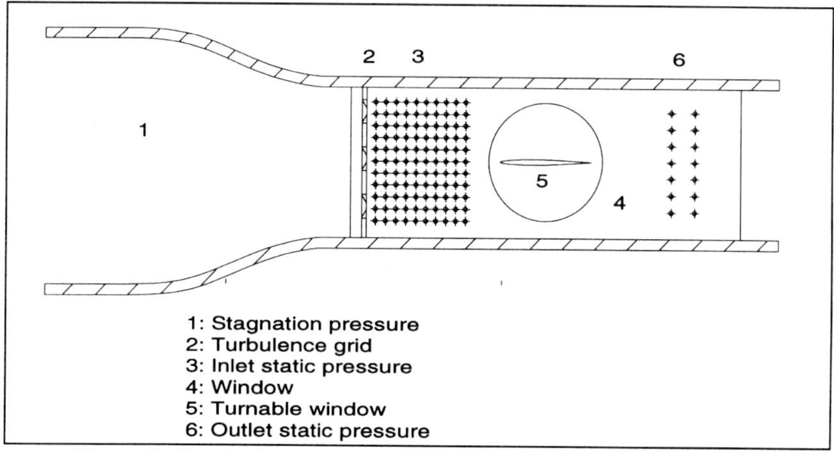

Figure 1: Sketch of the HPT VM100 tunnel with the NACA63 profile

For the presented tests, the inlet Mach number was varied between 0.35 and 0.7 with an inlet flow temperature of 303K. This gave inlet chord-based Reynolds numbers from 540 000 to 1 160 000. The inlet free-stream turbulence level was varied by means of grids, designed in accordance with well-known theories (6), (7). The turbulence quantities were measured with a hot wire probe at the position of the leading edge, but with the airfoil removed. The inlet flow conditions and the turbulence level in the tunnel can be seen in Table 1. With mesh-sizes of 10mm and bar-sizes of 2.1mm (grid1) resp. 12.3mm and 1.5mm (grid2), the grids were positioned 10 to 13 mesh-sizes upstream the airfoil leading edge. For the presented cases, it can be seen that the turbulence intensity is varying with inlet Mach number (Table 1).

The integral length scale of the turbulence (Table 2) was determined by first calculating the autocorrelation in time by an fft analysis and thereafter integrating the autocorrelation function to the first zero-crossing. This results in the integral time scale which is multiplied with the mean velocity to give the length scale in the stream-wise direction. For Grid 1 at M_i=0.35 and Grid 2 at M_i=0.5 (Table 2) the integral length scale seems suspiciously high. This depends on the fact that the integral time scale is an order of ten times larger than for the other cases. The reason for this has not yet been understood and since the turbulence measurements and length scale calculations have been performed identically for all cases, a possible error could not be determined. Thus, these values are kept in the table.

<table>
<tr><td colspan="5" align="center">Table 1:
Turbulence intensities in VM100</td><td colspan="4" align="center">Table 2:
Turbulence length scales in VM100</td></tr>
<tr><td colspan="5">Tu [%]</td><td colspan="4">Λ [mm]</td></tr>
<tr><td>M_1 [-]</td><td>Re [-]</td><td>No grid</td><td>Grid 1</td><td>Grid 2</td><td>M_1 [-]</td><td>No grid</td><td>Grid 1</td><td>Grid 2</td></tr>
<tr><td>0.35</td><td>540 000</td><td>0.2</td><td>3.7</td><td>2.9</td><td>0.35</td><td>3</td><td>105</td><td>52</td></tr>
<tr><td>0.5</td><td>780 000</td><td>0.3</td><td></td><td>4.9</td><td>0.5</td><td>40</td><td></td><td>379</td></tr>
<tr><td>0.7</td><td>1 160 000</td><td>0.6</td><td></td><td></td><td>0.7</td><td>42</td><td></td><td></td></tr>
</table>

A symmetrical 2D airfoil NACA63A006 with a chord length of 80mm and a span of 300mm was used for the tests. The model was instrumented with 66 hot film gauges, of type Senflex, along the mid span chord of the entire profile surface, as well as 13 pressure tappings on each side of the profile. In this way, both pressure and hot film data could be achieved at the same time. The angle of attack has been varied between -6° and 6°. The airfoil is clamped through the turnable windows and thanks to the additional length of the model, pressure tubings and hot film leads can easily be taken out through the window. The model has also been used to verify the two-dimensionality of the flow by being traversed in the spanwise direction. It was found that even for an inlet Mach number of 0.7 and an incidence angle of 6° the flow is 2D in the midspan of the test section (8).

The pressure measurements were performed with a PSI 8400 system equipped with pressure ranges of ±35kPa and ±100kPa + atmospheric. A barometer was used to record the atmospheric pressure and the surface and sidewall Mach numbers (M_x) were calculated according to:

$$M_x = \sqrt{\left(\left(\frac{P_0}{P_x}\right)^{\frac{\kappa-1}{\kappa}} - 1\right) * \frac{2}{\kappa-1}}$$

[Eq. 1]

where x denotes the position of the static pressure on the wall and blade surface, P_x is the pressure at x and P_0 is the inlet stagnation pressure. Since the working media was air at low temperature, κ (=C_p/C_v) was taken as 1.4.

The equipment used for the hot wire measurements was a 16 channel constant temperature anemometer system from TSI (IFA300). The hot film voltage signals were sampled at 50kHz, low-pass filtered at 20kHz and thereafter recorded on the PC hard disk. The voltage signals (e) were normalised with the mean voltage at zero flow conditions (e_{0mean}).

The information concerning the transition process was given by the skew values of the hot film signal, defined according to (9):

$$e_{skew} = \frac{\frac{1}{n}\sum_{1}^{n}(e - e_{mean})^3}{e_{rms}^3}$$

[Eq. 2]

From the skew can be determined the transition position, since the signal is zero during laminar flow, thereafter increasing at transition onset, until a positive maximum at an intermittency (γ) of about 0.25, decreasing through zero at $\gamma{\approx}0.5$ to a negative minimum at $\gamma{\approx}0.75$ and returning to zero when the intermittency is 1, i.e. fully turbulent flow (10).

2.1 Accuracy

The pressure ranges used for the presented tests have absolute pressure accuracies of ±20.9Pa (±35kPa range) and ±51.3Pa (±100kPa range). This gives a Mach number error of 0.23% of the measured value at M=0.5.

The total hot film anemometer accuracy is ±0.01V (±0.22% Full Scale), which includes a possible error of 10°C in temperature drift and temperature compensation in the AD card and the anemometer.

Combining the pressure and anemometer accuracy, the error of the measured free-stream turbulence level is considered to be ±0.1%, after in-situ calibration of the hot wire probe.

The total thickness of the hot film sheet and the adhesive is 0.16mm, which adds a total 0.32mm to the profile thickness (6.7% of maximum blade thickness). The leading edge radius increases from 0.2mm to 0.36mm.

The total accuracy in angle setting, is considered to be $\pm0.2°$, based on the accuracy of the angle measurement device ($\pm0.1°$) and the manual adjustment of the zero angle.

Due to the finite size of the hot film sensors, the behaviour of the boundary layer can only be studied for discrete points with a distance of about 3% chord. If the transition onset is determined to be between the last sensor that shows laminar flow and the first sensor that shows transitional flow, the transition onset is given with an accuracy of ±1.5% chord. The same is valid for the transition end, given by the last sensor showing transitional flow and the first sensor showing fully turbulent flow. For the higher free-stream turbulence cases, the fluctuations are higher also in the laminar boundary layer, making the detection of the transition onset somewhat less distinct. Thus the accuracy for these cases is considered to be ±3% chord.

3 NUMERICS

The numerical study was carried out by solving the Reynolds-averaged Navier-Stokes equations with a three-dimensional compressible flow solver. The algorithm is based on a cell-centred finite volume approximation using structured multiblock grids. The numerical inviscid flux is computed via a third-order-accurate upwind-biased scheme, and the compact central scheme is applied to the viscous terms. A three-stage explicit Runge-Kutta scheme is used for time integration to steady state. The start and the end of transition were predicted with the correlations of Abu-Ghannam and Shaw (11), and the turbulent closure is provided by the Baldwin-Lomax model (12). In the AGS model of transition onset, the turbulence intensity is based on the mean value of the initial upstream level and the local one at the start of transition. Computationally, assuming no turbulence decay in the flow path, the evolution of the Tu level around the airfoil can be estimated from the measured inlet conditions and the computed local free-stream velocity. More details of the present numerical method can be found in other publications (13), (14).

The computational grid is a five block H-C-H type mesh which contains 18941 nodes. A C-mesh is used around the airfoil to ensure a good grid quality in all the cases investigated, which allows approximately at least 20 nodes to resolve the boundary layer. The geometry of the airfoil was modified, in accordance with the measurements, to take the thickness of the hot film into account. Notice that the experiments are performed with the airfoil surface in thermal equilibrium with the surroundings, but are all conducted with the air at near ambient conditions. Two typical cases are calculated with the wall either in adiabatic condition or at a temperature of 288 K (based on the measurements). The results show that the computed transition onset and end are almost identical even for the highest Mach-number case, but they could be more clearly identified from the computed heat transfer than from the skin friction coefficient. In all the computations, therefore, the constant wall temperature is used. The transition onset is defined as the location with the minimum of the computed local heat transfer, and the end of transition is at the maximum thereafter. A fully turbulent calculation is assumed on the suction surface if the transition onset is detected experimentally at the leading edge. Otherwise, a transition calculation is performed.

It should be noticed that the definitions of the transition start are not the same for the numerical and the experimental methods. The skew value, used by the experiments, start increasing immediately when the first turbulent spots are seen, while the heat transfer, used by the numerics, reacts slower. Similarly, the last fluctuations towards the end of the transition process will be detected by the skew but not by the heat transfer. A comparison of transition detection between heat transfer and skew measurements has earlier been performed by Svensdotter and Fransson (15). It is thus expected that the experiments will detect a somewhat earlier transition and a longer process than will the numerical predictions.

4 RESULTS

The Mach number distribution for inlet Mach numbers 0.35, 0.5 and 0.7, at the low free-stream turbulence level and for a few of the investigated incidence angles, can be found in Figure 2 (a-c). It can be seen that the Mach number increases for increasing incidence on the forward part of the suction side, while decreasing on the pressure side. For inlet Mach number 0.35, the Mach number distribution shows that the leading edge separation seems to start between 5° and 6° incidence, while flow visualisation shows a separation bubble also at 5° angle of attack. This is also confirmed by the numerical calculation. The separation bubble is in general increasing in length with increasing incidence and increasing Mach number. At M=0.7 a separation is detected already at 3° incidence and, for the higher angles of attack, the boundary layer is separated over the entire blade chord until the shock.

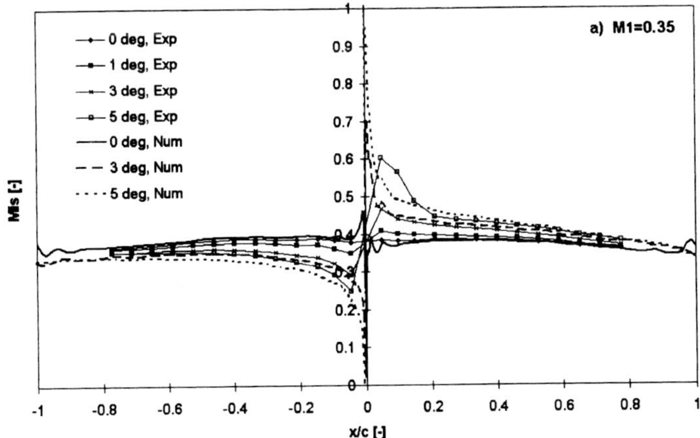

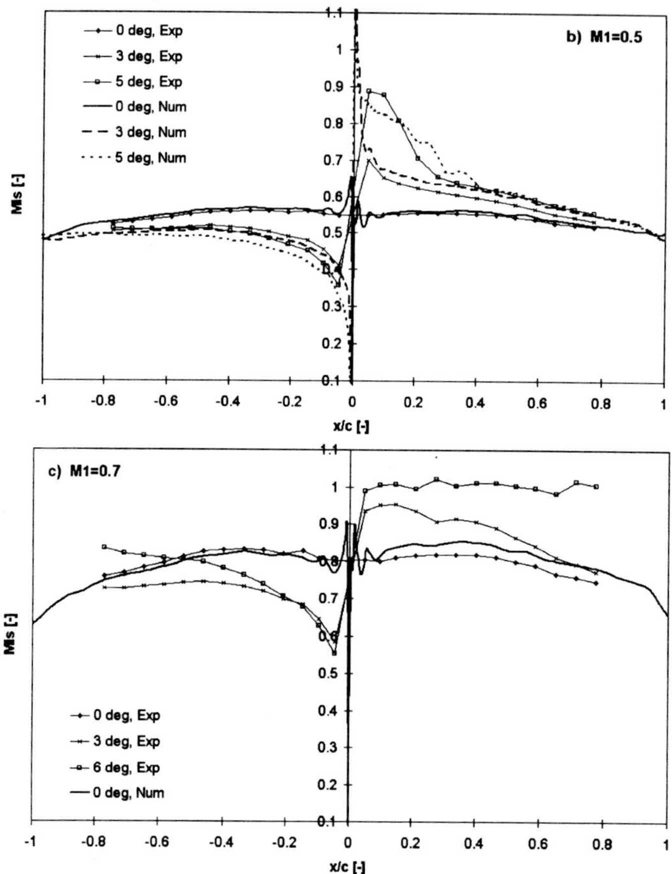

Figure 2: Mach number distribution for inlet Mach numbers of 0.35, 0.5 and 0.7

For M_1=0.7 a shock is detected at about 60% chord for 5° incidence, moving slightly towards the trailing edge for increasing incidence. The flow visualisation also shows a short laminar separation bubble at the leading edge for 2.5° angle of attack at all Mach numbers (not detected by the pressure measurements). In Figure 2 the experimental measurements are also compared with a number of the numerical results, and good agreements are generally obtained.

The transition onset and end, for inlet Mach number 0.35 and a free-stream turbulence of 0.2% at varying angles of attack, are depicted in Figure 3. It can be seen that the suction side (upper side at positive incidence) transition onset and end move rapidly towards the leading edge with increasing incidence and from 3° the boundary layer is turbulent from the leading edge. The length of the transition zone is decreasing with increasing incidence, although the behaviour does not seem to be linear. On the pressure side (lower side at positive incidence) the transition onset and end moves towards the trailing edge for increasing incidence and the length of the transition zone is increasing. The process is slower than on the suction side. For angles above 1.5° the boundary layer is still transitional at the trailing edge. This is why no

C557/159 © IMechE 1999

transition end is marked in the figure. For zero degrees incidence the transition process should be identical on both sides of the blade, however this is not the case. It was found, after measuring the profile, that the blade is not perfectly symmetric at the leading edge (16). Another possible reason is the error in the angle adjustment, which was estimated to be ±0.2°.

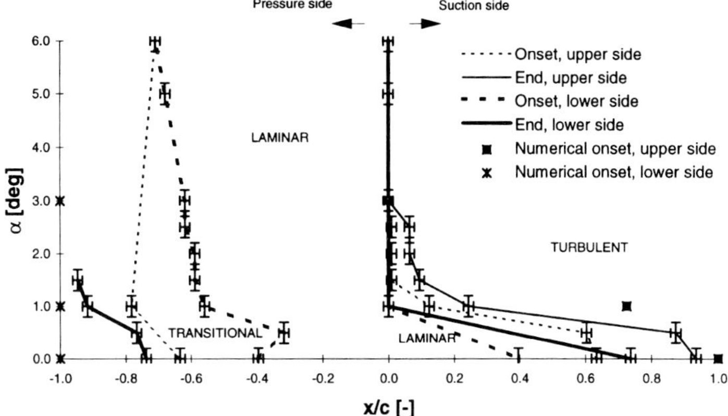

Figure 3: Transition regions for M=0.35 and Tu=0.2%

The numerical results depicted in Figure 3 do not show a very good agreement with the experimental results. This is discussed in the last paragraph of this section.

Increasing the Mach number to 0.5 (Figure 4) moves the transition process towards the leading edge on both suction and pressure side. The trends concerning increasing incidence are the same with the transition process moving towards the leading edge on the suction side and towards the trailing edge on the pressure side. At zero incidence, the transition onset has moved from about 60% chord for M=0.35 to about 40% chord at M=0.5. The transition end also moves towards the leading edge, but not as much. At M=0.35, the transition end is detected at about 95% chord and at M=0.5 the transition end has moved to 85% chord. Also for this Mach number can be noted the asymmetry in the transition onset and end positions on the two sides of the airfoil.

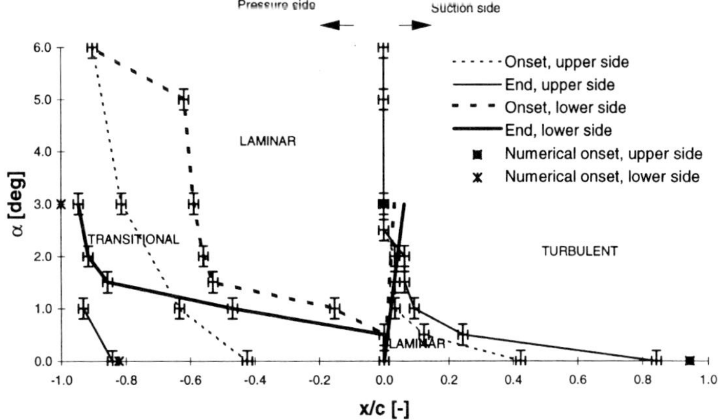

Figure 4: Transition regions for M=0.5 and Tu=0.3%

At M=0.7 (Figure 5) the transition onset is detected very close to or at the leading edge on both suction side, for all incidences, and pressure side, for incidences below 2°. The transition end on the suction side is moving towards the leading edge, compared with the case for M=0.5, for the lowest incidences (0° - 1°) but staying constant for the higher incidences. As was mentioned earlier, from 3° angle of attack a separation is forming and at higher incidences the suction side boundary layer is entirely separated and experiences a shock at about 60% chord. The pressure side transition onset positions for the higher incidences are not much changed compared with those at M=0.5, while the transition ends have moved towards the leading edge. This shows that an increasing Mach number will decrease the length of the transition zone.

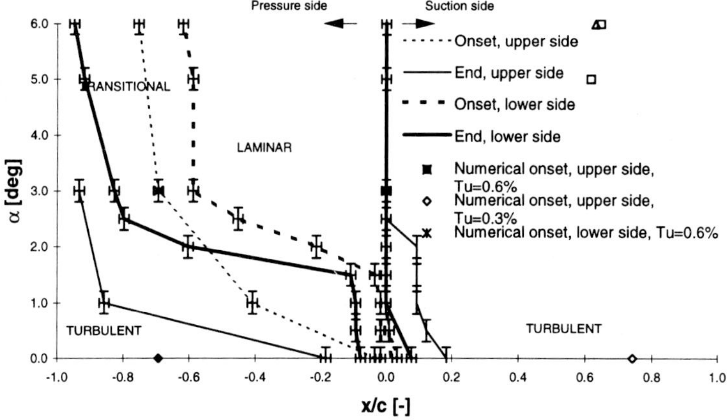

Figure 5: Transition regions for M=0.7 and Tu=0.6%

The leading edge laminar separation bubble, seen for the above three cases at 2.5° incidence, showed a reattachment with transitional flow at M=0.35 and a turbulent reattachment for the higher Mach numbers.

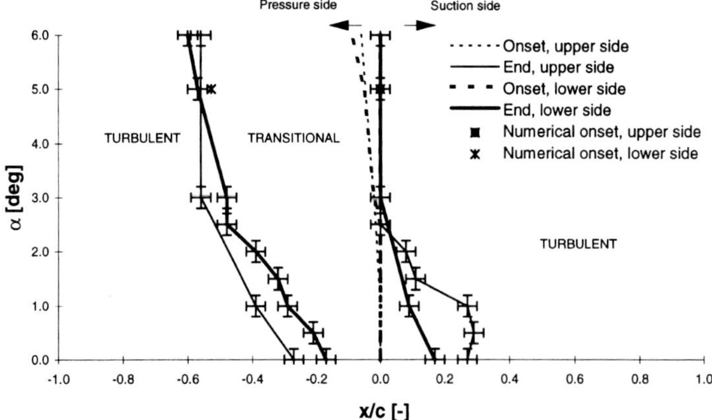

Figure 6: Transition regions for M=0.35 and Tu=2.9%

Increasing the turbulence level (Figure 6) to 2.9% from 0.2% (Figure 2) is shown to significantly move the transition onset and end towards the leading edge for low incidence angles. Fully turbulent flow at the entire suction side is reached at 2.5° incidence and the laminar separation bubble that was detected for the low free-stream turbulence case is not seen here. At the suction side the transition onset occurs at the leading edge for all incidences investigated. The transition process is developing faster at this turbulence level and even for small angles of attack, most of the blade surface experiences fully turbulent flow. The pressure side transition onset is also set at the leading edge until 3°incidence, where the transition onset starts moving slightly towards the trailing edge. The transition end is seen at about 60% chord for the highest incidence, which means that for this high turbulence level fully turbulent flow is developed at the aft part of the blade for all incidence angles.

Further increasing the turbulence intensity to 3.7% (not shown) does not significantly change the transition onset. On both pressure and suction side the transition end seems to occur somewhat closer to the leading edge for small incidence angles. Between 2° and 3° incidence, the hot films detect an instability at the suction side leading edge that does not seem to be associated with the transition process.

For a Mach number of 0.5 and a turbulence level of 4.9% (Figure 7) the transition onset is detected at the leading edge for all incidence angles investigated. This test case is performed with the same grid as for M=0.35 and the turbulence level of 2.9%, which means that the change in turbulence is associated only with the change in velocity. For this flow condition only a small fraction of the suction surface is transitional, from 10% chord at zero incidence decreasing until 1.5° incidence where the flow turns entirely turbulent. At the pressure side the fully turbulent flow starts at about 10% chord for zero incidence and is delayed to about 40% chord at maximum (6°) incidence. Comparing the M=0.35 flow case with the M=0.5 case, it appears that the transition process is decreasing in length for increasing Mach number although the effect can not be separated neither from the effect of the increased turbulence intensity nor from the Reynolds number effect.

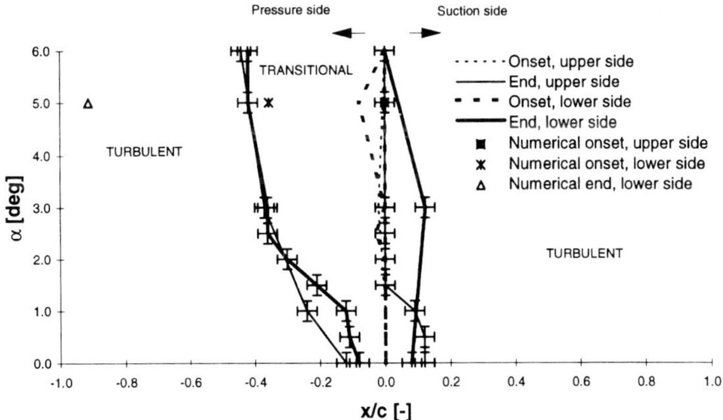

Figure 7: Transition regions for M=0.5 and Tu=4.9%

It is seen from Figure 3 to Figure 7 that some basic characteristics of the transition process on this airfoil, as discussed above, are also confirmed by the numerical results. For instance, the transition onset moves downstream with increasing incidence angle on pressure side (Figure 4) but moves upstream on suction side (Figure 3). However, the numerical transition generally starts and ends much later than the measured transition. This can be partly explained by the fact that the transition models used in the calculations are not accurate to predict the transition in strong pressure gradient flow conditions. Also the lowest free-stream turbulence levels are not in the recommended range of the AGS model.

5 CONCLUSIONS

Experimental and numerical investigations have been performed on a NACA63 isolated airfoil for Mach numbers up to 0.7 and turbulence levels up to 4.9% (for M=0.5). The Mach number has not been changed independently of the Reynolds number, why we are looking at a combined effect.

As expected from other studies, although performed for incompressible velocities (4), (5), (17), an increased incidence moves the suction side transition onset rapidly towards the leading edge. Also the end of the transition zone shifts towards the leading edge faster than the onset. This means that the length of the transition process is decreasing for increasing incidence.

The pressure side transition onset and end moves towards the trailing edge for increasing incidence and the transition process slows down, i.e. the length of the process is increasing.

It can clearly be seen how the transition process for the investigated Mach numbers is affected by the change in free-stream turbulence level. Both transition onset and end move quickly towards the leading edge and the length of the process decreases. A laminar-separated transition, detected for 2.5° incidence at the low free-stream turbulence level is not seen for the increased turbulence. Other researchers (18), although investigating lower Mach numbers, have also mentioned that an increasing free-stream turbulence level decreases the length of the leading edge separation bubble. There are also investigations mentioned (19) where an increasing Mach number is not shown to decrease the transition length.

Due to the fact that the turbulence level changed with varying Mach numbers during the experimental investigations, it could not be determined if it was the Mach number or the turbulence intensity that was of most importance for moving the transition process upstream. However, the numerical results indicates that the Mach number effect on transition onset is significant at a low turbulence level. For instance, increasing Mach number from 0.5 to 0.7 moves the transition onset approximately 13% and 20% of the chord distance upstream on the suction and pressure surfaces, respectively.

The non-symmetry discovered concerning the transition onset position was found to be caused by the fact that the profile was not perfectly symmetric at the leading edge and that the angle adjustment had a possible error of ±0.2°.

6 ACKNOWLEDGEMENTS

A large part of the experimental work was performed as a Diploma Thesis (16) by Mr Ruggero Gandolfi from the Department of Aeronautics and Aerospace Engineering, Polytecnico Di Torino, Italy, with financial support from the Erasmus student exchange program.

7 REFERENCES

(1) Abbott, I.; von Doenhoff, A.; 1959, "Theory of Wing Sections Including a Summary of Airfoil Data", *Dover Publications Inc., New York, 486-60586-8*

(2) Bullivant, K.; 1941, "Tests of the NACA 0025 and 0035 airfoils in the Full-Scale Wind Tunnel", *NACA Report No. 708*

(3) Goett, H.; Bullivant, K.; 1939, "Tests of NACA 0009, 0012 and 0018 Airfoils in the Full-Scale Wind Tunnel", *ASME 96-GT-113*

(4) Pucher, P.; Göhl, R.; 1987, "Experimental Investigation of Boundary Layer Separation With Heated Thin-Film Sensors", *Journal of Turbomachinery, April 1987, Vol. 109, pp. 303-309*

(5) Mateer, G.; Monson, D.; Menter, F.; 1996, "Skin-Friction Measurements and Calculations on a Lifting Airfoil", *AIAA Journal, Vol. 34. No. 2, February 1996, pp. 231-236*

(6) Roach, P.; 1987, "The Generation of Nearly Isotropic Turbulence by Means of Grids", *Journal of Heat and Fluid Flow, Vol. 8 No. 2, June 1987*

(7) Baines, W.; Peterson, E.; 1951, "An Investigation of Flow Through Screens", *Transactions of the ASME, July 1951, pp. 467-480*

(8) Baunot, A.; 1997, "Pressure Measurements on a NACA63A006 Single Airfoil in Transonic Flow", *Internal report KTH/HPT 97-24*

(9) Halstead, D.; Wisler, D.; Okiishi, T.; Walker, G.; Hodson, H.; Shin, H., "Boundary Layer Development in Axial Compressors and Turbines Part 1 of 4: Composite Picture", *ASME 95-GT-461*

(10) Halstead, D.; Wisler, D.; Okiishi, T.; Walker, G.; Hodson, H.; Shin, H., "Boundary Layer Development in Axial Compressors and Turbines Part 2 of 4: Compressors", *ASME 95-GT-462*

(11) Abu-Ghannam, B.J.; Shaw, R.; 1980, "Natural Transition of boundary Layer - The Effect of Turbulence, Pressure Gradient, and Flow History", *Journal of Mechanical Engineering Science, Vol. 22, No. 5, pp. 213-228.*

(12) Baldwin, B.S.; Lomax, H.; 1978, "Thin Layer Approximation and Algebraic Model for separated Turbulent Flow", *AIAA Paper 78-257*

(13) Hu, J.; Fransson, T.; 1997, "Transition Predictions for Turbomachinery Flows Using Navier-Stokes Solver and Experimental Correlation", *AIAA 97-2230.*

(14) Hu, J.; Fransson, T.; 1998, "On the Application of Transition Correlations in Turbomachinery Flow Calculation", *ASME 98-GT-460.*

(15) Svensdotter, S.; Fransson, T.; 1998, "Hot Film and Liquid Crystal Measurements on a Turbine Airfoil at Varying Re and Turbulence Intensities", *AIAA 98-3453, 34:th Joint Propulsion Conference, 1998, Cleveland*

(16) Gandolfi, R.; 1998
"Transonic Flow Over a NACA63A006 Isolated Airfoil, Flow Field and Transition Studies"
Diploma thesis No 533, Chair of Heat and Power Technology, The Royal Institute of Technology, Stockholm, Sweden

(17) Chen, H.; Platzer, M.; Cebeci, T.; 1995, "Analysis of Airfoil at Low Reynolds Numbers"
Proceedings of the 7:th International Symposium on Unsteady Aerodynamics and Aeroelasticity of Turbomachines, pp. 287 - 302, Tanida, Y and Namba, M editors

(18) Walraevens, R.; Cumpsty, N.; 1995, "Leading Edge Separation Bubbles on Turbomachine Blades", *Journal of Turbomachinery, January 1995, Vol. 117, pp 115-125*

(19) Boyle, R.; Simon, F.; 1998, "Mach Number Effects on Turbine Blade Transition Length Prediction", *ASME 98-GT-367*

C557/107/99

Experiments on by-pass boundary layer transition with several turbulence length scales

P JONAS, O MAZUR and **V URUBA**
Institute of Thermomechanics AS CR, Prague, Czech Republic

The role of free-stream turbulence on the by-pass transition of a flat plate boundary layer was experimentally investigated. Keeping the mean value of flow velocity constant at 5m/s and the intensity of longitudinal velocity fluctuation on 3% at the leading edge of the flat plate, the dissipation length scale was changed from 2.3 up to 34.5 mm. It has been found that the onset of the last stage of the by-pass transition is postponed and probably extends further downstream in a boundary layer perturbed by fine-grain free-stream turbulence.

INTRODUCTION

The boundary layer transition problem has been subjected to intensive exploration since Burgers and Hegge Zijnen´s time (1924), however, the problem has not been fully understood yet. The published observations as well as accurate prediction methods which still lack accuracy, e.g. Savill (**1**), show that the phenomenon depends on many influences which have not been fully considered so far. One of the underestimated factors is the length scale of the incoming turbulence. The studies of the by-pass transition investigate mainly the effect of the turbulence level at the leading edge of the flat plate. However, there are reasons to expect that the length scale of the incoming flow turbulence will control the transition as well.

This effect follows mainly from the fundamental features of the turbulence dynamics. From dimensional estimates it follows, e.g. Tennekes and Lumley (**2**), that two turbulent streams arriving at the leading edge of a rigid surface with the same turbulent velocity scale but with different length scales will excite the boundary layer on the surface in different ways because the „larger" turbulent disturbances will decay less rapidly then the „smaller" ones.

A further reason to consider the effect of the length scale on by-pass transition follows from the results of investigations of the grid turbulence passing over a wall moving at the mean velocity of the outer stream measured by Uzkan and Reynolds (**3**), Thomas and

Hancock (**4**) and Hunt and Graham (**5**). It has been observed that both the influence of the wall constraint and of the viscous boundary conditions depend on the length scale. These effects change the properties of turbulence in the region where transition occurs.

Motivation for studying the role of the length scale in the by-pass transition can be found in the research on turbulent boundary layers perturbed by free-stream turbulence. Hancock (**6**), Hancock and Bradshaw (**7**) and the present author (**8**) proved that the effects of free stream turbulence on turbulent boundary layer characteristics depend both on the intensity and the dissipation length parameter of turbulent disturbances. As the turbulent boundary layer is the termination of the transition process, the process itself must depend on the length scale of turbulence.

Preliminary confirmation of the effect of the length scale of the free-stream turbulence on flat plate boundary layer transition is presented in (**9**). The aim of the presented experiments was to contribute to better understanding the physics of by-pass transition. Experimental data were acquired for validation of the transition prediction methods for boundary conditions prescribed by the COST/ERCOFTAC Special Interest Group on Transition for the Test Case T3A+ (turbulence intensity Iu = 3% and several turbulent length scales Le = ~ in the leading edge plane at constant mean velocity 5 m/s). The first part of the results obtained is presented in this paper.

EXPERIMENTAL APPARATUS AND PROCEDURE

The investigation has been carried out in the closed circuit wind tunnel (0.9×0.5) m^2 of the Institute of Thermomechanics of the Academy of Sciences of the Czech Republic. The investigated boundary layer was developing on an aerodynamically smooth plate 2.75 m long and 0.9 m wide, made from a laminated wood-chip board 25 mm thick. It has a thin cylindrical leading edge designed by Kosorygin et al. (**10**) and it is equipped with devices suppressing the circulation (flap and screen). The entire arrangement is illustrated in Fig. 1 where also the co-ordinate system (x,y,z) is introduced.

The unperturbed turbulence level at the entrance of the test section is about 0.3%. The increased turbulence level is produced by means of turbulence generators - plane grids/screens of various geometries. Each grid is placed across the flow at proper distance $x = x_G$ upstream of the leading edge of the plate to achieve nearly homogeneous turbulence, close to isotropic, with the intensity of fluctuations of the longitudinal velocity component Iu = 3 % and different length scales in the leading edge plane (x = 0). Table 1 summarises the most important parameters of turbulence generators and the important features of turbulence produced in the leading edge plane.

The measurements were carried out mostly by means of two single hot wire probes

Table 1 Grid turbulence generators

Grid	mesh [mm]	diameter [mm]	porosity [%]	x_G [m]	Iu_e [%]	Iv_e/Iu_e [1]	$\left(\overline{uv}\right)_e / \overline{U}_e^2$	L_e [mm]
GT 1	20	3	72	-0.445	3.0	1.02	$-2.0 \cdot 10^{-5}$	6.8
GT 3a	40	6	72	-1.087	3.0	1.05	$-2.1 \cdot 10^{-5}$	16.1
GT 3b	40	6	72	-1.076	3.0	1.05	$-2.1 \cdot 10^{-5}$	15.7
GT 5	35	10	51	-1.350	3.0	1.03	$-1.6 \cdot 10^{-5}$	34.5
GT 8	5.75	1.65	51	-0.194	3.0	~ 1	$\sim 10^{-5}$	3.9
GT 9	5	1	64	-0.123	3.0	~ 1	$\sim 10^{-5}$	2.3

working in constant temperature anemometer mode. One is in a fixed position in the free stream, the second „profile probe", is supported by the x-y traversing system controlled by a personal computer. If necessary, the profile probe is replaced by a X-wire probe.

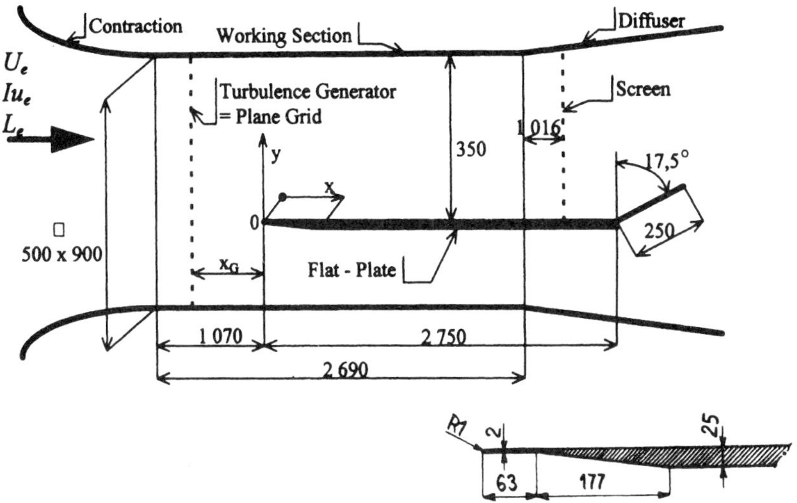

Fig. 1 Sketch of test section

The measuring system consists of anemometer bridges DANTEC 55M01, signal conditioners DANTEC 56N20 and a personal computer Pentium EISA 133 MHz. The computer is equipped with the Data Acquisition System of National Instruments (EISA A-2000) controlled by the software LabVIEW. As the prescribed value of the mean free-stream velocity is $U_e = 5$ m/s (see definition of Test Case T3A+), it is quite satisfactory to choose the sampling rate 25 kHz and from 30 up to 40 seconds long records of 12 bit samples; variable offset voltage is exercised.

With regard to the required low velocity, it has been necessary to develop a special calibration set-up with a metering nozzle (amplification rate 10) to reach a satisfactory accuracy of hot wire calibrations in the region from about 0.3 m/s to up to 10 m/s.

The distance y from the wall to the hot wire of the „profile probe" is measured with an accurate cathetometer. The uncertainty of the measurement of successive shift is less then 0.01 mm but the possible error of reading the initial hot wire position y_1 might be up to $y_0=0.05$ mm.

All hot wire probes have been carefully calibrated after every two or three hours of operation. The heat transfer from the hot wire is expressed in customary manner using the generalised Collis& Williams approximation of the cooling law and the directional sensitivity formula after Hinze (**11**). On the basis of these formulas the samples of the thermoanemometer output voltage are converted into samples of momentary velocity. The ensembles of velocity samples have been analysed in a customary way to obtain statistical moments, spectra and other statistical characteristics of the longitudinal component of the stream velocity. However, the results presented here are based only on the measured mean velocity profiles. The effect of the wall proximity on the hot wire readings is corrected by a procedure developed in (12).

Accuracy estimates are derived from the spread of repeated measurements. In the course of correcting the effect of the wall proximity the error of reading the initial hot wire distance from the wall is rectified (+/- 0.01 mm) and the derivative $\left(d\overline{U}/dy\right)_w$ at the wall is determined with the relative error of about 2 percent in average. The same estimate applies to the skin-friction coefficient C_f. The estimated error of the mean velocity $\overline{U}$ is less then 0.3 percent, while the error of turbulence intensities Iu and Iv is about 3 percent. The uncertainties of Taylor microscale λ_2 and dissipation length parameter L_e are estimated to be up to 6 percent. These estimates produce possible errors in the determination of Reynolds number of the free-stream turbulence Re_T (up to 10%), shape factor H_{12} (up to 2%) and intermittency factor γ (up to 5%).

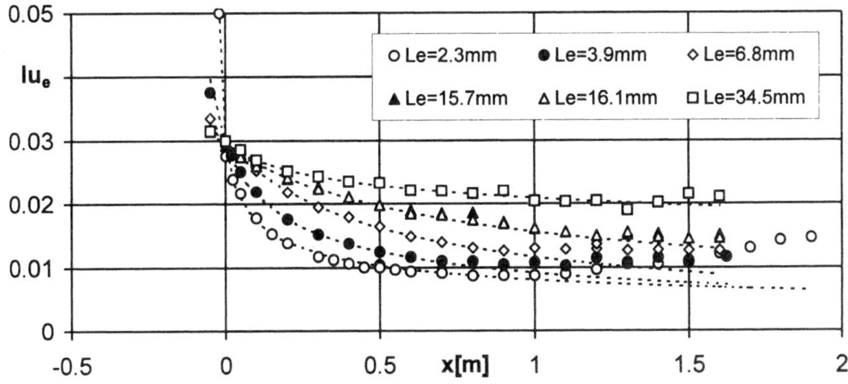

Fig. 2 Free stream turbulence intensity distributions

PROPERTIES OF THE OUTER FLOW

It has been required that the mean value of the outer flow velocity be $\overline{U}_e = 5$ m/s. Throughout the whole experimental program this requirement has been fulfilled with differences among individual profiles of less than 2 percent but the variation related to a single profile is less then 0.3 percent. It should be noted that only 1.6 m of the total length of the plate lies in the test section with a cross section of 0.9 m x 0.5 m. There, $\overline{U}_e$ is constant (in the limits of measuring accuracy ±0.5%) over the whole surface. However, the remaining part of the plate, the (longitudinal co-ordinate x > 1.6 m), is in a decelerated flow, and so the local mean flow velocity decreases down to $0{,}96\ \overline{U}_e$ in the distance x = 2 m from the leading edge. This might slightly accelerate the transition process, which is not terminated sooner.

The external flow turbulence is characterised by the intensities

$$Iu_e = \sqrt{\left(\overline{u^2}\right)_e}\Big/\overline{U}_e \quad \text{and} \quad Iv_e = \sqrt{\left(\overline{v^2}\right)_e}\Big/\overline{U}_e \tag{1}$$

of longitudinal and lateral components of velocity fluctuations u and v. In the Fig. 2, distributions of Iu_e along the plate downstream of the grid-generators are shown. It is obvious that in sections $x > 0.8$ m convection and production of turbulence become important and in fine grain turbulence (GT 8 and GT 9) turbulent fluctuations begin to amplify.

Two length scales are introduced in this paper, namely the dissipation length parameter L_e taken after Hancock and Bradshaw (7)

$$L_e = -\left(\overline{u^2}\right)_e^{3/2} \left/ \ \overline{U}_e \frac{d\left(\overline{u^2}\right)_e}{dx} \right. \tag{2}$$

and the lateral dissipation length scale (Taylor microscale) λ_2 derived from the auto-correlation function by applying Taylor's hypothesis and with the assumption on satisfactory isotropy (see Table 1).

The dissipation length parameter L_e is computed by means of an empirically determined decay law (a power-law fit of $\left(\overline{u^2}\right)_e$ as a linear function of $(x - x_0)^m$; x_0 and m are statistical estimates). It is obvious that the length L_e loses its physical meaning if the rate of turbulent energy dissipation would not dominate over convection and production of turbulent energy. This **problem** arises downstream of the grids GT 8 and GT 9. Hence, in analysing the results, the lateral dissipation length scale λ_2 is mostly considered, and Hancock's parameter L_e is applied (Table 1) to describe turbulence only in the vicinity of the leading edge of the plate ($x=0$).

The Reynolds number of outer stream turbulence defined as

$$Re_T = \lambda_2 \sqrt{\left(\overline{u^2}\right)_e} \left/ \nu \right. \tag{3}$$

is plotted in Fig. 3 as a function of the streamwise distance from the leading edge. As apparent, the streamwise distribution and the value of Re_T depend on the grid turbulence and cover the range roughly from 15 to 60.

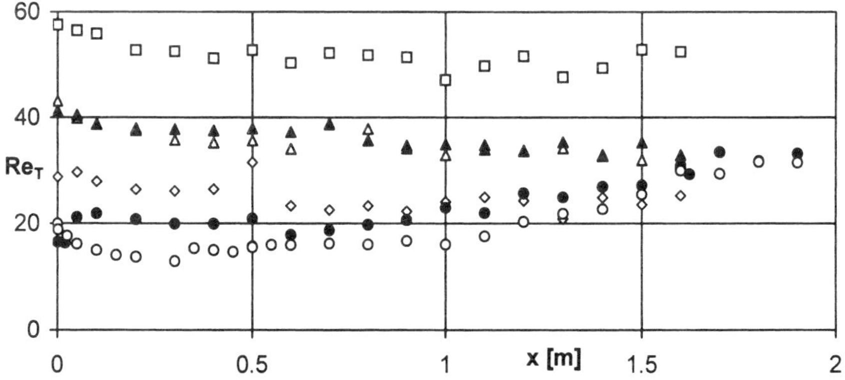

Fig. 3 Reynolds number of free stream turbulence; key as in Fig.2

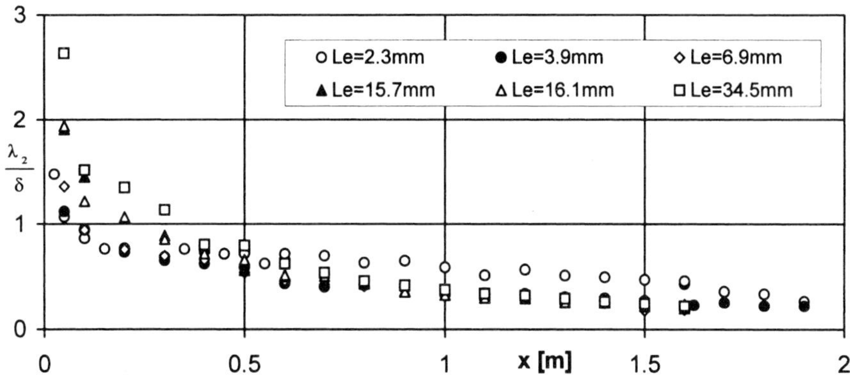

Fig. 4 Ratio of lateral dissipation length scale to boundary layer thickness

TRANSITIONAL BOUNDARY LAYERS

Downstream of each turbulence generator (see Table 1), an array of mean velocity profiles has been measured. The density of the investigated sections can be seen from the plot of the ratio of lateral dissipation length scale λ_2 to boundary layer thickness δ (half percent boundary layer thickness) as shown in Fig. 4. In the vicinity of the critical value of the momentum thickness Reynolds number, $Re_2=163$, the ratio varies from about 0.7 up to 1.5. The range of length scales in the leading edge plane x=0 is broader and is displayed in Table 1.

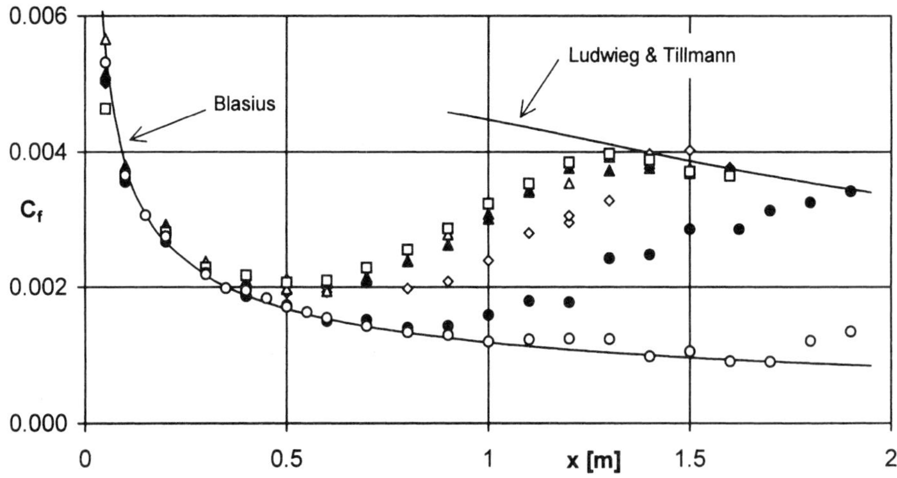

Fig. 5 Skin friction coefficient distributions along the plate; key as in Fig.4

The evolution of the boundary layer within the transition range is characterised first of all by the skin friction coefficient

$$C_f = 2\mu\left(d\overline{U}/dy\right)_w \Big/ \rho\overline{U}_e^2 \tag{4}$$

Fig. 5 shows the distributions of C_f along the plate. Though the intensity Iu_e in the leading edge plane is the same downstream of each turbulence generator, the onset of the last phase of transition interferes with the decreasing length scale of turbulence. With the finest grid-generator, the end of the transition process has not been achieved in the test section.

This observation is confirmed by the distributions of the shape factor H_{12}

$$H_{12} = \delta_1/\delta_2 \tag{5}$$

where δ_1 is the displacement thickness and δ_2 the momentum thickness calculated from the mean velocity profiles. In Fig. 6, the behaviour of H_{12} as a function of the momentum thickness Reynolds number is shown. Near the leading edge (x=0) all mean velocity profiles are very close to the one of Blasius (see also Fig. 5). Farther downstream the shapes of profiles change to turn ultimately into profiles typical of a turbulent boundary layer. This development is faster in outer-flow with large-scale turbulence. Because of the low value of the Reynolds number Re_2 the ultimate value of the shape factor is $H_{12} \cong 1.5$.

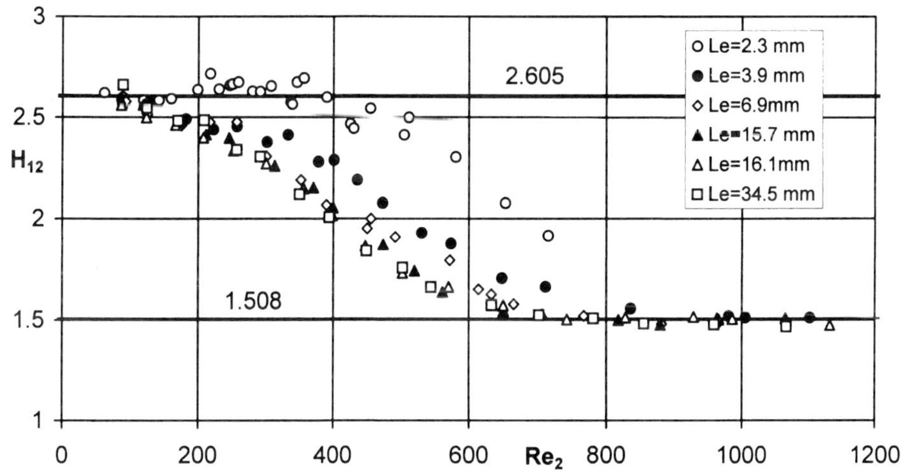

Fig. 6 Shape factor as a function of momentum thickness Reynolds number

The intermittency factor γ has been computed using the concept proposed by Emmons (**13**)

$$C_f(Re_2) = (1-\gamma)(C_f)_L + \gamma(C_f)_T \tag{6}$$

where $(C_f)_L$ is the value following from the Blasius solution at the same value of the momentum thickness Reynolds number Re_2 and $(C_f)_T$ is the skin friction coefficient according to the formula from Ludwieg and Tillmann (e.g. Schlichting (**14**)). The value

$H_{12}=1.5$ has been substituted into the Ludwieg and Tillmann formula. Again the plot of the intermittency factor distributions shown in Fig. 7 confirms the distinct effect of turbulence length scale on by-pass transition.

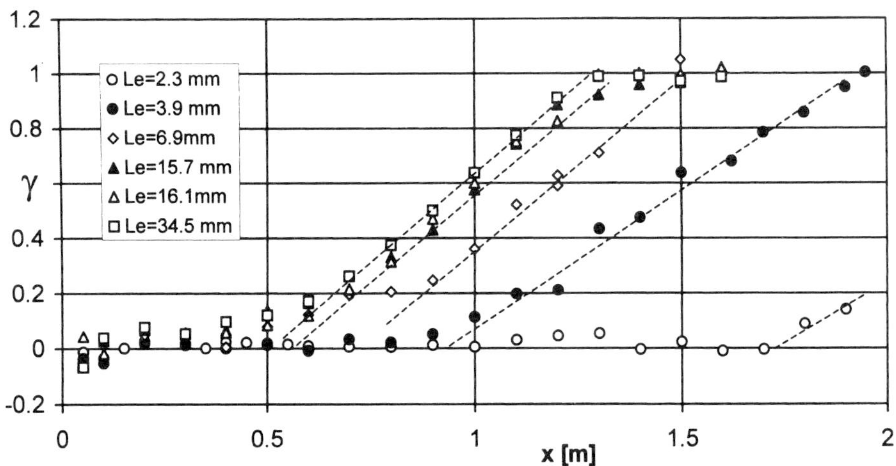

Fig. 7 Intermittency distributions downstream of various turbulence generators

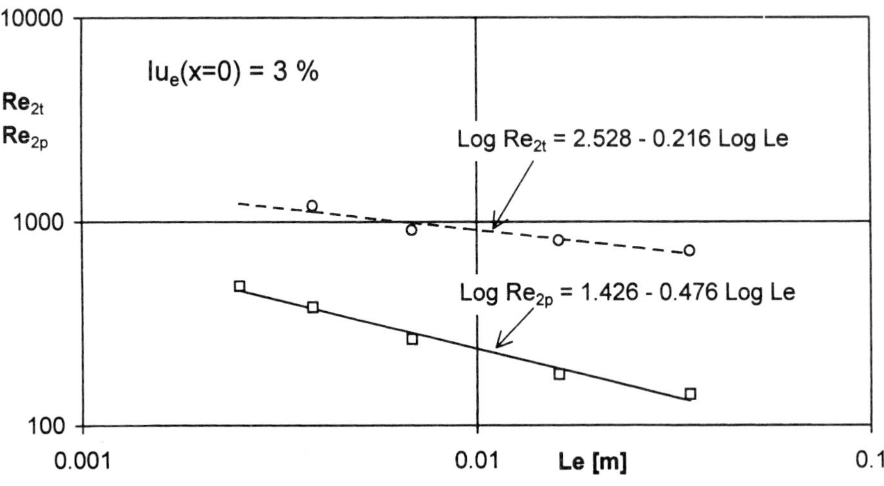

Fig. 8 The effect of the dissipation length parameter on the onset and the end of transition

CONCLUSIONS

The length scale of the turbulence in the incoming flow influences the onset and the end of laminar/ turbulent transition. This is confirmed by the distributions of integral properties as well as by the measured profiles of statistical characteristics.

In Fig. 8 it is shown that at the given intensity of streamwise turbulent fluctuations Iu_e=3% in the leading edge plane x=0 the onset of the last stage of the transition is postponed (higher values of Re_{2p}) and probably extends farther downstream (higher values of Re_{2t}; also see Fig. 7) in a boundary layer perturbed by a fine-grain free-stream turbulence than in a boundary layer perturbed by larger scale turbulence. An attempt has been made to describe these observations quantitatively, see Fig. 8.

ACKNOWLEDGEMENTS

The authors gratefully acknowledge the supports received from the Grant Agency of the Academy of Sciences (No: A2076602), from the Grant Agency of the Czech Republic (No:101/96/1696) and from the government via the Ministry of Education Youth and Physical Training of CR (No: OC.F1.10).

NOMENCLATURE

C_f = skin friction coefficient
H_{12} = shape factor
Iu = longitudinal (streamwise) component of turbulent velocity fluctuations intensity
Iv = lateral (cross stream) component of turbulent velocity fluctuations intensity
L_e = dissipation length parameter
$-\overline{uv}$ = turbulent shear stress
Re_2 = momentum thickness Reynolds number
Re_T = Reynolds number of outer stream turbulence
$\overline{U}_e$ = mean free-stream (bulk flow) velocity
u = fluctuating component of streamwise velocity
v = fluctuating component of cross stream velocity
x_G = distance of the grid upstream of the leading edge
δ_1 = displacement thickness
δ_2 = momentum thickness
γ = intermittency factor
λ_2 = lateral Taylor microscale
ρ = density
Subscripts
e = in the outer flow
L = from the Blasius solution of laminar boundary layer flow
p = at the onset of the last stage (region of the turbulent spots production) of transition
t = at the end of transition
T = from the Ludwieg and Tillmann skin friction coefficient formula
w = at the surface

REFERENCES

1   **Savill, A.M.** ERCOFTAC Special Interst Group on Transition/Retransition, *ERCOFTAC Bulletin*, 1991, **11**.

2   **Tennekes, H.** and **Lumley, J.L.** *A first course in turbulence*, (MIT Press, Cambridge-Massachusetts).

3   **Uzkan, T.** and **Reynolds, W.C.** A shear-free turbulent boundary layer, *J. Fluid Mech.*, 1967, **28**, 803.

4   **Thomas, N.H.** and **Hancock, P.E.** Grid turbulence near a moving wall, *J. Fluid Mech.*, 1977, **82**, 481.

5   **Hunt, J.C.R.** and **Graham, J.M.R.** Free-stream turbulence near plane boundaries, *J. Fluid Mech.*, 1978, **84**, 209.

6   **Hancock, P.E.** The effect of free stream turbulence on turbulent boundary layers, *PhD thesis*, 1981, (Imperial College, London).

7   **Hancock, P.E.** and **Bradshaw, P.** Turbulence structure of a boundary layer beneath a turbulent free stream, *J. Fluid Mech.*, 1989, **205**, 45.

8   **Jonáš, P.** On turbulent boundary layer perturbed by outer flow turbulence, *Power Eng. Journal*, 1992, No.3, 69.

9   **Jonáš, P.** On the role of the length scale in the by-pass transition, *ZAMM - Z. angew. Math. Mech.*, 1997, 77, S145

10   **Kosorygin, V.S.**, **Levchenko V.Ja.** and **Polyakov N.F.** On problem of the origin of waves in a laminar boundary layer (in Russian), *Preprint ITPM*, 1982, No. 12-82 (ITPM SO AN SSSR, Novosibirsk).

11   **Hinze, J.O.** Turbulence, 2nd edition, 1975 (McGraw-Hill, New York).

12   **Jonáš, P.**, **Mazur, O.** and **Uruba, V.** Statistical characteristics of the wall friction in a flat plate boundary layer through by-pass transition, *ZAMM - Z. angew. Math. Mech.*, 1998, in print.

13   **Emmons H.W.** The laminar turbulent transition in boundary layer, *J. Aero. Sci.*, 1951, **18**, 490.

14   **Schlichting H.** *Grenzschicht-Theorie*, 1965, (Verl. G. Braun, Karlsruhe).

C557/131/99

Modelling turbomachine-blade flows with non-linear eddy–viscosity models and second-moment closure

W-L CHEN and **M A LESCHZINER**
Department of Mechanical Engineering, UMIST, Manchester, UK

SYNOPSIS

The predictive characteristics of a non-linear eddy-viscosity model and a second-moment-closure are contrasted with those of a linear eddy-viscosity model by reference to flows around three blade profiles: a controlled diffusion (CD) compressor profile, a double-circular-arc (DCA) compressor profile, and a turbine profile, known as the "VKI blade". The flow over the VKI blade is at the design state, in which case all models return similar predictions which are in good agreement with the experimental data, provided transition is represented realistically. In contrast, the CD-blade flow is at off-design conditions, while that around the DCA blade contains a large separated region on the suction side. Hence, in both cases, the interaction between the suction-side boundary layer and the passage flow is strong, and turbulent transport is influential. In these circumstances, second-moment closure yields a distinctly superior representation of the flow, relative to the other models. While transition is not of decisive importance in the compressor-blade flows, it is observed that only the non-linear eddy-viscosity model captures this process, albeit only qualitatively correctly, by virtue of specific model features designed to render the turbulence production insensitive to irrotational straining associated with impingement at the leading edge.

INTRODUCTION

The flow in a turbomachine cascade operating close to its design condition is largely dominated by inviscid processes, whereas turbulent transport is influential, principally, in the relatively thin near-wall layers. In off-design conditions, however, the boundary layers on the blades and the end walls grow rapidly, separation can occur on both suction and pressure sides, and there arises a strong two-way interaction between the boundary layers and the main passage flow. In such circumstances, turbulence transport has a major effect on the primary flow properties that are of direct consequence to the operational parameters of the cascade.

Conventional eddy-viscosity models which employ linear relationships between the turbulent stresses and strains constitute the "industry standard" in computational schemes for turbomachine aerodynamics [1-3]. While these models are adequate in attached flows, in which the shear stress is the only dynamically active stress component, they often display serious defects in flows featuring strong curvature, impingement and separation - precisely the combination encountered in off-design conditions. These defects are rooted, principally, in the inability of the models to resolve turbulence anisotropy and the distinctly different contributions which different strain types make to the turbulence intensity through their interaction with specific stress components. The manifestations of these defects in blade flows include excessive levels of turbulent mixing in the leading-edge region and the suction-side boundary layer, a suppression of self-sustaining periodic vortex shedding behind blunt trailing edges, and a failure to predict bypass transition following leading-edge impingement.

The only practical modelling alternative to linear eddy-viscosity formulations is to adopt closures based on non-linear stress-strain relations or on evolution-equations for all Reynolds-stress components - the latter referred to as "second-moment closure". Both types of closure have been the subject of intensive research and validation efforts over the past two decades, by reference to numerous generic flows unrelated to specific areas of applications and, in some instances, to flows pertinent to combustor aerodynamics, cyclone separators, external aerodynamics, heated cavities and environmental problems. With few exceptions [4,5], they have not, however, been investigated in the context of demanding turbomachine flows.

This paper focuses on the predictive performance of both types of model in flows around compressor and turbine blade profiles.

TURBULENCE MODELLING

Attention is focused mainly on the performance of the second-moment closure of Gibson and Launder [6]. This is a *high-Re* model, applicable only to fully-established turbulence, and has been used here in combination with a *low-Re* linear or non-linear eddy-viscosity model to bridge the semi-viscous near-wall region. Matching of the two models is normally effected at a selected near-wall grid line with the distance from the wall y^+ ranging from 60 to100. Recent experience with *low-Re* variants of second-moment closure does not suggest distinct advantages over the hybrid methodology adopted here. Two eddy-viscosity models, one linear (Launder-Sharma, [7]) and the other cubic (Craft-Launder-Suga, [8]), have also been used, and related results are included in comparisons to follow.

The linear model employs the conventional Bousinesq relationship between stresses and strains,

$$\frac{\overline{u_i u_j}}{k} = \frac{2}{3}\delta_{ij} - \frac{v_t}{k}S_{ij} \tag{1}$$

where $S_{ij}=\partial u/\partial x_j+\partial u/\partial x_i$, while, the non-linear model uses a cubic relationship which arises from a general tensorial expansion in terms of the strain and vorticity tensors, subject to kinematic constraints arising from physical considerations,

$$\frac{\overline{u_i u_j}}{k} = \frac{2}{3}\delta_{ij} - \frac{v_t}{k}S_{ij}$$

$$+c_1\frac{v_t}{\tilde{\varepsilon}}(S_{ik}S_{kj}-\frac{1}{3}\delta_{ij}S_{kl}S_{kl})+c_2\frac{v_t}{\tilde{\varepsilon}}(\Omega_{ik}S_{kj}+\Omega_{jk}S_{ki})$$

$$+c_3\frac{v_t}{\tilde{\varepsilon}}(\Omega_{ik}\Omega_{kj}-\frac{1}{3}\delta_{ij}\Omega_{kl}\Omega_{kl}) \tag{2}$$

$$+c_4\frac{v_t k}{\tilde{\varepsilon}^2}(S_{ki}\Omega_{lj}+S_{kj}\Omega_{li})S_{kl}+c_5\frac{v_t k}{\tilde{\varepsilon}^2}(S_{kl}S_{kl}-\Omega_{kl}\Omega_{kl})S_{ij}$$

where $\Omega_{ij}=\partial u/\partial x_j-\partial u/\partial x_i$. The coefficients in equation (2) are given in table (1).

c_1	c_2	c_3	c_4	c_5
-0.1	0.1	0.26	$-10\,c_\mu^2$	$-5c_\mu^2$

Table (1) : Numerical values for the coefficients in equation (2).

Both eddy-viscosity models determine the turbulence energy and its rate of dissipation, needed to determine the eddy viscosity, from related differential transport equations which are applicable without Reynolds-number restrictions.

The second-moment model is based on a set of differential equation for all Reynolds-stress components,:

$$\frac{D\overline{u_i u_j}}{Dt} = \frac{\partial}{\partial x_k}(c_s \overline{u_l u_k}\frac{k}{\varepsilon}\frac{\partial \overline{u_i u_j}}{\partial x_l}) + P_{ij} + \Phi_{ij} - \frac{2}{3}\delta_{ij}\varepsilon \tag{3}$$

where

$$P_{ij} = -(\overline{u_i u_k}\frac{\partial u_j}{\partial x_k} + \overline{u_j u_k}\frac{\partial u_i}{\partial x_k}) \tag{4}$$

In equation (5), $\Phi_{ij2}=-0.6(P_{ij}-1/3P_{kk}\delta_{ij})$. The above system is closed by an equation for the dissipation

$$\Phi_{ij} = -1.8\frac{\varepsilon}{k}(\overline{u_i u_j}-\frac{2}{3}\delta_{ij}k) + \Phi_{ij2}$$

$$+0.5\frac{\varepsilon}{k}(\overline{u_m u_l}n_m n_l\delta_{ij} - 1.5\overline{u_i u_m}n_m n_j - 1.5\overline{u_j u_m}n_m n_i)f_y \tag{5}$$

$$+0.3(\Phi_{ml2}n_m n_l\delta_{ij} - 1.5\Phi_{im2}n_m n_j - 1.5\Phi_{jm2}n_m n_i)f_y$$

rate, which is not included here.

Most of the current low-Reynolds-number turbulence models cannot, without specific modification, predict transition with an acceptable degree or realism . For any but the simplest zero-pressure-gradient, flat-plate flow, they tend to return a premature onset of transition and to underestimate the length of the transitional region. There is a particularly serious problem with transition following impingement and laminar separation, such as often encountered in blade flows. Here, even a slight over-estimation of the turbulence level upstream of the leading edge, provokes much too early transition and a misrepresentation of the boundary-layer structure.

To improve the predictive accuracy of models, Schmidt and Patankar [9] proposed a Production Term Modification (PTM) which restricts the growth of the turbulence-production term along the streamwise direction, so as to delay the onset of transition and to prolong the transition process. Here, a similar concept is adopted, especially in relation to the turbine-cascade blade where transition occurs late, due to the strong acceleration of the boundary layer, and is therefore a highly influential process. The current PTM contains two limiters on the turbulence-production term. The first aims to delay the predicted onset of transition and is as follows:

$$P_k \rightarrow f_{p1}P_k, \quad Re_\theta < Re_{\theta b} \quad (laminar\ )$$

$$P_k \rightarrow [f_{p1}+(1-f_{p1})\frac{Re_\theta - Re_{\theta b}}{Re_{\theta e}-Re_{\theta b}}]P_k, \quad Re_{\theta b} < Re_\theta < Re_{\theta e} \quad (transitional\ ) \tag{6}$$

$$P_k \rightarrow P_k, \quad Re_\theta > Re_{\theta b}, \quad (turbulent\ )$$

where

$$Re_{\theta b} = 163 + e^{6.91 - Tu}, \quad Re_{\theta e} = 2.667 \, Re_{\theta b}$$

$$f_{p1} = -0.0047 \, Tu^2 - 0.087 \, Tu + 0.95 \quad (Launder - Sharma) \tag{7}$$

$$f_{p1} = -0.018 \, Tu^2 + 0.026 \, Tu + 1 \quad (Craft - Launder - Suga)$$

In equation (7), Tu is the turbulence intensity(%) at the inlet. The appearance of Tu in eq. (7) introduces a sensitivity to the specified inletdissipation-length scale l_e. This is not known, however, and should be so chosen as to correctly return the decay rate of turbulence energy in the free-stream region in accordance with the experimental data.

The second modification is designed to prolong the predicted length of transition, and has the form

$$P_{k(s)} = P_{k(s-\Delta s)} + \Delta P_k$$
$$\Delta P_k = \min \left(P_{k(s)} - P_{k(s-\Delta s)} \; , \; 1000 \, f_{p2} \nu_{t\infty} \Delta s \right) \tag{8}$$

where s is the streamwise coordinate along the blade surface and $f_{p2} = -20.45Tu + 262.25$ for both models.

NUMERICAL METHOD

Calculations reported below have been performed with the multi-block version of the general non-orthogonal fully collocated finite-volume scheme STREAM (Lien, *et al.*, [10]). In this, convection is represented by bounded (TVD) variant of the second-order QUICK scheme. The solution is iterated to convergence using a pressure-correction approach. Multi-block grids, combining O- and H-types, have been used to construct high-quality meshes around the blade, avoiding grid distortion near the leading and trailing edges. A fully implicit scheme, with the time variation approximated to second-order accuracy, is incorporated to solve unsteady problems.

RESULTS

Computations have been performed for a VKI turbine blade and two compressor blades, one a Controlled-Diffusion (CD) blade and the other a Double-Circular-Arc (DCA) blade. All three flows are virtually incompressible. Fig. 1 shows the blade geometries in terms of the associated multi-block meshes. Experimental data for the VKI blade at the design inlet angle have been reported by Ubaldi, *et al.* [11]. Data for CD blade at inlet angles 40°, 43.4° and 46°, relative to the axial cascade direction, have been obtained by Elazar and Shreeve [12], while measurements for the DCA blade flows at -8.5°, -1.5° and +5° incidence, relative to the design angle, have been made by Zierke and Deutsch [13]. The turbine-blade flow is attached, and the blade boundary layers are relatively thin. In contrast, the compressor-blade boundary layers on the suction side are subjected to adverse pressure gradient and are thick or separated. In the following computations, the PTM is only used for the VKI blade..

VKI blade

In this geometry, the passage flow is accelerating, and this occurring in combination with the blade operating at the design state, results in thin boundary layers. Not surprisingly, therefore, the only influential factor in terms of predictive accuracy is transition. With a production term modification introduced, all models yield similar results which are in good agreement with the data. Without this modification, all models predict pre-mature transition and therefore more or less excessive boundary-layer thickness. Fig. 2 shows that the pressure distributions around the blade returned by all models are almost identical. Here, there is no difference whether Gibson-Launder model is used either with linear or non-linear models. The displacement thickness and shape factor are shown in Fig. 3. Here, the predictions show a greater degree of variability, but are nevertheless close to one another and in good agreement with the experimental data. The shape factor suggests that transition in the suction-side boundary layer occurs within 0.3-0.5 of chord, while the pressure-side boundary layer is transitional over virtually the entire blade surface. Clearly, it is this behaviour which makes the modelling of transition modelling the dominant predictive factor. In terms of the vortex shedding in the wake, only the non-linear model captures this process with a shedding frequency of 1250 Hz, relative to the experimentally observed value of 1700 Hz.

CD blade

The case considered here is the flow at the highest inlet angle. This is the most challenging, involving the highest loading and adverse pressure gradient on blade surface, which drives the boundary layer on the suction side towards separation. However, apart from a leading-edge separation bubble, the boundary layer on the suction side remains attached because of the particularly favourable aerodynamic design of the blade.

Fig. 4a shows blade-surface Cp distributions, while velocity profiles in the suction-side boundary layer are given in Fig. 4b. The models are seen to behave differently especially near the leading edge and the region close to the trailing edge. The presence of the leading-edge separation bubble reduces the slope of Cp. This bubble is only captured by the non-linear model which gives a much better prediction of Cp and velocity profiles than the other two models. However, this model also overestimates the tendency towards separation near the trailing edge, resulting in a higher displacement thickness and an excessive shape factor, as seen in Fig. 5. This weakness may be rooted in, or at least aggravated by, too slow a recovery from the leading-edge separation, as is indicated by the poor agreement with the data at $x/C=54.3\%$. The prediction returned by the second moment closure is in good agreement with the data downstream of the leading-edge separation, but fails to represent correctly the transitional leading-edge flow which is very sensitive to the representation of the flow just upstream of the leading edge. the linear eddy-viscosity model fails to capture the leading-edge separation bubble and returns an incorrect growth of boundary layer towards the trailing edge. This behaviour arises from a combination of grossly excessive levels of turbulence energy in the impingement region, insufficient sensitivity to curvature and over-estimation of turbulent mixing in adverse pressure gradient. On the pressure side, the boundary layer is generally much thinner than that on the suction side, and is much less influential in terms of contribution to the overall performance of the blade. As seen in Fig. 5, the pressure-side displacement thickness returned by all models is in close agreement with the data. However, there are large discrepancies between the predicted and the measured shape-factor distributions, especially within the first 30% of the chord. This may reflect defects in the representation of transition, but there is also some doubt about the accuracy of the experimental data because of the thinness of the boundary layer on the pressure surface.

DCA blade

The flow over the DCA blade is highly challenging, even at near-design conditions, owing to the blade's highly curved surface which introduces strong adverse pressure gradient and high streamline curvature. This results in a very rapid growth of the boundary layer on the suction side, leading to a large trailing-edge separation. In such circumstances, the boundary layer thickness can exceed 20% of the pitch length and the interaction between the boundary layer and the passage flow becomes strong. Here, turbulence transport plays an especially important role and model quality becomes a crucial issue.

The case presented below is the flow at an inlet angle which departs by -1.5° from the design value. Even at this near-design condition, there is a massive trailing-edge separation which occupies approximately 20% of the chord length on the suction surface. Fig. 6 compares Cp distributions. The existence of the trailing-edge separation is evident from the pressure plateau located around 80% of the chord to the trailing edge. Only the second moment closure returns a credible prediction of the separation process, and this reflects its greater sensitivity to streamline curvature and the lower level of mixing it predicts in the boundary layer in the presence of adverse pressure gradient. Fig. 7 shows the suction-side velocity profiles at three different sections, while Fig. 8 gives the distributions of displacement thickness and shape factor. The velocity profiles close to the leading edge indicate that neither the second-moment closure nor the linear eddy-viscosity captures the transitional state of the leading-edge flow. However, this is not a crucially important issue in this flow, as transition occurs very close to the leading edge and does not have a major impact on the development of the downstream boundary layer. Here, then, it is the quality of the models in fully-turbulent conditions which is the key to a satisfactory prediction of the flow. Thus, as the flow progresses beyond the immediate leading-edge region, the linear model predicts an excessively turbulent boundary layer which is not sufficiently sensitive to adverse pressure gradient and which therefore remains attached. In contrast, the boundary layer returned by the second-moment closure grows rapidly, in accordance with experiment, and eventually separates at around 80% of the chord. As regards the non-linear model, despite the use of the elaborate stress-strain relation, the model fails to capture the trailing-edge separation, returning results which are no better than the linear model in the trailing-edge region. However, the root of its failure is different, being a consequence of an overestimation of the leading-edge separation bubble. In the experiment, this bubble only occupies a few percent of the chord, and the boundary layer reattaches in a fully turbulent state at 3% of the chord. In contrast, the predicted boundary layer reattaches just upstream of the first measurement section at 7.3%. This results in a serious underestimation of the initial displacement thickness and turbulence intensity. With the wrong initial profiles, there arises a wrong response of the boundary layer to the pressure field further downstream. Although the non-linear term is sensitive to streamline curvature, by virtue of its cubic fragments, this sensitivity is perhaps insufficiently strong, and may be a contributory factor to the model's failure. The superior performance of the second-moment closure relative to the other two models is strikingly conveyed by the distributions of displacement thickness and shape factor. In particular, the steep rise in both quantities on the suction side, towards the trailing edge, predicted by the closure reflects the presence of separation. On the pressure side, the boundary layer is transitional up to about 50% of the chord, and none of the models gives a credible prediction of this behaviour, a failure which made it imperative to introduce production-term modification when computing the VKI turbine blade.

CONCLUSIONS

For the VKI turbine blade operating at near-design conditions, the boundary layers are thin and attached, and all model return results which are in close agreement with the data, provided transition is captured correctly. In contrast, the boundary layers on the suction side of compressor-cascade blades are subjected to strong adverse pressure gradient and grow rapidly, possibly to the point of separation. In such circumstances, overall predictive accuracy is sensitive to turbulence modelling. Here, the performance of the second-moment closure is observed to be distinctly better than that of the other two eddy-viscosity models, especially in the case of the DCA compressor blade, where extensive separation occurs near the trailing edge. Despite the fact that the cubic stress-strain relationship is designed to reproduce the response of turbulence to streamline curvature and irrotational straining, the cubic model fails to capture suction-side separation. This failure appears to be rooted in the model's tendency to over-estimate the extent of the laminar leading-edge separation, returning a wrong initial boundary-layer profile after reattachment. The practice of limiting turbulence production, via the PTM, is not advocated as an universal proposal and was used here for the VKI blade mainly to demonstrate the extreme importance of predicting correctly the transition position and length to the overall predictive realism of the flow around turbine blades.

ACKNOWLEDGEMENTS

The authors gratefully acknowledge the assistance given by Prof. Zunino and his colleagues who provided the experimental data for the VKI turbine cascade. The financial support from the

Commission of the European Communities under the Brite-EuRam III Contract "TURMUNSFLAT" No. BE95-1698 is also greatly appreciated.

REFERENCES

[1] Lakshminarayana, B. and Govindan, T.R., (1981), "Analysis of Turbulent Boundary Layer on Cascade and Rotor Blades of Turbomachinery", *AIAA Journal*, Vol. 19, p. 1333.

[2] Ho, Y.K., Walker, G.J. and Stow, P., (1990), "Boundary-Layer and Navier-Stokes Analysis of a NASA Controlled-Diffusion Compressor Blade", *ASME paper 90-GT-263*.

[3] Hobson, G.V. and Lakshminarayana, B., (1991), "Prediction of Cascade Performance Using an Incompressible Navier-Stokes Technique", *ASME Journal of Turbomachinery*, Vol. 113. p. 561.

[4] Chen, W.L., Lien, F.S. and Leschziner, M.A., (1998), "Computational Prediction of Flow around Highly-Loaded Compressor-cascade Blades with Non-linear Eddy-viscosity Models", to be appear in *Int. Journal of Heat and Fluid Flows*.

[5] Chen, W.L., Lien, F.S. and Leschziner, M.A., (1997), "Computational Modelling of Highly-loaded Compressor Cascade Flow", *Proc.11th Symp. on Turbulent Shear Flows*, Grenoble, France. [6] Gibson, M.M. and Launder, B.E. (1978), "Ground Effects on Pressure Fluctuations in the Atmospheric Boundary Layer", *Journal of Fluid Mechanics*, Vol. 86, p. 491.

[7] Launder, B.E. and Sharma, B.I. (1974), Application of the Energy-Dissipation Model of Turbulence to the Calculation of Flow near a Spinning Disc", *Int. Journal of Heat and Mass Transfer*, Vol. 1, p.131.

[8] Craft, T.J., Launder, B.E. and Suga, K. (1993), "Extending the Applicability of Eddy-viscosity Model Through the Use Deformation Invariant and Non-linear Elements", *Proc. 5th Int. Symp. Refined Flow Modelling and Turbulence Measurements*, p. 125.

[9] Schmidt, R.C. and Patankar, S.V., (1991), "Simulating Boundary Layer Transition with Low-Reynolds-Number k-ε Turbulence Models, Part 1 & 2", *ASME Journal of Turbomachinery*, Vol. 113, p. 10.

[10] Lien, F.S., Chen, W.L. and Leschziner, M.A (1996), "A Multiblock Implementation of a Non-orthogonal, Collocated Finite Volume Algorithm for Complex Turbulent Flows", *Int. Journal for Numerical Methods in Fluids*, Vol. 23, p. 567.

[11] Ubaldi, M., Zunino, P., Campora, U. and Ghiglione, A., (1996), "Detailed Velocity and Turbulence Measurements of the Profile Boundary Layer in a Large Scale Turbine Cascade", *ASME paper 96-GT-42*.

[12] Elazar, Y and Shreeve, R.P. (1990), "Viscous Flow in a Controlled Diffusion Compressor Cascade with Increasing Incidence", *ASME Journal of Turbomachinery*, Vol. 112, p. 256.

[13] Zierke, W.C. and Deutsch, S. (1989), "The Measurement of Boundary Layers on a Compressor Blade in Cascade", *NASA CR-185118*.

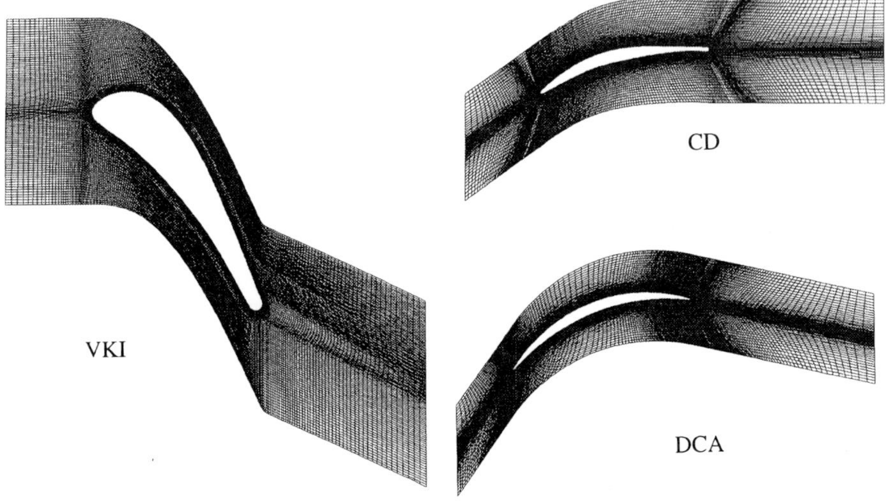

Fig. 1: Geometries and computational meshes for VKI, CD and DCA cascades.

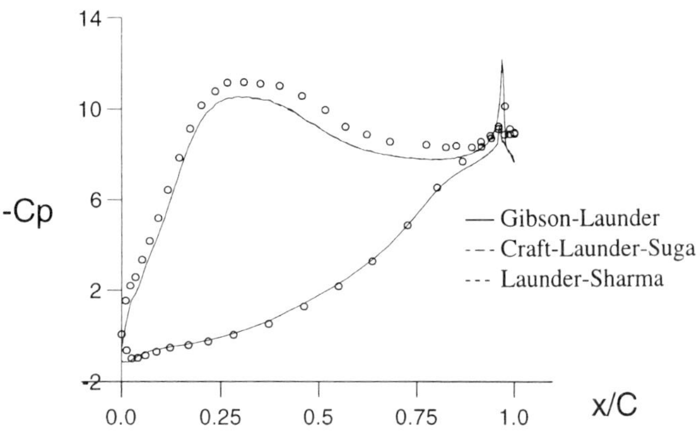

Fig. 2: VKI turbine cascade, Cp distributions.

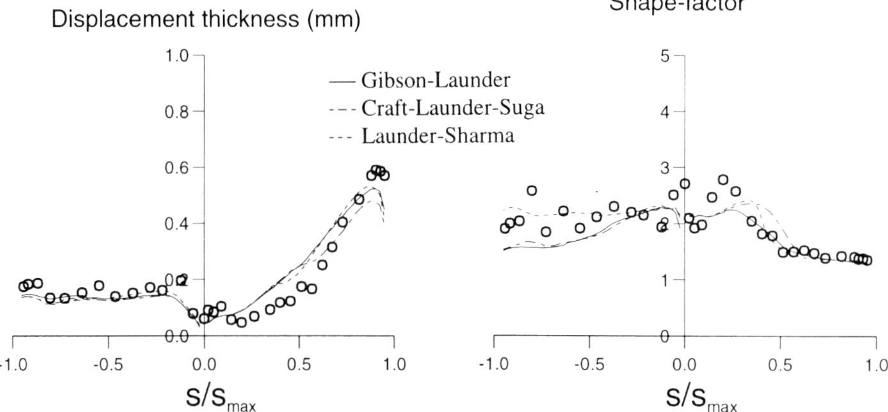

Fig. 3: VKI turbine cascade, distributions of momentum thickness and shape factor.

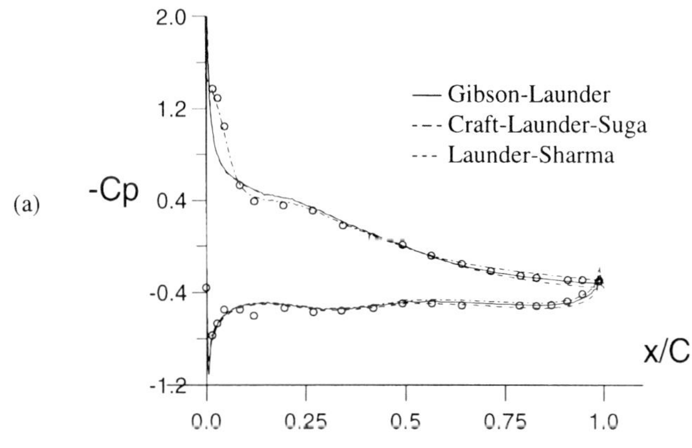

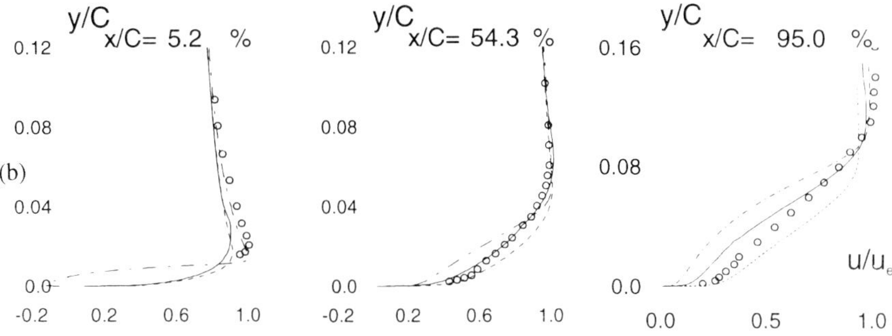

Fig. 4: CD compressor blade, (a) Cp distribution, and (b) suction side velocity profiles at $\beta=46°$.

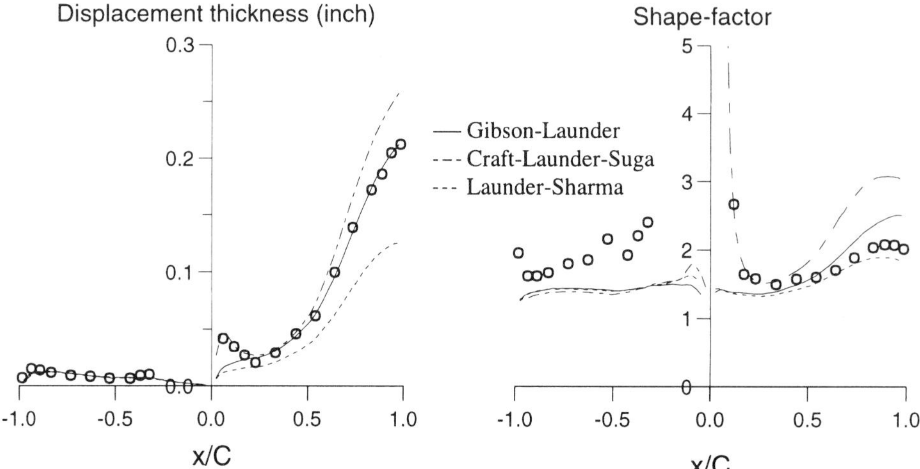

Fig. 5: CD compressor blade, distributions of momentum thickness and shape factor.

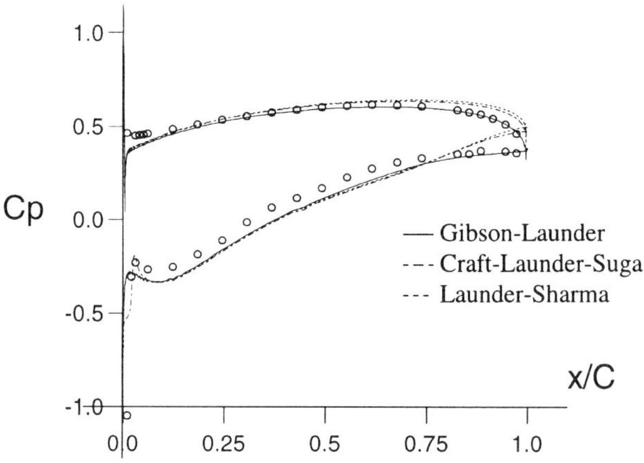

Fig. 6: DCA compressor blade, Cp distributions at -1.5° incidence.

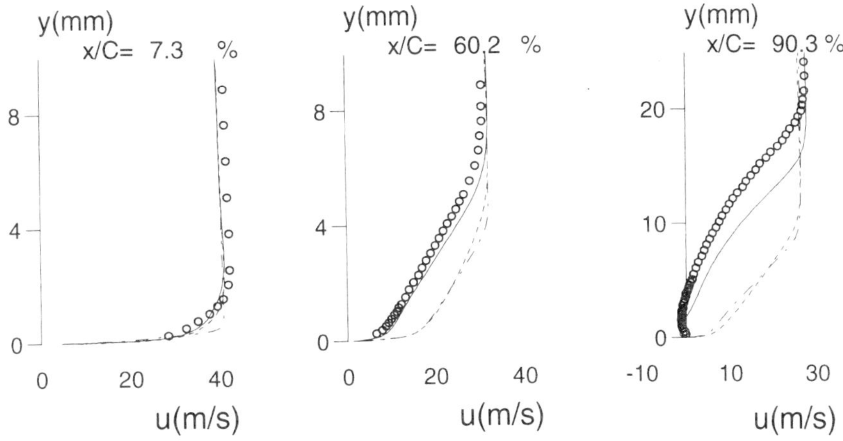

Fig. 7: DCA compressor blade, suction side velocity profiles.

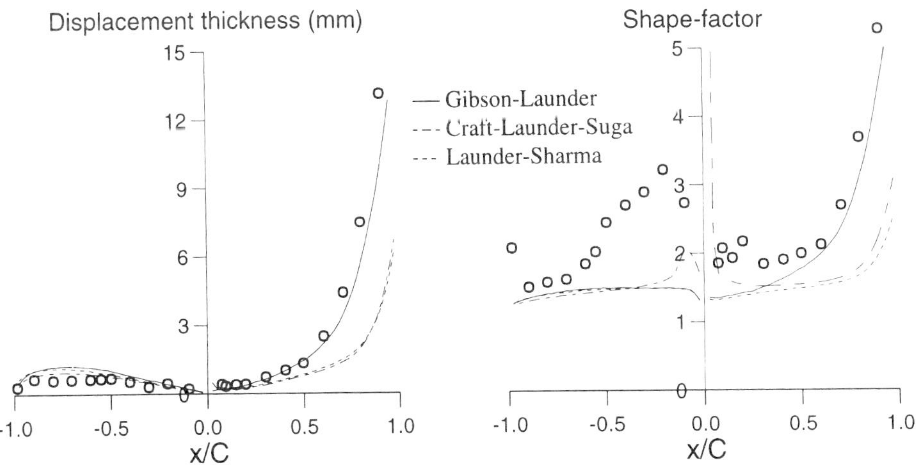

Fig. 8: DCA compressor blade, distributions of momentum thickness and shape factor.

Experiments on turbulent flow separation

B G B MUHAMMAD-KLINGMANN and **J P R GUSTAVSSON**
Department of Mechanics, Royal Institute of Technology, Stockholm, Sweden

A fully turbulent boundary layer was established on a flat plate under varying adverse pressures. This study should be relevant for the design of blading in large compressors and turbines, e g wind power turbines, where it is of interest to exploit the large load obtained at flow conditions near separation. Boundary layer profiles were measured with hot-wire anemometry in the attached flow case, and PIV (Particle Image Velocimetry) in the separated cases. The boundary layer evolution down to the position of separation was analysed using different scaling and separation prediction methods found in the literature.

LIST OF SYMBOLS

B	boundary layer parameter in Perry-Schofield scaling	U_s	velocity parameter in Perry-Schofield scaling
C_p	pressure coefficient	x	streamwise co-ordinate
H	boundary layer shape factor	y	wall-normal co-ordinate
Re_θ	Reynolds number based on θ	Γ	pressure gradient parameter in Buri's integral method
S	Stratford parameter; skewness of U-distribution	δ	physical boundary layer thickness
u_{rms}	rms of U		
u_τ	friction velocity	δ^*	displacement
U	mean velocity within the boundary layer	θ	momentum thickness
		κ	von Karman's constant $= 0.41$
U_o	mean velocity at the boundary layer edge	ν	fluid kinematic viscosity
		ρ	fluid density
U_{ref}	reference velocity, equal to velocity in the smallest cross-section of the duct	τ_w	wall shear stress
		χ	back-flow coefficient

1. INTRODUCTION

The character of the boundary layer on the blade affects both heat transfer rate and acceptable load on the blade of a turbine. The boundary layer on the blade will experience laminar-turbulent transition at some position along the chord, sometimes after a local flow separation near the leading edge. The Reynolds number based on θ at transition varies between 200 and 1000 depending on surface and free stream conditions. On high load blading, flow separation may again occur towards the rear of the blade, giving a loss of efficiency, due to the wake past the blade. However, the lift continues to increase, and this is expoited in wind power turbines, which can be operated at conditions near stall.

The objective of the present experiment is to study the development of a turbulent boundary layer under varying pressure gradients, in order to make an evaluation of existing methods of flow separation prediction and scaling. A pressure gradient similar to that over a turbine blade is obtained on a flat plate in a wind tunnel by expanding the test section area. When scaled with the maximun dynamic pressure, the maximum pressure gradient in case 3 was about 2.5 and in case 1a it was about 6.5 m^{-1}. As a comparison, a wind tunnel test (9) of a wind power turbine blade on the verge of separation, had a normalised maximum pressure gradient of 4.5 m^{-1}. The chord length of the model was 0.45 m, which is similar to that of the "bump" used in the present experiment. Measurements were made using mainly CTA hot wire anemometry and PIV (Particle Imaging Velocimetry).

2. METHODS FOR PREDICTION OF FLOW SEPARATION

2.1. Diagnosing flow separation

Turbulent flow separation is defined in modern literature to occur when the back-flow coefficient χ (i.e. the time portion during which the flow is reversed) reaches 50%, see Simpson (1). Alternatively, the point where the mean value of the wall shear stress vanishes is taken to be the point of separation. In 2D flows, these two definitions are equivalent. Another criterion, established by Dengel & Fernholz (2) by measuring χ at the wall, is the shape factor $H=\delta^*/\theta$, which reaches a value of 2.85 at separation.

2.2. The Momentum integral equation

The evolution of the momentum thickness θ and shape factor H is given by the momentum integral equation

$$\frac{d\theta}{dx} + (2 + H)\frac{\theta}{U_o}\frac{dU_o}{dx} = \frac{\tau_w}{\rho U_o^2} \tag{1}$$

Here, U_o is the velocity at the boundary layer edge, and τ_w the wall shear stress. Different empirical relations have been suggested for τ_w in turbulent boundary layers. The current work uses the relation cited by Schlichting (1960):

$$\frac{\tau_w}{\rho U_o^2} = 1.2Re_\theta^{-0.268}10^{1.5H} \tag{2}$$

That leaves us with two unknowns, θ and H, hence a second equation is needed. The equation derived by Duncan, Thom and Young (4) was used, which represents the conservation of mass flow through the boundary layer in the form

$$\frac{1}{U_o}\frac{d}{dx}(U_o\theta\,\frac{2H}{H-1}) = F(H) \tag{3}$$

Here F is a linear function of H, $F(H)= C{\cdot}(H\text{-}1.1)$, where the constant C is chosen to fit the present experimental data. To complete the method, values of H and θ must be provided at the starting point. These values were also taken from the present measurements.

2.3. Empirical separation criteria for turbulent boundary layers

One of the empirical separation criteria for turbulent flow derived from the momentum equation is the method of Buri (5, also quoted in 4), where the quantity

$$\Gamma = \frac{\theta}{U_o}\frac{dU_o}{dx}\,Re_\theta^{\,0.25} \tag{4}$$

is to reach a value of -0.06 at separation. θ is computed using a simplified version of the momentum integral equation, obtained by assuming both $\tau_w/(\rho\,U_o^{\,2})$ and H to be functions only of Γ. This leads to an integral expression for Γ.

$$\frac{d}{dx}(\theta\,Re_\theta) = 0.016 - 4.05\Gamma$$

The early laminar part of the boundary layer is handled separately, but the method is quite robust against slight errors in the initial value of θ of the turbulent boundary layer.

Stratford (6) suggested a correlation based on the pressure gradient, which at separation obeys the following equality:

$$S = C_P\sqrt{(x\frac{dC_P}{dx})} = 0.39\sqrt[10]{(10^{-6}\,Re_x)}. \tag{5}$$

When $C_P \le 4/7$ and $d^2p/dx^2 \le 0$ (for $d^2p/dx^2 > 0$, the numerical coefficient 0.39 on the right hand side of the above relation should be replaced with 0.35).C_P is the pressure coefficient defined as $(p\text{-}p_{ref})/(p_o\text{-}p_{ref})$, where p, p_o and p_{ref} are the static, total and reference pressures. The reference pressure is here taken to be the static pressure at the duct throat.

3. EXPERIMENTAL SETUP

The experiments were performed in a low speed wind tunnel with a 1 m long test section (cross section 0.5 m x 0.4 m). The upper ceiling of the test section was replaced with a flexible sheet the shape of which could be adjusted by means of long screws (see figure 1). A 1.08m long flat plate was placed in the test section and aligned so as to obtain a straight inflow into the duct. The boundary layer was removed from the ceiling by suction through small holes, in order to be able to study the effect of the adverse pressure gradient on the flat plate surface. About 10% of the flow through the duct was hence removed. The velocity in the throat section of the duct, denoted U_{ref}, was about 40 m/s, and free stream turbulence about 1%. The Reynolds number based on this velocity and the maximum distance downstream of the throat is approximately 12 million.

In order to assure fully turbulent boundary layer flow, a transition trip was inserted on the plate at a position 0.18 m from the leading edge. Although this introduces a somwhat artificial feature as compared to a real turbine blade case, it has the advantage of providing a well defined starting point for calculations.

The plate spanned the entire wind tunnel width and hence the wall boundary layers will influence the flow over the plate. While this means that the free-stream will be somewhat more accelerated than the height distribution in itself indicates, the fact that the width of the tunnel (0.4m) is significantly larger than the height and boundary layer thicknesses up to separation should assure the two-dimensionality of the flow of the studied region, which was close to the centre line of the plate.

3.1. Flow configurations

Three different cases were studied as indicated in figure 1, namely

- Case 1a and b: separated flow
- Case 2: separation bubble
- Case 3: attached flow

(The photograph in figure 1 shows case 1a.) Completely separated flow was studied in two different configurations, 1a and b. In the latter, a 3mm trip was used to keep the separation front steady at x=0.68m. Case 2 (separation with reattachment) was obtained in the same geometry as case 3, but with a 7 mm high separation trip introduced at x=0.67 m. This was the smallest trip found to be able to persistently hold the sepation bubble in place on the plate. Smaller trips allowed the separated region to move around on the plate, or in the latter case, to move to the ceiling.

3.2. Measurement techniques

In all configurations, 37 static pressure taps in the plate and a large number of tufts taped to the plate were used to assure 2D flow and give some idea about the stability and position of the separated flow region.

A Preston tube was used in cases 1a and 3 to measure the wall shear stress.

PIV measurements were made in the x-y-plane plane in cases 2 and 3. The measurements where made with a 400 mJ Nd:YAG laser (Spectraphysics) and a Kodak ES 1.0 high-resolution digital camera, controlled by Dantec's so-called Flowmap system. 64x64-pixel interrogation areas with 75% overlap were used, giving averages over 2.6x2.6-mm squares with c/c distances of 0.65 mm. The closest used position was about 1.5 mm away from the wall. At each position 100-300 frame pairs were taken. The smoke was inserted through slits in the plate at the positions x=0.3 and 0.6 m and the smoke particle size is in the order of 2-3 µm. In order to get output data of acceptable quality, the measured velocity vectors had to be validated in several steps checking the feasibility and reliability of each measurement typically leading to over 50% of them being discarded. Comparisons with LDV measurements made later in the same set-up showed differences between LDV and PIV measurements of in the order of 0.5 m/s (10). These are probably due to the relatively low number of samples taken by PIV and the averaging over large areas done in PIV.

CTA hot wire measurements were performed in case 3, using a traversing mechanism placed directly on the plate, a 5 µm Pt hotwire at 70% overheat and a DISA anemometer. Approximately 100 000 samples were taken at each position to allow accurate computation of mean velocity and perturbation velocity moments. Due to the relatively large sample, the mean values should not suffer from significant errors apart from calibration errors which may be of importance at low velocities. The y-position of the probe could not be determined exactly, but should not be more in error than 0.05 mm. A comparison with hot-wire measurements in the attached flow case indicated that the deviations from these measurements were less than 1 m/s. Most of this deviation was found to be a result of the retardation of the flow close (<7mm) to the plate caused by the hot-wire mount. Farther away from the plate, the deviations were found to be less than 0.5 m/s in the examined case.

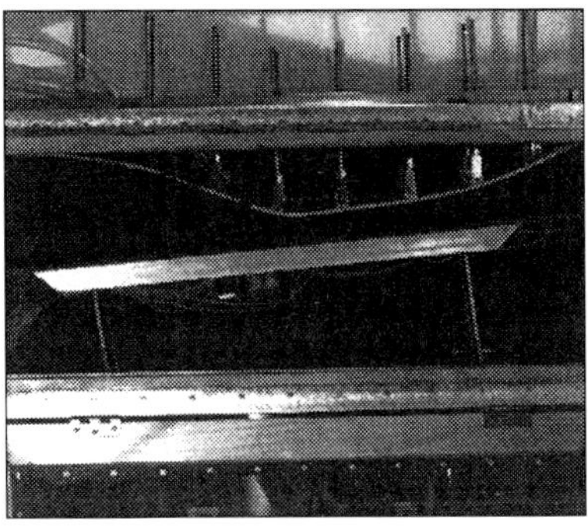
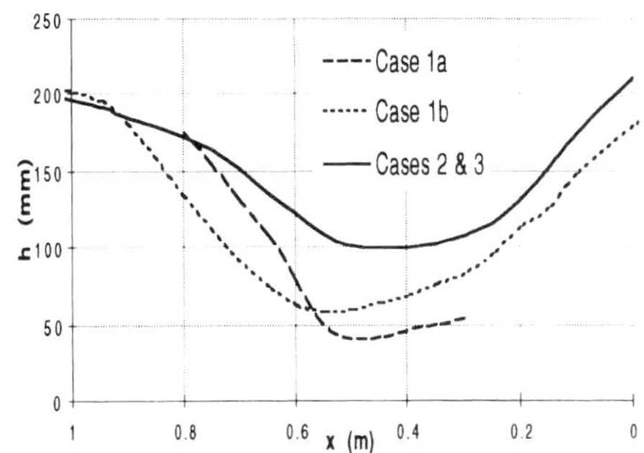

Figure 1. Left: Photograph of the set-up. Right: Height distribution for different flow configurations.

4. RESULTS

4.1. Laminar versus turbulent boundary layer separation

Figures 2 and 3 show the pressure and wall shear stress distributions obtained in case 1a at different wind tunnel speeds. τ_w was measured with a Preston tube, and the measurements show where the wall shear stress becomes zero. The lowest speed case is qualitatively different from the others, because the boundary layer was laminar, and hence separation occurred close to the throat (x=0.525 m). The three high-velocity cases have the same shape of the pressure distribution, and separate close to x=0.65.

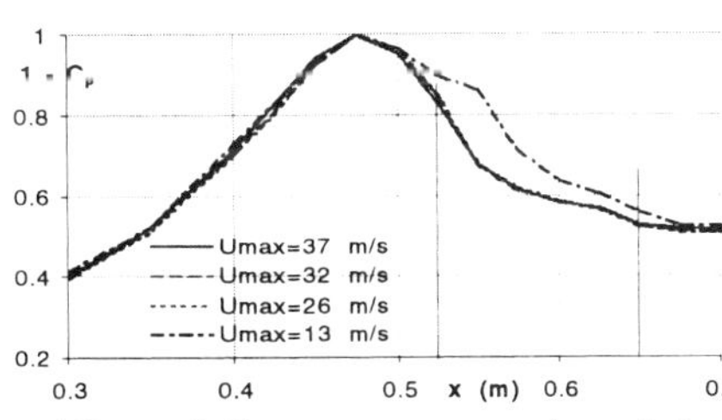

Figure 2. Pressure recovery (case 1a)

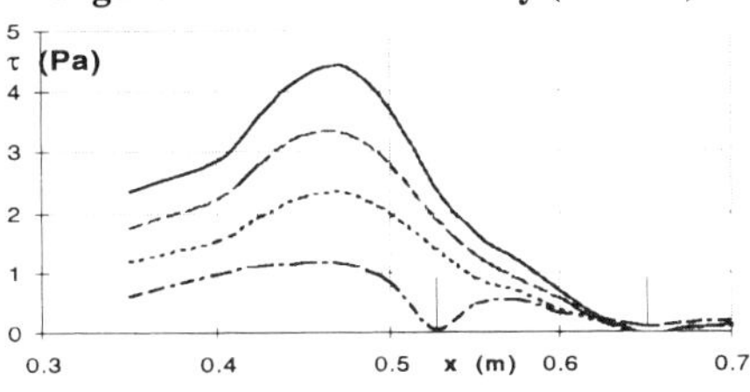

Figure 3. Wall shear stress (case 1a).

4.2. Boundary layer development

Figures 4 and 5 show the boundary layer development for the cases 2 and 3 (separation with reattachment vs. attached flow). Also shown is the evolution of the shape factor H. The Reynolds number based on theta ranges from 2000 at x=0.5 m to about 8000 at x=0.9 m. Figure 4 uses PIV data since there is significant flow reversal, while figure 5 is based on higher-quality hot-wire measurements. Unfortunately, the hot-wire mount used could not traverse farther away from the plate than 19 mm, so in the two most down-stream positions, the free-stream was not reached. In the attached case, the shape factor reaches a value of 2.4 at the most down-stream position and in the separated flow case, the flow reattaches when H reaches a value of approximately 3. A typical feature of adverse pressure gradient flow is that the peak of u_{rms}, which is at first located near the wall, decreases as the flow decelerates, while a broad peak develops further out in the boundary layer. An inherent problem with PIV is the inability to resolve both the wall region and the outer flow region. If the high velocities found near the free-stream are to be captured, interrogation areas large enough to make most particles remain in their respective interrogation area between exposures must be chosen. Since the distance traveled by the particles between the exposures can only be determined with an accuracy comparable with the pixel size of the pictures, the velocity resolution will be relatively coarser for small flow velocities, where particles only move a few pixels between exposures. In the present case, we chose to capture most of the boundary layer, accepting a rather poor resolution near the wall. The reattachment point was estimated from the distribution of the back-flow coefficient χ, shown in figure 6.

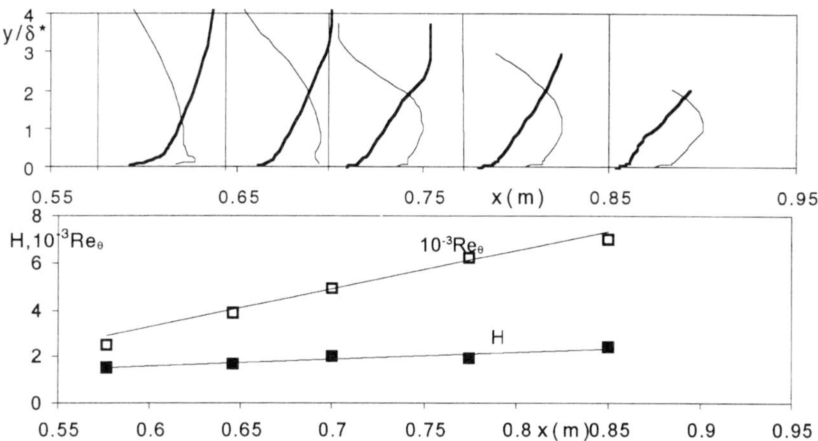

Figure 4. Evolution of the boundary layer in case 3 (attached flow).
Upper: $0.05\cdot U(y)$ and $0.05\cdot u_{rms}/u_{rms,max}$ at different x-positions, lower: H and $10^3\cdot Re_\theta$.

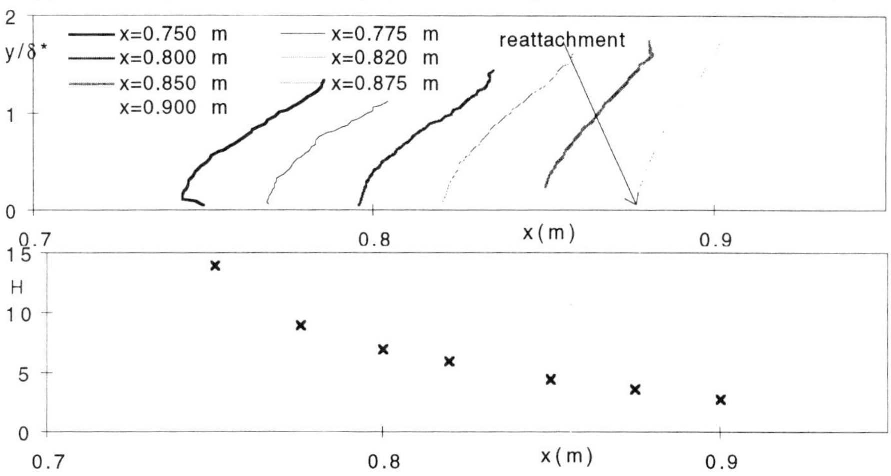

Figure 5. Evolution of the boundary layer in case 2 (separation with reattachment).
Upper: $0.05\cdot U/U_{ref}(x)$, lower: H(x).

4.3. Perry-Schofield scaling (case 3)

Perry & Schofield (7) suggested a similarity scaling of the mean velocity profiles near separation, which is based on a length scale B and a velocity scale U_s. The idea is that the mean velocity profiles in an adverse pressure gradient flow should collapse onto a single function f when scaled according to

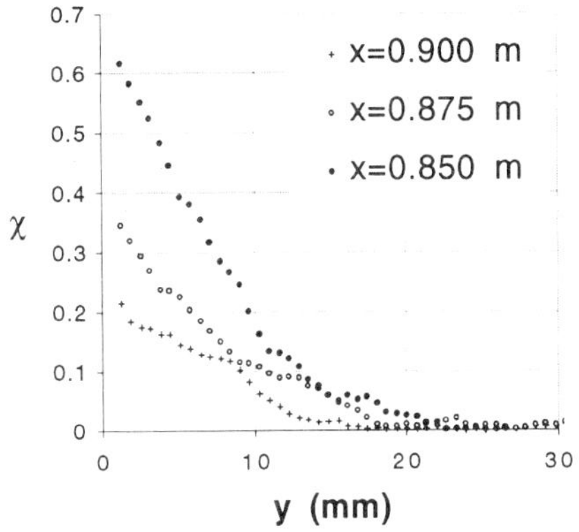

Figure 6. Back-flow coefficient χ (case 2).

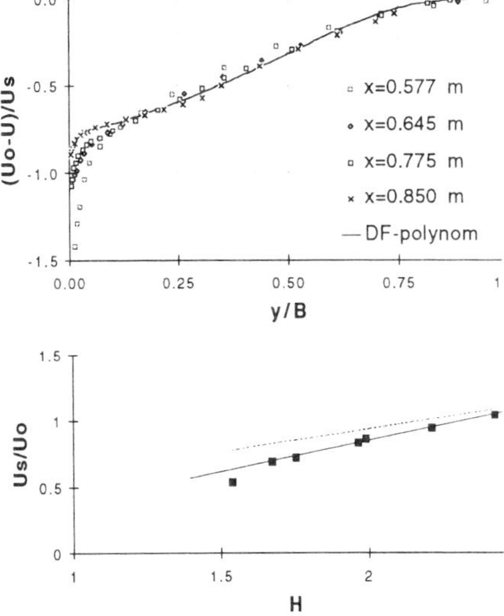

$$\frac{U_o - U_s}{U_s} = f\left(\frac{y}{B}\right)$$

where B is defined as $B=2.86\delta^* U_o/U_s$. The non-dimensionalised velocity profile f has been approximated by Dengel and Fernholz (2) with a 7th order polynomial in $\zeta=y/B$:

Figure 7. Upper: boundary layer profiles plotted in Perry-Schofield co-ordinates. Lower: variation of U_s with H:

    ——— **straight line fit to data**
    - - - - **line based on DF's data.**

$$f(\zeta)=0.781 - 0.535\zeta - 0.739\zeta^2 - 2.352\zeta^3 + 13.81\zeta^4 - 33.178\zeta^5 + 36.502\zeta^6 - 14.324\zeta^7.$$

The velocity scale U_s is chosen to give the optimum fit between the measured velocity profile and the Perry-Schofield profile in the central part of the boundary layer. Fernholz and co-workers (2, 8) found that there is a linear relation between U_s/U_o and H. This was also found in the present experiment, albeit with a somewhat different slope, see figure 7. Note how the measured profile approaches that given by $f(\zeta)$ as H increases. This is in agreement with Dengel and Fernholz' (2) observation that Perry-Schofield co-ordinates are only applicable at $H \geq 2.2$.

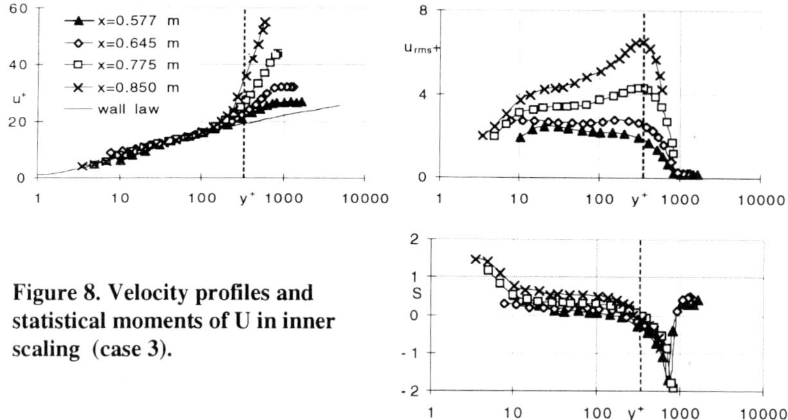

Figure 8. Velocity profiles and statistical moments of U in inner scaling (case 3).

4.4. Scaling with wall variables (case 3)

The attached boundary layer profiles were also plotted in wall co-ordinates u^+ and y^+ (i.e. y and U are normalised with u_τ and v, where u_τ is defined as $\sqrt{(\tau_w/\rho)}$, τ_w is the wall shear stress and v the kinematic viscosity). u_τ was evaluated from a best fit of the measured profiles to the universal wall law $u^+ = 1/\kappa \ln y^+ + 5$. The wall shear stress thus obtained was in excellent agreement with values obtained from Preston tube measurements. Figure 8 shows U, u_{rms} and the skewness (S) of U for the different downstream positions in case 3. An interesting feature is that the new peak in u_{rms} lies approximately where the U-profile departs from the log-law, i.e. where the wake region begins. In inner variables, the maximum of u_{rms}^+ coincides with the zero-crossing of S, occurring roughly at $y^+ = 300$.

4.5. The momentum integral equation

The parameters in equation (1) were calibrated to give a best fit with the experimental data from case 3, and the method was then applied to case 1b, where flow separation occurs at $x=0.68$ m. First, the initial value of H was adjusted to yield the correct values of θ. The experimental values of H were then used to adjust the constant C in equation (3). The result is shown in figure 9. Also shown in the figure are the results obtained with the method of Buri.

Figure 9a shows the pressure distributions for the two cases. In case 3, the pressure coefficient is evaluated from wall pressure measurements. For the separated case, however, the wall pressure is no longer the same as that in the free stream. In this case, the wall shape was used to compute the free stream velocity based on conservation of mass through the duct. The pressure distribution derived in this way was in good agreement with the measured pressure distribution up to the position of the separation.

Figures 9b and c show the development of the momentum thickness, θ, and the shape factor H. It can be seen that Buri's method gives quite a good prediction of θ, all the more valuable, since no adjustments have to be made. To obtain a correct value of H is more difficult - the results are quite sensitive both to changes in the constant C and to the modelling of the wall shear stress term.

In case 1b, H and θ both begin to increase quickly past $x=0.7$, and the wall shear stress drops to zero at x=0.70 m. Figure 9e shows the development of Γ, which is similar to the driving term $\theta/U_o \, dU_o/dx$ in equation (1). The minimum value of Γ reached according to (4) is –0.025 in case 3, which is below the critical value for separation (Γ=-0.06). In case 1b, this critical value is reached at x=0.66. No wall-shear stress measurements were made in this case, but the position where a minimum separation trip (3mm) was required to give stable separation (as indicated by tufts and static pressure taps in the plate) was at x=0.68 m. The separation is hence accurately predicted in this case.

The critical value of the Stratford parameter S according to formula (5) occurs at $x\approx0.65$ for case 3, i.e. the criterion of Stratford predict separation where in fact the flow remains attached. Also in case 1b, separation is predicted much too early, $x\approx0.55$.

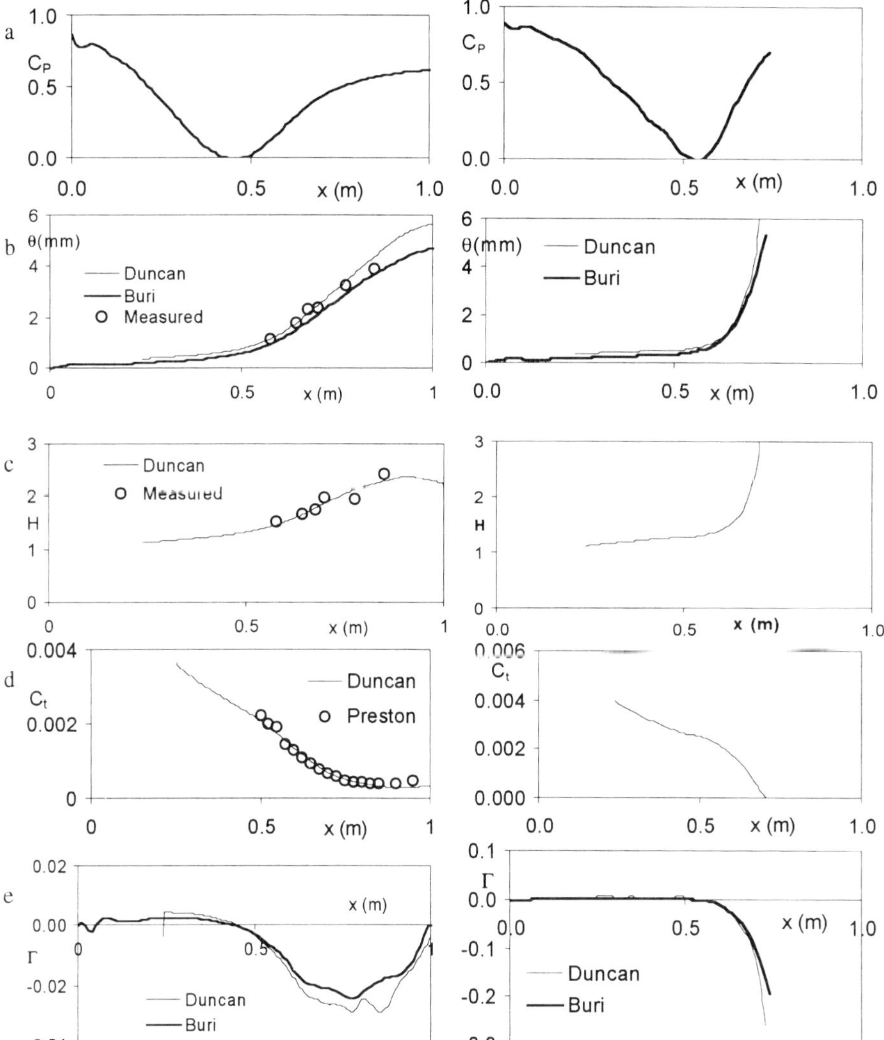

Figure 9. Boundary layer development. Left column: case 3; right column: case 1b.

5. CONCUDING REMARKS

In the present study, the development of a boundary layer under varying pressure gradients has been studied, and the results have been compared with different scaling laws known from the literature. In the attached flow, the value of the shape factor H reached 2.4, while in the case with separation, the reattachment occurred at $H \approx 3$. This is in fair agreement with the results previously obtained by Dengel & Fernholz (2) in a rather mild separation bubble, where H was found to be 2.85 both at separation and reattachment. Hence, the value H may be used directly as separation criterion. An interesting observation is that the boundary layer fluctuations (u_{rms} and S of U) appear to scale with inner variables. In contrast to this is the observation that H can be directly related to the Perry-Schofield velocity parameter U_s, which represents an outer variable scaling. Hence the boundary layer profile can then be reconstructed once H is known.

Ideally, one would wish to possess a criterion, which relates separation directly to the shape of the blade, or to the velocity field between the blades in a particular blade layout. However, methods trying to relate separation directly to some parameter based on the pressure gradient fail to produce the desired result. The problem of predicting separation can then be viewed as how to obtain a reliable value of H without having to perform detailed measurements of the boundary layer, or well resolved boundary layer computations. In this study we have tried to apply different methods based on the momentum integral equation. These attempts show that it is relatively easy to predict the development of the momentum thickness (θ), while the calculated value of H is quite sensitive to how the terms in the equation are modelled. While the methods proposed by Buri and Duncan et al predict separation relatively well, Simpson's method appears to underestimate the ability of a turbulent boundary layer to withstand a strong deceleration without separating.

REFERENCES

1. R. L. Simpson, "Turbulent boundary-layer separation", *Annual Review of Fluid Mechanics* **21** (1989)
2. P. Dengel, P. & H. H. Fernholz," An experimental investigation of an incompressible turbulent boundary layer in the vicinity of separation", *Journal of Fluid Mechanics* **212**, p. 615-636 (1990)
3. H. Schlichting, "Boundary Layer Theory", pp 676-679. McGraw-Hill Book Company (1960).
4. W. J. Duncan, A. S. Thom and A. D. Young, "Mechanics of Fluids". Edward Arnold Printers Ltd, London (1970).
5. A. Buri, "Eine Berechnungsgrundlage für die turbulente Grenzschicht bei beschleunigter und verzögerter Strömung". Diss. Zürich (1931). see also (3)
6. B. S. Stratford, "The prediction of separation of the turbulent boundary layer", *Journal of Fluid Mechanics* **5**, p. 1-16 (1959)
7. A. E. Perry & W. H. Schofield, "Mean velocity and shear stress distributions in turbulent boundary layers". *Phys. Fluids* **16**, p. 2058 (1970)
8. A. Alving & H. H. Fernholz, "Mean-velocity scaling in and around a mild, turbulent separation bubble. *Phys. Fluids* **7**(8), p. 1956-1969 (1995)
9. Wind tunnel test of FFA-W3-211 profile, private communication Torkel Hambræus and Anders Björck, FFA (The Aeronautical Research Institute of Sweden)
10. Kristian Angele, KTH (The Royal Institute of Technology), private communication

C557/143/99

An effect of the curved passage depth on the shock-induced separation flow structure

J CZERWINSKA and **P DOERFFER**
Institute of Fluid Flow Machinery, Polish Academy of Sciences IMP PAN, Gdansk, Poland

Normal shock wave induced separation in a curved channel shows a significant influence of the channel spanwise depth upon the flow structure in a separated area. Measurement possibilities of separated flows are rather limited, so the numerical simulations have been carried out. They have supported the experiment and allowed much better insight into the 3-D details of the flow structure.

1 INTRODUCTION

The flow in a curved passage is the basic element in turbomachinery aerodynamics. Passages in turbines are sometimes not very high along the blade span. Also wind tunnels used for experimental investigations have strong limits in respect to spanwise dimension. Often transonic conditions appear in such passages due to a large pressure drop, when a local supersonic area is formed at the convex wall. This local supersonic area is always terminated by a normal shock wave.

An existence of the shock wave and its interaction with a boundary layer at the convex wall introduces a high tendency towards separation. If the separation occurs the spanwise depth of the passage may control the separated flow structure. It is important to analyse such flow conditions because they influence the passage flow and may be responsible for differences in experiments between wind tunnels. Due to the channel curvature a secondary flow is always induced in the side wall boundary layers. This flow interferes with separated flow structure in the way that may be dependent on the passage depth. In order to investigate this phenomenon independently a passage flow has been chosen instead of a cascade flow in order to avoid an uncontrolled influence of the horse shoe vortices.

2   EXPERIMENT

The curved passage used for the investigations has a convex wall of radius R = 300 mm and a typical boundary layer thickness in front of the shock wave boundary layer interaction about $\delta = 3.5$ mm (Fig.1). Investigated Mach number range, upstream the shock wave, was M = 1.35 to M = 1.47, covering the induction and development of separation.

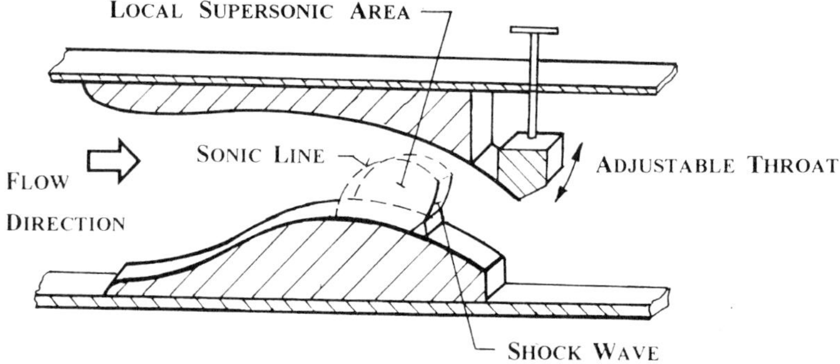

Fig.1. Configuration of the measurement passage

Two passage depth have been investigated H = 50 mm and H = 150 mm. The first of these cases has been studied in the University of Karlsruhe. Wind tunnel used was working intermittently where vacuum tanks sucked in the air through a test section. The separation flow structure has been analysed and presented in [1, 2, 3].

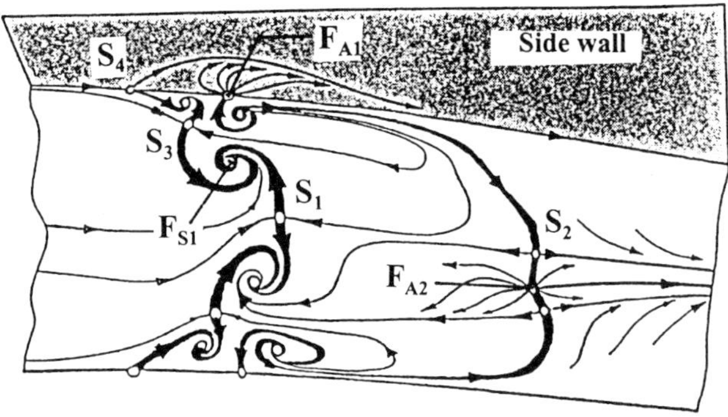

Fig.2. Flow structure at the test wall, narrow wind tunnel

The second one has been investigated in the Aviation Institute in Warsaw [4]. The wind tunnel there was driven by pressure tanks and the air was discharged to the atmosphere through the test section.

Carried out measurements have reviled that the shock induced separation displays globally different flow structures in both cases. These structures at M = 1.47 are sketched in Fig.2 and 3. They have been extracted from the oil visualisations. Reynolds number range is of the same order 10^5 in both wind tunnels. It is defined by the undisturbed boundary layer thickness upstream the interaction and the upstream flow parameters. Downstream the shock wave, in the area of separation, the boundary layer is about three times thicker than upstream the interaction (where it is about 3.5 mm thick). These dimensions indicate that the streamwise extent of separation structures is an order of magnitude larger than the corresponding sizes normal to the wall.

The two structures presented are globally different. This is well displayed by the separation line which starts at saddle S_4 in the test wall - side wall corner and ends up on the centreline as the saddle S_1 in case of the narrow test section and as the node N_1 in case of the wide test section. The differences concern first of all the sense of rotation of the separation focus F_{S1}. Secondly in the wide tunnel there is enough spanwise space for the node N_1 to appear in the test wall centre. Due to this something reminding 2-D separation bubble is formed with separation at N_1-S_1 and reattachment at S_2-N_2. Such area is not present in a narrow wind tunnel.

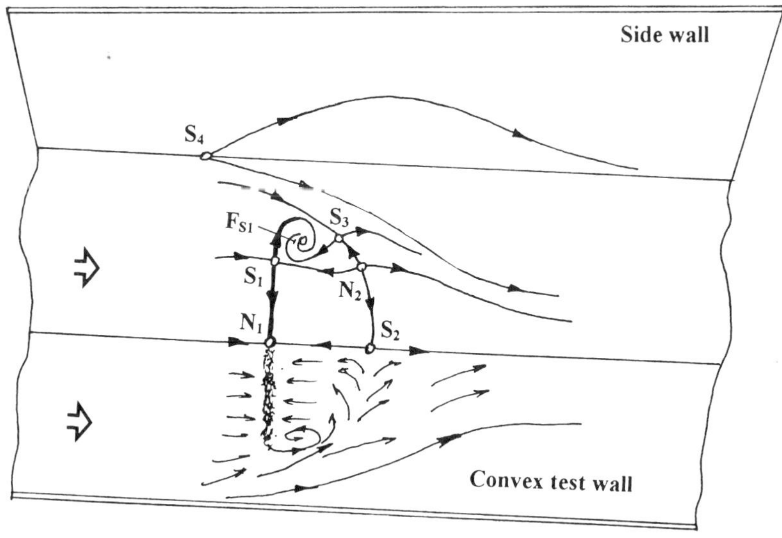

Fig.3. Flow structure at the test wall, wide wind tunnel

The sense of rotation of F_{S1} in the narrow test section (Fig.2) and a direct neighbourhood of the same focal point on the other side of the centreline cause that the separation saddle point S_1 is shifted downstream the F_{S1} location. This particular example illustrates the difficulty of separation definition in 3-D flow. The shock wave and the onset of separation is located upstream the F_{S1} but in the centre line there is no change of flow direction there. Due to the F_{S1} sense of rotation the flow direction at the test plate middle is still streamwise within the separated area.

The reattachment is similar in both cases. In a narrow wind tunnel it takes place in a form of attachment focus F_{A1} and a saddle S_2. In case of a wide wind tunnel the reattachment node N_2 which corresponds to F_{A1} and a saddle point S_2.

Typical experimental measurements are carried out in the middle plane of the test section. Along this middle plane skin friction coefficient values (c_f) have been determined from the measured boundary layer profiles. The typical distributions along the centre line of the test wall are presented in Fig.4. A qualitative difference of the distributions has been observed upstream the separation point. In case of a wide wind tunnel the c_f drop goes directly to zero value and the separation line (Fig.3) is just downstream the shock wave. In case of the narrow wind tunnel the c_f drop does not reach zero but runs into a plateau of very low c_f value before it indicates separation. It means that the reverse flow in the middle of the wall appears rather far downstream the shock induced skin friction drop.

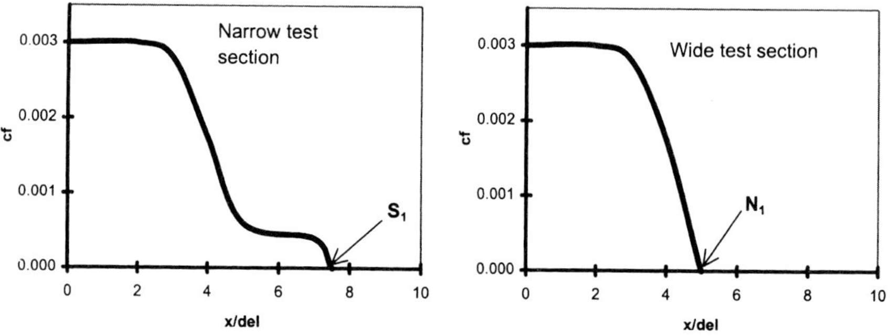

Fig.4. Skin friction coefficient distribution along the test wall centre line

In the case of a narrow passage the separation flow structure is controlled by the secondary motion developed within side wall boundary layers. The main vortices of separated flow at the convex wall are forced to rotate in opposite direction than in a case of wide passage. This leads to a very complicated pattern of the separation line showing the domination of 3-D effects.

In case of a wide wind tunnel the separation structures close to side walls are distinctly apart from each other. An area of a more uniform flow zone sets in between, where separation looks more 2-D like.

Our experimental observations showed that in a narrow wind tunnel the increase of separation by high Mach numbers leads to asymmetry of the separation pattern [3]. In case of the wide test section this effect has not been observed.

3 NUMERICAL SIMULATIONS

The abilities of experimental measurements within 3-D separated flow areas are very limited. In our case they consisted of pressure measurements in a very few locations and oil visualisations on the wall which deliver an important information on the flow structure. However, the flow structure at the wall does not provide a unique solution of a 3-D flow pattern above it.

In order to understand the topology of separated flow and especially the global structural changes which are controlled by passage depth it is useful to employ numerical simulations. The numerical simulations of a viscose, turbulent and compressible flow have been carried out by means of the solver KAPPA [5] used in our research group. 3-D flow through the above mentioned passages has been computed at Mach number around M=1.47 at which the experiment showed structures of Fig.2 and Fig.3. In order to access the flow structure within separated area the number of grid points has to be very large. A structural grid of 129×97×65 (streamwise × height × span) has been used. This grid covered half span with a symmetry condition in the channel middle. The grid point closest to the wall was at $y^+ - 0.03$ for the test wall, where the separation structures are investigated, and $y^+ = 0.5$ at other walls. In order to reduce the calculation time the full multigrid and parallel computing are employed in our solver. It has been observed that in lower multigrid levels, hence for less grid points the structures don't change but the resolution of structures is worse and the scope of separation diminishes.

It should be reminded that at present state of numerical simulations there are no turbulence models that provide a good modelling of the separated flows. Therefore one of the useful tasks of presented here work is to find out if any of available numerical schemes together with a number of turbulence models gives a possibility to obtain separation structures of at least similar qualitative layout as those observed in experiment. If this is successful than one could grasp the chance to use field data to understand the main features undergoing global structural changes induced by the variation in a spanwise depth of the channel.

The flow simulations carried out at present used the non-linear κ-τ turbulence model of Craft-Launder-Suga. This is the non-linear eddy viscosity model incorporating invariant parameters of the mean strain and vorticity. Applications of the model are generally in good agreement with experiment [6] and accuracy is comparable with current second-moment closures even though computing requirements for this latter type are about 3-5 times grater than model of Craft-Launder-Suga.

It could be generally said that the numerical results agree very well with experiment. It concerns the flow structure in the passage. Even the "after expansion" downstream the interaction is much stronger and closed with a weak shock wave in the narrow wind tunnel. This is a typical effect of the side wall boundary layer which was very well displayed by 3-D simulations. The only result we would like to present here is the static pressure distribution along the test wall in Fig.5.

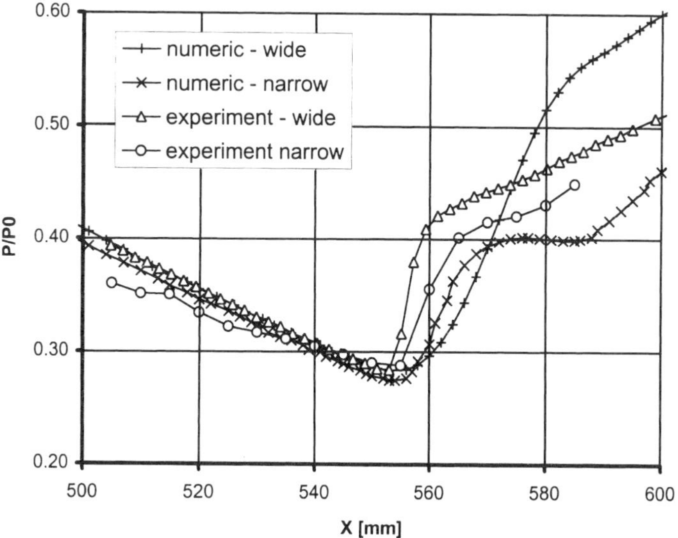

Fig.5. Static pressure distributions along the test wall for all considered cases

The figure above shows that the pressure gradients upstream the interaction are nearly identical for numerical simulations and experiment. The shock wave location in the test section are not exactly the same and the presented curves are shifted slightly in relation to each other in order to overlap the upstream part of pressure distribution. The maximum gradient of pressure jump under the shock wave is the same for all cases except the experiment in the wide wind tunnel. It is a bit steeper for this case. The characteristic tendency to form a plateau downstream the pressure jump in the narrow wind tunnel is shown in experiment also. The only noticeable difference is that in numerical simulation of wide wind tunnel the pressure rises more significantly than in experiment far downstream the shock wave. This may be explained by the different realisation of the outflow boundary condition in the experiment and in the numerical simulations. This boundary condition allows to control the characteristic upstream Mach number value. In the experiment the outlet from the test section is flexible and allows for cross section area variation. In numerical simulations the passage geometry is fixed and the flow is controlled by the outlet static pressure. This difference may be responsible for some differences in pressure distribution close to the outlet but should not have any effect on the shock boundary layer interaction.

C557/143 © IMechE 1999

The numerical simulations confirm the existence of structures observed in experiment to certain extend. Fig.6 and Fig.7 show a plan view on the test wall between the middle line (lower) and the side wall (upper line) for the narrow and the wide channel respectively. Plotted arrows present velocity vectors of unit length in 3-D at the plane close to the test wall. So, what is seen is a projection of these vectors on the test wall. Therefore it may be considered as a representation of the shear lines as in oil visualisation. Some additional arrows have been added in order to better indicate the flow structure.

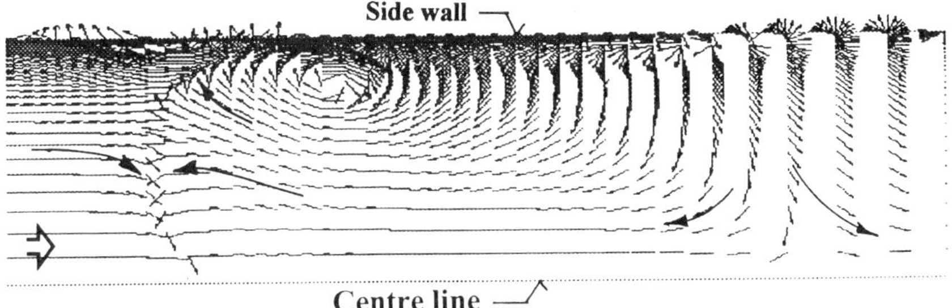

Fig.6. Plan view on the velocity vectors near the test wall in the narrow test section

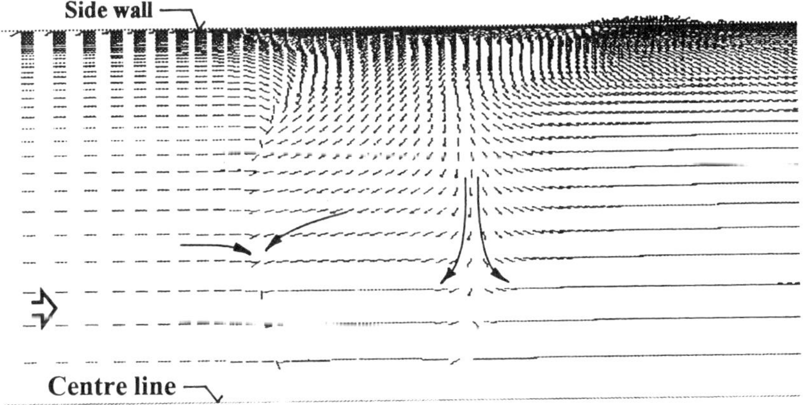

Fig.7. Plan view on the velocity vectors near the test wall in the wide test section

Flow direction is from the left to the right side. The first general observation is that the proportion between separation length and channel width is in agreement with experiment. In case of a narrow test section three elements of flow structure are very distinct. To the left the separation line, then the reattachment focus close to the side wall and to the right the reattachment saddle on the middle line. A complicated 3-D structure along the test wall-side wall corner upstream of the separation is also very distinct.

The existence of the large separation focus F_{s1} is however not so evident as in experiment (Fig.2). There is a strong downstream shift of the separation saddle in the wall centre. This fact may be induced by the existence of a rotation with the same sense of rotation as F_{s1} which is not displayed so well by presented vector plots. In order to see if such rotation exists a single streamline is plotted in Fig.8a. This is a 3-D streamline shown in plan view (upper) and side view (lower). The streamline starts at the marked point. Its shape indicates the existence of an anti-clockwise rotation as in F_{S1} and further on it lifts up from the test wall into a separated area.

In case of wide test section Fig.7 the flow close to the test wall-side wall does not show so strong 3-D effects as in the other case. It is noticeable that in a large part of the test wall the separation looks 2-D like having the separation and reattachment lines nearly straight and perpendicular to the oncoming stream.

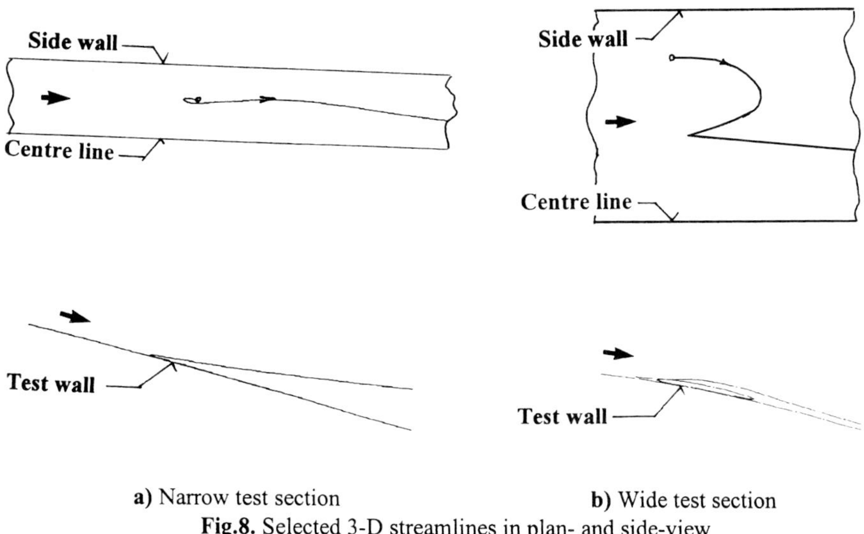

a) Narrow test section   **b)** Wide test section

Fig.8. Selected 3-D streamlines in plan- and side-view

The streamline generated in the side wall region (Fig.8b) shows an existence of rotation in the opposite direction to the narrow test sections what is in agreement to the separation focus F_{S1} in Fig.3.

One of the interesting issues that we hoped to be able to answer with the help of numerical simulations is the 3-D character of the reattachment saddle S_2. In the streamwise middle plane, normal to the test wall it is possible that this point is a nod or a saddle point. It is essential because it decides if the S_2 is the reattachment or the separation point. It could not be answered experimentally due to the limited measurement possibilities. The convergence of streamlines downstream the S_2 may induce separation, which in 3-D is a local phenomenon and may happen on streamwise planes. So, our previous claim that S_2 is a reattachment saddle needs to be proved.

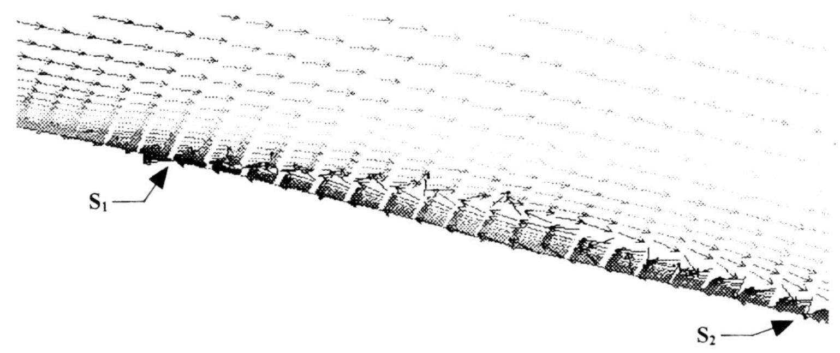

a) Unit velocity vectors at the streamwise middle plane, narrow wind tunnel

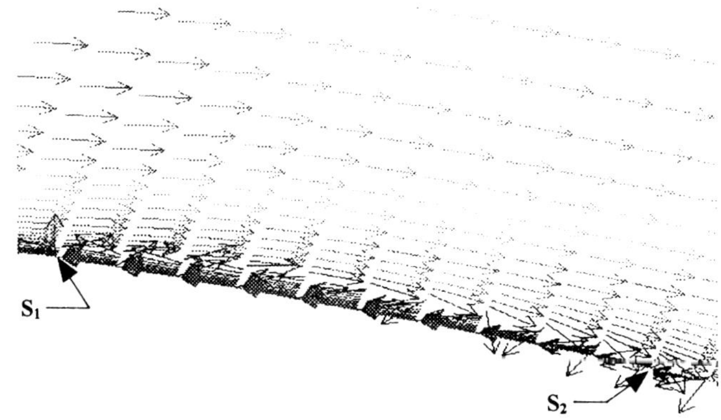

b) Unit velocity vectors at the streamwise middle plane, wide wind tunnel
Fig.9. Flow structure at the streamwise middle plane

Thanks to the field data of numerical simulation one is able to answer this question easily. In Fig.9 velocity unit vectors are plotted at the streamwise middle plane perpendicular to the test wall. In Fig.9a the narrow test section case is presented and in Fig.9b the wide test section case is displayed. The S_2 in the presented plain is a saddle point in both cases. It means that the S_2 is really a reattachment point. It is also indicated once again that the separation length is much larger in case of the narrow test section.

4 CONCLUSIONS

Experimental results indicate that the shock induced separation in a curved duct may be influenced considerably by the spanwise channel depth. It concerns 3-D structure of separated flow.

The used numerical scheme and turbulence model have provided a tool which allows to obtain separation structures similar as in the experiment. Obtained results have confirmed that the carried out numerical simulations are able to predict separation structures rather well. Certain details have to be improved still in order to obtain even better coincidence of numerical results and experiment.

Results obtained allowed already to make conclusion that was beyond existing experimental ability. Namely it has been proved that S_2 is a reattachment point in 3-D sense.

Further analysis has to be continued in order to resolve structure details at the streamwise corner separation development.

REFERENCES

[1]   P. Doerffer "An experimental investigation of the Mach number effect upon a normal shock wave - turbulent boundary layer interaction on a curved wall", Acta Mechanica 76, pp.35-51 (1989).

[2]   P. Doerffer, U. Dallmann "Mach number dependence of flow separation induced by normal shock wave - turbulent boundary layer interaction at a curved wall", AIAA Paper 89-0353.

[3]   P. Doerffer, U. Dallmann "Reynolds number effect on separation structures in normal shock wave - turbulent boundary layer interaction", AIAA Journal, Vol.27, No.9, 1989.

[4]   P. Doerffer, W. Kania, M. Bejm "Interaction between a shock wave and a turbulent boundary layer" in Polish, IMP PAN Gdańsk, Internal Report, 269/97.

[5]   F. Magagnato, "KAPPA" TASK Quarterly, Volume 1, No4., Gdańsk 1998, ISSN 1428-6394.

[6]   T.J. Craft, B.E.Launder, K. Suga, "Extending the applicability of eddy viscosity models through the use of deformation invariants and non-linear elements", 5th IAHR Conference on Refined-Flow Modelling and Turbulence Measurement, Paris 7-10 September 1993

[7]   S.Tatsumi, L.Martinelli, A.Jameson, " Flux-limited schemes for the compressible Navier-Stokes equations", AIAA Journal, vol. 33, No. 2, 1995

[8]   J.Czerwińska, "Computations of the 3-D transonic flow in a curved channel", IMP PAN Gdańsk, Internal Report, 260/98.

C557/139/99

Unsteady flow past a turbine blade using non-linear two-equation turbulence models

F MAGAGNATO
Institute für Strömungslehre, Universität Karsruhe (TH), Germany

Synopsis

The flow past turbine blades with thick trailing edges is characterized by periodically separating large scale vortices. The failure of two-equation turbulence models to predict the near wake of a turbine blade correctly appears to be mainly due to the inability of accounting properly for the large scale statistics. The investigation presented here should reveal if a time-dependent calculation in conjunction with a linear as well as a non-linear two-equation model can prescribe these flow fields better than a steady state calculation since the large scale vortices are resolved by the numerical scheme and the turbulence model must account only for the small scales.

1. Introduction

The flow past a turbine blade designed by Cicatelli and Sieverding (1) and measured by Ubaldi et al. (2) has been numerically investigated under the framework of the Brite-EURAM project called TURMUNSFLAT. The code used is a 3-D block structured finite volume cell centered scheme called KAPPA (3) developed at the University of Karlsruhe, with an explicit Runge-Kutta type integration in time (4). The convergence to steady state calculations is accelerated by a full multigrid method and an implicit residual averaging technique (5). The original artificial dissipation scheme of Jameson et al. (3) as well as the SLIP and USLIP scheme after Tatsumi et al. (6) and the AUSM-scheme according to Liou/Steffen (7) are implemented in the code. Several turbulence models ranging from the simple mixing length model of Baldwin /Lomax (8) to the linear eddy viscosity two-equation models of Launder/Sharma (9) and Speziale et al. (10) and non-linear eddy viscosity models proposed by

Speziale (11) and Craft et al. (12) are implemented. The code is parallelized with the MPI message passing library.

The simulation of unsteady flow fields on turbine stages with low Reynolds number models requires a very efficient time integration scheme in order to calculate these flow fields with an acceptable amount of computer time. The necessity of using parallel solvers for this application gives an additional requirement for these schemes. The author chose a formulation of the dual-time-stepping scheme proposed by Arnone et al. (13), which is very efficient and parallelizable. The basic idea is to reformulate the governing equations so that it can be treated as a modified steady state problem in a fictitious time τ. If the time discretization is made implicit, stability restrictions are removed and acceleration techniques such as multigrid, local time stepping etc. can be used instead of traditional time-consuming factorization (i.e. alternate direction implicit and lower/upper schemes).

An important aspect of the calculation past turbine blades is the type of turbulence model used to represent the unresolved turbulent fluctuations. Linear two-equation eddy viscosity models are the simplest turbulence models which can handle such flows. A new type of non-linear eddy viscosity model offers now a good compromise between the robustness and efficiency of the traditional linear eddy viscosity model and the in principal more accurate but computationally expensive Reynolds stress models. This type of models has a non-linear stress-strain relation in contrast to the linear models and are capable to model anisotropic normal Reynolds stresses. The last term on the right hand side of equation eq. (1) is the non-linear part of the above mentioned stress-strain relation.

$$ -\rho\,\overline{u_i u_j} = -\frac{2}{3}\rho\,k\delta_{ij} + \mu_t D_{ij} + F_{ij}(D_{ij}^2, D_{ij}^3,..) \quad \text{eq. (1)} $$

where D_{ij} is the strain tensor

Since there are several non-linear models proposed in the past, one has to choose the one that offers the best compromise with respect to stability, economy, usefulness, portability and accuracy for the intended application. The model proposed by Speziale (11) offers only minor improvements in accuracy because it adopts only quadratic terms in the formulation of the stress-strain relation (for details see Launder (14) and Magagnato (15)). Another disadvantage is the inclusion of the Oldroyd derivative in the stress-strain relation creating some numerical stability problems and higher computational effort. The non-linear model proposed by Craft/Launder/Suga (12) to date seems to offer the best compromise for the intended application because of the inclusion of cubic terms in the stress-strain relation as well as the expected numerical stability of this model. Preliminary computations on steady as well as unsteady flow fields fully confirmed this expectation. The additional computational effort of the non-linear model with respect to its linear counterpart is in the range of only 5-10% overhead whereas the model of Speziale needs about 30-40% more cpu-time compared to a linear model. All prescribed turbulence models based on k-ε have been transformed into an equivalent k-τ formulation. The reason for that is the numerical robustness of the τ-equation due to the linear distribution of the turbulence time scale τ in the boundary layer and the numerically robust boundary condition of τ=0 at the wall. In order to suppress negative turbulent kinetic energy or turbulent time scales during the transient phase, the solution of these quantities were not allowed to fall below 0.1 % of the freestream value.

2. Calculations

Since the aim of this work focuses on unsteady flow fields the flow around a circular cylinder at large Reynolds number is suitable for a first test. The well documented experiment of Cantwell and Coles (16) has been adopted. The Reynolds number Re_D is 140 000 based on the diameter of the cylinder and the freestream velocity $u = 21.2 m/s$. The turbulence level Tu is 0.6%. An O-type grid has been generated with a total of 52 000 points on the finest grid (see Fig. 1). Calculations with app. 13 000 points showed almost the same results as with 52 000 points indicating mesh independence. All calculations have been performed with the switch scheme of Jameson et al. (4). Since the characteristic turbulence length scale at the boundary is not known, the common practice is to choose the turbulence length scale such that the resulting eddy viscosity in the freestream is in the order of the laminar viscosity. Eddy viscosity models are not able to predict natural transition to turbulence no matter if they are formulated with wall functions (low Reynolds number modeling) or log law of the wall, therefore the transition location must be prescribed explicitly. In the case of the circular cylinder the results depend strongly on the appropriate location of the transition region. In these calculations the transition point has been fixed at $\alpha = 80°$ since it is known from experiments that transition of sub-critical flow past cylinder occurs in the separated free shear layers. The velocity component in streamwise-direction is shown in Fig. 2 for the non-linear model. The linear model predicts a Strouhal number of $St_{cl} = 0.226$ and a mean drag coefficient of $c_D = 0.95$, whereas the non-linear model predicts $St_{cl} = 0.215$ and $c_D = 0.91$. Compared with the experimental findings of Cantwell and Coles $St_{cl} = 0.179$ both turbulence models overpredict considerably this value. But they are in better agreement with the Strouhal number range of $St_{cl} = 0.16 - 0.225$ measured by Lienhard (17). The drag coefficients predicted by both models are considerably lower than the measurements of Cantwell and Coles of $c_D = 1.237$. But calculations with transition locations imposed more down stream ($\alpha = 90°$) gave drag coefficients as high as $c_D = 1.55$ for the linear model and $c_D = 1.65$ for the non-linear model.

Transition fixed at $\alpha=80°$ for all calculations	c_D	St_{cl}
Experiment Lienhard	0.77 – 1 4	0.16 – 0.225
Experiment Cantwell/Coles	1.237	0.179
Launder/Sharma Model (linear)	0.95	0.226
Craft et al. Model (non-linear)	0.91	0.215

The surface pressure distribution compared to the measurements is shown in Fig. 3 ($\alpha = 80°$). The result calculated with the linear model is displayed by a dashed line whereas the non-linear one is displayed with a solid line. Both models predict an overshoot in the $\alpha = 90°$ region followed by a higher pressure recovery in the wake of the cylinder $90° < \alpha < 270°$. The pressure distribution of this test case depends largely on the turbulence length scale specified in the freestream and the resulting transition region. Calculations with different turbulence length scales showed large mean total drag variations of more than 50% in both directions. Even a complete suppression of the vortex shedding could be observed by using a very high turbulence length scale.

Next the flow around a turbine blade has been calculated. The measurements were conducted by Ubaldi et al. (2) . The Reynolds number $Re_{2is}= 1.6 \cdot 10^6$ is based on the chord length s = 0.3m and the isentropic exit Mach number $Ma_{2is} = 0.23$. The measured turbulence level was Tu = 3% . No turbulence length scale has been measured. In contrast to the cylinder flow the decay of the turbulent kinetic energy along the turbine blade has been measured. In order to match these values a turbulence length scale of $L_{tu} = 0.018m$ was applied at the upstream boundary. A very fine grid has been generated for this test case with about 72 000 points for one blade. Calculations with about 18 000 points showed again almost the same results as with the fine grid. Experiments indicated a dominant periodic nature of the wake with a frequency of about 1700 Hz for the test case with natural transition. A fully turbulent calculation with the linear two-equation model of Speziale et al. (10) predicted an absolutely steady flow field (Fig. 4) caused by a high level of eddy viscosity (Fig.5) which suppressed all flow instabilities. A calculation without turbulence model was done next in order to verify if the suppression is dominated by numerical dissipation. This calculation showed a vortex shedding frequency of about 1652 Hz indicating that the results are not contaminated by numerical dissipation. In contrast to the linear model a fully turbulent calculation with the non-linear model of Craft et al. (12) was capable to predict an unsteady flow field but with a reduced vortex shedding frequency of about 1080 Hz . The reason for that lies in a considerable lower (approximately 60%) eddy viscosity distribution in the near wake (see Fig. 6) compared to the linear model allowing the flow instabilities to grow. The contours of velocity magnitude are displayed in Fig 7 (please note only one blade has been calculated using periodic boundary conditions). The comparison of the velocity profiles calculated with the non-linear model at the suction side shows a very good agreement with the measurements at $s/s_{max} = 0.35$ as well as $s/s_{max} = 0.95$ (see Fig.8).

	Frequency	St
Experiment Ubaldi et al.	≈1700 Hz	0.34
Laminar calculation	1652 Hz	0.33
Speziale et al. model (linear)	0 Hz	0.0
Craft et al. model (non-linear)	1080 Hz	0.216

The vortex shedding frequency of the laminar calculation is very close to the experimental value. We believe that the dissipative nature of an eddy viscosity model may not adequately prescribe the formation of the vortices in the near wake. Numerical experiments with lower turbulence length scale than the above used values results in a lower eddy viscosity distribution in the near wake and an increased vortex shedding frequency of about 1500 Hz.

3. Conclusions

A linear as well as a non-linear eddy viscosity model has been validated against flow past a circular cylinder and a flow around a turbine blade. In the former the pressure distribution predicted with both models are quite different from the measurements and depends strongly on the prescribed transition location whereas the Strouhal number is almost independent for both models. Calculations from Beaudan/Moin (18) past circular cylinder at a lower Reynolds number (Re_d = 3500) demonstrate that the flow is highly three-dimensional in the near wake. And a two-dimensional calculation can not reproduce the pairs of counter-rotating streamwise vortices observed in the experiment. It is believed that the same is true for higher Reynolds numbers which would explain the above mentioned disagreement.

In contrast to the circular cylinder the flow past the VKI-turbine blade is only modestly three-dimensional in the wake and therefore a two-dimensional calculation can capture the dominating vortices of this flow. It has been shown that the linear eddy viscosity turbulence model can not predict the experimentally observed vortex shedding from the trailing edge whereas the non-linear eddy viscosity model can predict a vortex shedding but with a frequency 37% below the measured value. The agreement of the velocity profiles between measurement and the calculation with the non-linear model are very good.
The conclusion is therefore that the non-linear model offer a somewhat better modeling of unsteady flow field around turbine blades.

Acknowledgements
The author gratefully acknowledges the financial support of the European Community under the Brite-EuRam III Contract No. BE95-1698 and the extensive discussions with Prof. Zunino from University of Genoa.

REFERENCES
[1]   Cicatelli, G. and Sieverding, C. H.: The Effects of Vortex Shedding on the Unsteady Pressure Distribution around the Trailing Edge of a Turbine Cascade. ASME paper no. 96-GT-359, 1996

[2] Ubaldi M., Zunino P., Campora U., Chiglione A.: Detailed Velocity and Turbulence Measurements of the Profile Boundary Layers in a Large Scale Turbine Cascade. ASME-Journal 96-GT-42, 1996

[3] Magagnato, F.: KAPPA-Kompressibel 2.0. Technical Report 97/4 , Institut für Strömungslehre, 1997

[4] Jameson, A., Schmidt W., Turkel E.: Numerical Solution of the Euler Equations by Finite Volume Methods using Runge-Kutta Time-Stepping Schemes. AIAA-paper 81-1259, 1981

[5]   Jameson, A.: Multigrid Algorithms for Compressible Flow Calculations. Technical Report 1743, MAE-Report, 1985

[6]   Tatsumi S., Martinelli L., Jameson, A.: Flux-Limited Schemes for the Compressible Navier-Stokes Equations. AIAA-Journal Vol. 33, No. 2, 1995

[7]   Liou M. S., Steffen C. J.: A New Flux Splitting Scheme. Journal of Computational Physics 107, 23-39, 1993

[8]   Baldwin B. S., Lomax H.: Thin Layer Approximation and Algebraic Model for Separated Turbulent Flows. AIAA-paper 78-257, 1978

[9]   Launder B., Sharma B. I.: Application of the Energy Dissipation Model of Turbulence to the Calibration of Flow near Spinning Disc. Letter Heat Mass Transfer, pages 131-138, 1976

[10] Speziale, C. G., Abid R., Anderson E. C. : A Critical Evaluation of Two-Equation Models for Near Wall Turbulence. ICASE Report No. 90-46, 1990

[12] Craft, T. J., Launder, B. E., Suga, K.: A Non-linear Eddy Viscosity Model including Sensitivity to Stress Anisotropy. 10th Symposium. Turbulent Shear Flows, Pennsylvania State University, 1995

[11] Speziale, C. G.: On Non-linear k-l and k-ε Models of Turbulence. Journal of Fluid Mechanics, pp. 458-475, 1987

[13] Arnone A., Liou M. S., Povinelli L. A.: Integration of Navier-Stokes Equations Using Dual Time Stepping and a Multigrid Method. AIAA-Journal Vol. 33, No. 6, 1995

[14] Launder, B. E.: Turbulence and Transition Modelling. Kluwer Academic Publishers, 1996

[15] Magagnato, F.: Untersuchung von linearen und nichtlinearen Wirbelviskositätsmodellen. Dissertation TH-Darmstadt, 1995

[16] Cantwell, B. and Coles, D.: An Experimental Study of Entrainment and Transport in the Turbulent near Wake of a Circular Cylinder. Journal of Fluid Mechanics, pp. 321-374, 1983

[17] Lienhard, J. H.: Synopsis of Lift, Drag and Vortex Frequency for Rigid Circular Cylinders. College of Engineering Research Division bulletin 300, Tech. Extension Service, Washington State University, 1966

[18] Beaudan, P. and Moin, P.: Numerical Experiments on the Flow Past Circular Cylinder at Sub-critical Reynolds Number, Report No. TF-62, Stanford University, California, 1994

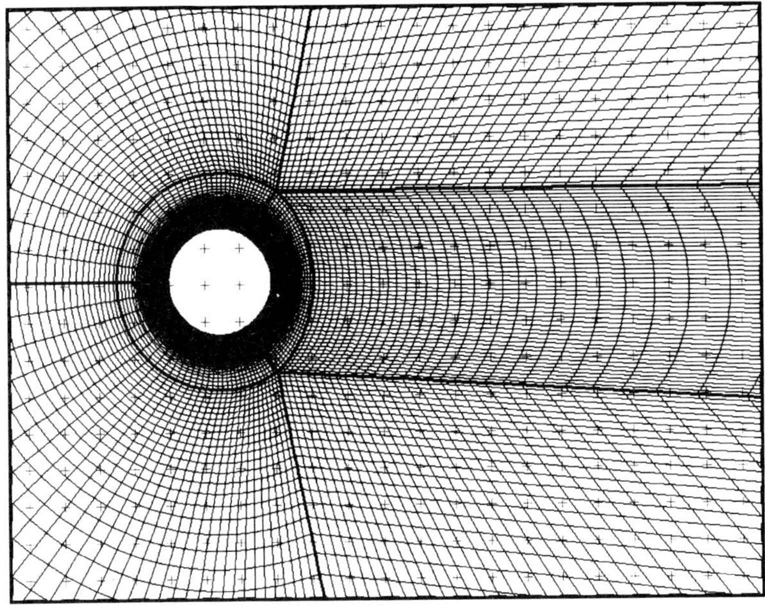

Figure 1 : Computational grid for circular cylinder

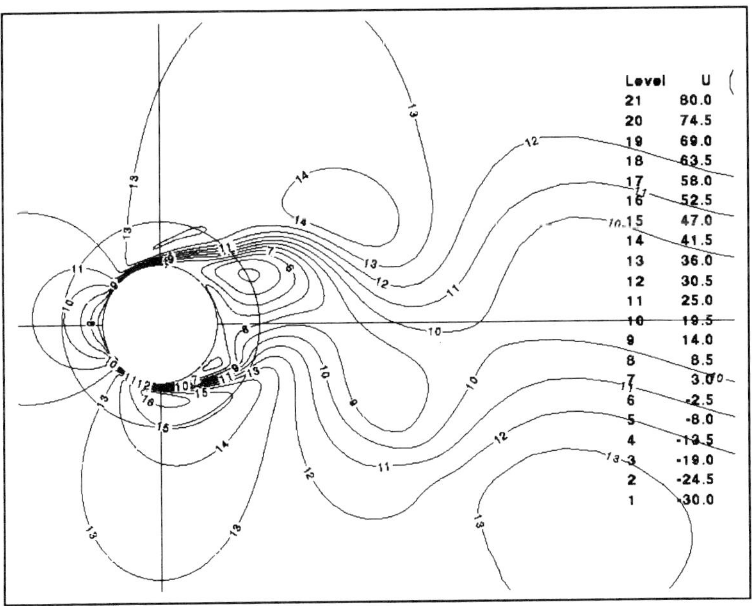

Figure 2 : Streamwise velocity distribution around the circular cylinder

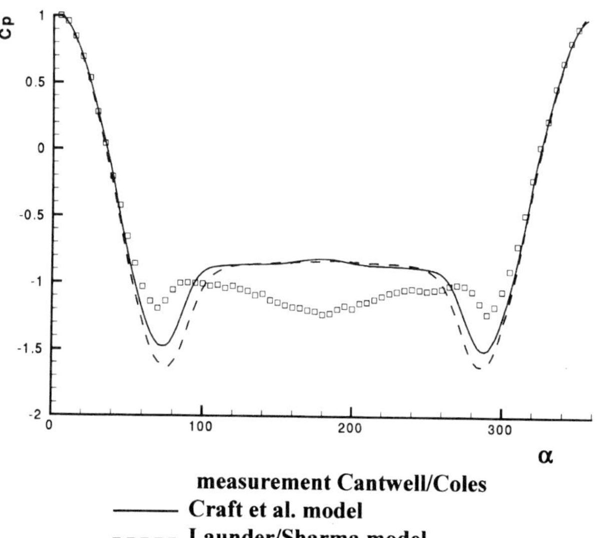

measurement Cantwell/Coles
——— **Craft et al. model**
- - - - - **Launder/Sharma model**

Figure 3 : Surface pressure distribution on circular cylinder

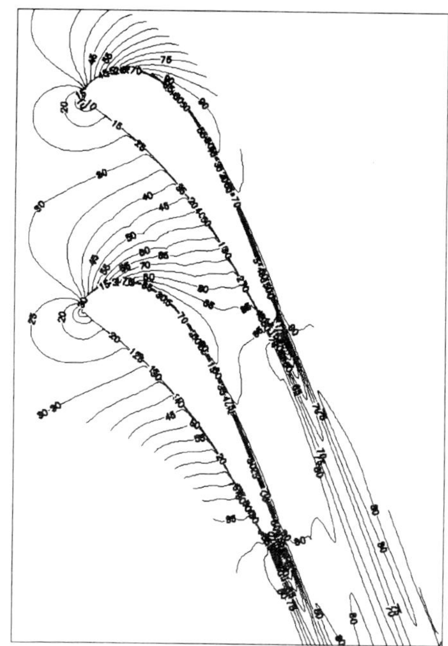

Figure 4 : Contours of velocity magnitude in of VKI turbine blade, Speziale et al. model
Units in m/s

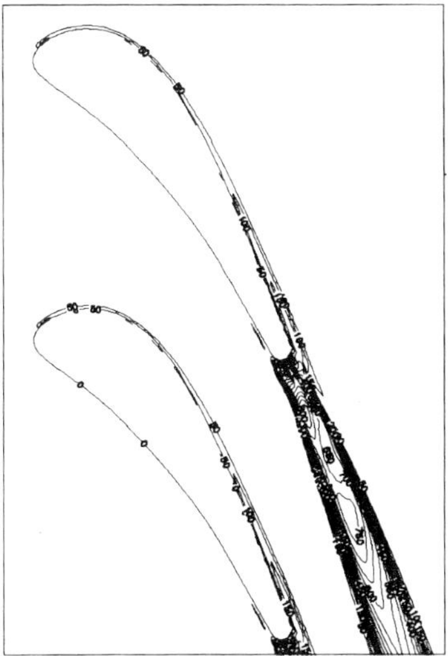

**Figure 5 : Contours of eddy viscosity ratio of VKI turbine blade, Speziale et al. model
Units in m/s**

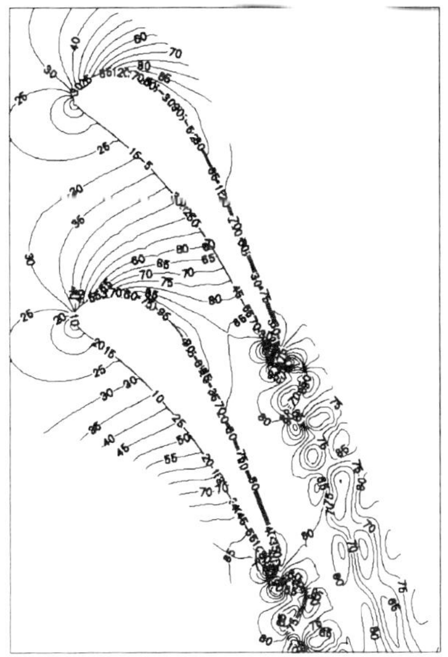

**Figure 6 : Contours of velocity magnitude of VKI turbine blade, Craft et al. model
Units in m/s**

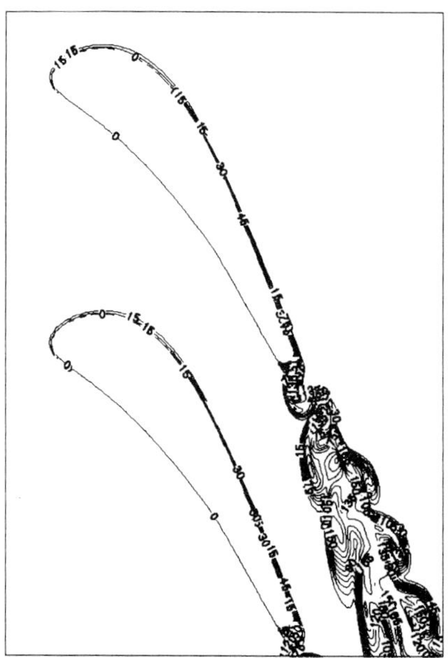

Figure 7 : Contours of eddy viscosity ratio of VKI turbine blade, Craft et al. model

$s/s_{max}=0.35$ $s/s_{max}=0.95$

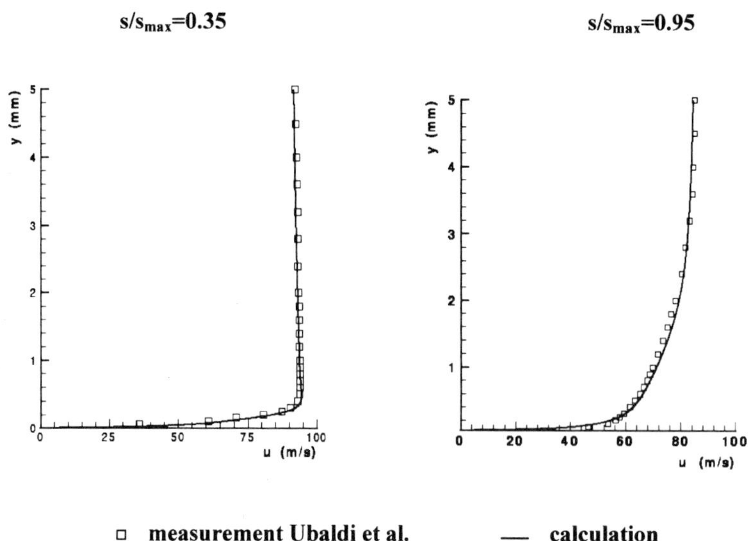

Figure 8 : Velocity profile at $s/s_{max} = 0.35$ and at $s/s_{max} = 0.95$ on the suction side

Investigation of wake-induced transition on a highly loaded low-pressure turbine cascade

S BRUNNER and **L FOTTNER**
Institut für Strahlenantriebe, Universität der Bundeswehr München, Neubiberg, Germany
V SCHULTE and **G KAPPLER**
BMW Rolls-Royce GmbH, Dahlewitz, Germany

SYNOPSIS

In turbomachines periodically unsteady flow is caused by the relative motion of rotor and stator rows. In order to simulate a moving blade row upstream of a highly loaded low pressure turbine linear cascade a wake generator has been designed and built in a high speed cascade wind tunnel. The wakes are generated with cylindrical bars moving with a velocity of up to 40 m/s in the test section upstream of the cascade inlet plane.

Measurements have been performed at varying Reynolds-numbers at steady and unsteady inlet flow conditions. For the unsteady inlet flow conditions the frequency (Strouhal-number) of the wake passing has been altered by varying the speed of the bars. The turbulence intensity and the velocity deficit of the bar wakes have been measured with a 1D hot-film probe. Wake-induced transition is semi-quantitatively mapped by employing a simultaneous surface hot-film anemometry system. Measurements of the surface pressure distribution and wake traverses have been performed.

If the effects of rotor-stator interaction are taken into account a new design of turbine blades with an increased blade loading (high lift) is possible. A significant reduction of blade numbers for constant stage loading can be achieved, which decreases the weight and the cost of the turbine module.

NOMENCLATURE

c_{ax}	[m/s]	axial velocity
E	[V]	anemometer voltage
f	[Hz]	frequency
l	[m]	chord length
s	[m]	position along surface length

g	[m]	surface length of profile suction side
t	[m], [s]	pitch, time
T	[s]	time period between two wakes
U	[m/s]	bar velocity

Greek

μ_3	[-]	skewness
ω	[-]	total pressure loss, $\omega = \Delta p_t / q_{2th}$
Θ	[m]	momentum thickness

Abbreviations

CTA		Constant Temperature Anemometry
Ma		Mach-number
Re		Reynolds-number
RMS		Root-Mean-Square
Sr		Strouhal-number $Sr = f*l/c_{ax}$

Subscripts

is		isentropic
2th		downstream conditions for isentropic flow
b		bar
c		cascade
~		ensemble averaged

1. INTRODUCTION

Recent research has revealed positive effects of unsteady flow on the development of profile boundary layers in turbine cascades at conditions with a laminar suction side separation bubble. Compared to steady flow a reduction of total pressure loss over a broad range of Reynolds-numbers has been shown. A new design of turbine blades with an increased blade loading (high lift) is possible if the effects of rotor-stator interaction are taken into account. A significant reduction of blade numbers for constant stage loading can be achieved. This decreases the weight and the cost of the turbine module.

Positive effects of rotor-stator interaction mainly occur with high lift low pressure turbines, where a laminar suction side separation bubble exists especially at low Reynolds-numbers (1), (2). Incoming wakes force the boundary layer of the suction side to undergo transition, whenever the Reynolds-number based upon momentum thickness exceeds the range of Re_Θ = 90-150 (3), (4). Following the wake-induced transitional regions calmed regions appear, which show laminar like boundary layer behaviour (2), (5), (6). With a full velocity profile and low entropy generation the calmed region combines the positive properties of laminar and turbulent boundary layers. It is able to inhibit turbulent spot generation and withstand larger negative pressure gradients than conventional laminar boundary layers. The calmed region is therefore able to suppress or delay a laminar suction side separation bubble. The replacement of a thick laminar separation bubble that is present at steady inflow conditions with intermittently becalmed or transitional attached flow at unsteady inflow conditions leads to a time mean reduction of loss generation. Extra benefit may be gained by setting an optimal sequence of transitional, separated and becalmed regions, hence a favourable reduced frequency (2) of the rotor wakes. In general the positive effect of rotor-stator interaction increases at higher loading (7).

Several parameters influence the loss generation of the low pressure turbine blade: the Reynolds-number Re, the Strouhal-number Sr, the wake strength (turbulence intensity or velocity deficit within the wake and the wake width) and the loading of the turbine cascade.

Up to date only experiments on rotor-stator interaction on high lift low pressure turbines are known to the authors which were performed in cold-flow turbine rig tests (8) or from tests at low speed cascade wind tunnels (1), (2), (3), (9), (10), or low speed rotating rigs (5), (9). So the logical step is to close the gap between turbine rig testing and testing at low speed by using a high speed cascade wind tunnel with moving wake simulation. Data taken primarily in low speed cascade testing can be verified and supplemented at high speed in order to optimize design criteria for high lift low pressure turbines taking into account the effects of rotor-stator interaction.

2. TEST FACILITY AND MEASUREMENT TECHNIQUE

2.1 The High Speed Cascade Wind Tunnel

The High Speed Cascade Wind Tunnel (HGK) of the Universität der Bundeswehr München is an open loop facility which can operate continuously and reach Mach-numbers up to Ma = 1.05 in the test section (Fig. 1). Being built inside a large pressure tank the wind tunnel offers the possibility to vary the Mach- and the Reynolds-number in the test section independently (11) in order to correctly simulate the flow conditions inside turbomachines.

The air is supplied by a six-stage axial compressor, driven by a 1.3 MW a.c. electric motor which is situated outside the pressure tank. The air enters through a diffuser into a settling chamber where it is cooled down to an adjustable constant temperature (30°C to 60°C). A turbulence generator and a nozzle are positioned upstream of the 300 mm wide test section. It has variable height.

<table>
<tr><td colspan="2">test section data:</td></tr>
<tr><td>- Mach number</td><td>: $0.2 \leq Ma \leq 1.05$</td></tr>
<tr><td>- Reynolds number</td><td>: $10^6 \leq Re/l \leq 1.5\ 10^7$</td></tr>
<tr><td>- degree of turbulence</td><td>: $0.3\% \leq Tu_1 \leq 6\%$</td></tr>
<tr><td>- upstream flow angle</td><td>: $25° \leq \beta_1 \leq 155°$</td></tr>
<tr><td>- blade height</td><td>: 300 mm</td></tr>
</table>

<table>
<tr><td colspan="3">wind-tunnel data:</td></tr>
<tr><td>- a.c. electric motor</td><td>: P = 1300 kW</td><td></td></tr>
<tr><td colspan="3">- axial compressor (six stages):</td></tr>
<tr><td>air flow rate</td><td>: V = 30 m^3/s</td><td>(max.)</td></tr>
<tr><td>total pressure ratio</td><td>: Π = 2.14</td><td>(max.)</td></tr>
<tr><td>rotational speed</td><td>: n = 6200 rpm</td><td>(max.)</td></tr>
<tr><td>- tank pressure</td><td>: 0.05 bar $\leq p_k \leq 1.2$ bar</td><td></td></tr>
</table>

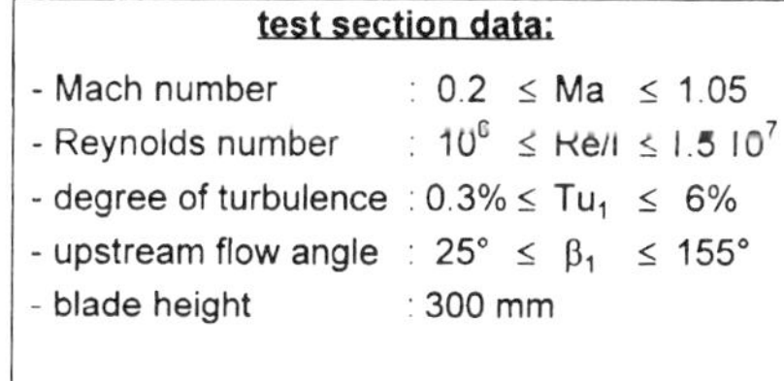
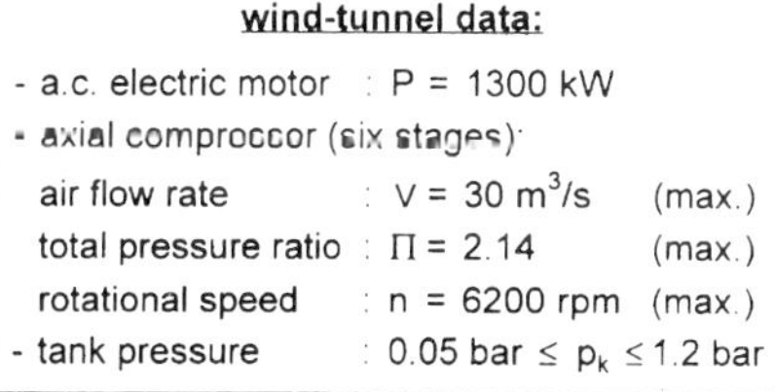
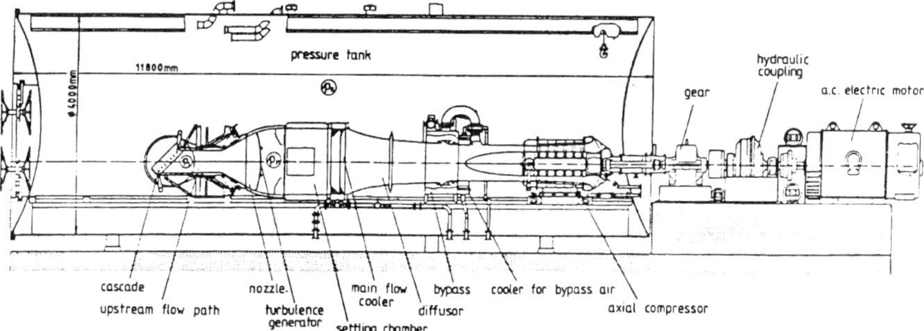

Figure 1: The High Speed Cascade Wind Tunnel

2.2 The wake generator

In turbomachines periodically unsteady flow is caused by the relative motion of rotor and stator rows. A wake generator has been designed and built in order to simulate a moving blade row upstream of the linear turbine cascade in the high speed cascade wind tunnel of the Universität der Bundeswehr München (fig.2), (12). The wakes are generated with cylindrical bars moving with a velocity of up to 40 m/s in the test section upstream of the cascade inlet plane (table 1). As shown in (13) the far wake of a cylinder is similar to that of an airfoil assumed that both cause the same total pressure losses. With the wake generator an angle of the incoming wake (see fig.3) of about 55° (Sr = 1.06) can be generated, whereas a representative turbine has a wake angle of about 65° (c_{ax}/U = 0.85). Despite this deviation of the velocity triangle basic investigations on the effects of wake-induced transition on the suction side boundary layer can be performed.

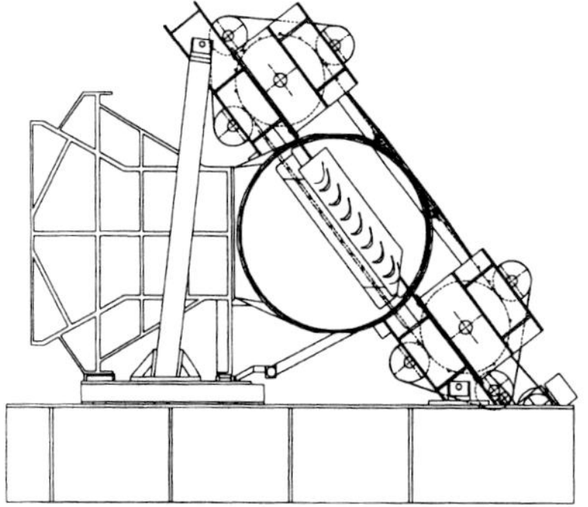

Figure 2: The generator of unsteady inlet flow conditions

Table 1: Design data for wake generator

bar diameter	2 mm
bar pitch	min. 10 mm
bar velocity	1 - 40 m/s
maximum number of bars	400
Strouhal-numbers in tested cases	Sr = 0.26 - 1.06
axial gap between bar plane and cascade inlet plane	$0.75 * l_{ax}$

The principle of the wake generator is based on cylindrical bars fastened to two rubber timing belts moving in front of and behind the cascade. The timing belts, having a length of 4000 mm, span over two main pulleys with a diameter of 400 mm positioned above and below the cascade. The distance between the trailing edge of the cascade and the returning bars is large enough that the flow in the blade passages is not disturbed and to allow the traverse of all probe types usually employed during the tests. The driving pulley is positioned on the lower main shaft and is connected by a timing belt to the a.c. electric motor using a 1:1 transmission ratio. The water-cooled motor has a nominal maximum

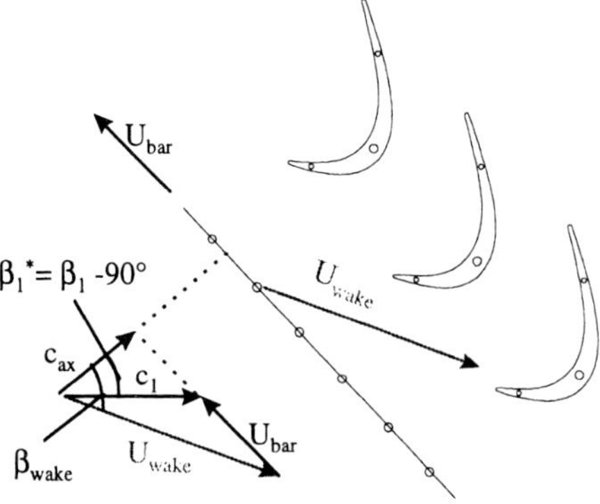

Figure 3: Angle of the incoming wake

power of 10 kW at 3000 rpm. The motor can move the bars in both directions for simulating the rotor-stator interaction of compressor as well as of turbine stages. The velocity of the motor can be digitally controlled with an accuracy of 0.1 rpm from 100 rpm to 3000 rpm.

In order to avoid disturbances of the inlet flow due to the timing belts in front of the cascade, it was necessary to reduce the width of the test section to 176 mm. Thus the blade aspect ratio for the used cascade was reduced to 1.76. Due to flow visualizations with the oil-and-dye technique on the blade surface and with a flow field traverse downstream ($x/l_{ax}=1.5$) of the cascade a two dimensional flow at mid-span of the turbine cascade had been assured.

2.3 Measurement Techniques

2.3.1 Pneumatic Measurements

Measurements of the surface pressure distribution and wake traverses in a plane $x/l_{ax} = 1.5$ downstream of the cascade entry plane have been performed. All measurements have been taken at mid-span of the turbine cascade. The surface pressure distribution has been measured utilizing a Scanivalve SDIU-System, whereas the wake traverses have been conducted with a five-hole probe and a PSI DPT6400-System (14), (15).

2.3.2 1D Hot-Film Measurements

The turbulence intensity and the velocity deficit due to the bars in the cascade inlet plane have been measured with a standard 1D hot-film probe (DANTEC HF-55R01) connected to a constant-temperature bridge of the type 55M01. The sensor calibration has been performed in the cascade inlet plane of the HGK test section and a 4th order polynomial has been used for the approximation of the calibration curve.

2.3.3 Glue-On Hot-Film Measurements

Wake-induced transition is semi-quantitatively mapped by employing a simultaneous surface hot-film CTA anemometry system (DANTEC-Streamline), (16). Twelve SENFLEX (array 9102) hot-film sensors, covering about 40% of the suction surface length, have been logged simultaneously. In total 24 hot-film sensors each spaced 2.5mm apart have been used for unsteady inlet flow conditions. In order to obtain the desired information about wake-induced transition on the suction surface no calibration of the hot-film sensors is necessary (17). The location and the mode of transition can be efficiently detected just by the normalized RMS value of the anemometer output signal.

2.3.4 Data Acquisition and Processing

The 1D hot-film probe measurements have been controlled by a HP 3565 workstation, which acquired data with a sample frequency of 65 kHz. The analog data have been filtered by an anti-aliasing filter and then have been digitized by a 14-bit A/D-converter.

With the simultaneous surface hot-film anemometry system 12 channels with a sample frequency of 30-40 kHz each have been obtained. The low-pass filter has been set to 10 kHz and no high-pass filter has been used. Additionally one channel was employed for a once per revolution trigger and another channel for the detection of each bar. The well established ensemble averaging technique has been applied to evaluate the raw data, using 300 ensembles each consisting of 5 wake passing periods.

The ensemble average of a time-dependent quantity b is given by:

$$\tilde{b}(t) = \frac{1}{N} \cdot \sum_{j=1}^{N} b_j(t)$$

where N is the number of ensembles and the ensemble RMS is given by:

$$\sqrt{\tilde{b}'^2(t)} = \sqrt{\frac{1}{N-1} \cdot \sum_{j=1}^{N} \left(b_j(t) - \tilde{b}(t) \right)^2}$$

The skewness (third-order moment)

$$\mu_3(t) = \frac{1}{\left(\tilde{b}(t) \right)^3} \cdot \frac{1}{N-1} \cdot \sum_{j=1}^{N} \left(b_j(t) - \tilde{b}(t) \right)^3$$

has also been calculated in order to gain a deeper insight into the boundary layer characteristics.

In the case of the 1D hot-film data, b is set to the velocity and for the glued-on hot-film data to the anemometer output voltage.

3. EXPERIMENTAL RESULTS

The measurements have been performed on a highly loaded low pressure turbine linear cascade (see fig.4), which, representing a current high lift LP turbine design, is specifically designed in order to take advantage of the positive effect of rotor-stator interaction on the suction side boundary layer. In order to obtain a deeper insight into the wake induced transition phenomena on the suction side, variations of the Reynolds-number and Strouhal-number (velocity of the bars) have been undertaken. Therefore a Reynolds-number range from $Re_{2th} = 70000$ to $Re_{2th} = 300000$ and a Strouhal-number range from $Sr = 0.26$ to $Sr = 1.06$ have been investigated. With the variation of the Strouhal-numbers the angles of the incoming wakes have been changed from about 48° ($Sr = 0.26$) to about 55° ($Sr = 1.06$).

Data shown in this paper have been acquired with a total of 100 bars, spaced 40 mm apart from each other. In order to examine the unsteady inlet conditions to the turbine cascade, caused by the moving bars, measurements with a 1D hot-film probe in the cascade inlet plane have been carried out. In figure 5 an example of such a 1D hot-film probe result is displayed, showing as well the ensemble averaged inlet velocity distribution as the ensemble averaged inlet

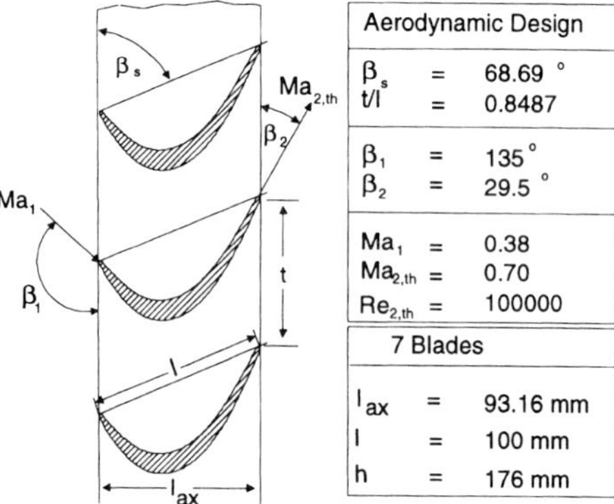

Figure 4: Aerodynamic design

turbulence level distribution both over the non-dimensional wake passing period (6 periods).

As the results of the 1D hot-film measurements at different conditions have been very similar to each other, only one representative result has been chosen. Considering an inlet turbulence intensity in the freestream of about $Tu_1 = 1\%$ at conditions of figure 5, it is clearly indicated by the 1D hot-film results that the bar wakes have already grown together in the cascade inlet plane. The turbulence level between two bar wakes can be detected at a level of about $Tu_{min} = 2\%$, whereas the peak turbulence levels within the bar wakes reach

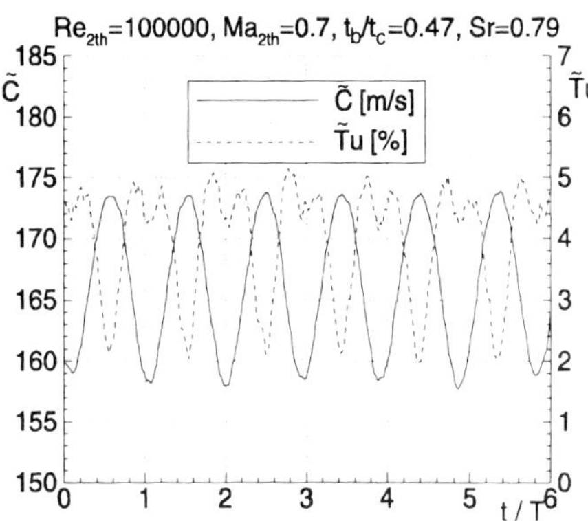

Figure 5: Result of 1D hot-film probe data

values of about $Tu_{max} = 5\%$. The ensemble averaged velocity distribution shows a velocity deficit of about 8.6% inside the bar wake, compared to the ensemble averaged velocity between two wakes. This velocity deficit is representative of a real engine situation.

In figure 6 the isentropic Mach-number distributions at various conditions are presented (27 suction side and 22 pressure side measurement points). By using the wake generator gaps in the wind tunnel test section upstream of the cascade inlet plane had to be provided to enable the traversing of the bars in front of the cascade. These gaps caused a change in incidence of about 5 degrees compared to the design condition, changing the inlet flow angle from $\beta_1=135°$ to about $\beta_{1flow}=140°$. Therefore all isentropic Mach-number distributions reveal a deceleration in the leading edge region of the suction side (fig. 6a) at steady inlet flow. The Mach-number distributions in the trailing edge region of the suction side in figure 6a show the presence of a

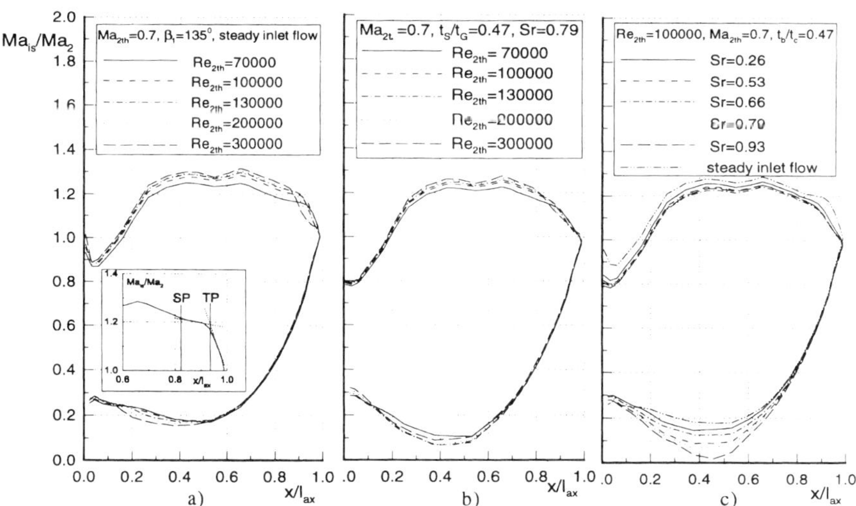

Figure 6: Isentropic Mach-number distributions; a) at steady inlet flow, at unsteady inlet flow with various b) outlet Reynolds-numbers and c) Strouhal-numbers

laminar suction side separation bubble at all investigated outlet Reynolds-numbers. The detection of a laminar separation bubble is been presented as a zoom-out in fig. 6a. The start of a plateau-like distribution of the isentropic Mach number distribution describes the location of the separation point (SP) of a laminar separation bubble. The point of intersection of the tangents on the isentropic Mach number distribution thereby detects the transition point (TP) in the region of the laminar separation bubble. Compared to figure 6a the laminar separation bubbles in figure 6b seem to be reduced in size or even suppressed at unsteady inlet flow conditions. This leads to the conclusion that the time-mean flow on the suction side at unsteady inlet flow is not separated to the same extent as at steady inlet flow conditions. Figure 6c shows Mach-number distributions at unsteady inlet flow conditions with various Strouhal-numbers. Due to the high mechanical load of the wake generator no time-mean pressure measurements and loss measurements have been carried out at Sr=1.06. The variation of Strouhal-numbers do not show any significant influence on the time-mean isentropic Mach-number distributions in the trailing edge region of the suction side.

In figure 7 glue-on hot-film measurement results are shown at an outlet Mach-number of $Ma_{2th} = 0.7$, an outlet Reynolds-number of $Re_{2th} = 100000$ and a variation in Strouhal-numbers from $Sr = 0.26$ to $Sr = 1.06$. The time-space diagrams of the RMS values are shown with the non dimensional wake passing period over the non dimensional axial chord length. The black regions thereby denote the location of the global maxima of RMS values, and therefore the

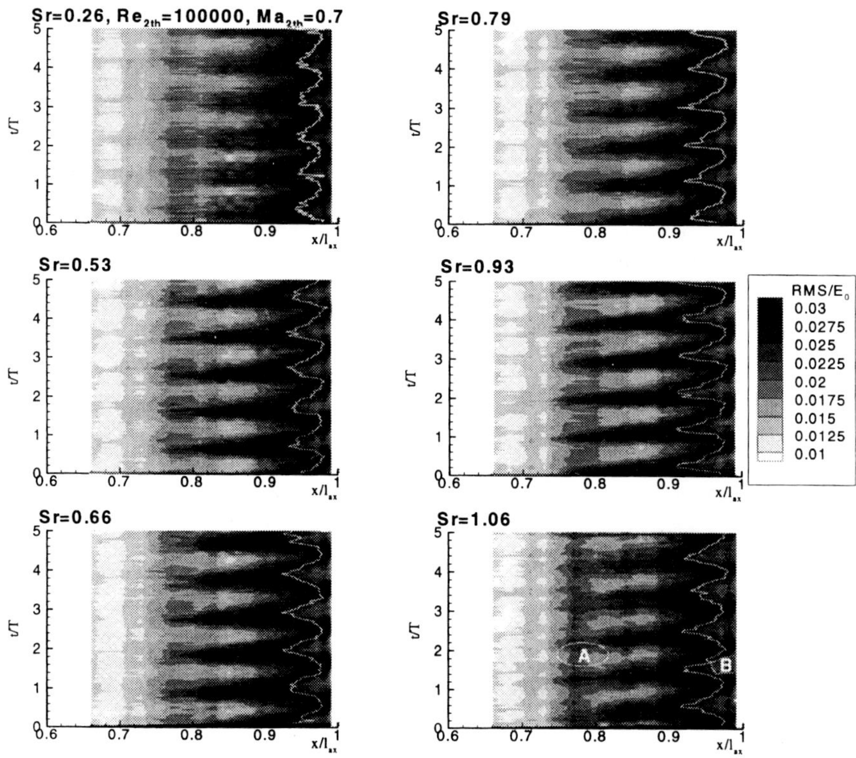

Figure 7: Glue-on hot-film results at various Strouhal-numbers

position where the intermittency factor is about 0.5. The intermittency is the probability that a point P in space and time is covered by turbulent flow (1). The periodic movement of the transition region by the passing of the wakes can be clearly identified. In addition, the location of zero skewness is indicated by the fine white line in figure 7. Zero skew is also an indicator for approximately 50% intermittency. It is observed, that the zero-crossing of the skewness also moves upstream whenever a wake is present on the suction side boundary layer. But the movement of the skewness zero-crossing is not as strong as the movement indicated by the RMS levels. At various Strouhal-numbers the onset of forced transition remains at the location of about x/l_{ax} = 0.76. Between two successive regions of wake-induced transition a laminar separation bubble seems to re-establish. The onset of a laminar separation bubble can be indicated by the local high RMS values (region A). The isentropic Mach-number distribution at steady inlet flow (fig. 6a) shows the onset of a laminar separation bubble at about x/l_{ax} = 0.8, which corresponds to the high RMS levels in region A (fig. 7). This could lead to the conclusion that a laminar separation bubble between two wake-induced turbulent patches occurs. Between the wake-induced transition the transition moves downstream, to a location x/l_{ax} = 0.98, where transition takes place without any wakes at the inlet. Wake-induced transition leads to a fully turbulent boundary layer upstream of the trailing edge (region B), while the boundary layer is still transitional between the wake-passing where it undergoes natural or separated boundary layer transition. As can be seen in figure 7, at this low Reynolds-number of Re_{2th} = 100000 there always exists a laminar separation bubble between the wake induced transitional paths with different Strouhal-numbers.

In figure 8 also time-space diagrams of RMS levels are shown, but now with a variation of the outlet Reynolds-numbers, whereas the outlet Mach-number and the Strouhal-number are kept constant at Ma_{2th} = 0.7 and Sr = 0.79. At outlet Reynolds-numbers of Re_{2th} = 70000 and Re_{2th} = 100000 a laminar suction side separation bubble is visible between the wake induced turbulent paths (region C and D). At even higher outlet Reynolds-numbers, as the transitional region moves upstream with increasing Reynolds-numbers, the laminar

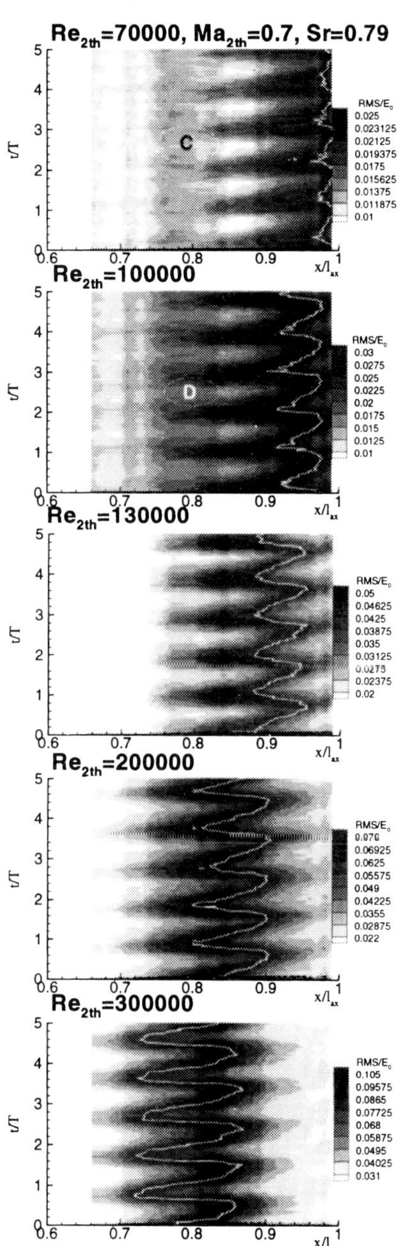

Figure 8: Glue-on hot-film results at various outlet Reynolds-number

separation bubble can no longer be detected. The increasing level of the hot-film signals with rising outlet Reynolds-number corresponds to the higher pressures inside the wind channel at higher Reynolds-numbers. At $Re_{2th} = 70000$ a wind tunnel pressure was adjusted to 4.2 kPa, whereas at $Re_{2th} = 300000$ a pressure of 18.4 kPa existed. With increasing outlet Reynolds-numbers not only the transition region moves upstream (see also fig. 6a, b), but also the extent of the transition region (detected by about 50% of the appropriate maximum RMS level) changes. The extent of the transition region increases from 20% of suction surface length at $Re_{2th} = 70000$ to about 33% at $Re_{2th} = 100000$. With even higher outlet Reynolds-numbers the extension decreases, however, to about 26% of the suction surface length at $Re_{2th} = 300000$. The zero-crossing of the skewness seems to agree better with the movement of the global RMS maxima with increasing Reynolds-numbers. There is an assumption that the wake induced transitional paths have already grown together at outlet Reynolds-numbers higher than $Re_{2th} = 130000$. Thus no natural transition between the wake-induced transitional paths can be detected at these outlet Reynolds-numbers.

Time-mean pressure measurements (wake-traverses) have been carried out with a five-hole probe. Hereby calculated bar losses (2) have been separated from the measured total losses to gain the total losses of the cascade. The results of the variation in Strouhal-numbers are shown in figure 9a and the results at various outlet Reynolds-numbers are displayed in figure 9b.

Figure 9a reveals, that at the Strouhal-number of about $Sr = 0.8$ there seems to be an optimal sequence of wake-induced transitional patches and becalmed regions, thus an optimal reduced frequency, though the minimum is very flat. The total pressure loss reduction amounts to about 34.1% ($Sr = 0.79$) compared to the total loss at steady inlet flow (which is shown as the data point for zero Sr). Due to the high mechanical load of the wake generator no loss measurements have been carried out at $Sr = 1.06$. In figure 9b total loss data of unsteady and steady inlet conditions are represented. In the entire Reynolds-number range the total pressure loss at unsteady inlet conditions is significantly below the loss at steady inflow, which is in line with the time-mean reduction of a laminar separation bubble at unsteady inlet flow conditions (fig. 6b). The data in figure 9b clearly exhibit the design potential for new high lift bladings especially at very low Reynolds-numbers. Even at the relatively high outlet Reynolds-number of $Re_{2th} = 300000$ there is still a reduction of total pressure loss at unsteady inlet conditions.

This is in line with previous findings for a low speed version of the current high lift LP profile (2).

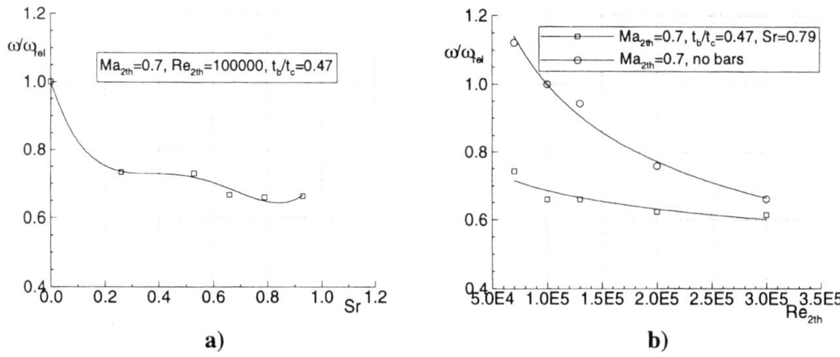

**Figure 9: Total pressure loss at various a) Strouhal-numbers and
b) outlet Reynolds-numbers**

4. SUMMARY

The investigation on wake-induced transition on the suction side of a turbine cascade and its dependency on loss production has shown an optimal Strouhal-number of Sr = 0.8 with a minimum in the total pressure loss. At Sr = 0.8 a reduction of total pressure loss of about 34% compared to steady inlet conditions has been achieved. Also beneficial effects of rotor-stator interaction have been revealed with respect to the total pressure loss over a wide Reynolds-number range, especially at very low Reynolds-numbers. Even at an outlet Reynolds-number of Re_{2th} = 300000 a total pressure loss reduction of about 7% has been observed. At Re_{2th} = 100000 a periodic re-establishing of a laminar separation bubble at unsteady inlet conditions has been detected at all measured Strouhal-numbers from Sr = 0.26 to Sr = 1.06. The Strouhal-number variation at Re_{2th} = 100000 has not shown any influence on the wake-induced transitional length; it seems to be constant at 33% suction surface length.

Results from (2) have been confirmed for high speed flow. This shows that the boundary layer behaviour previously outlined in low speed moving bar cascade experiments seems to be applicable to high sub-sonic turbine exit Mach-numbers. This provides more confidence into the design of high lift turbines under consideration of unsteady inflow caused by rotor-stator interaction.

A significant reduction of blade numbers for constant stage loading can be achieved. This decreases the weight and the cost of the turbine module.

5. ACKNOWLEDGEMENTS

The reported work was performed within a research project that is part of the national research cooperation „AG TURBO". The project has been supported by the German Ministry of Education, Science, Research and Technology (BMBF) and the BMW Rolls-Royce GmbH. The permission for publication is gratefully acknowledged.

REFERENCES
(1) Schulte, V.; *Unsteady Separated Boundary Layers in Axial-flow Turbomachinery;* PhD Dissertation, Cambridge University, 1995
(2) Schulte, V.; Hodson, H.P.; *Unsteady wake-induced boundary layer transition in high lift LP turbines;* ASME 96-GT-486, 1996, Journal of Turbomachinery, Vol. 120, pp. 28-35, Jan. 1998
(3) Hodson, H.P.; *Blade Row Interference Effects in Axial Turbomachinery Stages: Bladerow Interactions in Low Pressure Turbines;* VKI Lecture Series 1998-02, 1998
(4) Mayle, R.E.; *The Role of Laminar-Turbulent Transition in Gas Turbine Engines;* ASME 91-GT-261, 1991
(5) Halstead, D.E.; Wisler, D.C.; Okiishi, T.H.; Walker, G.J.; Hodson, H.P.; Shin, H.; *Boundary Layer Development in Axial Compressors and Turbines;* Part 3: LP-Turbines, ASME 95-GT-463, 1995
(6) Schulte, V.; Hodson, H.P.; *Prediction Of The Becalmed Region for LP Turbine Profile Design;* ASME 97-GT-398, 1997
(7) Curtis, E.M.; Hodson, H.P.; Banieghbal, M.R.; Howell, R.J.; Harvey, N.W.; *Development of blade profiles for low pressure turbine applications;* ASME 96-GT-358, 1996

(8) Arndt, N.; *Blade Row Interaction in a Multistage Low-Pressure Turbine;* Journal of Turbomachinery, Vol. 115, pp. 137-146, Jan. 1993

(9) Banieghbal, M.R.; Curtis, E.M.; Denton, J.D.; Hodson, H.P.; Huntsmann, I.; Schulte, V.; *Wake Passing in LP Turbine Blades;* AGARD CP 571, Paper 23, 1995

(10) Schobeiri, M.T.; Pappu, K.; *Experimental Study on the Effect of Unsteadiness on Boundary Layer Development on a Linear Turbine Cascade;* Experiments in Fluids 23/4, S. 306-316, 1997

(11) Sturm, W.; Fottner, L.; *The High-Speed Cascade Wind-Tunnel of the German Armed Forces University Munich;* 8th Symposium on Measuring Techniques in Transonic and Supersonic Flow in Cascades and Turbomachines, 1985

(12) Acton, P.; Fottner, L.; *The Generation of Instationary Flow Conditions in the High Speed Cascade Wind Tunnel;* 13th Symposium on Measuring Techniques in Transonic and Supersonic Flow in Cascades and Turbomachines, 1996

(13) Pfeil, H.; Eifler, J.; *Turbulenzverhältnisse hinter rotierenden Zylindergittern;* Forschung im Ingenieurwesen, Vol. 42, pp. 27-32, 1979

(14) Amecke, J.; *Auswertung von Nachlaufmessungen an ebenen Schaufelgittern;* Bericht 67A49, AVA Göttingen, 1967

(15) Ganzert, W.; Fottner, L.; *WINPANDA - an Enhanced PC-Based Data Acquisition System for Wake and Profile Distribution;* 13th Symposium on Measuring Techniques in Transonic and Supersonic Flow in Cascades and Turbomachines, 1996

(16) Brunner, S.; Teusch, R.; Stadtmüller, P.; Fottner, L.; *The Use of Simultaneous Surface Hot Film Anemometry to Investigate Unsteady Wake Induced Transition in Turbine and Compressor Cascades;* 14th Symposium on Measuring Techniques in Transonic and Supersonic Flow in Cascades and Turbomachines, Sept. 1998

(17) Schröder, Th.; *Investigations of Blade Row Interaction and Boundary Layer Transition Phenomena in a Multistage Aero Engine Low-Pressure Turbine by Measurements with hot-Film Probes and Surface-Mounted Hot-Film Gauges;* Von Karman Institute for Fluid Dynamics, Lecture Series 1991-06, 1991

C557/084/99

A new intermittency model incorporating the calming effect

O N RAMESH and **H P HODSON**
Department of Engineering, University of Cambridge, UK

Synopsis: In this paper, a transitional intermittency model is presented, which incorporates the stabilising effect of the calmed region trailing behind the turbulent spot. In order to estimate the intermittency, the flow is allowed to break down into turbulent spots throughout the transition zone with equal probability. The calming effect is then added to this to derive an expression for the modified spot formation distribution. This expression is used to arrive at an intermittency distribution that can be used to model the boundary layer parameters in the transitional zone. Even though this formulation is mainly valid for constant pressure flows, it is shown to be applicable to flows with pressure gradient as well.

List of Symbols

g	Spot formation rate per unit area per unit time
g_c	Corrected spot formation rate per unit area per unit time
g_n	Spot formation rate (per unit distance per unit time) in concentrated breakdown model
G	Non-dimensional spot formation rate
G_c	Corrected non-dimensional spot formation rate
G_N	Non-dimensional spot formation rate in concentrated breakdown model
N	'Crumble' $= G_N R_\theta^{\,3}$
q	Freestream turbulence level

R_θ Reynolds number based on momentum thickness θ

x,z,t   Streamwise, spanwise and time co-ordinates

x_t Transition onset streamwise location

X $= Ux/\nu$

X_t $= Ux_t/\nu$

U Freestream velocity

w Width of the rectangular calmed region

α Half propagation angle of the turbulent spot = 10.5° for constant pressure flow

γ Intermittency

λ Transition extent

λ_θ Polhausen pressure gradient parameter

ν Kinematic viscosity

σ Spot propagation parameter

σ' Propagation parameter for spot and the calmed region

τ Ratio of σ' and σ

1. Introduction

The study of laminar-turbulent transition in attached boundary layer flows is a problem of tremendous importance both from the fundamental and applied perspectives. This has led to a spate of studies of this problem ever since Emmons [1] discovered turbulent spots and characterised the transition zone as being made up of these constituent spots - islands of turbulence in a laminar sea.

The intermittency (γ), is defined as the fraction of time for which the flow is turbulent at a point in the flow in the transitional zone. It is one of the most important parameters to model the transitional boundary layer parameters. The turbulent spots, which are formed due to the breakdown of the laminar flow, are seen to grow bigger till they coalesce to form the fully turbulent region. Hence the intermittency factor varies gradually from zero to unity across the transition zone.

The distribution of intermittency depends on the spot formation distributions across the transition zone. Over the years, various proposals for the spot formation rate and intermittency distribution have been put forward. These vary from the so - called continuous breakdown hypothesis of Emmons [1] and its variants (e.g., [2]) to the concentrated breakdown hypothesis of Narasimha [3]. Opinion has been divided as to the appropriateness of these hypotheses and they have led to many debates, especially in the field of turbomachinery, where transition modelling remains to be one of the most pressing problems.

In this work, we seek to reconcile the different modelling philosophies in vogue and this is expected to be achieved by incorporating two new factors - the stabilising effect of the so-called calmed region and that of the fully turbulent fluid in inhibiting further spot production.

In the following sections, after explaining the background of intermittency model in the transition zone, the effects of the calmed region and the turbulent spot in modifying the spot production distribution are quantified to arrive at an expression for intermittency distribution.

The analyses presented here are mainly valid for steady flows but the basic formulation has more general applicability for unsteady flows also, as shown by Schulte & Hodson [4]. The consideration of unsteady effects is especially important in LP turbine applications where wakes from the upstream blade rows makes the flow inherently unsteady and hence the transitional dynamics is quite different from the steady situation. However, as long as the time scale of unsteadiness of the mean flow is much larger than the time scale of the turbulence in the transitional boundary layer, one can tackle this problem as a quasi-steady flow (as far as transition modelling is concerned). Hence steady transitional modelling ideas can be brought to bear on the problem at hand. A recent work on the transitional boundary layer measurements in axial compressors and turbines, is given by Halstead et al. [5].

2. Background

2.1 Bypass Transition

If we consider the boundary layer development over a flat plate, the initially laminar flow, through a sequence of instabilities, breaks down into turbulent spots. The turbulent spots grow and merge as they propagate downstream to eventually encapsulate the whole flow field downstream. In turbomachinery flows, where the freestream disturbance levels are quite high, most of the initial stages of instabilities could be completely bypassed. This bypass transition is primarily the transition route of concern to us here. Fig. 1 shows the situation when breakdown occurs at only one streamwise location. This is therefore an example of concentrated breakdown.

The prediction of the transition onset location, is a formidable (and open) problem in hydrodynamic stability. Hence, it is usual to describe it in terms of a critical Reynolds number based on the boundary layer momentum thickness. For typical LP turbine situations, the peak freestream turbulence level in the neighbourhood of transition is typically of the order of about 4% [6]. Under these circumstances, the corresponding momentum thickness based Reynolds number would be about 168 at the transition onset location according to the correlation of Mayle [7]. In many LP turbines, this means that the transition sometimes begins before peak suction on the suction surface, but almost always ends after peak suction in the adverse pressure gradient region. It is this aspect, viz., that transition begins and ends in very different regimes that has led to the debates regarding the breakdown hypotheses.

2.2 Turbulent spots

Fig. 2 shows the schematic of a turbulent spot in bypass transition in a two-dimensional constant pressure flow. As Schubauer and Klebanoff [8] found, the spots are roughly heart shaped in planform with a pointed leading edge; the spots are often approximated by a triangular planform shape (following [9]). In the elevation, the spot can be seen to have an overhang in its shape which travels with a celerity (velocity) of about 88% of the freestream velocity; the trailing edge of the spot travels with a celerity of about half that of the freestream velocity. The planform shape varies slowly across the height of the boundary layer such that the planform at the boundary layer edge becomes a point.

The difference in the leading and trailing edge celerities of the boundary layer means that the spot grows in spatial extent as it moves downstream. Kinematically, since the spot is seen to have a self-similar form, this would mean a constant angle of propagation of the wedge enveloping the spot. This angle is about 21° for constant pressure flows.

Schubauer and Klebanoff [8] also found that a region of non-turbulent calmed region trails behind the turbulent spot. This calmed region is characterised by fuller velocity profiles than the laminar boundary layer which makes it less susceptible to destabilisation. Recent experiments (e.g., [10]) indicate that the calmed region is nearly rectangular in planform and its trailing edge is seen to travel with about 25% of the freestream velocity.

2.3 Intermittency Models

Following [1], consider the spatio-temporal evolution of a turbulent spot in planform. Together with the streamwise *(x)* and spanwise *(z)* co-ordinates, by considering the evolution in time *(t)*, the three-dimensional x-z-t space is constructed. A point $P(x,z,t)$ in this space gives the relative location in space and time of different points in the planform of a turbulent spot. A spot generated at $P(x, z, t)$ will sweep out a volume called the propagation volume in this space; a slice of this volume at a constant value of time (t) would reveal the planform of the spot at that instant. By extrapolating the generators of the propagation volume upstream of point P, we can construct the so-called dependence volume for the point P. Only spots which are created within this volume can make the point P turbulent.

The dependence volume of $P(x,z,t)$ is shown in Fig. 3. Here, the spot planform is taken to be triangular for the sake of simplicity following [9]. By working out the probability that a spot will be formed in the propagation volume, Emmons[1] derived the expression for intermittency at any point x as

$$\gamma = 1 - \exp\left\{ -\iiint_V g(x, z, t)\,dV \right\},$$

$$2.1$$

where $g(x,z,t)$ is the spot formation rate per unit time per unit area and dV is an elemental volume in the propagation cone; the product $g\,dV$ is the probability that a spot is formed in the volume dV. Note that, even though the equation (2.1) is valid even when $g(x, y, t)$ is non-stationary in all the variables, we will consider here only steady two-dimensional transition so that g is only a function of x. Further, if the spot propagation wedge is linear, as it is for constant pressure flows, it can be shown ([1], [11]) that

$$\iint dz_0 dt_0 = \frac{\sigma}{U}\left(x_0 - x\right)^2,$$

$$2.2$$

where

$$\sigma = Ut \iint_S \frac{dx dz}{x^3},$$

$$2.3$$

is the non-dimensional spot propagation parameter and it is equal to 0.25 for flat plate flow [11]; the integration is carried out over the spot planform (S). Since the integrand in equation (2.3) is weighted by $1/x^3$, it is biased more towards the trailing edge (or the lower limit of integration). Hence,

$$\gamma = 1 - \exp\left\{ -\int_{x_t}^{x} \frac{\sigma}{U} g\left(x_0 - x\right)^2 dx_0 \right\},$$

$$2.4$$

As a first approximation, Emmons assumed g to be a constant i.e., the spots are assumed to be formed with equal probability everywhere. This 'continuous breakdown' hypothesis leads to the expression

$$\gamma = 1 - \exp\left\{-\frac{\sigma}{3U} g(x - x_t)^3\right\} \qquad\qquad 2.5$$

for intermittency. Here x_t is the transition onset location from which the integration is started.

Schubauer and Klebanoff [8] measured intermittency distributions on a flat plate and they found that Emmons' hypothesis of 'continuous breakdown' did not agree with their measurements. To resolve this, Narasimha [3] proposed that the spots were born in a narrow belt around the onset location, and the length of this belt is so small in comparison to the transition zone length that the spots could be assumed to be formed only at the location x_t. This could be modelled by a Dirac Delta function centered at x_t, such that this 'concentrated breakdown' hypothesis would lead to

$$g = g_n \qquad x = x_t$$

$$\qquad\qquad\qquad\qquad\qquad\qquad\qquad 2.6a$$

$$g = 0 \qquad x \neq x_t$$

$$\gamma = 1 - \exp\left\{-\frac{\sigma}{U} g_n(x - x_t)^2\right\} \qquad\qquad 2.6b$$

The spot formation rate in Narasimha's model (g_n) has the dimensions of no. of spots per unit distance per unit time; whereas g in Emmons' model has the dimensions of no. of spots/area/time.

Further, by defining $\lambda = x(\gamma = 0.75) - x(\gamma = 0.25)$, which is a measure of the transition zone extent, and the non-dimensional streamwise distance $\xi = (x - x_t)/\lambda$ in equation (2.6b), the 'universal intermittency distribution'

$$\gamma = 1 - \exp\left\{-0.41\xi^2\right\} \qquad\qquad 2.7$$

is obtained. This expression has been found to be in very good agreement with measurements in constant pressure flows.

Even though the universal intermittency distribution (2.7) has been mainly derived for zero pressure gradient flows, it is found to be a good description of many pressure gradient flows as well [12]. However, there has been a lot of debate, especially in the field of turbomachinery where strong adverse and favourable pressure gradients are encountered at high freestream disturbance levels, as to the appropriateness of this hypothesis. There have even been suggestions that a hypothesis like Emmons' 'continuous breakdown' hypothesis is more relevant for bypass transition. (e.g., [2],[7],[13]).

Originally, it was suggested by Emmons that during continuous breakdown, the spot formation rate should increase downstream. It was thought that with the increase in Reynolds number, the flow becomes more unstable. However, the mean velocity profile becomes fuller, and hence more stable, downstream as the intermittency rises. This would imply a decreasing spot formation rate with increasing x. In reality, the breakdown of the laminar flow is neither 'continuous' nor 'concentrated' strictly but it is somewhere in between these two extreme pictures. In practice, for steady constant pressure flows, the belt over which the breakdown occurs is so small in comparison to the length of the transition zone, that the breakdown could be safely assumed to be concentrated at a preferred streamwise location for purposes of modelling.

It is only in recent years that the importance of the calmed region in transition zone has begun to be realised (see [4]). Since the calmed region is a very stable region with a full velocity profile, it was suggested by Schulte & Hodson [4] that the birth of spots is inhibited in these regions; also, a spot cannot be formed in a region which is already turbulent. These factors are now believed to explain the success of the 'concentrated breakdown hypothesis' in describing two-dimensional constant pressure steady flows. A reconciliation is hence effected between the 'concentrated' and 'continuous' breakdown hypotheses by incorporating the role of calming and the turbulent regions.

In this paper, following [4], the calming effect and the role of turbulent spots in inhibiting further spot production are quantified, to arrive at expressions for spot formation distributions and a closed form expression for the intermittency.distribution.

3. Transitional Intermittency with modified spot formation distribution

3.1 Propagation parameter for the spot and the calmed zone

In this section, the calmed region is quantified by a non-dimensional spot propagation parameter for use in later analyses. To do this, consider the turbulent spot with the calmed region trailing behind it (Fig. 2). The calmed region is seen to be fairly approximated by a rectangle in the planform. Then, following [1], we can define a propagation parameter (σ'') for the calmed region. This is given by (cf. Equation (2.3))

$$\sigma'' = Ut\iint_{C} \frac{wdx}{x^3},$$

3.1

where w is the width of the calmed region and the integration is from the spot trailing edge to the trailing edge of the calmed region (C). For the half angle of the propagation wedge $\alpha = 10.5°$, it can be shown that $\sigma'' = 1.11$.

The propagation parameter for the spot and the calmed region is the sum $\sigma' = \sigma + \sigma'' = 1.36$. With this, we can define another parameter(τ)

$$\tau = \sigma' / \sigma = 5.44$$

3.2

This ratio is a measure of the relative significance of a spot in terms of its calming vs. its contribution to intermittency.

3.2 Spot formation with calming and turbulence

In the present intermittency model, the spot formation rate is modified by the calming effect and the already turbulent fluid. To indicate this, we define the corrected spot formation rate by g_c. Note that this quantity is, in effect, a modified form of the spot formation rate g used in Emmons' model. Then the corrected intermittency is given by [4]

$$\gamma = 1 - \exp\left\{ -\iiint_{V} g_c(x_0, z_0, t_0)dV_0 \right\},$$

3.3

where $dV_0 = dx_0\, dz_0\, dt_0$. Physically, the expression (3.3) gives the probability that a flow is turbulent at a point since the integration is over the dependence volume V of the spot.

Since new spots cannot be born in a region which is already turbulent, or in a calmed region of an earlier spot (since the calmed region stabilises the flow and inhibits any flow breakdown), the probability of spot formation in an elemental volume dV is given by

$$P(S) \;=\; g_c\, dV \;=\; P(S/L)\, P(L) \;=\; g\;dV\, P(L),$$

In the above expression, $P(L)$, is the probability that the flow is laminar, and $P(S/L) = g$ is the probability that the spot is formed provided the flow is laminar (this is similar to the spot formation rate used in Emmons' theory i.e., independent of any calming effect etc.,). Here, $g = g(x)$, i.e., the uncorrected spot formation rate can be, in general, a function of x. For constant value of $g(x) = g$, this is given by

$$g_c(x) = g(x)\exp\left\{-\iiint\limits_{V+V'} g_c(x_0)\,dx_0 dz_0 dt_0\right\}. \qquad 3.4$$

Since the propagators of the dependence volume for the spot as well as the calmed region are linear for a constant pressure flow, we could write

$$\iint\limits_{S+C} dz_0 dt_0 = \frac{\sigma'}{U}\left(x_0 - x\right)^2.$$

Hence, equation(3.4) becomes

$$g_c(x) = g(x)\exp\left\{-\frac{\sigma'}{U}\int\limits_{x_t}^{x} g_c(x_0)\left(x_0 - x\right)^2 dx_0\right\}. \qquad 3.5$$

This expression means that the corrected spot production rate at any location is obtained by starting with an uncorrected value at the onset location and as we march downstream allowing the calming and the turbulent zone effects to modify the spot formation rate. Notice, from equation (3.5), that its integrand vanishes as $x_0 \to x$. This implies that the corrected spot formation rate at any location x is influenced only the corresponding values upstream of x, but not by its value at x itself.

Equation(3.5) can be solved analytically to get a closed form solution. To do this, substitute the functional form

$$g_c(x) = g(x)\varphi(x)$$

in equation (3.5).   This leads to(after rearranging)

$$-\ln[\varphi(x)] = \int_{x_t}^{x} \frac{\sigma'}{U} g(x_0)\varphi(x_0)(x_0 - x)^2 \, dx_0$$

By inspecting both the sides of this expression, we can immediately see that

$$\frac{d}{dx_0}\left[\frac{1}{\varphi(x_0)}\right] = \frac{\sigma'}{U} g(x_0)(x_0 - x)^2,$$

and by integrating the last expression and rearranging,

$$\varphi(x_0) = \frac{1}{\displaystyle\int_{x_t}^{x_0} \frac{\sigma'}{U} g(x_0)(x_0 - x)^2 \, dx_0 + C},$$

where C is an arbitrary constant which becomes unity by virtue of the initial condition $\varphi(x_0 = x_t) = 1$. Hence

$$\varphi.(x_0) = \frac{1}{1 + \displaystyle\int_{x_t}^{x_0} \frac{\sigma'}{U} g(x_0).(x_0 - x)^2 \, dx_0},$$

so that, for $g(x) = g = $ constant (which is considered here for the sake of simplicity)

$$\varphi \ (x_0) = \frac{1}{1 + \dfrac{\sigma'}{U} g\left\{(x - x_t)^3 - (x - x_0)^3\right\}}.$$

Hence, the corrected spot formation rate at any streamwise location x can be written as

$$g_c(x) = \frac{g}{1 + \dfrac{g\sigma'}{3U}(x - x_t)^3} . \tag{3.6}$$

Some observations are in order with regard to the expression (3.6). The factor σ' in (3.6) denotes the effect of both the calmed region and the already turbulent spot in modifying the spot production rate. Accordingly, if we put $\sigma' \rightarrow 0$ in this expression, we get $g_c \rightarrow g$, i.e., the spots are allowed to breakdown continuously as in Emmons' model. To go to the other limit, if we allow $\sigma' (= \sigma + \sigma'') \rightarrow \infty$ by allowing σ'' to go to ∞, then

$$g_c \ (x) \ = \ g \qquad\qquad at \ x = x_t$$
$$g_c \ (x) \ = \ 0 \qquad\qquad x \neq x_t$$

The first condition (at $x = x_t$) comes about because $(x - x_t)^3$ goes to zero faster than σ' goes to infinity and hence the denominator in (3.6) becomes unity. This is same as the concentrated breakdown model (equation 2.6a).

It should be noted that, even if we do not include calming in the model, and take $\sigma' = \sigma$, the denominator in the expression (3.6) is greater than unity (for $x > x_t$). This means that, even without the calming effect, the breakdown rate is a decreasing function of x, and hence different from the continuous breakdown picture. This is due to the inhibition of further spot production in an already turbulent region. On top of this, if we also take calming into account, the breakdown rate tends more and more towards the concentrated breakdown model as the amount of calming is increased.

A schematic of the spot formation rate with the concentrated and continuous breakdown models are shown along with that given by the present model in Fig. 4.

An estimate of the streamwise extent over which the spots are formed can be calculated as follows. If we take the location at which the spot formation rate drops to 10% of the initial value as the last location of spot birth (denoted by x_s), then from (3.6) it can be seen that

$$ x_s - x_t = 3\left(\frac{U}{g\sigma'}\right)^{1/3} \qquad\qquad 3.7 $$

This expression gives the belt over which spots are born. This belt will shrink to zero width (as in the concentrated breakdown model) if either g or σ' is large or if U is small. This belt is about half the length of the transition zone for constant pressure flows (this estimate is arrived at by making use of the relation (3.11) between the spot formation rates of concentrated breakdown and the present theory as used in the next section).

3.3   A new intermittency distribution

The expression (3.6) can be cast in terms of non-dimensional variables as

$$ G_c = \frac{G}{\left[1 + \tau G\left(X - X_t\right)^3\right]} , \qquad\qquad 3.8 $$

where

$$ G_c = \frac{g_c \sigma v^3}{3U^4}, \quad X = \frac{Ux}{v}, \quad G = \frac{g\sigma v^3}{3U^4} \quad \text{and} \quad X_t = \frac{Ux_t}{v} . $$

$\tau = \sigma'/\sigma$, as in the previous sections. Since the non-dimensional spot formation rates (denoted by G etc.,) are likely to be microscopic in value, these non-dimensional G's by themselves have no physical significance. They are used here for the sake of convenience as they have been used by other workers (e.g., [2]) in the past. In terms of these non-dimensional variables, the expression for intermittency becomes

$$\gamma = 1 - \exp\left\{ - \int_{X_t}^{X} 3G_c \left(X_0 - X \right)^2 dX_0 \right\},\qquad\qquad 3.9$$

which, by using equation (3.8) becomes

$$\gamma = 1 - \exp\left\{ - \int_{X_t}^{X} \frac{3G}{1 + \tau G\left(X_0 - X_t \right)^3} \left(X_0 - X \right)^2 dX_0 \right\}.\qquad\qquad 3.10$$

This integral can be evaluated to give the final intermittency variation (after some algebra) and this expression is shown in Appendix A.

As mentioned earlier, the concentrated breakdown model has been found to be an excellent description of transition in constant pressure steady flows and it has been validated against different experiments over the years. Hence, in order to validate the new intermittency distribution, we compare it with the concentrated breakdown model. In order to do this, we need a relation between the spot formation rates at the onsets for both the theories, i.e., between g and g_n, or between the non-dimensional variables G and G_n (see Appendix B). Here, the concentrated and the present models are matched such that they give the same intermittency over most of the transition zone. This gives

$$G = 0.45 G_N^{3/2},\qquad\qquad 3.11$$

Also a relation is needed between the transition onset locations in both the models as these locations could be expected to be quite different from physical considerations. Here, we find that

$$X_t(Conc) - X_t(present) = 27{,}000.\qquad\qquad 3.12$$

The relation (3.12) above has been found numerically for constant pressure flows. Numerical experiments with different pressure gradients indicate that this constant pressure situation is likely to be an upper bound for difference between onset location (according to concentrated breakdown and the present theory).

Using expressions (A1, 3.11 and 3.12), the variation of intermittency with $X = \dfrac{Ux}{v}$ is shown plotted in Fig. 5, where it is compared with the concentrated breakdown model. It can be seen that the agreement between these models is excellent thereby validating the current model for constant pressure steady flows.

One of the main things to be noted in Fig. 5, apart from the excellent agreement between these theories over most of the transition zone, is that the present model has a faster pickup close to the origin than the concentrated breakdown model. Close to the onset, many experiments have in fact shown that the intermittency is non-zero even upstream of the onset location inferred by the concentrated breakdown model. This bears out the trend shown by the present model. This is a clear demonstration that the effect of calming and the turbulent fluid are crucial factors in determining the intermittency distribution realistically in constant pressure flows and more so in pressure gradient flows where the calming/turbulent factor could be expected to have strong influence on the intermittency distribution [4].

3.4 Pressure gradient flows

For pressure gradient flows, the calculation of intermittency is performed numerically as follows. The spot planform is assumed to be triangular (as in [9]) and the calmed region is taken to be of rectangular planform. The momentum thickness θ is calculated by the Thwaites method and from this the Polhausen pressure gradient parameter $\lambda_\theta = \dfrac{\theta^2}{v}\dfrac{dU}{dx}$, is obtained. Gostelow et al. [14] have correlated the spot formation rate G_n as a function of the pressure gradient parameter λ_θ at the transition onset location and the freestream turbulence level q. The correlation for G_n is given by

$$G_n = 0.86 X 10^{-3} \exp\left\{2.134\lambda_\theta \ln(q) - 59.23\lambda_\theta - 0.5641\ln(q)\right\}.$$

The corresponding G is obtained by using a modified form of the relation (3.11) and this is given by

$$G = G_N^{\,3/2}.$$

The change in the constant of proportionality is required because of the assumed triangular planform for the spot which reduces σ from 0.25 to 0.17. Then by using equations (3.3-3.4), the intermittency expression is obtained by numerical integration. This calculation has been performed for different pressure gradients as shown in Fig. 6. The leading edge, trailing edge and calmed region celerities are 88%, 50% and 25% of the freestream velocity respectively. The Reynolds number for the calculation is typical of a low pressure turbine.

Fig. 6 presents a comparison of calculated steady flow intermittency distributions. The two breakdown models are compared for a range of pressure distributions. In each case, transition begins at $\mathrm{Re}_{\theta,trans} = 168$ which corresponds to an inlet turbulence intensity of approximately 4 percent. It can be seen that in mildly accelerating and decelerating flows ($\lambda_\theta = 0.04, -0.04$), the agreement is very good. In a strong adverse pressure gradient ($\lambda_\theta = -0.08$), which is most relevant to low pressure turbines and compressors, the agreement is excellent. In a zero pressure gradient flow, the form of the curves is identical but there is a displacement of the distributions. This was noted earlier. In practice, this displacement could be allowed for. However, it amounts to no more than about 10 percent of the length of the transition zone. On an LP turbine blade, transition often starts near peak suction and probably occurs over approximately 10-15 percent of the surface length. In practice, therefore, the differences between the two models are relatively small in all circumstances. Therefore, it is concluded that the present model may be used in situations where the pressure gradient changes significantly within the transition zone. Such a situation arises in many low pressure turbines.

4. Conclusions

A new transitional intermittency model incorporating the calming effect is presented. The spot formation rate at any location is allowed to be modified by the calmed region as well as the already turbulent fluid. The intermittency distribution obtained from this model is compared with the concentrated breakdown and in doing so correlations between the spot formation parameters between these models are obtained for constant pressure flow. By numerical experiments for different pressure gradients, the validity of this correlation for pressure gradient flows has also been demonstrated.

Acknowledgement: This work has been carried out with the support of Rolls Royce plc. and the Defence Evaluation and Research Agency (MOD and DTI), Pyestock. The authors would like to thank them for funding it and for their permission to publish this paper.

References

[1] Emmons, HW, 1951,"The laminar-turbulent transition in a boundary layer- Part 1", Journal of Aerospace Science, Vol. 18, No7, pp490-498

[2] Soundranayagam, S, and Potti, MGS, 1991, "Transition in laterally convergent and divergent flows", Conf. on Boundary layer transition and control, Cambridge, April 1991.

[3] Narasimha, R, 1957, "On the Distribution of Intermittency in the Transition Region of a Boundary Layer", Journal of Aerospace Science, Vol. 24, pp 711-712.

[4] Schulte, V, Hodson, HP, 1997,"Prediction of the becalmed region for LP profile design", ASME-paper 97-GT-XXX, 1997

[5] Halstead, DE, Wisler, DC, Okiishi, TH, Walker, GJ, Hodson, HP and Shin, HW, 1997, "Boundary Layer development in axial compressors and turbines :part 1 of 4- composite picture", Jl. Of Turbomachinery, Vol. 119, No. 1.

[6] Hodson, HP, 1998, "Blade row interference effects in axial turbomachinery stages", Lecture series 1998-02, *Blade row interactions in low pressure turbines*, VKI, Belgium.

[7] Mayle, RE, 1991, "The role of laminar-turbulent transition in gas turbine engines", ASME paper 91-GT-261.

[8] Schubauer, GB, and Klebanoff, PS, 1955,"Contributions on the Mechanics of Boundary Layer Transition", NACA TN 3489 (1955) and NACA Rep. 1289 (1956)

[9] Chen, K.K. and Thyson, N.A., 1971, "Extension of Emmons' Spot Theory to Flows on Blunt Bodies", AIAA Journal, Vol 9, No 5, pp 821-825.

[10] Gostelow, JP, Walker, GJ, Solomon, WJ, Hong, G, Melwani, N, 1996, "Investigation of the calmed region behind a turbulent spot", ASME-paper 96-GT-489

[11] Narasimha, R. 1978, "A note on certain turbulent spot and burst frequencies", Report 78 FM 10, Department of Aeronautical Engineering, Indian Institute of Science, Bangalore.

[12] Narasimha, R, and Dey, J, 1989, "Transition-zone models for two-dimensional boundary layers: a review", Sadhana, Vol. 14, 93-120.

[13] Abu-Ghannam, B.J. and Shaw, R., "Natural transition of boundary layers - the effects of turbulence, pressure gradient and flow history," J. Mech. Eng. Sci., Vol. 22, No. 5, 1980, pp 213-228.

[14] Gostelow, JP, Blunden, AR, and Walker, GJ, 1992, "Effects of free-stream turbulence and adverse pressure gradients on boundary layer transition". ASME Gas Turbine Conference, Cologne, June.

Appendix A

The solution to equation (3.10) is

$$\gamma = 1 - \exp\{-I\}$$

where

$$I = \frac{1}{\tau}\ln\left\{1 + \tau G\left(X - X_t\right)^3\right\} + 3G(X - X_t)I_1$$

$$I_1 = \frac{A}{2(\tau G)^{2/3}}\ln\left\{(\tau G)^{2/3}(X - X_t)^2 - (\tau G)^{1/3}(X - X_t) + 1\right\} +$$

$$\frac{2}{\sqrt{3}(\tau G)^{1/3}}\left(\frac{A}{2(\tau G)^{1/3}} + B\right)\left\{\tan^{-1}\left[\frac{2(\tau G)^{1/3}(X - X_t) - 1}{\sqrt{3}}\right] + \frac{\pi}{6}\right\} + \frac{C}{(\tau G)^{1/3}}\ln\left\{(\tau G)^{1/3}(X - X_t) + 1\right\}$$

and

$$A = -\frac{\left(X - X_t\right)(\tau G)^{1/3} + 2}{3},$$

$$B = \frac{2}{3}\left[(X - X_t) - (\tau G)^{1/3}\right],$$

$$C = \frac{\left(X - X_t\right) + 2(\tau G)^{-1/3}}{3}.$$

$$(A1)$$

Appendix B

The objective is to obtain a functional form of the relation between G and G_n. To do this, we match the λ's obtained from the concentrated and continuous breakdown theory (without any calming). This is done as the analysis is tractable this way. The relation between G and G_n obtained this way is likely to be different from the present model including calming but the functional form of the relation is nevertheless the same. Hence by matching λ's between equations (2.5) and (2.6b), we get

$$\frac{0.641}{G_n^{1/2}} = \frac{0.4545}{G^{1/3}}.$$

This leads to the functional form

$$G = kG_n^{3/2}$$

If we include calming also, then for the present theory $k = 0.45$ is seen to give good agreement with the concentrated breakdown model. The general form of the above expression is consistent with the fact that the intermittency expressions for the concentrated and continuous breakdown models vary as $\exp(-x^2)$ and $\exp(-x^3)$ respectively (see equations (2.5) & (2.6b)).

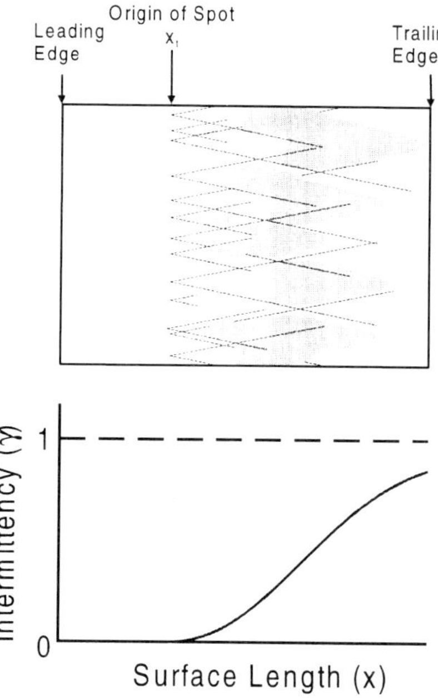

Fig. 1 Schematic of bypass transition

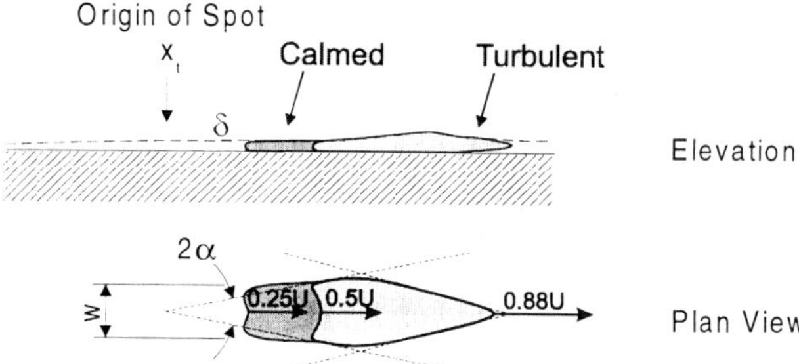

Fig. 2 Schematic of a turbulent spot

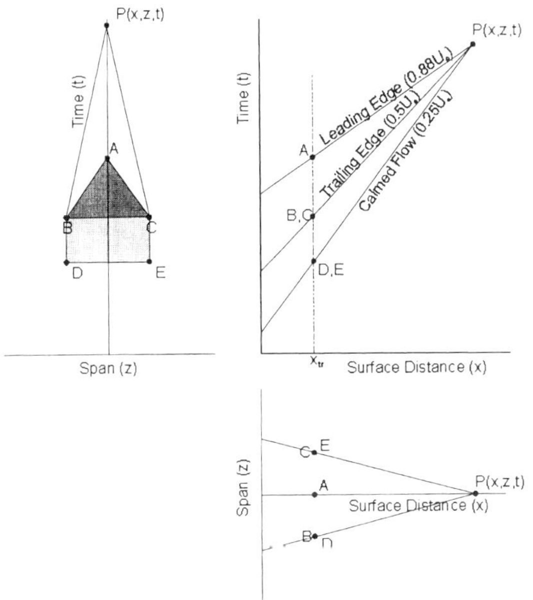

Fig. 3 Dependence volume of a turbulent spot and its calmed region

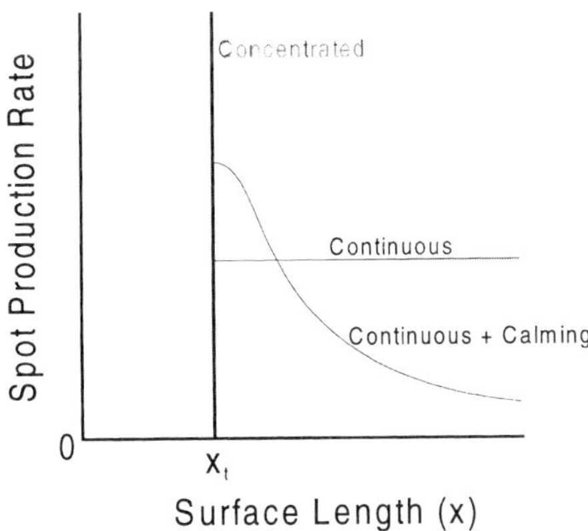

Fig. 4 Schematic of spot formation rates according to different breakdown models

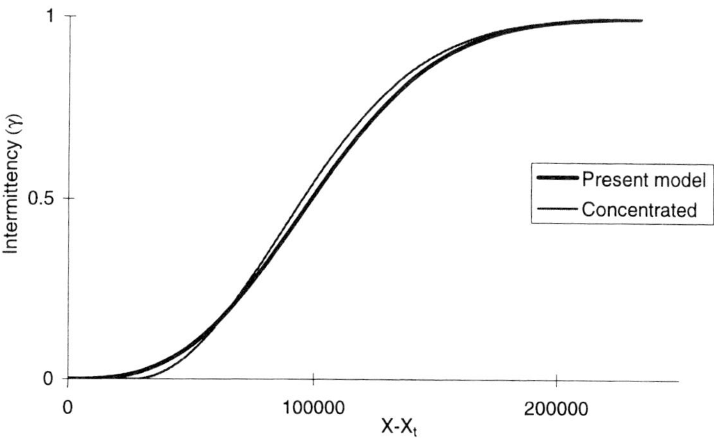

Fig . 5 Intermittency distribution for a constant pressure flow

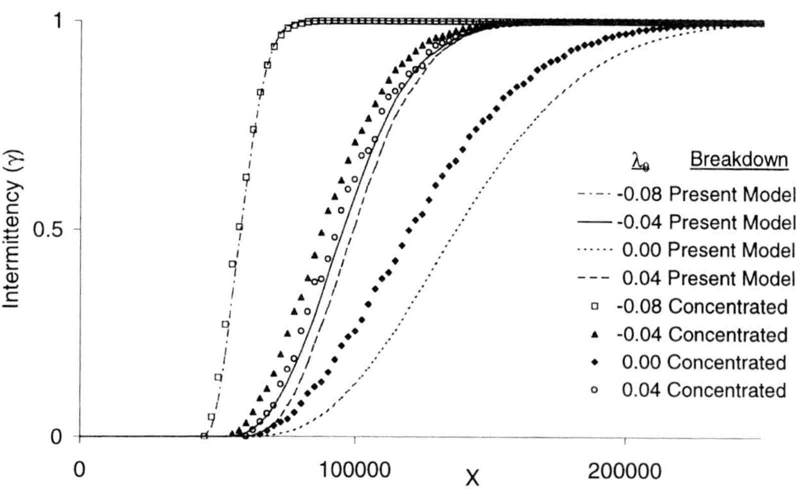

Fig. 6 Intermittency distributions for pressure gradient flows

C557/130/99

Turbulence modelling of rotor-stator interaction with linear and non-linear eddy–viscosity models

W-L CHEN and **M A LESCHZINER**
Department of Mechanical Engineering, UMIST, Manchester, UK

SYNOPSIS

The flow over the first 1.5 stages of a 2.5-stage compressor, consisting of an inlet-guide-vane (IGV) stage and two rotor-stator stages, is calculated using linear and non-linear, low-Reynolds-number k-ε eddy-viscosity models. A sliding-grid system is used to simulate the motion of the rotor cascade. The computation is two-dimensional, as this case is said to have been specifically designed to be treated as such, at least in terms of wake-blade-interaction phenomena. Measurements are only available at two streamwise planes, one behind the IGV stage and the other downstream of the first rotor. The computed solutions indicate that the flow, including the wake structure, is not materially sensitive to turbulence modelling, except for the anisotropic turbulence intensity for which the non-linear model gives a much better result. However, to investigate and identify the role of turbulence modelling with greater rigour, more data are required, especially in the blade-surface boundary layers.

INTRODUCTION

Turbomachinery flows are highly unsteady because of the relative motion of rotor and stator stages. Excluding 'random' turbulence, unsteadiness may be viewed as a combination of three components: the inviscid potential effect, the periodic passing of wakes generated by the upstream blade rows, and the shedding of vortices from the blade trailing edges, which cause the wakes themselves to be unsteady. The first process is due to pressure surges associated with the relative motion of the rotors with respect to the stators, and this effect becomes stronger with a decreasing gap between any rotor and a neighbouring stator. The second arises from the interaction of the blade boundary layers with the wakes, specifically from the variation of the angle of attack at the leading edge due to the velocity deficit of the impinging wake. The third process, the scales of which are much smaller than those of the other two, may be regarded as provoking high-frequency perturbations to the second. All these unsteady features and the interaction between them makes the simulation of unsteady turbomachinery flows a highly challenging task.

The predictive representation of rotor-stator interaction is clearly sensitive to the quality of the turbulence model used to generate the prediction. The requirement is for a model that represents correctly the response of the flow's turbulence characteristics, and hence its mean properties, to blade curvature, leading-edge impingement, transition and wake-induced unsteadiness. The present "industry standard" is to employ linear eddy-viscosity models, the most complex of which involve differetial transport equations for turbulence-energy and its rate of dissipation or a related length-scale-determining parameter [1-3]. While some studies demonstrate that such models resolve adequately the response of turbulence in boundary layers to unsteady perturbations at frequencies representative of those of wakes interacting with blades, it is well known that these models display serious defects in complex flow conditions. Specifically, they tend to underestimate the response of turbulence to curvature and to seriously overestimate its response to normal straining. Both defects combine, unless counteracted by *ad-hoc* corrections, to give excessive turbulent mixing, to suppress separation and vortex shedding and to provoke premature transition following impingement.

While no alternative modelling strategy appears to exist which cures all the above ills, it is generally accepted that second-moment closure and some non-linear eddy-viscosity models go some way towards capturing correctly the distinctly weak response of turbulence to normal straining, in contrast to shear straining, the attenuation of turbulence due to suction-side curvature and the presence of vortex shedding generated at thick trailing edges. The last effect, in particular, reflects the lower levels of turbulent mixing and thinner boundary layer predicted by these models relative to linear eddy-viscosity formulations in the boundary layers from which the wakes emanate. All these attributes are likely to contribute, in some measure, to the quality with which wake-blade interaction is predicted in the context rotor-stator flows.

The purpose of this paper is to report a numerical study of the flow in a rotor-stator configuration, in which the performance of linear and non-linear low-Reynolds number eddy-viscosity models is contrasted. A sliding-grid, multi-block methodology is introduced, which allows the relative motion between rotors and stators to be included and which provides the vehicle for examining the performance of the turbulence models.

TURBULENCE MODELLING

Two eddy-viscosity models feature in the present study, one linear (Launder-Sharma, [4]) and the other cubic (Craft-Launder-Suga, [5]). The linear model is known to generate much too large turbulence energy upon the impingement of blade's leading edge, thus, a modification proposed by Kato and Launder [6] is adopted here to suppress the excessive impingement turbulence energy. Both linear and non-linear models are applicable without any Reynolds-number restrictions and thus allow integration down to the wall through the semi-viscous sub-layer. The linear model employs the conventional, linear relationship between stresses and strains,

$$\frac{\overline{u_i u_j}}{k} = \frac{2}{3}\delta_{ij} - \frac{v_t}{k}S_{ij} \tag{1}$$

where $S_{ij} = \partial u_i/\partial x_j + \partial u_j/\partial x_i$, while the cubic model employs a general cubic tensorial expansion in the strain and vorticity tensors,

$$\frac{\overline{u_i u_j}}{k} = \frac{2}{3}\delta_{ij} - \frac{v_t}{k}S_{ij}$$
$$+ c_1\frac{v_t}{\tilde{\varepsilon}}(S_{ik}S_{kj} - \frac{1}{3}\delta_{ij}S_{kl}S_{kl}) + c_2\frac{v_t}{\tilde{\varepsilon}}(\Omega_{ik}S_{kj} + \Omega_{jk}S_{ki})$$
$$+ c_3\frac{v_t}{\tilde{\varepsilon}}(\Omega_{ik}\Omega_{kj} - \frac{1}{3}\delta_{ij}\Omega_{kl}\Omega_{kl}) \tag{2}$$
$$+ c_4\frac{v_t k}{\tilde{\varepsilon}^2}(S_{ki}\Omega_{lj} + S_{kj}\Omega_{li})S_{kl} + c_5\frac{v_t k}{\tilde{\varepsilon}^2}(S_{kl}S_{kl} - \Omega_{kl}\Omega_{kl})S_{ij}$$

where $\Omega_{ij} = \partial u_i/\partial x_j - \partial u_j/\partial x_i$. The coefficients in equation (2) are given in Table (1).

c_1	c_2	c_3	c_4	c_5
-0.1	0.1	0.26	$-10\,c_\mu^{\,2}$	$-5c_\mu^{\,2}$

Table (1) : Numerical values for the coefficients in equation (2).

In equation (1) and (2), v_t is the turbulence eddy viscosity,

$$v_t = c_\mu\, f_\mu\, \frac{k^2}{\tilde{\varepsilon}} \tag{3}$$

where

$$c_\mu = 0.09, \quad f_\mu = \exp\left[\frac{-3.4}{(1+\tilde{R}_t)^2}\right] \quad (for\ Launder\ -Sharma\ )$$

$$c_\mu = \frac{0.3}{1+0.35\,[\max(\tilde{S},\tilde{\Omega}]^{1.5}}\{1 - \exp\left[\frac{-0.36}{\exp(-0.75\max(\max(\tilde{S},\tilde{\Omega}))}\right]\}, \tag{4}$$

$$f_\mu = 1 - \exp\left[-(\frac{\tilde{R}_t}{90})^{0.5} - (\frac{\tilde{R}_t}{400})^2\right], \quad (for\ Craft\ -Launder\ -Suga\ )$$

and

$$\tilde{S} = \frac{k}{\tilde{\varepsilon}}\sqrt{0.5\,S_{ij}S_{ij}}, \quad \tilde{\Omega} = \frac{k}{\tilde{\varepsilon}}\sqrt{0.5\,\Omega_{ij}\Omega_{ij}}, \quad \tilde{R}_t = \frac{k^2}{v\,\tilde{\varepsilon}} \tag{5}$$

Both eddy-viscosity models determine the turbulence energy and its rate of dissipation from related differential transport equations which are applicable without Reynolds-number restrictions.

As is demonstrated in ref [5], the inclusion of quadratic terms in the stress-strain relations allows normal-stress anisotropy to be captured, while the cubic terms establish the appropriate sensitivity to streamline curvature. The additional dependence of the coefficients on the strain and vorticity invariants allows the cubic model to respond selectively to shear and irrotational straining.

In the computations to follow, no specific transition-related practices have been used. Low-Re models of the type used herein are able, in principle, to predict (bypass) transition. While the quality of this prediction is variable (as is the case with all models used in complex conditions), the absence of experimental data for boundary-layer properties in the flow considered below, does not allow transitional features to be examined, and the inclusion of *ad-hoc* transition relations was not, therefore, felt to be justified. Although the influence of such additional practices may be observed by references to changes in the wake profiles, such observations would not permit conclusions to be drawn on the quality with which transition is represented on the blade surfaces.

NUMERICAL METHOD

Calculations reported below have been performed with the multi-block version of the general non-orthogonal fully collocated finite-volume scheme STREAM (Lien, *et al.*, [7]). In this, convection is represented by bounded variants of second-order schemes. The solution is iterated to convergence for each time step using a pressure-correction approach. Multi-block grids, combining O- and H-types, have been used to construct high-quality meshes around the blade, avoiding grid distortion near the leading and trailing edges. A fully implicit scheme, with the time variation approximated to second-order accuracy, is incorporated to solve unsteady problems.

The present sliding-grid strategy consists of the following key elements:
(i) **Grid sliding**: In the case of a single rotor/stator configuration, shown in Fig. 1a, the whole rotor grid is sliding together with the attached coordinate frame at a speed **u** relative to the stationary stator grid

to simulate the rotor motion. At boundaries AB, CD, EF and GH, periodic conditions are applied. At the end of one blade cycle - that is, when point E moves to the position opposite point D - the whole rotor grid is shifted a pitch length upwards (point G is opposite point B), and a new blade cycle begins.

(ii) Treatment of interface between rotor and stator: The interface between stator and rotor is identified by boundaries BD and DG. Here, boundary segment BE aligns with DF because of periodicity. At this common interface, the rotor and stator meshes do not overlap. However, two arrays of additional near-interface cells (the shaded region in Fig. 1b) are needed for both rotor and stator to perform the second-order interpolation for convection. The property values at these cells are obtained by bi-quadratic interpolation.

(iii) Bulk correction at the rotor-stator interface: Although the bi-quadratic interpolation practice is numerically accurate, it does not ensure unconditional mass-conservation. A bulk correction is therefore effected on the first array of the near-interface cells of both stator and rotor. This is achieved by first calculating the mass-flow rate at the interface, then comparing this to the inlet mass-flow rate, and finally multiplying the interface-normal velocity component by the ratio of interface-to-inlet mass-flow rate. This practice ensures mass conservation at the interface and is crucially important for achieving overall accuracy and realism.

RESULTS

Before attention is directed to the rotor-stator configuration, it is instructive to highlight the sensitivity of one particular flow feature, generally pertinent to rotor-stator flows, to turbulence modelling. It has already been pointed out in the Introduction that linear eddy-viscosity models tend to give rise to excessive turbulence mixing and hence tend to suppress trailing-edge vortex shedding. The ability of different turbulence models to capture vortex shedding and hence wake unsteadiness, is one issue which was investigated in a study of the flow around a VKI turbine blade in a linear cascade, for which experimental data have been obtained by Ubaldi, *et. al.* [8]. Computations relating to the flow around the blade itself are reported in another paper in these proceedings (Chen & Leschziner [9]). Here, we merely note that the linear model (1) failed to capture shedding, while the cubic model (2) returned a pronounced shedding process at a frequency of 1250 Hz, relative to the experimentally observed value of 1700 Hz. Although this difference is substantial, the frequency is sensitive to a whole host of issues, including inlet turbulence characteristics and the entire history of the boundary layers as they develop along the suction and pressure sides.

The principal geometry of the present paper is the flow over a two-and-half-stage compressor cascade, examined experimentally by Sentker and Riess [10]. The geometry consists of one row of Inlet Guide Van (IGV) and two identical stages of rotor-stator cascades. The number of IGV, rotor and stator blades is 20, 30 and 26, respectively. In the computation, only the IGV and the first stage have been included, as experimental data are restricted to this geometric zone. In addition, the number of stator blades has been reduced from 26 to 25 to allow the computational domain to be restricted circumferentially to 4-IGV, 6-rotor and 5-stator blade passages. This reduction of stator-blade number might result in a slight change of mass-flow rate over each stator-blade passage but the overall flow rate should not be affected. However, the stator is located downstream of E2 where a slight change of flow rate in an individual stator passage should pose little effect on E2. More significantly, the reduction alters the passing frequency of the stator blades for each rotor passage. This, in turn, influences the pressure fluctuation at E2. Nevertheless, this difference will be less than 4%. Fig. 2 shows the computational mesh which consists of 180 blocks and 260,000 nodes. There are 12 blocks for each blade passage to allow a quality O-type grid to be inserted around the blade surface, and the grid is fine enough to ensure that the y^+ value at the wall-nearest node is below 1. Reference is made below to rotor velocity $\mathbf{u}$, absolute velocity $\mathbf{c}$ and relative velocity $\mathbf{w}$, and these are defined in the vector diagram in Fig. 3. In some comparisons to follow, only the solution obtained with the linear model is included, in which case that returned by the non-linear model is essentially identical.

Fig. 4a and 4b shows the variation of the absolute velocity across one rotor pitch and over the duration by which the rotor moved by one pitch, while Fig. 4c gives comparisons of absolute velocity at three pitchwise positions and three rotor positions. As seen, the predicted behaviour is not sensitive to turbulence modelling, both inside and outside the wake. One reason is that both models incorporate mechanisms (albeit different in nature) to suppress excessive impingement-generated turbulence energy near the leading edge. Additionally, the boundary layers on the blade surfaces are rather thin, in the design condition investigated, so that the boundary layers are not substantially sensitive to normal stresses or curvature. The variations outside the wake, for which experimental data are available, merely reflect the potential effect of the passing rotor blades and should not be affected by turbulence modelling. Confirmation is offered by Fig. 5 which shows the pitch-. and time-wise variation of the streamwise normal stress. Both experiment and calculation indicate that the moving rotor blades generate low levels of turbulence, and that turbulence activity is confined to the IGV wakes. Fig. 5 also suggests, albeit vaguely, that models underestimate the turbulence intensity of the IGV wakes. However, the substantial and irregular temporal variations in the experimental data do not really allow a firm statement to be offered. To investigate modelling issues with greater confidence, more experimental data, especially within the IGV boundary layers, are needed.

The flow at section E2 is much more complex than that at E1, and is punctuated by the presence of two sets of wakes, as shows schematically in Fig. 6. Comparisons of fields of absolute and relative-velocity at E2 are shown in Fig. 7 and Fig. 8, respectively. The presence of the IGV wakes cannot be clearly identified in Fig. 7 as they are diffuse and weak at this location. The distinctly peaky behaviour in the experimental profile are, as acknowledged in [10], due to the lack of resolution rather than reflecting real flow phenomena. The rotor wake is distinctly brought out in the relative-velocity field shown in Fig. 8. Again, the periodic gaps in the experimental data reflect lack of measurement resolution. It is difficult to make categorical statements about the agreement between experiment and calculation. What may be said is that the calculation faithfully resolves the wake behaviour as well as the undulations between the wakes, which reflect the combination of potential effects and the embedded IGV wakes. A more detailed comparison at specific locations is not possible here due to the limited availability of experimental data. In terms of model performance, both models again return very similar results in terms of the mean flow quantities. However, this is not the case in relation to normal stress, as is conveyed in Fig. 9. The experimental variation suggests a maximum value of around 14 (m/sec)^2. Also, here, the presence of the IGV wakes is clearly seen. For example, that at $0°$ is the one shifted from -7° at the location E1, due to the displacement effect caused by the rotor suction side. Both turbulence models return a weaker turbulence footprint of the IGV wake than the experiment, and this gives strength to the tentative suggestion, offered in relation to Fig. 5, that the real IGV wakes are more turbulent than predicted by either model. However, the level returned by the non-linear model is closer to the experimental data. The non-linear model also predicts distinctly closer agreement with experiment in respect of the maximum level of rotor-wake intensity, at around 14 (m/sec)^2, relative to 9 (m/sec)^2 returned by the linear model. There is no vortex shedding predicted due to the thinness of trailing edge and the high level of inlet-flow turbulence intensity. Therefore, to a large extent, the observed differences in wake turbulence reflect the fact that the non-linear model is able to represent, in contrast to the linear form, the anisotropic nature of sheared turbulence - in particular, the especially high level of streamwise intensity.

CONCLUSIONS

The flow in a 1.5-stage compressor has been calculated with linear and non-linear eddy-viscosity models. Since the compressor operates at its design condition, it is not surprising to observe that both models return similar results which are in fairly good agreement with the data at the first measurement station downstream of the IGV. Larger discrepancies between the measurements and the computations are observed at the second station downstream the first rotor. At this station, both models again give similar results in terms of mean-flow quantities. However, the non-linear model returns markedly better agreement with the data for the streamwise turbulence intensity, reflecting its ability to resolve normal-stress anisotropy. Due to the limited availability of experimental data, especially for the blade boundary layers, it is not possible, at this stage, to identify the origin of discrepancies between the predictions and the experimental data in the wakes.

ACKNOWLEDGEMENTS

The authors gratefully acknowledge the assistance given to them by Prof. Riess and Dr. Sentker from the University of Hannover, who provided the experimental data for this test case. The financial support from the Commission of the European Communities under the Brite-EuRam III Contract "TURMUNSFLAT" No. BE95-1698 is also greatly appreciated.

REFERENCES

[1] Cho, N.H., Liu, X., Rodi, W. and Schonung, B., (1993), "Calculation of Wake-Induced Unsteady Flow in a Turbine Cascade", *ASME Journal of Turbomachinery*, Vol. 115, p. 675.

[2] Arnone, A. and Pacciani, R., (1996), "Rotor-Stator Interaction Analysis Using the Navier-Stokes Equations and a Multigrid Method", *ASME Journal of Turbomachinery*, Vol. 118, p. 679.

[3] Michelassi, V. and Martelli, F. (1998), "Blade Row Interference Effects in Axial Turbomachinery Stages", *VKI Lecture Seeies* 1998-02.

[4] Launder, B.E. and Sharma, B.I. (1974), Application of the Energy-Dissipation Model of Turbulence to the Calculation of Flow near a Spinning Disc", *Int. Journal of Heat and Mass Transfer*, Vol. 1, p.131.

[5] Craft, T.J., Launder, B.E. and Suga, K. (1993), "Extending the Applicability of Eddy-viscosity Model Through the Use Deformation Invariant and Non-linear Elements", *Proc. 5th Int. Symp. Refined Flow Modelling and Turbulence Measurements*, p. 125.

[6] Kato, M. and Launder, B.E. (1993), Proc 9th Symp on *Turbulent Shear Flows*, Kyoto, p. 10.4.1.

[7] Lien, F.S., Chen, W.L. and Leschziner, M.A. (1996), "A Multiblock Implementation of a Non-orthogonal, Collocated Finite Volume Algorithm for Complex Turbulent Flows", *Int. Journal for Numerical Methods in Fluids*, Vol. 23, p. 567.

[8] Ubaldi, M., Zunino, P., Campora, U. and Ghiglione, A., (1996), "Detailed Velocity and Turbulence Measurements of the Profile Boundary Layer in a Large Scale Turbine Cascade", *ASME paper 96-GT-42*.

[9] Chen, W.L. and Leschziner, M.A., (1998), "Modelling Turbomachine-blade Flows with Non-linear Eddy-viscosity Models and Second-moment Closure", paper submitted to the *3rd European Conference on Turbomachinery*.

[10] Riess, W. and Sentker, A. (1998), "Measurement Techniques for Unsteady for Unsteady Flow and Turbulence in a Low-Speed Axial Compressor", *Proc. ERCOFTAC Workshop on Turbomachinery Flow Predictions VII*, Aussois, France.

FIGURES

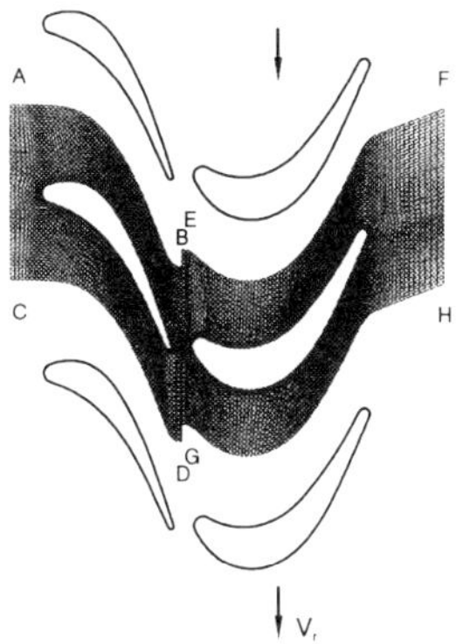

Fig. 1a: Single roto/stator passage.

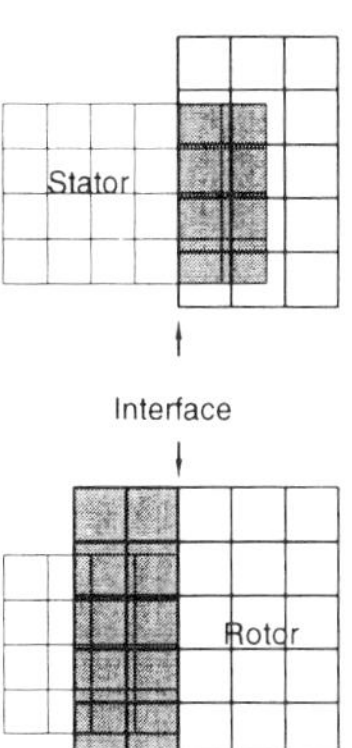

Fig. 1b: Interface between rotor and stator.

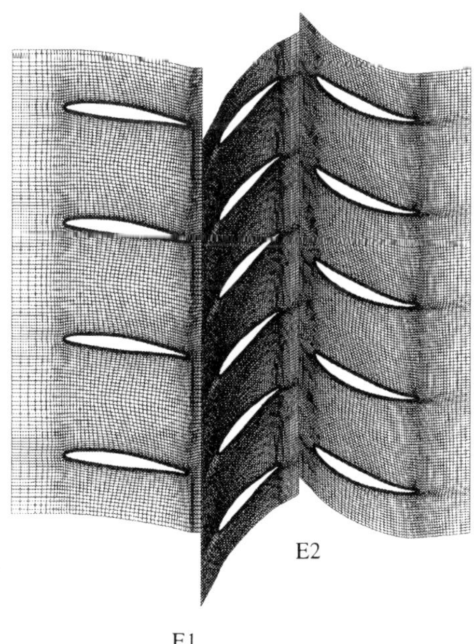

Fig. 2: Computational grid for the Hannover compressor (shown 1/4 of the actual size).

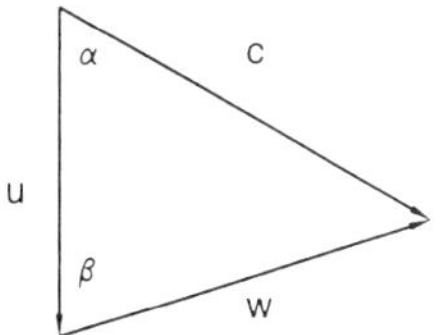

Fig. 3: Definition of velocity vectors.

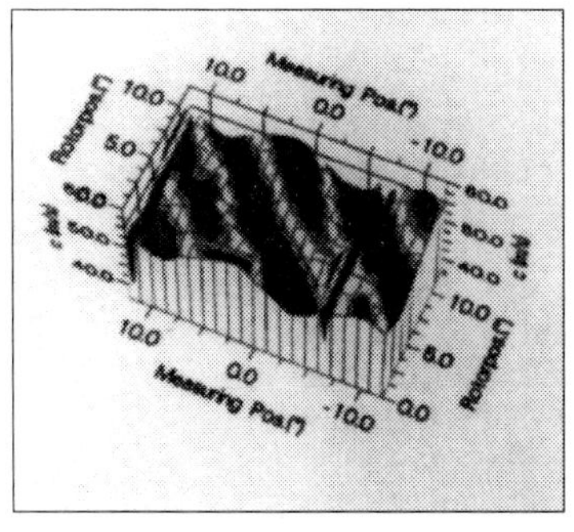

a. Exp. data

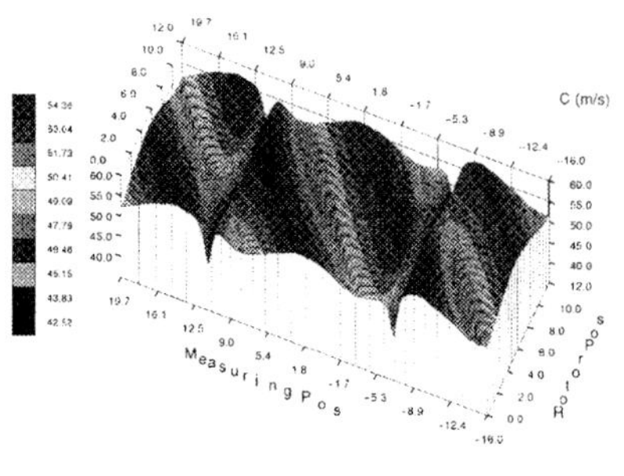

b. Kato-Launder

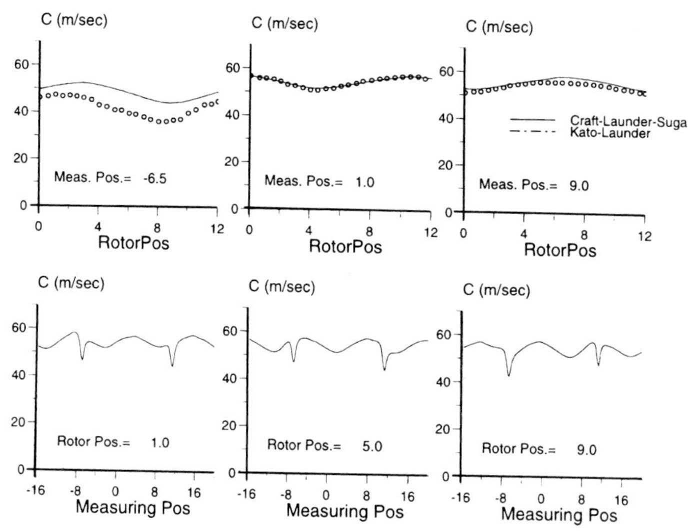

c. Absolute velocity profiles at different locations
Fig. 4: Hannover compressor, absolute velocity at E1.

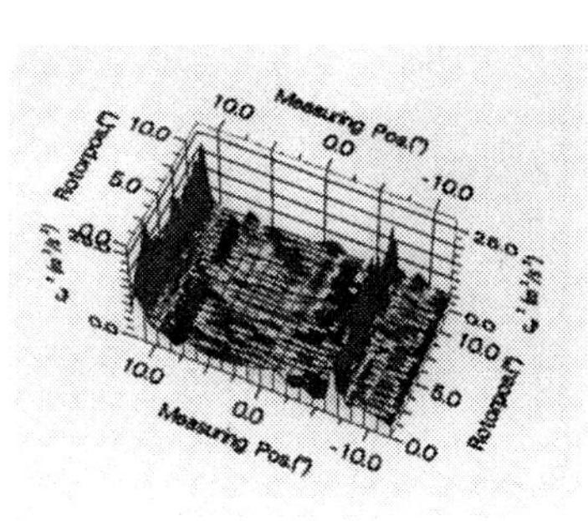
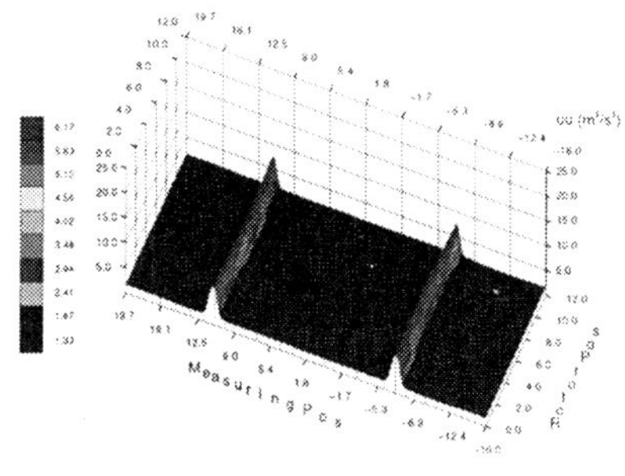

a: Exp. data

b. Kato-Launder

Fig. 5: Hannover compressor, normal stress at E1.

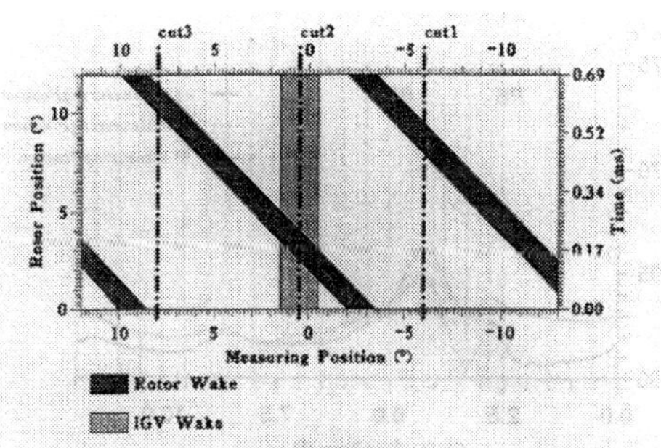

Fig. 6: Hannover compressor, wake pattern at E2.

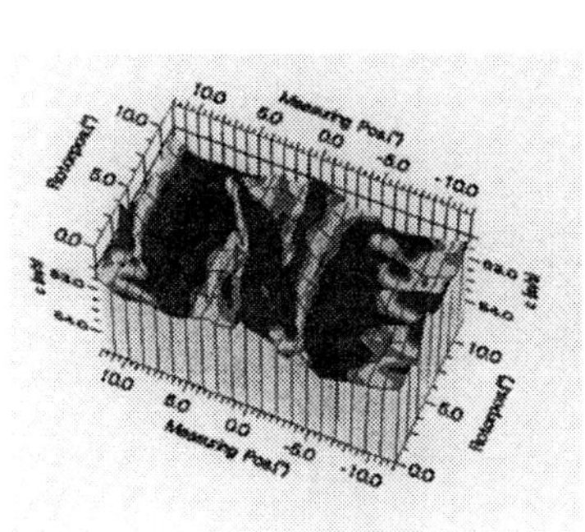

a. Exp. data

b. Kato-Launder

Fig. 7: Hannover compressor, absolute velocity at E2.

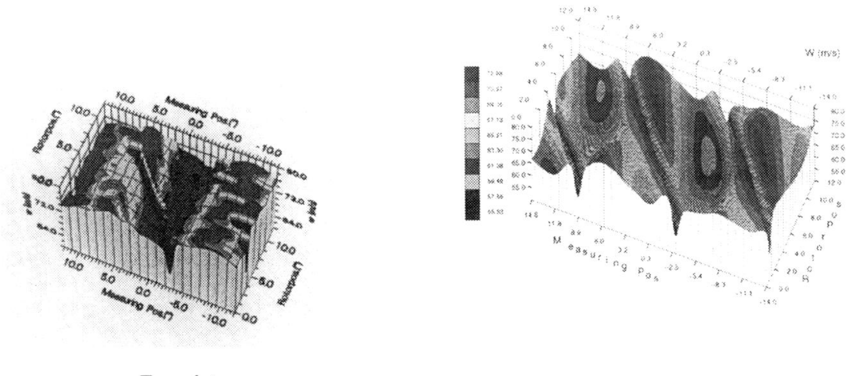

a. Exp. data   b. Kato-Launder

Fig. 8: Hannover compressor, relative velocity at E2.

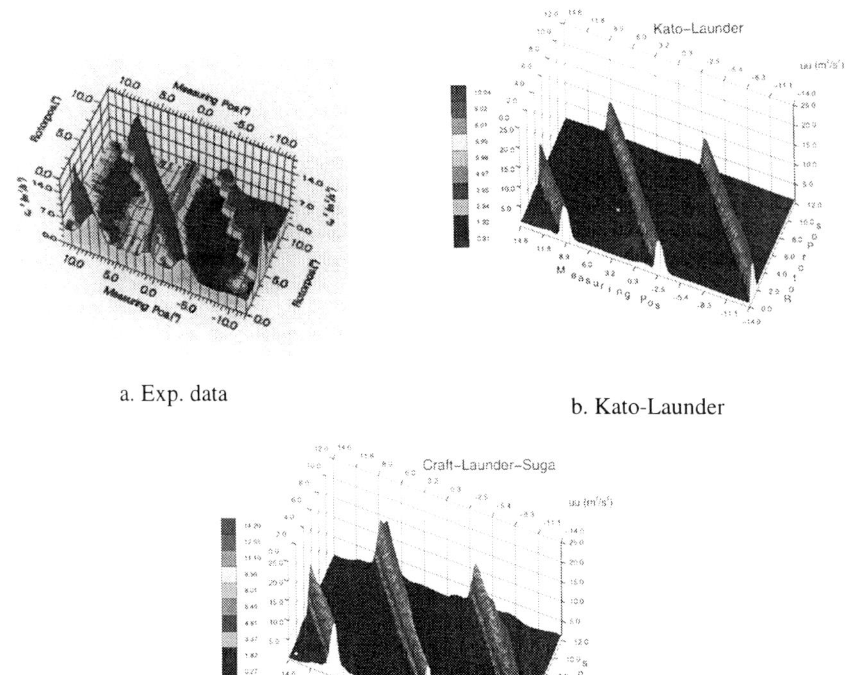

a. Exp. data   b. Kato-Launder

c. Craft-Launder-Suga

Fig. 9: Hannover compressor, normal stress at E2.

Blade–Row Interaction
and Aeroelasticity

C557/057/99

Experimental investigation of the unsteady rotor aerodynamics of a transonic turbine stage

R DÉNOS, C H SIEVERDING, T ARTS, J F BROUCKAERT, and **G PANIAGUA**
von Karman Institute for Fluid Dynamics, Rhode Saint Genèse, Belgium
V MICHELASSI
Energetics Department 'Sergio Stecco', University of Florence, Italy

ABSTRACT

The paper describes some results of a large experimental program on the unsteady flow through the rotor of a transonic turbine stage in the large VKI Compression Tube Turbine Facility. The tests were carried out as part of a BRITE EURAM project. The test program covered the investigation of the effects of a variation of the rotational speed of the rotor, the axial stator-rotor distance and the stator trailing edge coolant flow ejection. The paper aims at presenting the measurements of the relative inlet total pressure and the rotor blade surface pressure at rotor mid-span.

1 INTRODUCTION

The outlet flow field of stator blade rows is characterized by a large number of non-uniformities. Pitchwise and spanwise static pressure gradients result respectively from different accelerations of the flow along the pressure side and the suction side and from radial equilibrium. In the case of transonic flow, additional strong static pressure gradients are associated with the trailing edge shocks. The accumulation of the blade boundary layer in the wake causes a total pressure drop and a total temperature drop in the case of cooled blades. In blade rows with low aspect ratio, strong secondary flows are encountered. This complex flow field is traversed by the downstream rotor which sees periodically changes of static pressure, inlet relative total pressure and temperature and relative inlet angle resulting from the interaction of the rotor with the upstream flow field. Both the pressure distribution and the boundary-layer characteristics undergo periodic changes in the rotor passage. As a result, stage losses estimated from isolated blade row correlations are often underestimated (Hodson [1]) and large differences in heat transfer rates are observed between cascade experiments and stage experiments (Garside et al. [2]). Blades are submitted to unsteady lift causing vibrations while unsteady heat transfer regime increases thermal fatigue. The understanding of such complex interaction is necessary to improve the blade design and prediction in terms of efficiency as well as the evaluation of mechanical and thermal fatigue.

Most of the experiments encountered in the literature were carried out at low speed. In this case, the pitchwise pressure gradient associated with the stator blade flow in the stator-rotor gap is rather smooth and decays exponentially with axial distance (Hodson [1]). This pressure field is modulated by the rotor passage events and periodic fluctuations propagate upstream and downstream. However, for spacings above 5% $c_{s,ax}$, the wake interaction becomes predominant (Doorly and Oldfield [3]). Considering the wake in the relative frame, the local velocity deficit causes a smaller incidence angle. As a result, the low momentum wake material tends to migrate from pressure side towards suction side. Moreover, when the wake approaches the rotor channel inlet, it is distorted due to higher velocities in the middle of the passage compared with that in the vicinity of the blade leading edge. These two effects were observed both in experiments (Hodson [4]) and calculations (Hodson [5], Giles [6]). The effect of the wake impinging on a rotor blade can be separated in two contributions: the modification of the circulation propagates at acoustic velocity (*wake effect*) whereas the perturbation associated with the wake passage on the rotor surface should appear at a velocity close to the convection velocity (*wake distortion effect*). Both types of propagations were observed (Hodson [4], Dietz and Ainsworth [7], Hilditch et al.[8]). The influence of the wake passing over the blade boundary layer was carefully studied in LP turbines and led to the observation of different propagation speeds for the leading edge and the trailing edge of turbulent spots, the detection of wake induced transition alternating with a becalmed region causing in the latter part of the blade a by-pass transition followed by a fully turbulent flow (Orth [9], Halstead et al [10]...).

The effect of the stator trailing edge shock interfering with the rotor flow received much less attention in the literature. A sweeping phenomenon of the shock from the blade crown down to the leading edge was put into evidence in both experiments (Doorly and Oldfield [3], Ashworth et al [11]) and calculations (Saxer and Giles [12], Giles [6]). Complex shock patterns were put into evidence like periodic stator trailing edge shock reflections on the rotor blade and periodic emergence of shocks between stator trailing edge and rotor leading edge (Giles [6]). The measurements of the unsteady pressure field and heat transfer coefficient in highly loaded transonic stages were carried out by several researchers (e.g. Hilditch et al.[8],

Moss et al. [13], Rao et al. [14], Guenette et al. [15]...) but there are still many features to be understood.

The von Karman isentropic compression tube facility CT3 was recently converted from an annular cascade test rig into a full turbine stage test rig. The experimental program comprised rotor mid-span measurements of pressure and heat transfer coefficient at 2 axial stator-rotor spacings, 3 different RPM, zero and 3% stator trailing edge coolant flow ejection. This research was made possible through a large European collaboration effort involving Alfa Romeo Avio, Fiat Avio, MTU, SNECMA, Turbomeca and the universities of Oxford, Limerick and Florence. The current paper is restricted to the description of the unsteady rotor pressure field.

2 NOMENCLATURE

a	speed of sound
c	chord
$\dot{m}$	mass flow
M	Mach number
P	pressure
Re	Reynolds number $\rho v c / \mu$
T	temperature
α	flow angle with respect to the axial direction

subscripts

c	coolant
0	total quantity
s	static quantity
1	stator inlet
2	stator outlet, rotor inlet
3	rotor outlet
r	relative to the rotor frame
s	stator
r	rotor
ax	axial direction (turbine stage axis)

3 TURBINE TEST RIG, TURBINE STAGE, INSTRUMENTATION, OPERATING CONDITIONS

The tests were performed in the VKI short duration compression tube facility CT-3. A detailed description of the wind tunnel and its operating principle was provided by Sieverding and Arts [20]. The facility is capable to simulate correctly Reynolds number, Mach number, gas/wall and gas/coolant temperature ratio of modern aero-engines HP turbines. From 1990 to 1994, the facility served for annular cascade testing (Arts and Heider [21], Sieverding et al. [22]). Recently, major modifications were undertaken to convert this annular cascade facility into a full single stage turbine test rig as shown in Fig. 1. The development of fast response instrumentation techniques (Dénos [16]) and an opto-electronic wireless transmission (Sieverding et al. [17], Byrne and Davies [18]) make this facility well suited for blade row interference studies.

The particular transonic stage under investigation consists of 43 vanes and 64 blades; Fig. 2 shows a 3-D view of the turbine stage geometry. The rotor was designed with a meridional flow channel divergence of 10 % to reduce the axial outlet velocity and to minimize secondary losses. The channel height is 50.7 mm at inlet with a tip diameter of 790.4 mm. Chords and axial chords of stator and rotor are 72.3 and 43.2, respectively 48.4 and 39.6 mm. The ratio between the number of rotor and stator blades is very close to 1.5 (1.49), which allows to limit the numerical simulations to a domain made of two vanes and three rotor blade channels.

The rotor mid-section is instrumented with 24 surface mounted miniature piezo-resistive pressure transducers (14 on the suction side and 10 on the pressure side) from Kulite distributed over 3 rotor blades. The implementation of the pressure chips of 1.2 x 1.2 mm^2 was realized at the University of Oxford, UK. The Fast Fourier Transform of a signal delivered by one of these sensors, located in the vicinity of the rotor leading edge, reveals at least 11 harmonics of the fundamental blade passing frequency (4.6 kHz). The relative total pressure is measured by means of a standard Kulite pressure transducer placed with a small retreat in a 2 mm cylinder inserted in the rotor blade leading edge. The signals of the transducers were corrected during the run-up of the rotor as well as during the blowdown for the centrifugal force with an in-house method and for the temperature according to a procedure developed by Ainsworth [23]. The overall uncertainty on the rotor surface pressure measurements is evaluated to be of the order of 1% of the relative inlet total pressure.

Full details on the turbine test rig, the turbine stage and the instrumentation are reported by Dénos in [16].

The main similarity parameters maintained in the present investigation are the stage pressure ratio ($P_{01}/P_{S3} = 3.05$), the Reynolds number ($Re_{2c} = 10^6$) as well as the temperature ratios (T_{gas}/T_{wall} and $T_{gas}/T_{coolant} = 1.5$). With wall temperatures near ambient during the test run, the values of total upstream pressure and temperature are respectively T_{01}=440 K and P_{01}=1.620 bar. The nominal rotor speed is 6500 RPM. Tests are run at 6000, 6500 and 6800 RPM with axial stator-rotor spacings of 35 and 50 % of the axial chord of the vane. The cooling mass flow ejected through the stator trailing edge slot is fixed at 0 and 3 % of the stage mass flow (10.8 kg/s). At 6500 RPM, the measured conditions are very close to the design conditions. Design stator exit Mach number is M_2=1.03. The amount of flow, which leaks through the hub seals between stator and rotor, was evaluated from the increase of the pressure during the test in the volume underneath the seal. It results that the leakage mass flow into this volume is less than 0.6 % of the overall mass flow (10.8 kg/s)

The test rig provides constant conditions of inlet and outlet pressure and temperature during about 0.2 s (the largest variation was observed on P_{03} and is of the order of 2.0 % of its mean value). During the blowdown, the rotor speeds-up at a rate of about 800 RPM/s but the time period used to reduce the unsteady periodic data being of the order of 0.03 s (3 rotor revolutions), the corresponding variations in rotational speed, relative rotor inlet angle and relative rotor inlet total pressure are respectively limited to 24 RPM, 0.5 deg. and 3 mbar.

4   ROTOR MID-SPAN TIME-AVERAGED MACH NUMBER DISTRIBUTION

The time-averaged pressure measurements at 6000, 6500 and 6800 RPM for a stator-rotor distance of 0.35 $c_{s,ax}$ with 3% coolant ejection are presented under the form of Mach number distribution in Fig. 3. At nominal rotational speed (6500 RPM), the Mach number is computed with the design relative total pressure $P_{02,rel}/P_{01} = 0.572$. At 6000 and 6800 RPM this value becomes respectively $P_{02,rel}/P_{01}$=0.586 and 0.565. The results of steady 3-D Navier-

Stokes stage computations performed with the code of Arnone et al. [24] for 6000, 6500 and 6800 RPM are also presented. Measured inlet total pressure and total temperature profiles were imposed at the stage inlet. A mixing plane approach is used to transfer the information from stator exit to rotor inlet. A Baldwin-Lomax algebraic turbulence model is implemented.

The measurements show an acceleration on the front suction side which is much larger than that calculated with the 3-D Navier-Stokes code and what was predicted during the design (a continuous acceleration was prescribed). As a result, a plateau is observed between s/c=0.4 and 0.5 at 6800 and 6500 RPM while a deceleration takes place at 6000 RPM. As the deceleration occurs above M=1, it is likely to be associated with a weak shock. The most probable explanation for this discrepancy is a difference of incidence angle between predictions and measurements. The 3D Navier-Stokes computation which fits at best the experimental results at 6500 RPM is the distribution at 6000 RPM. This suggests that the relative inlet angle would be 50 deg instead of 44.8 predicted at 6500 RPM. This would imply an absolute stator exit angle of 74.3 deg whereas the same angle measured in the isolated stator case was 72.3 deg (see Sieverding et al. [22]). Since the uncertainty on this measurement was estimated to +/- 0.8 deg, the remaining difference (1.2 deg) would have to be ascribed to the stator-rotor interaction.

The front suction side Mach number distribution is very sensitive to a change of the rotational speed. This effect can be entirely attributed to the change of incidence which results from the change of the peripheral speed. With a design inlet angle of 45.8 deg. and assuming a constant stator exit Mach number of 1.03, the variation of incidence with respect to design at 6000 and 6800 RPM are respectively 4.15 deg. and -2.9 deg. The pressure side is sensitive to incidence only in the leading edge region.

The overall effect of the stator coolant flow ejection on the pressure distribution at the stator/rotor distance 0.35 $c_{s,ax}$ is small. On the front suction side, a slightly faster acceleration is observed. An increase of the axial rotor distance from 0.35 $c_{s,ax}$ to 0.5 $c_{s,ax}$ has no significant effect on the pressure distribution. This observation is in agreement with results reported by Dring et al. [25].

5   UNSTEADY ROTOR MID-SPAN PRESSURE MEASUREMENTS

5.1   Phase reference, phase-locked averaging

The rotational speed is recorded at high frequency (64 TTL pulses per rotation) in order to have a very precise measurement and to correctly identify the position of the rotor with respect to the stator at any time. The position of a rotor blade with respect to a stator blade is defined by the phase φ. For a given instrumented blade, $\varphi = 0$ defines the position for which the stacking axis of the instrumented blade is aligned with the stacking axis of the vane i.e. the blade leading edge, see Fig. 5. At $\varphi = 1$ the instrumented blade has completed an angular displacement equal to one stator pitch. All signals were corrected for the phase delays induced by the instrumentation and transmission boards.

All time resolved data were processed by means of a phase-locked averaging technique, including 3 full revolutions of the rotor (129 periods). The phase-locked averaging technique takes into account the evolution of the rotational speed as a function of time and uses the raw signal without any interpolation. The output of the phase-locked averaging routine provides a high quality statistical analysis of the time resolved signals and is composed of:
- the mean period across a pitch (phase locked average)
- the minimum and maximum of the mean period
- the RMS over one pitch (for each point of the phase-locked average)

- a linear regression correlation coefficient which relates how close the raw signal is to the phase-locked averaged signal.

5.2 Rotor inlet relative total pressure

The phase- locked averaged signals of the total pressure variation $\Delta P_{02r}/ P_{01}$ at the axial stator-rotor distance x = 0.35 $c_{s.ax}$ for 6500 RPM with and without stator trailing edge coolant ejection are shown in Fig. 4 together with the RMS distribution. A sharp pressure rise associated with a peak in RMS can be seen at φ=0.97. This large fluctuation can be associated with the traverse of the probe across the trailing edge shock. To aid interpretation, Fig. 5 presents schematically the interference of the stator trailing edge shock and wake patterns with the rotor blades as derived from Schlieren photographs by Kapteijn et al. published in [26] for the same flow conditions in a stationary linear cascade.

In Fig. 5, the position of the interference of the shock with the plane corresponding to the probe nose can be located at φ = 0.91. This simple prediction, which does not account for any unsteady effect, is quite close to the measurements. Notice that the peak in the RMS does not mean that high static pressure fluctuations occurs within the shock but most probably that the shock position varies slightly from one stator passage to another resulting in artificial fluctuations in the phase-locked averaging process.

The bump in the RMS distribution just to the right of the peak, with a local maximum at φ = 1.15, probably points to the impact of the stator wake. This interpretation is supported by the difference in the RMS distribution between the pressure traces with and without trailing edge coolant flow ejection. The path of the wake in a steady, interaction free hypothesis, is also plotted in Fig. 5, using the pitchwise averaged outlet angle measured in a previous test campaign with the stator alone (Sieverding et al. [27]). Notice that the wake and the shock intersect in the vicinity of the measuring plane. The predicted wake impact on the total pressure probe sitting at leading edge occurs at φ=0.88 instead of φ=1.15 ($\Delta\varphi$=0.27) found in the measurements. The wake slip effect, as mentioned in the introduction for low speed turbines, results in a lower incidence angle and causes the wake to hit the leading edge sooner than expected. An opposite effect is observed here. In contrast with low speed turbines, transonic turbines exhibit large pitchwise static pressure gradients downstream of the vane. If the relative flow angle inside the wake is computed in a zone of low static pressure, it can be locally higher than the mean flow angle. The wake path and the pressure discontinuity propagation line being different, the over or under deflection of the wake would depend on how often the wake traveled in a low or high static pressure zone. As a result it is very difficult to anticipate if the wake will hit the rotor leading edge sooner or later than expected. In any case, the variation of angle caused by a total pressure drop of 8% would not result in a phase difference larger than 0.03 in the present case. The most likely explanation is the distortion of the path observed when the wake approaches the rotor leading edge. An unsteady blade row computation using rotating inlet boundary conditions (total pressure wake, total temperature wake) with a 2D Euler code showed that the distortion of the wake, at the time it hits the rotor leading edge, was causing a phase delay of $\Delta\varphi$=0.25 with respect to what is expected from the main flow relative angle (Dénos [16]). An additional explanation could be that the vane absolute exit flow angle is higher than expected. The observed difference of $\Delta\varphi$=0.27 means that the absolute exit angle would be 75.7 deg. instead of 72.3 deg measured downstream of the vane alone. In the author's opinion, such a large difference is unlikely.

As a next step, let us analyze the effect of a pitchwise variation of the absolute flow variables on the relative inlet total pressure. The variations of P_{02}, T_{02}, α_2 are known from a previous test campaign. The stator pitchwise static pressure distribution at mid-span P_{s2} was

C557/057 © IMechE 1999

computed with a 2D-Euler code. The relative inlet total pressure trace obtain in this manner is very close to the measured one (Dénos [16]). The agreement is further improved when changing the phase of the total pressure wake with respect to other traces in order to take into account the wake slip effect. Table 1 presents the effect of the variation of each of the above parameters around their mean value on the resulting relative inlet total pressure.

P_{02} 10%	P_{s2} 30 %	T_{02} 5 %	α_2 2 deg
4.27 %	17 %	0.75%	0.98 %

Table 1: Contributions of typical variations of P_{02}, P_{s2}, T_{02}, α_2 around their mean values on the relative inlet total pressure P_{02r}.

It appears clearly that the static pressure field dictates the shape and the amplitude of the relative inlet total pressure. This can easily be understood when comparing the static to total pressure ratio corresponding to the absolute stator exit Mach number with the ratio corresponding to the relative rotor inlet Mach number. In the first case ($M_{2is\ mid-span}$=1.03) the static pressure represents only 51% of the total pressure, whereas in the second case (M_{2r}=0.45) it amounts to 87%.

The influence of the rotational speed is shown in Fig. 6 a. The amplitude of the fluctuation clearly increases with the rotational speed. This can be explained by two effects:
- for fixed stator outlet conditions, the increase of the rotational speed causes a decrease in the relative inlet speed which reinforces the contribution of the static pressure variation (an increase of rotational speed from 6000 to 6800 RPM causes a change of 3.6% in P_{02r});
- the measurements of the intermediate static pressure at tip across one pitch revealed that the amplitude of the static pressure variation increases by an amount of 10 % when the rotational speed rises from 6000 to 6800 RPM.
The second contribution is clearly dominant here.

It is surprising to note that the coolant flow ejection has no significant influence on the pressure distribution in spite of significant differences in the trailing edge shock patterns revealed in the Schlieren photographs (Kapteijn et al. [26]). As regards the RMS distribution the only, but weak, effect of the trailing edge coolant ejection occurs in the region of the wake interference.

The influence of spacing is shown in Fig. 6 b. The patterns are quite different with the emergence of an additional peak of non-negligible amplitude. According to the geometrical prediction of a steady shock at 0.50 $c_{s,ax}$ downstream of the stator, the sharp pressure rise should be observed at φ=0.85 (0.91 predicted at 0.35 $c_{s,ax}$). In reality, the sharp pressure rise appears at the same phase than in the first plane suggesting that between 0.35 $c_{s,ax}$ and 0.50 $c_{s,ax}$ the stator trailing edge shock propagates nearly in axial direction. This observation is in quite good agreement with the Schlieren picture taken in the isolated stator case. At 0.35 $c_{s,ax}$, the shock traverses the wake and is deflected to a more axial direction.

The amplitude of the total pressure fluctuations at 6500 RPM amounts to 6.5% of the upstream total pressure P_{01} or 11.2% of the design relative inlet total pressure P_{02r}.

5.3   Rotor mid-span unsteady pressure field

The unsteady pressure traces at each of the 24 measurement locations are presented as $(P - P_{pitch-average})/P_{01}$ over two pitches in Fig. 7 a, b and c. The mean value of the phase-locked average of all signals is zero due to high pass filtering. However, a difference of mean level between two successive traces was introduced in the graphs for clarity. The static pressure fluctuation measured in gauge 14 is very close to the measured relative inlet total pressure fluctuation.

The largest fluctuations are observed in the leading edge region (Fig. 7 b and Fig. 9-a) and can be associated with the stator trailing edge shock hitting the crown of the blade (gauges 5 and 4) and sweeping the front suction side (gauges 3,2,1) towards the leading edge (gauge 14). For all these gauges, a similar pattern is observed: one large fluctuation per period beginning with a steep pressure rise, immediately followed by a short double peak pattern. Then, after a short decrease, a plateau is observed. This steep pressure rise is associated with a peak in the RMS traces across the pitch as shown in Fig. 10. The path of the shock in this region can be determined by plotting the phase corresponding to the steep rise as a function of the curvilinear abscissa of the gauge. This is compared in Fig. 8 with the path of a steady state stator trailing edge shock. The intersection of the two curves suggests that the stator trailing edge shock impact oscillates around the trajectory predicted with a steady shock. In the computation of Michelassi [19], the stator trailing edge shock inclination is changing depending on the position of the rotor blade with respect to the stator blade.

The steep pressure rise due to the shock impact is always followed by a double peak which is not fully understood. It might be due to a shock induced separation bubble observed by Doorly and Oldfield [3] in their cascade tests with shocks produced by rotating bars. Indeed, this steep adverse pressure gradient occurs in the front suction side of the rotor blade where a large time-averaged favorable pressure gradient dominates due to the local high curvature. This periodic adverse pressure gradient is likely to cause a periodic separation followed by a recirculation bubble and a reattachment of the flow. Due to this, the curvature of the time-averaged blade contour seen by the flow is increased which could also explain the over acceleration observed on the front rotor suction side in the time-averaged Mach number distribution. A similar double peak feature was observed by Dietz and Ainsworth [7] but in their case, the first peak appears within the steep pressure rise.

The amplitude of the fluctuations is increasing rapidly from 8 % of P_{01} at gauge 14 to a maximum peak to peak pressure difference of 24% on the suction side at gauge 4. Referred to the theoretical inlet relative total pressure P_{02r}, this pressure amplitude amounts to 40 %. The static pressure fluctuation at gauge 4 exceeds by far the predicted pitchwise variation of static pressure behind the stator (15% of P_{01}). This large change of the static pressure fluctuation could be attributed to a different blockage effect depending on the location of the shock impact in the region between the crown and the leading edge. However, an inverse tendency is observed in the calculation of Giles and in the measurements of Rao et al. [14]. In their case the stator exit Mach number was slightly higher, $M_{2,is} = 1.12$ instead of 1.05 here and the stage geometry was different.

On the front pressure side (gauges 15 and 16, Fig. 7 b), the overall pressure modulation is smooth except for the appearance of two small pressure peaks attributed to weak shock events. The observation of two distinct spike at the same phase in the RMS traces reinforces this conjecture (Fig. 10). In his calculation Giles [6] also observed the occurrence of two peaks, the first being due to the stator trailing edge shock, the second due to the impact of a reflected shock. This second shock originates at the trailing edge of a stator blade and is reflected across the rotor passage after impinging on the suction side of the precedent rotor blade.

Once the shock has left the leading edge region on the pressure side (Fig. 7 a), the pressure fluctuations become very small and the nearly flat curve at gauge 18 is possibly explained by the fact that the gauge is located in a concave part of the blade which the shock can not access. In the rear part of the convergent channel, the gauges on both the pressure and suction sides, respectively gauges 22-24 and 7-8, exhibit a double fluctuation over one period with similar amplitude and phase. This suggests that a unique phenomenon affects simultaneously the entire passage in this region. A careful examination of the traces 19-24 (Fig. 7 c) reveals that while the first pressure wave occurs simultaneously for all traces, the phase of the second pressure wave is slightly shifted from one trace to the successive trace, showing a perturbation travelling upstream. Pressure perturbations travelling upstream are also mentioned by Giles [6].

On the rear suction side, the amplitude of the fluctuations at points 7 to 13 is much smaller than in the leading edge region (Fig. 9-a). The time-averaged RMS and the correlation of the raw signal with the phase-locked averaged signal (Fig. 9-b and c) bring more information on the nature of the fluctuations. In these locations, the time-averaged RMS is of the same order as at the leading edge but the correlation coefficient decreases dramatically down to 0.3 at the trailing edge. This means that, while the amplitude of the periodic component decreases, random fluctuations of increasing importance appear. Unlike at the leading edge, most of the fluctuations in the signal are not correlated with the shock passing events. Notice that this part of the rotor blade is hidden from a direct influence of the shock. The FFT analysis did not reveal any other frequency than the blade passing frequency meaning that the fluctuations are mostly random. This region corresponds with a small deceleration in the Mach number distribution (Fig. 3) which could cause a thickening of the boundary layer associated with turbulent fluctuations of increasing importance. On the late pressure side, the RMS stays at a constant low level after point 18 but with a high level of correlation. This means that most of the fluctuations are correlated with the blade passing frequency. This part of the pressure side seems accessible to the stator trailing edge shock and could maybe reached by reflected shocks.

As mentioned in the introduction, two effects of the wake can be distinguished: the wake effect travelling at acoustic speed (v+a and v-a) and the wake distortion effect travelling in the flow at the local convective speed. In some cases, pressure perturbations travelling close to the convective speed were observed (Hodson [4], Hilditch et al.[28]); in other experiments (Dietz and Ainsworth [7]) pressure fluctuations travelling at acoustic speed were identified. An attempt was done to track the propagation of the perturbations associated with a wake along the blade. This was performed assuming the injection of a perturbation at point 14 (close or at stagnation point) at $\varphi=1.15$ (from observations in relative total pressure traces). Then, three propagation lines were computed: one with the local convective speed v, the second with v+a, the third with v-a plotted as dots on top of the RMS traces on Fig. 10. The v-a trace stops at gauge 4 because the local convective speed becomes almost equal to the speed of sound resulting in a zero propagation velocity. No clear correlation between the traces and the peaks in RMS appear. The same dots plotted on the static pressure signal do not reveal any trend either. As mentioned earlier, the wake influence is at least one order of magnitude lower than the shock influence due to the transonic regime of the nozzle.

The absence of coolant ejection causes a small decrease in the amplitude of the fluctuations, mainly in the front suction side. As the rotational speed increases, the pressure traces showed that the sweeping phenomenon of the stator trailing edge shock occurs faster and faster. On most of the blade portion, the amplitude of the fluctuations tends to increase slightly with increasing rotational following the trend observed on the intermediate static pressure field P_{s2}. The shape of the RMS curves and correlation curves are quantitatively conserved.

Most of the qualitative features described for the first spacing are also encountered for the second spacing (see Fig. 11). The amplitude of the fluctuations are noticeably smaller in the leading edge region (gauge 2, 3, 4, Fig. 9 a) showing that the shock intensity has probably decreased strongly with increasing axial distance. Additional peaks are seen in the traces probably due to a different unsteady shock pattern.

5.4 Unsteady blade force

The unsteady blade pressure distributions were used to calculate the time varying tangential blade force. Fig. 12 presents these variations non-dimensionalized by the time averaged blade force. The total amplitude of the fluctuations of the blade force is 10 %. In addition, the variation of the load distribution along the blade induces a time varying torsional moment. In many turbines the axial stator-rotor distance is even smaller than in the present turbine with as result even higher blade force fluctuations.

6 CONCLUSIONS

The careful instrumentation of the rotor and elaborate calibration procedures provided high quality rotor data for three rotational speeds (6000, 6500, 6800 RPM), 2 coolant configurations (3%, 0%) and 2 spacings between stator and rotor (0.35 $c_{s,ax}$, 0.50 $c_{s,ax}$) at mid-span. This large set of experimental data should provide a useful base for code validation. Some results deserve particular attention.

The rotor was designed to fit the experimental outlet flow conditions of an existing guide vane. Based on the measured time averaged rotor Mach number distribution, it is concluded that the real relative inlet flow angle is considerably higher than could have been expected from the measured outlet flow angle behind the stator in absence of the rotor. The difference is at least partially to be attributed to 3-D blade row interference effects not accounted for in the rotor blade design. Another source for this discrepancy could also be a periodic separation of the flow on the front suction side resulting in an increased time-averaged acceleration on this part. The influence of coolant ejection and spacing on the time averaged pressure distribution is weak. The effect of the rotational speed on the time-averaged Mach number distribution can be explained by the change of relative inlet angle.

In this transonic stage, the unsteady relative inlet total pressure fluctuations is mainly dictated by the static pressure field which contains steep gradients due to the stator trailing edge shock; the amplitude of the fluctuations is of the order of 10 % of the relative total pressure. The effect of the wake is one order of magnitude smaller. The amplitude and the shape of the fluctuations change with rotational speed and stator-rotor axial distance. No significant difference is observed due to coolant ejection.

The interaction of the stator trailing edge shock with the rotor blade leads to extremely strong pressure gradients in the rotor leading edge region and the front suction side with a maximum pressure amplitude of nearly 40% of the relative inlet total pressure at the point near the maximum suction side curvature. The large difference between the amplitude of the fluctuations at the leading edge and at the crown was attributed to a changing blockage effect faced by the stator trailing edge shock. The amplitude of the fluctuations decreases strongly on the rear suction side and pressure side; on the first, random fluctuations of increasing amplitude appears while on the second, fluctuations remain mainly correlated with blade passing events. Pressure perturbations travelling upstream were identified on the rear pressure side. The influence of the wake on the pressure traces could not be identified. The coolant ejection does not modify significantly the unsteady pressure field while the increase of the rotational speed leads to slightly larger fluctuations. When increasing the spacing, the amplitude of the fluctuations decreases significantly on the front suction side probably due to the smearing of the stator trailing edge shock with axial distance.

The calculation of the tangential blade force based on the instantaneous pressure distributions shows a maximum amplitude of the variation of the blade forces of 10% over one period.

Work is underway to improve the understanding of the complex unsteady shock pattern and of the wake influence with the help of the quasi 3-D unsteady Navier&Stokes code of Michelassi [19].

ACKNOWLEDGEMENTS

The above research was carried out under contract for the European Commission as part of the BRITE EURAM AER 2-92-044 turbine project "Investigation of the Aerodynamics and cooling of Advanced Engine Turbine Components". The authors wish to acknowledge this financial support as well as the contributions of the industrial partners ALFA AVIO, FIAT, MTU, SNECMA and TURBOMECA who made this research possible.

REFERENCES

[1] Hodson H.P.: "Boundary layer and loss measurements on the rotor of an axial flow turbine", ASME paper 83-GT-4, 1983.

[2] Garside T., Moss R.W., Ainsworth R.W.- Dancer S.N., Rose M.G.:"Heat transfer to rotating turbine blades in a flow undisturbed by wakes", ASME paper 94-GT-94.

[3] Doorly D.J. and Oldfield M.L.G.: "Simulation of the effect of shock wave passing on a turbine rotor blade" Journal of Engineering for Gas Turbines and Power, 107, pp998-1006, ASME Paper 85-GT-112, 1985.

[4] Hodson H.P.: "Measurements of wake-generated unsteadiness in the rotor passages of an axial flow turbine", Journal of Engineering for Gas Turbine and Power vol.107, pp 467-476, 1985.

[5] Hodson, H.P.: "An inviscid blade-to-blade prediction of a wake-generated unsteady flow", ASME 84-GT-43, 1984.

[6] Giles M.B.: "Stator/rotor interaction in a transonic turbine" AIAA 88-3093,1988.

[7] Dietz A. J. and Ainsworth R. W.: "Unsteady pressure measurements on the rotor of a model turbine stage in a transient flow facility" ASME Paper 92-GT-156, 1992.

[8] Hilditch M.A., Smith G.C. and Singh U.K.: " Unsteady flow in a single turbine stage", ASME paper 98-GT-531.

[9] Orth U.: "Unsteady boundary layer transition in flow periodically disturbed by wakes", ASME paper 92-GT-283, 1992.

[10] Halstead D.E., Wisler D.C., Okiishi T.H., Hodson H.P. and Hyoun Woo Shin: "Boundary layer development in axial compressors and turbines. Part 1: Composite picture" ASME Paper 95-GT-461, 1995.

[11] Ashworth D.A., LaGraff J.E., Shultz D.L. and Grindrod K.J.: "Unsteady aerodynamic and heat transfer processes in a transonic turbine stage", Journal of Enginnering for Gas Turbines and Power, Vol 107, pp1022-1030, 1985.

[12] Saxer, A.P and Giles, M.B. "Predictions of 3-D Steady and unsteady inviscid transonic stator/rotor interaction with inlet radial temperature non-uniformity", ASME 93-GT-10, 1993

[13] Moss R.W., Sheldrake, C.D., Ainsworth, R.W. - Smith A.D., Dancer S.N.: "Unsteady pressure and heat transfer measurements on a rotating blade surface in a transient flow facility" 85th propulsion and energetic panel symposium on loss mechanism and unsteady flows in turbomachines, Derby, UK, 1995. AGARD CP No. 571, 1996.

[14] Rao K.V., Delaney R.A. and Dunn M.G.: "Vane-blade interaction in a transonic turbine. Part 1: Aerodynamics. Part 2: Heat transfer" AIAA Journal of Propulsion and Power, Vol. 10, No. 3, pp 305-317, 1994.

[15] Guenette G.R., Epstein A.H., Giles M.B., Haimes R.- Norton R.J.G.: "Fully scaled transonic turbine rotor heat transfer measurements" ASME Journal of turbomachinery, Vol. 111, pp 1-7, 1989.

[16] Dénos. R.: "Investigation of the unsteady aero-thermal flow field in the rotor of a transonic turbine", PhD thesis, IVK-University of Poitiers. 1996.

[17] Sieverding C.H., Vanhaeverbeek C., and Schulze G.: "An opto-electronic data transmission system for measurements on rotating turbomachinery components" ASME Paper 92-GT-337, 1992.

[18] Byrne C. M. and Davies M. R. D.: "Data transmission systems for a transient gas turbine rotor" ASME 96-GT-514, 1996.

[19] Michelassi V., Martelli, F.- Dénos, R., Arts T., Sieverding C.H.: "Unsteady heat transfer in stator-rotor interaction by two equation turbulence model" ASME paper 98-GT-243.

[20] Sieverding C.H. and Arts T.: "The VKI Compression Tube Annular Cascade Facility CT3" ASME Paper 92-GT-336, 1992.

[21] Arts T. and Heider R.: "Aerodynamic and thermal performance of 3D annular transonic nozzle guide vane. 1-Experimental investigation" 30th AIAA/ASME/SAE/ASEE Joint Propulsion Conference, Indianapolis, USA, 1994.

[22] Sieverding C.H., Arts T., Dénos R., Martelli F. :"Investigation of the flow field downstream of a turbine trailing edge cooled nozzle guide vane" Journal of Turbomachinery, Vol. 118, pp 291-300, 1996, ASME Paper 94-GT-209, 1994.

[23] Ainsworth R.: "Recent development in fast response aerodynamic technology" Von Karman Institute Lecture Series 1995-01 on Measurement Techniques, 1995.

[24] Arnone A., Meng-Sing L. and Povinelli L.: "Multigrid Calculation of three-dimensional viscous cascade flows" Journal of Propulsion and Power, Vol. 9, No.4, pp 605-614, July-August 1993.

[25] Dring R.P., Joslyn, H.D., Hardin and L.W., Wagner, J.H.: "Turbine rotor-stator interaction" Journal of Engineering for Power, vol. 104:No 2, pp 729-742, ASME Paper 82-GT-3, 1982.

[26] Kapteijn C., Amecke J.- Michelassi V.: "Aerodynamic performance of a transonic turbine guide vane with trailing edge coolant ejection. Part 1-Experimental approach" Journal of Turbomachinery, Vol. 118, No. 3, pp 519-528, July 1996.

[27] Sieverding C. H., Arts T., Dénos R., Amecke J., Kapteijn C., Martelli F., Michelassi F., Coluantoni S., Santoriello G., Schröder T., Bernard J., and Lapidus Y.: "Advances in engine technology, chapter 3: Investigation of the wake mixing process behind a transonic turbine inlet guide vanes with trailing edge coolant flow ejection". New York, John Wiley & Sons Ltd., 1995. Editor: Dunker R.

[28] Hilditch M.A., Smith G.C., Anderson J.S. and Chana K.S.: "Unsteady measurements in an axial flow turbine" 85th Propulsion and Energetic Panel Symposium on Loss Mechanism and Unsteady Flows in Turbomachines, Derby, UK, 1995. AGARD CP No. 571, 1996.

7 FIGURES

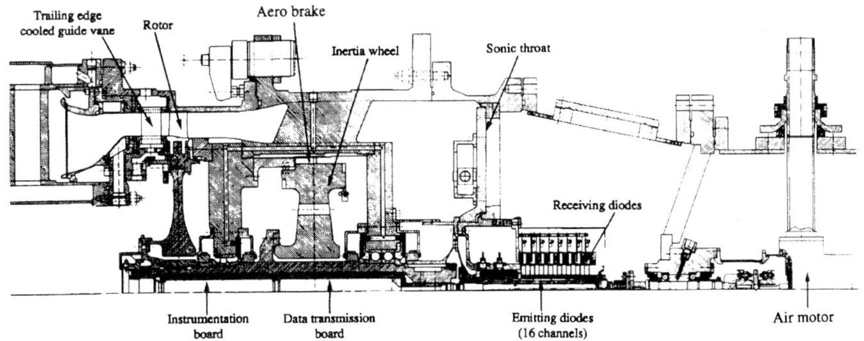

Fig. 1:Turbine test section

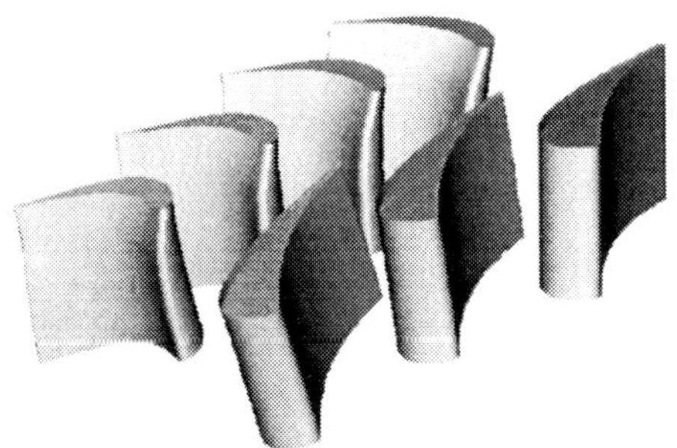

Fig. 2: 3D view of the Brite turbine stage

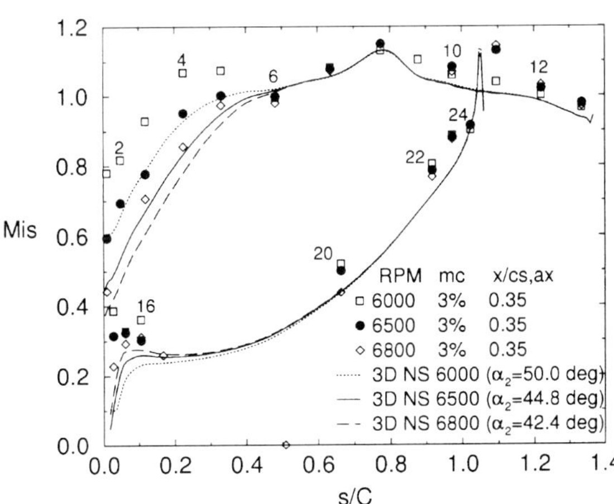

Fig. 3: Time averaged rotor isentropic Mach number distribution

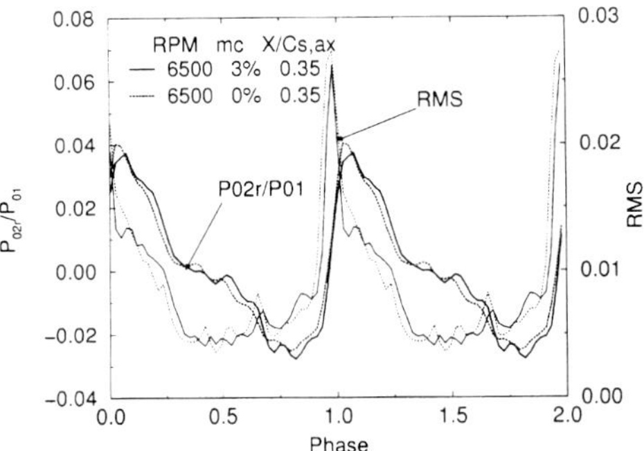

Fig. 4: Rotor relative inlet total pressure phased-locked average signals and RMS at 6500 RPM, stator-rotor spacing 0.35 $c_{s,ax}$; influence of trailing edge coolant ejection

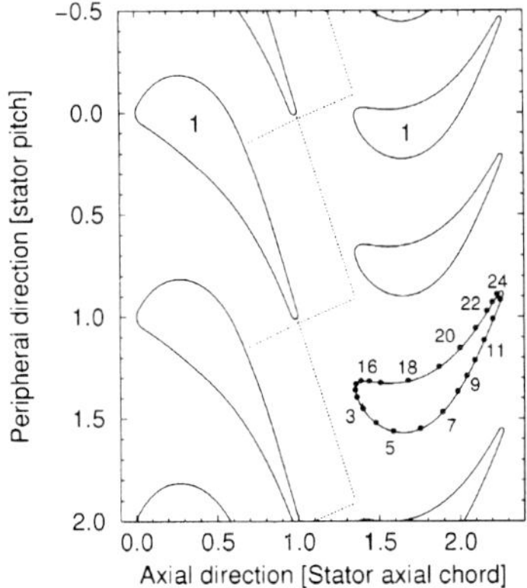

Fig. 5: Rotor/stator position at phase 0; schematic presentation of the stator shock and wake interaction with rotor blade for a stator-rotor spacing of 0.35 $c_{s,ax}$.

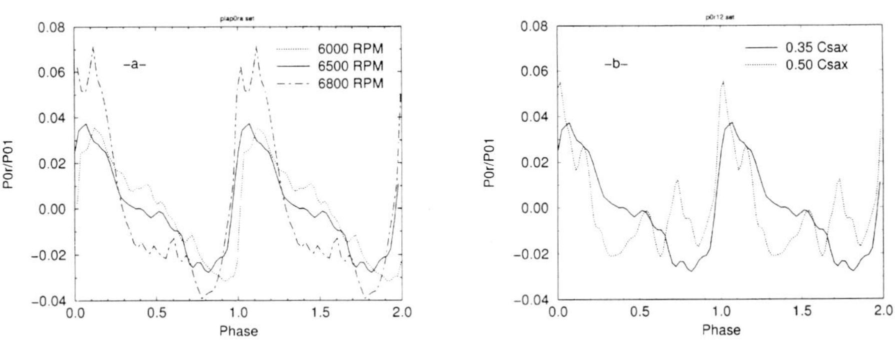

Fig. 6: Influence of a) the rotational speed (0.35 $c_{s,ax}$, $\dot{m}_c$ =3%) and b) the stator-rotor spacing (6500 RPM, $\dot{m}_c$ = 3%) on relative inlet total pressure fluctuations.

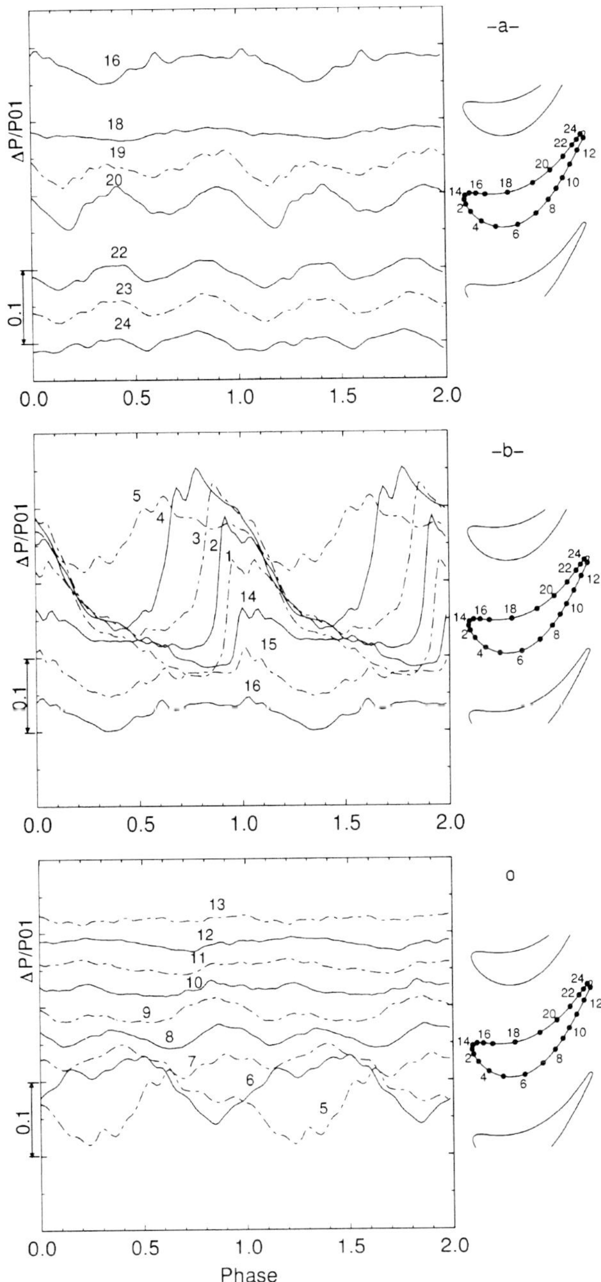

Fig. 7: Pressure traces (6500 RPM, 0.35 $c_{s,ax}$, $\dot{m}_c$ =3%) a) gauge 24 to 16, b) gauge 16 to 5, c) gauge 5 to 13.

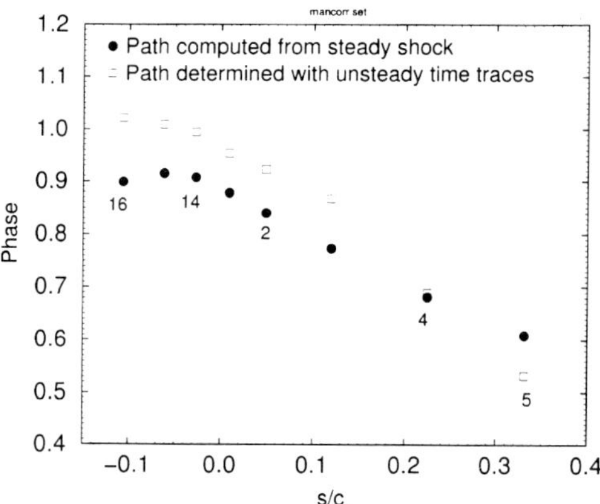

Fig. 8: Shock path estimated from unsteady measurements and from steady shock pattern.

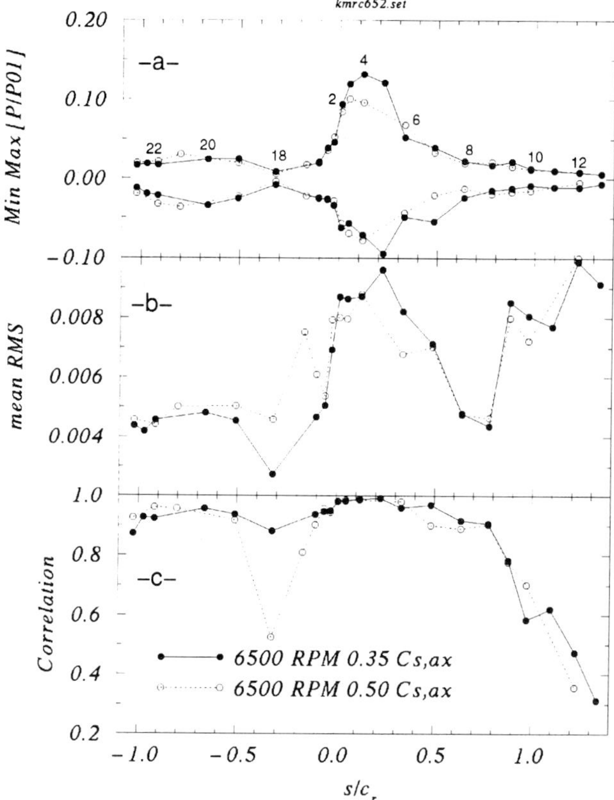

Fig. 9: a) Minimum, maximum of mean period, b) time-averaged RMS and c) correlation coefficient of pressure traces for spacings 0.35 $c_{s,ax}$ and 0.50 $c_{s,ax}$ (6500 RPM, $\dot{m}_c = 3\%$)

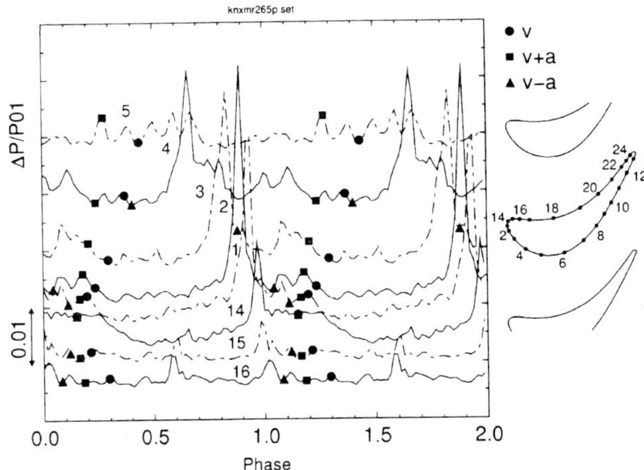

Fig. 10: RMS traces (6500 RPM, 0.35 $c_{s,ax}$, $\dot{m}_c = 3\%$) in the blade nose region; propagation directions at v, v+a, v-a for a perturbation injected at point 14 at $\varphi=0.15$

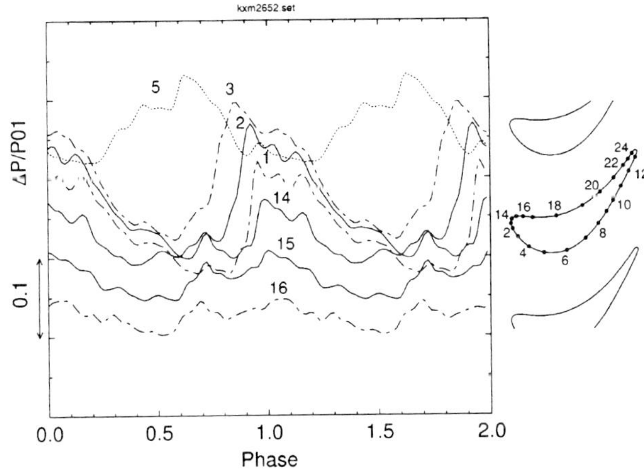

Fig. 11: Pressure traces: gauge 16 to 5 (6500 RPM, 0.50 $c_{s,ax}$, $\dot{m}_c = 3\%$)

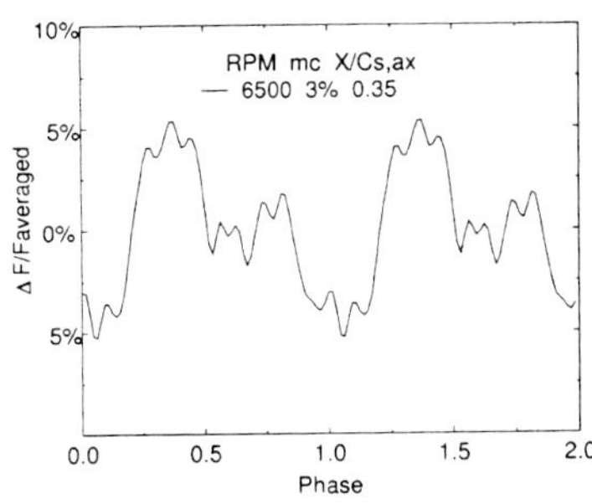

Fig. 12: Unsteady tangential blade force

C557/049/99

Three-dimensional unsteady Navier–Stokes analysis of stator–rotor interaction in axial-flow turbines

L HE
School of Engineering, University of Durham, UK

SYNOPSIS

A 3D full Navier-Stokes method is developed and applied to calculations of unsteady flows through multiple bladerows in axial flow turbomachinery. The solver adopts the cell-centred finite volume discretization and the 4-stage Runge-Kutta time-marching scheme. Unsteady calculations are effectively accelerated by using a time-consistent multi-grid technique, resulting in a speed-up by a factor of 10~20 with adequate temporal accuracy. The computational efficiency and validity of the present multi-grid technique is illustrated by comparisons with the results of the conventional dual time-stepping scheme. Calculated unsteady pressures on blade surfaces for a turbine stage test rig compare well with the corresponding experimental data. A computational study of turbine stage performances at different stator-rotor axial gaps reveals a marked three-dimensional behaviour of the interaction between incoming wakes and rotor passage-vortex structures. The time-averaged losses from unsteady calculations show a considerable spanwise redistribution compared to the steady results. 2D and 3D calculations indicate opposite trends in stage efficiency variation when the stator-rotor gap is reduced.

1. INTRODUCTION

Modern designs of axial-flow turbomachinery are driven toward highly compact configurations. The trend of increase in aerodynamic loading and decrease in machine size and weight makes it important to understand and evaluate unsteady bladerow interaction effects. This is reflected by the rapid increase in research and development activities in the area of unsteady flows in turbomachinery. There have been many efforts in computational fluid dynamics modelling of unsteady flows due to bladerow interactions. Two-dimensional calculations have been shown to be very useful in providing detailed information to further our understanding of the basic unsteady interaction mechanisms. For instance, the kinematic wake transportation model (1) and the associated energy separation mechanism was examined at a low speed condition (2), whilst the unsteady vane shock-rotor interaction was revealed in a transonic turbine (3).

Given the progress in 2D modelling and understanding of the unsteady bladerow interaction mechanisms, one would naturally ask, to what extent 2D results and arguments can be extended to a 3D blade row environment? This is highly relevant since modern blading designs are often intent to take advantages of three-dimensional flows. Although fully 3D unsteady viscous computations for turbomachinery stage configurations have started to emerge (4), the differences between 2D and 3D solutions in the context of the bladerow interaction effects on time-average performances have rarely been addressed.

In the present work, a previously developed Navier-Stokes solver for blade flutter calculations has been modified and extended to calculate 3D unsteady flows in multi-blade rows. There are two main emphases. Firstly, efforts were made to enhance computational efficiency by employing a new time-consistent multi-grid scheme. Secondly, the effects of the bladerow axial gap on stage performances were examined with particular attention paid to the difference between 2D calculations and 3D calculations.

2. COMPUTATIONAL METHOD

2.1. Unsteady Viscous Flow Model

For convenience of simulating flows in multiple bladerow turbomachinery, a cylindrical co-ordinate (x, θ, r) in an absolute system is adopted. For a rotor bladerow, this would avoid the extra terms related to the Coriolis term etc, and is preferred for convenience of the treatment at a rotor-stator interface. A mesh around a rotor row will be seen as a moving mesh in the absolute system. An integral form of the 3-D Navier-Stokes equations over a moving finite volume is:

$$\frac{\partial}{\partial t}\iiint_{\delta V} U\, dV + \oiint_{\delta A}[\ F\mathbf{n_x} + (G - Uv_{mg})\mathbf{n}_\theta + H\mathbf{n_r}]\cdot\ \mathbf{dA}\ = \iiint_{\delta V} S_i\, dV + \oiint_{\delta A}[\ V_x\,\mathbf{n_x} + V_\theta\,\mathbf{n}_\theta + V_r\,\mathbf{n_r}]\cdot\ \mathbf{dA}$$

(1)

where U, F, G and H are the standard primitive variable and inviscid flux vectors. $v_{mg} = \omega r$ is the mesh grid moving velocity and the extra inviscid flux term $U\,v_{mg}$ accounts for the contribution to the fluxes due to moving grids. S_i is the inviscid source term to include the centrifugal effect for maintaining a circumferential movement in the radial momentum equation. Different levels of viscous terms can be used. In the present work, the full viscous stress terms are adopted:

$$
V_x = \begin{pmatrix} 0 \\ \tau_{xx} \\ r\tau_{\theta x} \\ \tau_{rx} \\ u\tau_{xx} + v\tau_{\theta x} + w\tau_{rx} - q_x \end{pmatrix}, \quad
V_\theta = \begin{pmatrix} 0 \\ \tau_{x\theta} \\ r\tau_{\theta\theta} \\ \tau_{r\theta} \\ u\tau_{x\theta} + v\tau_{\theta\theta} + w\tau_{r\theta} - q_\theta \end{pmatrix}, \quad
V_r = \begin{pmatrix} 0 \\ \tau_{xr} \\ r\tau_{\theta r} \\ \tau_{rr} \\ u\tau_{xr} + v\tau_{\theta r} + w\tau_{rr} - q_r \end{pmatrix}
$$

The detailed formulations for these stress terms can be found elsewhere (6).

For turbulent flows, random unsteadiness will generate extra stress terms when ensemble-averaged at a known frequency. It is recognised that turbulence modelling for closure will inevitably involve uncertainties, especially for unsteady flows. Here for simplicity, the mixing length model by Baldwin and Lomax (7) widely used for steady flow calculations is used in a quasi-steady manner, assuming that a typical length scale of turbulence is still much smaller than that of the periodic unsteadiness to be dealt with.

2.2 Boundary Conditions

The computation is carried out in a multi-passage and multi-row domain, as shown in Fig.1 for a mid-span section of a turbine stage.

The relatively moving rotor and stator meshes are patched together at the interface. There are two options for the interface treatment. The first one is the so-called 'mixing-plane' firstly proposed by Denton, which was used for single passage multi-bladerow steady flow calculations (8). The present mixing-plane treatment follows the non-reflective procedure by Saxer and Giles (9). At each spanwise section, the 'mixed-out' variables at both the rotor and stator sides are flux-averaged. The difference in the mixed-out variables across the interface represents a jump in characteristics. The procedure is to drive characteristics jumps to zero in a nonreflective manner. It should be noted that the mixing-plane treatment is always associated with an entropy jump across the interface due to the mixing loss. The second method is for unsteady flow calculations. A 2nd order interpolation and correction method (10) enables local instantaneous information to be transferred directly across the interface.

On blade/endwall surfaces, the log-law is applied to determine the surface shear-stress and the tangential velocity is left to slip. This slip wall condition is preferred for unsteady 3D multi-passage calculations because of the relatively coarse meshes to be used. The solver includes a simple tip-leakage model similar to that previously used blade flutter calculation (11). For zero tip clearance cases, the rotor blades are effectively shrouded with a moving casing wall between the leading-edge and the trailing-edge.

At the circumferential periodic boundaries, the direct periodic (repeating) condition is currently used. This means that the numbers of rotor and stator passages included in the domain need to be such as to have the same total circumferential length.

At the inlet, stagnation parameters and flow angles are specified. At the exit, the pitchwise mean static pressure at each spanwise section is specified, and the local upstream-running characteristic is formulated to drive the pitchwise average pressure to the specified value, while the local pitchwise nonuniformity is determined by the downstream-running characteristic.

2.3 Discretization and Solution Methods

The governing equations are discretized in space using the cell-centred finite volume scheme, together with the blend 2nd and 4th order artificial dissipation (12) to damp numerical oscillations. The baseline temporal integration of the discretized equations is carried out using the explicit four-step Runge-Kutta scheme. It is well known that the explicit time-marching scheme is subject to a limitation on the length of time-step due to the numerical stability requirement, and this is very restrictive on unsteady viscous flow calculations. In order to overcome this limitation, the multi-grid technique described as follows is adopted.

<u>Time-Consistent Multi-Grid</u>

An effective way to relax the stability limitation is to use a time-consistent multiple grid technique. Consider that once the basic fine mesh is generated, the whole computational domain can be divided into small coarse mesh blocks for the multi-grid acceleration. For a fine mesh cell which has an allowable time-step less than the uniform global time-step for unsteady calculations, the equations will be firstly time-marched using the fine-mesh net flux at the allowable time step. Then, the deficit in the time-step length will be made up by using the net flux from the coarser mesh block which this fine-mesh cell belongs to. The conservative flux relation between the coarse and fine meshes means that the spatial accuracy is still controlled by the fine mesh. But the temporal accuracy is now subject to the coarse mesh. In the present method, the coarse blocks are taken to be essentially of the same spatial size. An adequate temporal accuracy is expected to be obtained as long as the size of the coarse block is much smaller than the spatial wave length of the physical unsteadiness to be modelled.

The simplest implementation of the time-consistent multi-grid is to use just two-grids as in the author's previous work (13), where the effectiveness of the technique was examined for unsteady transonic flows around oscillating cascades or due to a purely self-excited shock oscillation. Because the coarse blocks are more or less of equal spatial size and the basic fine viscous flow mesh is normally quite non-uniform, the numbers of fine mesh cells in each coarse block can thus be very different. In the present work, several levels of intermediate meshes between the basic fine mesh and the coarse mesh are introduced. This modification gives more weight to the contributions from finer meshes, and therefore improves the accuracy. Consider that we use M levels of intermediate meshes. The general multi-grid formulation for the temporal change of flow variables on the fine mesh is:

$$\delta U_f = \Delta t_f \, \frac{R_f}{\Delta V_f} + \sum_{i=1}^{i=M} \Delta t_i \, \frac{R_i}{\Delta V_i} + \Delta t_c \, \frac{R_c}{\Delta V_c} \tag{2}$$

where subscripts 'f', 'i' and 'c' denote the fine-mesh, the intermediate-mesh (level i) and the coarse mesh respectively. R is the net flux; ΔV is the cell or block volume. Δt_f and Δt_i are the allowable time-step lengths on the fine-mesh and the intermediate-mesh of level i ($1 \leq i \leq M$) respectively. In order to maintain a uniform global time step length Δt for unsteady calculations, the time-step length for the coarse block Δt_c needs to satisfy the time-consistence condition:

$$\Delta t_c = \Delta t - \Delta t_f - \sum_{i=1}^{i=M} \Delta t_i \tag{3}$$

Dual Time-Stepping

The dual time-stepping scheme was also implemented as an alternative acceleration option. This scheme was proposed by Jameson (14) and has been widely used for unsteady flow calculations. It starts with an implicit backward temporal difference of the discretized equations:

$$\frac{3}{2\Delta t} U^{n+1} - \frac{2}{\Delta t} U^{n} + \frac{1}{2\Delta t} U^{n-1} = \frac{R^{n+1}}{\Delta V} \tag{4}$$

A pseudo time τ is then introduced:

$$\frac{\partial U}{\partial \tau} = \frac{R^{n+1}}{\Delta V} - [\frac{3}{2\Delta t} U^{n+1} - \frac{2}{\Delta t} U^{n} + \frac{1}{2\Delta t} U^{n-1}] \tag{5}$$

The above equations can be integrated in the pseudo time by any time-marching methods with standard acceleration techniques for steady flows. Here the multi-grid method is used. When the residual is driven to zero, the second order temporal accuracy is fully recovered, independent of the multi-grid acceleration technique.

It should be mentioned that most of the calculations to be presented were carried out with the time-consistent multi-grid technique. The dual time-stepping scheme was mainly used to check the validity and effectiveness of the multi-grid technique for unsteady flows.

3. RESULT AND DISCUSSION

3.1. A Turbine Test Stage

Firstly, a test turbine stage was calculated for validation purposes. This is a subsonic turbine stage case for which experimental measurements of unsteady pressure on rotor blade surfaces are available (15). In the test rig configuration, the turbine stage was inverted, i.e. the rotor bladerow was kept stationary and the vane bladerow was rotated in the opposite direction. This was arranged in order to avoid the difficulty in unsteady data acquisitions in a rotating system. The hub/tip ratio of both blade rows is 0.8. The aspect ratio for the vane blades is 0.75 while that for the rotor is 1.34. All blades are of non-twist 2D profiles representative of mid-span I. P. turbine blading. High frequency response pressure transducers were embedded at different chordwise positions on the suction pressure surfaces (8 on each surface) at the mid-span section in one blade passage of the rotor row.

For a direct comparison, the calculations for this case were carried out with the same inverted stage of the rotating stator and stationary rotor. The test stage has 32 vane blades and 48 rotor blades, so the stage computational domain only needs to contain 2 vane blade passages and 3 rotor passages to achieve the same rotor/stator blade count. For all stage calculations, a periodic solution could normally obtained in 10 blade passing periods. Fig.2 shows typical time traces of tangential forces on the two bladerows when the unsteady calculation was started from a single passage steady flow solution.

A 3D calculation was carried out with a mesh density of 35x81x35 per passage. The tip clearance of the rotating blades was not included. Time-averaged pressure distributions at the rotor mid-span sections are shown in Fig.3. Unsteady pressure distributions on the suction and pressure surfaces at the mid-span sections are compared with the experimental data, as shown in Fig. 4. It should be noted that the unsteady pressures presented are the fluctuation components from the time-averaged values. The time traces at different chordwise positions are vertically shifted for clarity, hence the scale in absolute unsteady pressure values is only correct for the top trace. The reference phase angle is chosen to match the calculated phase with the experimental one for the leading edge point on the pressure surface (i.e. the first experimental tapping position on the pressure surface). It can be seen that the regions with appreciable unsteady pressures are the first half of the suction surface and a small region near the leading edge on the pressure surface, both of which are featured by large mean velocity gradients. Overall, the calculated magnitudes and relative phase angles of unsteady pressure are in good agreement with the experimental data.

Further calculations were then carried out to clarify several modelling issues. Fig.5 shows the unsteady pressure at the mid-span suction surface from a 3D and a 2D calculation. Although some difference can be noticed especially at the 50% axial chord position, the comparison seems to suggest that the unsteady pressures at the mid-span are not significantly affected by the 3D effects. The results for 2D configurations on two different mesh densities, 35x81 per blade passage and 61x81 per passage, are shown in Fig.6 for the three chordwise positions with large unsteady pressures. The comparison indicates a small mesh-dependence.

All the above calculations were carried out using the time-consistent multi-grid scheme, which enables to use a much larger time step. For the present calculations, the number of time-steps in one rotor blade-passing period is 200, corresponding to the maximum CFL number being 15 times bigger than that determined by the stability on the basic fine mesh. For this multi-grid method, there is a question concerning the timewise accuracy. To verify the validity and effectiveness of the multi-grid method, calculations for the 2D case were also carried out using the dual time-stepping method, in which the multi-grid is only applied in the pseudo time acceleration at the each real time step and the 2nd order timewise accuracy is fully maintained on the fine mesh. The dual time stepping calculation was carried out at the same number of the real time steps in one period as the time-consistent multi-grid. At each real time step, 20-30 pseudo time-marching iterations were performed to drive the residual down by two order of magnitudes. The calculated results by the two different methods are shown in Fig.7. The good agreement between the two calculations suggests that when the time-step length is enlarged by a factor of 15, the loss of temporal accuracy by the time-consistent multi-grid is negligible. Further increase in the time-step length (e.g. by a factor of 20 or more) would result in noticeable discrepancies between the two methods, suggesting an upper limit of the time-step length for the present time-consistent multi-grid scheme. It is noted significantly that because of the computing effort of the sub-iterations at each real time-step, the CPU time for the dual time-stepping method is about a factor of 20 more than that required by the time-consistent multi-grid. It is recognised that this difference in computing efficiency might significantly change if finer meshes are used. In that case, many more time steps per period will have to be used for the time-consistent multi-grid method in order to keep the loss of temporal accuracy under control.

Overall the calculated results for this turbine stage case are favourable and serve well as a satisfactory validation of the present method, given that the baseline method had been previously validated for unsteady flows around oscillating blades.

3.2. Effect of Axial Gap on Turbine Stage Performances

One of the important aspects of bladerow interaction is the effect of axial gap distance between rotor and stator. For compressors, reducing the axial gap should have a beneficial effect because a wake shed from an upstream rotor would be stretched in the stator passage and consequently generate a smaller mixing loss than that if the wake is completely mixed out before entering the stator. However, the effect of axial gap between vane and rotor rows for turbine is less clear. There does not seem to be a consistent link between the gap distance and the stage aerodynamic performance, though a degradation in turbine efficiency at a smaller gap is often suspected (5).

The aerodynamic performances for the afore-mentioned turbine stage configuration were examined at different axial gaps. In this case, the test stage was reverted back to a normal turbine stage so that the first vane blade row was kept stationary, and the second rotor row was rotating. Three gap distances were examined: 10 %, 27% and 44 % of the rotor axial chord. All the calculations were carried out at an inlet stagnation pressure of 158 kPa, an inlet stagnation temperature of 314 K, an exit static pressure of 101 kPa, a rotor speed of 4411 R.P.M. and a Reynolds number of 10^6. It is noted that the calculations were at the same back pressure condition rather than a fixed mass flow which is practically difficult for time-marching calculations without further iterations. The calculated results however show a small variation of mass flow in a range of 0.5% at different axial gaps. The stagnation parameters are mass-averaged and time-averaged. The stage total-to-total isentropic efficiency was calculated based on the mass and time averaged stagnation pressure and temperature values at the stage inlet and exit planes. It is not expected that a CFD calculation like the present one can give trustworthy results in the absolute values of stage efficiency. The main interest here is to see if any trend in relative variations of stage performance can be identified.

Firstly 2D calculations of the mid-span section were carried out with a mesh density of 50x81 per passage. Table 1 shows the calculated stage total to total efficiencies by both the unsteady solutions and the steady mixing-plane solutions. The unsteady results indicate a very slight decrease in efficiency as the gap is reduced. These 2D unsteady results are similar to those obtained by Venable etc (5). The mixing-plane treatment however produces a very different trend. Although at the large gap, both approaches give almost identical answers, 1% drop in the efficiency was produced by the mixing-plane treatment at the small gap. The feature that the mixing-plane treatment overpredicts stage losses in comparison with unsteady calculations was also observed by Fritsch and Giles (16). The 1% drop in stage efficiency is quite significant and therefore care must be taken when using the mixing-plane approach at small inter-row gaps.

Method	*10% C_{ax}*	*27% C_{ax}*	*44% C_{ax}*
Unsteady (multi-passages)	0.946	0.948	0.949
Steady (mixing-plane)	0.936	0.945	0.948

Table 1. Calculated steady and time-averaged stage efficiency at different gaps (2D)

The 3D calculations were performed at two different mesh densities. The coarse mesh consists of 35x81x36 points per passage (a total of 510,300 points for the stage). The fine mesh consists of 50x81x50 points per passage (a total of 1012,500 points for the stage). The results are shown in Table 2. Probably the most significant observation from the 3D results is that the time-average stage efficiency variation against the gap is in an opposite trend compared to the 2D results. Although the solutions are not mesh-independent, the same trend is produced by the two different mesh-densities. The coarse and fine mesh solutions produce an efficiency gain by 0.6% and 0.5% respectively when the axial gap was reduced from 44% to 10% chord.

Mesh Density	*10% Cax*	*27% Cax*	*44% Cax*
35x81x36x5passages	0.920	0.917	0.915
50x81x50x5passages	0.925	0.921	0.919

Table 2. Calculated time-averaged stage efficiencies with different mesh densities (3D)

We will use the results for the middle gap (27% Cax) as a representative to discuss the mechanisms. Let's first look at the flow structures from steady flow solutions with the mixing-plane treatment. Fig.8 shows the steady entropy contours at an axial plane 5% downstream of the stator trailing-edge. Fig.9 shows the steady entropy contours at 5% downstream of the rotor trailing-edge. Since there are no inlet endwall boundary layers specified and also the stator is very lightly loaded, the stator passage secondary flow is quite weak. There is only a small indication of the passage-vortex associated loss accumulation toward the stator suction surface/endwall corners (Fig.8). Thus, the stator wakes are essentially of a 2D pattern. However, there is a very strong secondary flow structure in the rotor passage (Fig.9). This is partially due to the fact that the non-twist rotor blades lead to considerable off-design conditions in the endwall regions with a negative incidence near the tip and a positive incidence near the hub. Though no noticeable blade surface boundary layer separation is indicated, the low momentum fluids due to the thickened surface boundary layers are expected to enhance the building up of the loss cores. The larger loss core from the hub side than that from the casing side is consistent with the high turning near the hub.

Given the strong rotor secondary flow structure (Fig.9) and the largely 2D incoming stator wakes (Fig.8), it is of interest to see to what extent the interaction would be of a 2D nature. Fig.10 shows the instantaneous entropy contours at the same rotor axial plane as that for the steady results (Fig.9). The 2D entropy contour plot on the mid-span section at the same time instant is shown in Fig.11. It can be seen from Fig.10 that the unsteady interaction is highly 3 dimensional. For a simple interpretation, just image two separate activities which could behave linearly and thus be superimposed. One is an unsteady 2D stator wake, and the other is a steady rotor passage vortex. It is known that for turbine blading a 2D incoming wake would act as a negative jet transporting high loss fluid from a pressure surface to a suction surface. The key point to note is that this cross-passage fluid movement due to the stator wake can be in the same or the opposite direction of that due to the rotor passage-vortex. In the very near wall regions, the low energy fluid is transported by the passage-vortex from the pressure to the suction surface, and thus is in the same direction as the wake transportation. On the other hand, in a region about 20- 30% span away from the endwall, the cross-passage movement due to the passage-vortex will have an opposite sense to that due to the wake. Consequently the resultant flow structure when a '2D' wake is interacting with a passage-vortex will have to be strongly 3 dimensional. Fig.11 gives a good indication of the positions of the 2D wakes in each of the three passages. At this instant, a wake is just passing through the trailing-edge plane of passage "A". We can see some kind of cancellation effect at around 20% span from the hub wall, where there appears to be a low loss core (Fig.10). A similar but smaller low loss core is also present at about 20% span from the casing wall. We can also see that the high loss region is much closer to the endwalls compared the steady solution (Fig.9), indicating the enhancement of the two cross-passage activities in the same direction in the local near wall regions. These depression and enhancement of cross-passage movements at different spanwise positions appear to be responsible for the shape of time-averaged entropy contours featured by the losses being redistributed towards the end-walls as shown in Fig.12.

The 2D calculations do not include the wake/passage-vortex interaction. Two opposing 2D unsteady mechanisms affecting the stage efficiency may be expected: the interaction between a stator wake and a rotor boundary layer (detrimental), and the stator wake stretching in a rotor passage (beneficial). As the inter-row gap is reduced, both effects would be enhanced but they seem to be very much balanced, resulting in essentially unchanged stage efficiency. For 3D situations, the stage efficiency is strongly affected by the wake/passage-vortex interaction. When the gap is reduced, there might be two causes for a stage efficiency gain. Firstly a smaller gap would result in thinner endwall boundary layers which not only reduce the endwall friction loss but also reduce the passage-vortex associated secondary losses. Secondly the enhanced unsteady interaction at a smaller gap would also lead loss-cores more confined toward endwalls as discussed earlier, which is again beneficial. The second cause appears to be more influential since the 3D steady mixing-plane results, which should include the first cause, do not significantly differ from their 2D counterparts.

Fig.13 shows the temporal variations of rotor blade tangential force at the three different axial gaps. When the gap is reduced from 44% to 27 % chord, the 1st harmonic force is increased by a factor of 2.7. But a further reduction of the gap by the same amount only increases the 1st harmonics of the force by a factor of 1.3. Given the possible efficiency gain and the non-linear variation of blade unsteady forces as the inter-row gap is reduced, there should be an optimum gap to balance both aerodynamic and mechanic requirements.

Finally, with regard to computing time, the fine mesh (50x81x50x5passages) calculation using 4 levels of time-consistent multi-grids consumed 85 CPU hours for 15 rotor passing periods on a single processor of a S.G.I./RS-10000 workstation.

4. CONCLUDING REMARKS

A 3-dimensional unsteady Navier-Stokes method is adopted for unsteady flows in multi-bladerows of turbomachines. Calculated unsteady surface pressures compare well with the experimental data for a turbine test stage case. The effectiveness of a multi-grid method is shown by comparison with the conventional dual time-stepping calculations.

The results reveal that the interaction between an essentially 2D stator wake and a rotor passage vortex is strongly 3 dimensional, highlighted by the opposite directions of the two cross-passage movements at different spanwise positions. In comparison with the steady results, the time-averaged entropy is redistributed with the loss cores closer to the endwalls.

In terms of the effect of axial inter-row gap distance, 2D unsteady calculations produce a very slight efficiency loss with a reduced gap. However, 3D unsteady results indicate around 0.5% efficiency gain when the gap is reduced from 44% to 10% of the rotor chord length.

Both 2D and 3D results show that the steady mixing-plane method underpredicts the stage efficiency, compared to unsteady calculations. The difference is more marked at a small inter-row gap and can be more than 1%.

ACKNOWLEDGEMENT

The author wishes to thank Drs P. Walker and B. Haller of ALSTOM Energy Ltd for providing the turbine stage test data and for useful discussions.

REFERENCES

1. Kerrebrock J. L. and Mikolajczak A.A. (1970) " Intra-stator Transport of rotor Wakes and Its Effect on Compressor Performance", J. of Engineering for Power, Vol. 92, pp.359-368.

2. Hodson H.P. and Dawes W.N. (1998) "On the Interpretation of Measured Profile Losses in Unsteady Wake-Turbine Blade Interaction Studies", ASME J. of Turbomachinery, Vol.120, pp276-284.

3. Giles, M. B. (1990) "Stator/Rotor Interaction in a Transonic Turbine", AIAA J. of Propulsion and Power, Vol.6, No.5.

4. Dawes W.N. (1995) "A Simulation of Unsteady Interaction of a Centrifugal Impeller and Its Vaned Diffuser: Flow Analysis", ASME J. of Turbomachinery, Vol.117, pp.213-222.

5. Venable B. L., Delaney R.A., Busby, J.A., Davis, R. L. Dorney, D.J., Dunn, M.G., Haldeman, C.W. and Abhari, R.S. (1998) "Influence of Vane-Blade Spacing on Transonic Turbine Stage Aerodynamics, Part 1: Time-Averaged Data and Analysis", ASME Paper 98-GT-481.

6. Hoffmann, K, Chiang, S., Siddiqui, S, and Papadakis, M (1996) "Fundamental Equations of Fluid Mechanics", Engineering Education Systems, U.S.A.

7. Baldwin, B.S. and Lomax, H. (1978) "Thin Layer Approximation and Algebraic Model for Separated Turbulent Flows", AIAA Paper, 78-0257, Jan., 1978.

8. Denton J.D. (1992) "The calculation of Three-Dimensional Viscous Flow Through Multistage Turbomachines", ASME J. of Turbomachinery, Vol. 114.

9. Saxer, A. P. and Giles, M. B (1993) "Quasi 3D Nonreflecting Boundary Conditions for Euler Calculations", AIAA J. of Propulsion and Power, Vol.9, No.2.

10. He, L. (1997) "Computational Study of Rotating Stall Inception in Axial-Flow Compressors", AIAA J. of Power and Propulsion, Vol.13, No.1.

11. He, L. and Denton, J. D., (1994), "Three-Dimensional Time-Marching Inviscid and Viscous Solutions for Unsteady Flows Around Vibrating Blades", ASME J. of Turbomachinery, Vol. 116, No.3.

12. Jameson, A., Schmidt, W. and Turkel, E. (1981) "Numerical Solutions of the Euler Equation by Finite Volume Method Using Runge-Kutta Time-Stepping Scheme", AIAA Paper 81-1259,

13. He, L. (1993) "New Two-Grid Acceleration Method for Unsteady Navier-Stokes Calculations", AIAA J. of Power and Propulsion, Vol.9, No.2, 1993.

14. Jameson, A. (1991) "Time-Dependent Calculations using Multi-Grid, with Applications to Unsteady Flows past Airfoil and Wings", AIAA Paper 91-1596.

15. Hall D. M., McGuire, P. M. and Price, D. W. (1981) "Excitation of Blade Vibration", GEC Turbine Generators, Report No. AGR 131/F.

16. Fritsch and Giles (1995) "An Asymptotic Analysis of Mixing Loss", ASME J. of Turbomachinery, Vol. 117, pp.367-374.

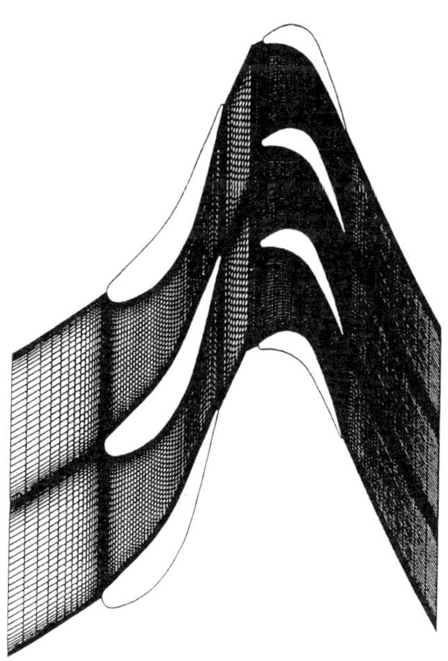

Fig.1 Computational Mesh for a Turbine Stage

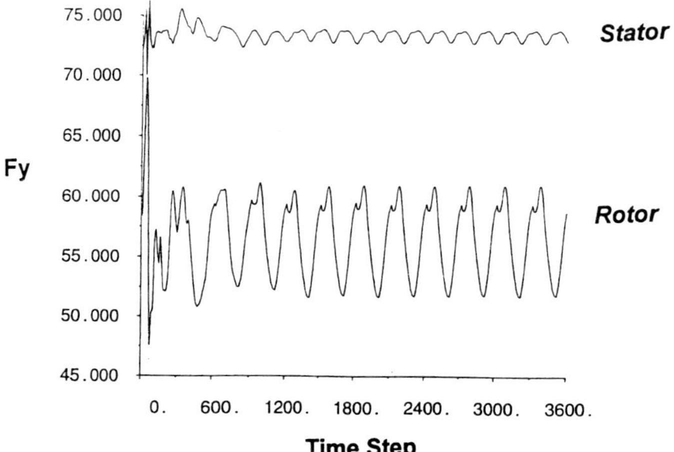

Fig.2 Time Histories of Tangential Blade Forces
(multi-passage unsteady run starting from a single passage steady solution) .

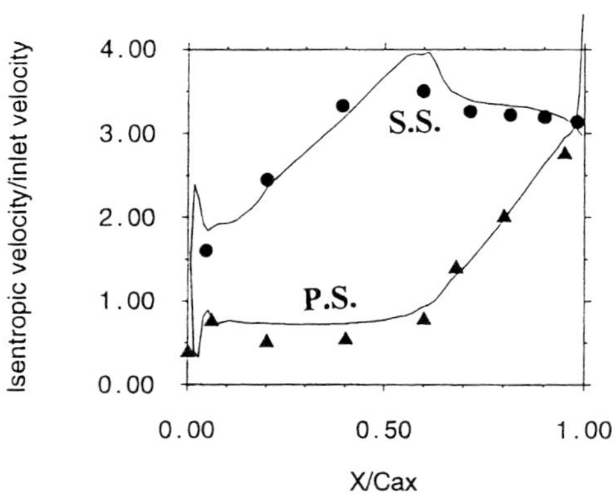

Fig.3 Time-Averaged Pressure Distributions on Rotor Mid-span Section
(-- Calculation;   ● ▲ Experiment)

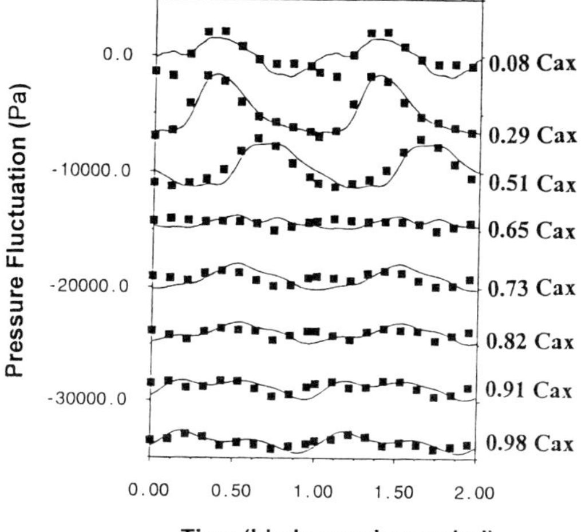

(a) SUCTION SURFACE

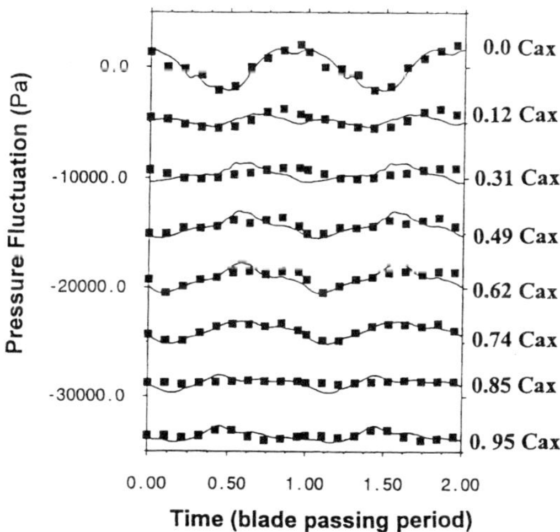

(b) PRESSURE SURFACE

Fig.4 Unsteady Pressure Variation in Two Periods at Different Chordwise Positions (— Calculation; ■■ Experiment)

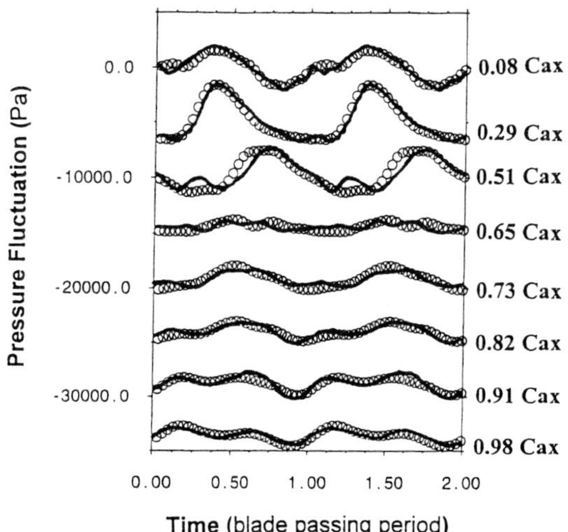

Fig.5 Calculated unsteady pressure on Suction Surface by 2D and 3D methods
(-- 2D ;   o o 3D)

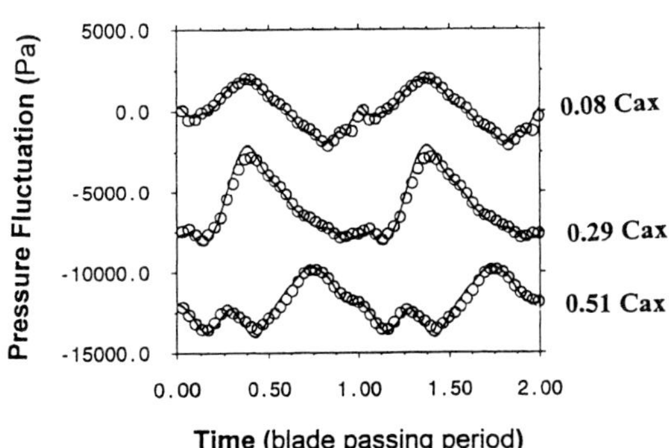

Fig.6 Unsteady Pressure Calculated with Different Mesh Densities
(-- 25x81 per passage;   o o 61x81 per passage)

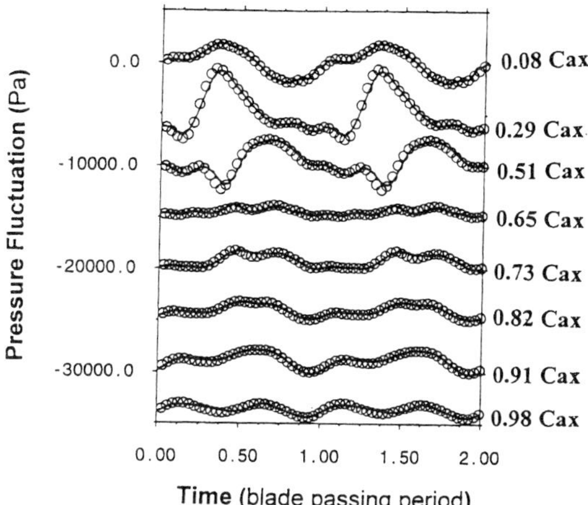

Fig.7 Unsteady Pressure on Suction Surface Calculated with Different Acceleration Schemes (-- Time-Consistent M-G; o o Dual Time Stepping)

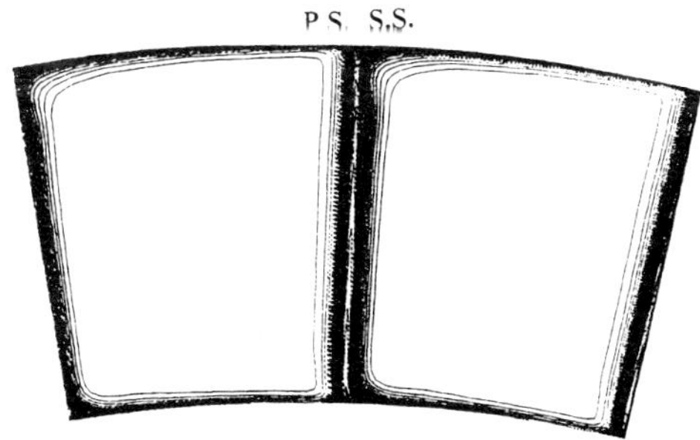

Fig.8 Entropy (exp(-S/R)) Contours at Stator Exit (Steady mixing-plane solution) (5%C downstream of trailing edge, interval=0.005)

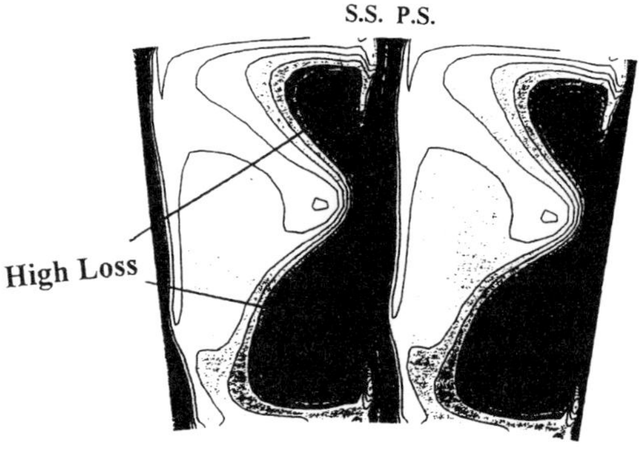

Fig.9 Entropy (exp(-S/R)) Contours at Rotor Exit (Steady mixing-plane solution)
(5%C downstream of trailing edge, interval=0.005, Gap=27% Cax)

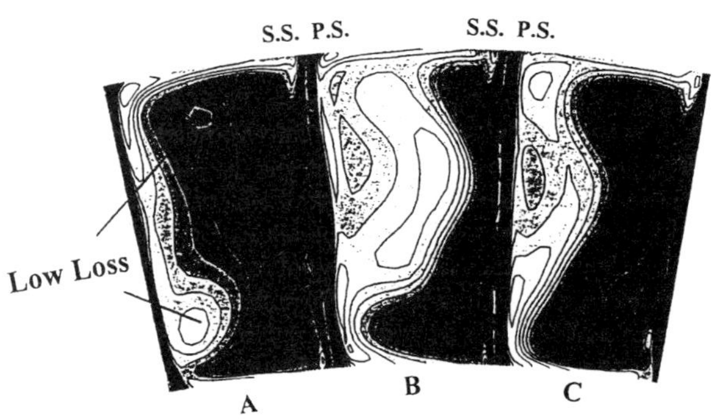

Fig.10 Instantaneous Unsteady Entropy (exp(-S/R)) Contours at Rotor Exit
(5%C downstream of trailing edge, interval=0.005, Gap=27% Cax)

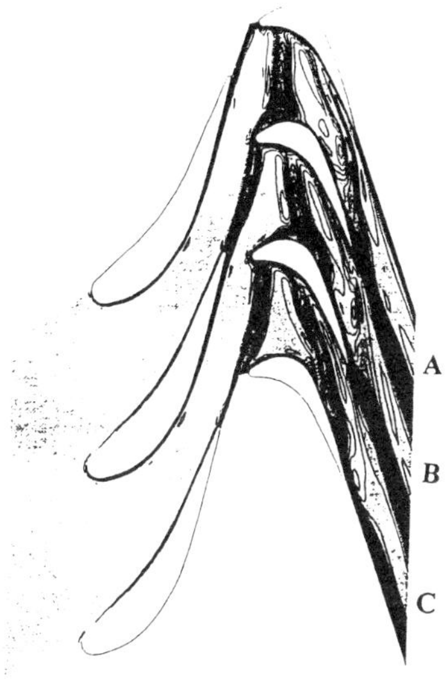

Fig.11 Unsteady Entropy (exp(-S/R)) Contours at 2D Mid-span Section at the Same Instant as that for Fig.10. (interval=0.005, Gap=27% Cax)

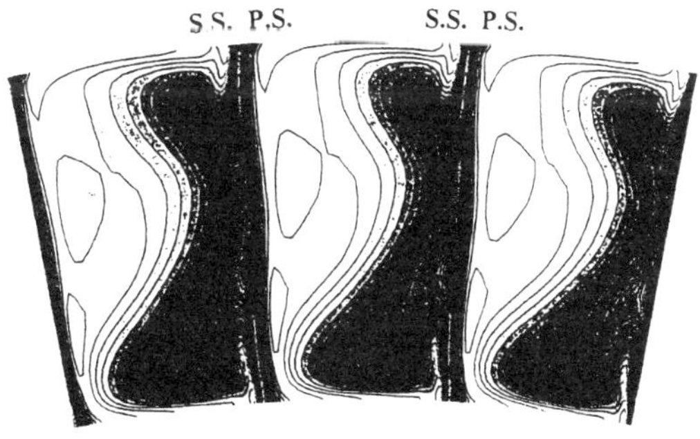

Fig.12 Time-averaged Unsteady Entropy Contours at Rotor Exit (5%C downstream of trailing edge, interval=0.005, Gap=27% Cax)

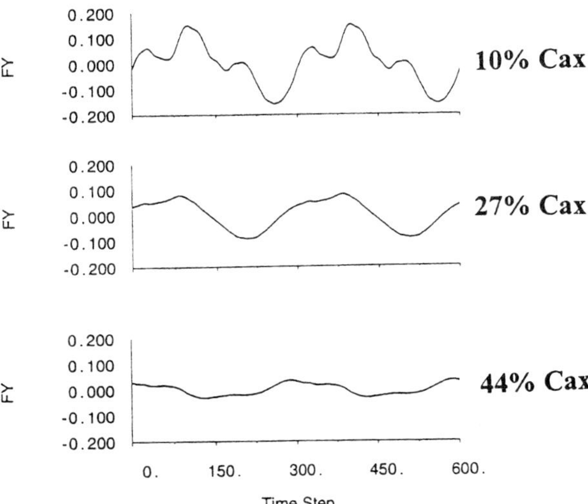

Fig.13 Unsteady Tangential Forces on a Rotor Blade at Three Inter-Row Gaps (non-dimensionalized by time-averaged values)

C557/018/99

Comparison of three approaches to model stator–rotor interaction in the turbine front stage of an industrial gas turbine

M VON HOYNINGEN-HUENE and **J HERMELER**
Siemens Power Generation (KWU), Mülheim an der Ruhr, Germany

SYNOPSIS

One method to improve the efficiency of state-of-the-art gas turbines even further is to take into account the interaction between stator and rotor blades in the compressor and the turbine. The focal point of the study presented here is to evaluate the degree of modelling complexity needed to capture all design-relevant blade-row interference effects in an axial turbine stage.

To this end the study compares three approaches, viz: a classical "stage interface" calculus with a steady-state mixing plane between the stator and the rotor, "frozen rotor" calculations and finally a fully unsteady calculation. Grid dependency effects are studied as well. The calculations are peformed using the commercially available Finite Volume Code CFX-TASCflow which is an important aerodynamic design tool at SIEMENS. It has been validated by different authors against experimental results for applications in turbomachinery and showed very good results [1, 3, 5, 2].

The gas turbine market is very dynamic and competitive thus requiring fast responses in engineering. From the designer's point of view it is therefore of vital importance to use tailor-made software tools for the various tasks that not only provide the right answers to the engineer's questions but are also highly efficient in terms of computational time, modelling and post-processing effort. In order to enable the different tools provided by commercial software companies and universities to be selected, a detailed study of their options and limitations is presented in this paper. The aim of this paper is to distinguish between design problems that can be solved using comparatively fast stage interface and frozen rotor calculations, and design problems that can only be tackled performing a time-resolved analysis. A better understanding of these tools and their efficient use enables engineers further to improve gas turbine efficiency and cooling concepts in an even faster moving R&D environment. One important objective highlighted in this paper is to determine the maximum pressure over the seal air injection gap and the blade surface in order to prevent hot air from inflating the coolant air channels at any time and thus to enhance the service life duration and reliability of the components.

1 INTRODUCTION

In this study, the limitations and options offered by three numerical approaches with different degrees of complexity are compared with regard to some of the most important flow field data needed for a good turbine design. In order to speed up the design process, the use of the fastest available tool that correctly predicts the phenomena of interest has become an absolute must.

In stage interface calculations of the flow field in turbomachine stages, the flow quantities are circumferentially averaged in a mixing plane at the interface between the stator and the rotor grids. Such methods have reached a high degree of maturity [12]; they perform quite well with regard to predicting the pressure distribution over the profile but they are not able accurately to simulate the entire flow field due to the model-inherent restrictions.

In addition to the above, another type of steady-state calculaton is performed: the frozen rotor calculation is a hybrid method in which the position of the rotor relative to the stator remains fixed throughout the calculation (11 different positions are chosen). The flow variables are exchanged directly via an interpolation procedure rather than being circumferentially averaged prior to the exchange as in the stage interface calculation. This represents a cost-efficient method to capture some of the flow characteristics associated with the stator wake. The third and most complex type of calculation is an unsteady calculation using the "transient sliding interface" option of the CFX-TASCflow Solver. In this case, the rotation of the rotor relative to the stator is captured correctly.

In the current hardware environment, full 3D calculations still require too much time to be used for the extensive parameter studies that are necessary for design purposes. The authors therefore have decided to perform 2D calculations at midspan which proved to be a useful tool for optimizing design parameters. Similar approaches were chosen by Dorney[4] and Michelassi[10]. The latter shows for untwisted blades that the results of 2D calculations agree well with measurements in a transonic flow environment. In the test case chosen the blades are untwisted in the stator and only slightly twisted in the rotor.

The limitation to 2D calculations means that 3D effects like secondary flow structures cannot be modelled.

2 TEST CASE

2.1 Description

The first stage of the 170 MW, 60 Hz SIEMENS V84.3A gas turbine with a single cycle efficiency of 38 % and a combined cycle efficiency of 58 % was chosen for the test.

2.2 Flow regime

Due to the high Reynolds number of 10^6 (Re with chord length), the flow is assumed to be completely turbulent. The flow is entirely subsonic having a Mach number at the rotor exit of 0.81 and a Mach number of the rotation of 0.43. The Strouhal number is defined to

$$Str = n \cdot f \cdot p_r / c_{ax} \qquad (1)$$

where n denotes the number of blades, f the rotation frequency, p_r the rotor pitch and c_{ax} the axial chord length. It is slightly lower than 2. The axial spacing/chord ratio equals 0.3. The chord length of the stator and the rotor are almost the same.

3 NUMERICAL APPROACH

3.1 Grid

A structured mesh in multi-block topology is created with the AEA tool TurboGrid [15]. As shown in the block border sketch in Fig. 1, the same type of grid is created for both stator and rotor: a slender O-grid surrounds the blade, and from the inflow boundary on, a C-grid is created around it. From the trailing edge region to the outflow boundary, an H-grid is created each for the pressure side and the suction side. The close-up view of the trailing edge region in Fig. 1 shows how the different blocks are connected. In pitchwise direction, the grids are periodically connected in a many-to-one manner. The stationary stator grid and the rotating rotor grid are attached without any overlaps.

For the unsteady calculation, a transient sliding interface option is used to ensure that at every time step, the rotor is in the right position relative to the stator. The flux-conservative second order interpolation algorithm used for data exchange between the two grids in the frozen rotor and the transient sliding interface cases is the same. It is described in [6].

In order to account for the variation of the stream tube thickness, the calculation is performed with two radially-oriented grid cells which enlarge in line with the variation of the stream tube thickness previously calculated with a Q3D tool. In the example given here, the thickness of the stream tube is almost constant in the stator whereas in the rotor it widens by 20 %. This is modelled using a corresponding linear widening of the rotor grid cells. After having performed a number of analyses to study convergence, the two grid cells are chosen to have a radial extension at the stator entrance of 1 % chord length each. Two grid cells correspond to three grid planes, the interior of which is situated on a surface of revolution at midspan.

In pitchwise direction, a simple periodic boundary condition can only be applied if the flow field exhibits periodicity after the same distance in the stator and the rotor grid. In order to limit the simulation to one channel in the stator and the rotor, this distance has to be equal to one blade pitch of both the stator and the rotor thus requiring a blade count ratio of 1:1. In this test case, the stator has 78 blades and the rotor 80. The stator is therefore scaled up by the factor $\frac{80}{78}$ maintaining a constant stagger angle in order to reach a blade count ratio of 1:1 (similarly performed by [11, 10]). Since this factor is close to unity, only minor errors are expected as a result of this scaling.

To judge grid dependency effects, three grids of different fineness are compared. Their grid point numbers parallel and perpendicular to the blade surface as well as the total number of grid points are shown in Table 1. The medium grid is shown in Fig. 1.

Table 1: **Grid resolution of the three grids used**

	coarse	middle	fine
around stator	157	192	192
⊥ stator	18	18	18
around rotor	158	184	290
⊥ rotor	22	34	54
grid pts. × 1000	26	33	76

3.2   Governing equations

The equations solved are the fully three-dimensional, unsteady Favre-averaged Navier-Stokes equations. For the chosen grid and boundary conditions, they automatically reduce to the equations governing the flow field in a stream tube of variable thickness. The fluid is assumed to behave like an ideal gas with a constant ratio of specific heat capacities.

3.3   Discretisation

The transport equations are discretised using a conservative finite volume method that is finite-element based in order to represent the geometry. The standard k-ϵ model [9] is used to account for the effects of turbulence. Near the walls, a wall function approach is applied which is based on the logarithmic law of the wall. The dimensionless wall distance y^+ is appropriately set to 30–50 for stage calculations. A second-order accurate skew upwind differencing scheme with physical advection correction is employed. In each implicit time step, an algebraic linear multigrid method iteratively solves the coupled system of the mass and momentum equations. Afterwards in this time step, the equations for k and ϵ are subsequently solved implicitly in a matrix formulation using the same multigrid method. More details about the theoretical basis of the software are reported in [16].

3.4   Boundary conditions

Total pressure and total temperature are used as boundary conditions at the inlet. For the turbulence model, the turbulence intensity and the eddy length scale are specified at the inflow boundary. In addition, the flow is assumed to be parallel to the machine axis at the inflow, i.e. the flow leaves the combustion chamber without any swirl. At the exit, a grid-face-averaged value of the static pressure is imposed in a way that the pressure at each node may follow the pressure fluctuations in order to allow pressure distortions to exit the computational domain. Extrapolation boundary conditions are applied to both k and ϵ at the outflow boundary. The pitchwise boundaries are symmetric and the upper and lower boundaries containing the stream tube are reflecting boundaries. All solid walls (i.e. the blade surfaces) are assumed to be adiabatic and perfectly smooth. The no-slip condition is applied.

4 RESULTS

For the unsteady calculation, one blade passing period corresponding in space to the pitch angle of 4.5°, is divided into 40 time steps. The distance between two time-steps corresponds to a relative motion of the rotor of $\frac{1}{40}$ pitch angle or 0.1125°. Frozen rotor calculations are performed for four different positions, characterised by the shift of the rotor compared with its reference position: 0 pitch angle (i.e. the relative position depicted in Fig. 1), $\frac{1}{4}$ pitch angle (1.125°), $\frac{1}{2}$ pitch angle (2.25°), and $\frac{3}{4}$ pitch angle (3.375°). The rotor blades rotate the respective angle upwards. Seven supplementary frozen rotor calculations at different positions are performed. The unsteady results at the respective rotor positions are compared with the frozen rotor calculations. Additionally, the time-averaged values of the unsteady calculation are compared with the stage interface calculation.

4.1 Grid dependency

Two grid dependency studies are carried out. While the first grid dependency study is performed on the basis of a frozen rotor calculation at a relative rotor shift of 0°, the second is based on the unsteady calculation. In both cases, the three grids described in Table 1 are compared. As shown in the upper left corner of Fig. 2, the time-mean of the static pressure distribution over the rotor for the unsteady calculations shows a considerable deviation between the coarse grid and the medium grid, whereas the results obtained using the medium grid and the fine grid come very close to each other differing by less than 0.32 % of the total pressure at the inlet ($p_{tot,in}$). The differences between the coarse and the fine grid locally exceed 1 % of $p_{tot,in}$. The static pressure distributions for the steady frozen rotor calculations show a similar sensitivity to grid refinement. The pressure distributions on the stator obtained with the medium and the fine grid differ even less than on the rotor. For the unsteady calculations, the effect of grid refinement on the relative flow angle distribution at the grid interface is also analyzed, and it is shown in the upper right corner of Fig. 2. Again, the deviation between the solutions obtained with the coarse grid and the medium grid is considerable although the latter has only 27 % more grid points than the coarse grid. Compared to that, the discrepancy between the medium and the fine grid is rather small with a maximum flow angle difference of 0.36°, even though the fine grid contains more than double the number of grid points than the medium grid. For time-resolved calculations, it is mandatory to additionally perform a grid refinement study of some representative instantaneous values. The time-resolved pressure contours over the stator and the rotor are very similar for the medium and the fine grid. The most critical unsteady parameter investigated is the instantaneous pressure distribution over the grid interface. Therefore, in the lower part of Fig. 2, this pressure variation is depicted for two representative instants in time. Again, the difference in the results obtained using the coarse and the medium grid, respectively, is much bigger than between the medium and the fine grid. The maximum deviation in instantaneous pressure at any location and any time amounts to less than 0.4 % of $p_{tot,in}$ between the medium and the fine grid while it exceeds 1 % of $p_{tot,in}$ between the coarse and the fine grid.

The medium grid is chosen for further studies because it offers a good compromise between a fine resolution in the leading edge and trailing edge regions and the number of grid points. Even though less than half the points of the fine grid are needed, the results are sufficiently accurate for the qualitative and quantitative studies of the time-resolved flow field intended here.

4.2 Computational time

The calculations are performed on a HP C 200 workstation. The steady calculations are performed until the residual drops below 10^{-5}. Severe convergence problems are not observed. A stage interface calculation or a frozen rotor calculation takes approx. 6 hours starting from scratch. The unsteady calculations start from a converged frozen rotor calculation. After about two periods with 15 internal iterations imposed per time-step, the residual to be obtained for each time step in the subsequent periods is set to 10^{-4}. This corresponds to about 10 internal iterations per time step. As described above, one blade passing period is divided into 40 time steps. Finally, about 10 periods are necessary to obtain a solution that is fully periodic in time. Since one iteration takes approx. 60 s on the HP C 200, the total theoretical computing time for the unsteady calculation under ideal conditions is approximately 3 days. Due to convergence problems however this time is likely to be exceeded. In fact, more than 10 periods are typically needed so that, in all, an unsteady calculation takes 1–2 weeks until full convergence is achieved. However, no extensive studies are undertaken how far the residual requirements could be lowered or if a calculation with less time steps per period would also be sufficiently accurate. Additional calculations have shown that a division of one blade passing period into 80 time steps does not significantly alter the solution. In general, emphasize is rather laid on accuracy then on solver speed-up. The periodicity of the unsteady solution is monitored regarding the pressure, velocity and entropy values at several points in the flow field.

Since the computational effort is proportional to $n \cdot \log n$ where n is the grid point number, a fully 3 D calculation of the same problem would take 1–2 months on a modern single processor work station such as the HP C 200 which is far too long for extensive parameter studies. In most practical cases, even more computational time would be needed because it is often necessary to simulate more than only one stator and one rotor passage.

4.3 Pressure distribution

In Fig. 3, the time-averaged static pressure distribution on the rotor is depicted together with the range of instantaneous unsteady static pressure values ("pressure envelope"). Additionally, the static pressure distribution resulting for two different frozen rotor position is shown (1.125° and 2.25°).

The static pressure values are normalized with the total pressure at stator inlet and are depicted over the axial chord length. The range of pressure values in the unsteady calculation clearly demonstrates that the pressure fluctuations are most distinct on the first quarter of the suction side, amounting to between 4–6 % of the local static pressure. In close vicinity to the leading edge, the fluctuations are damped by the velocity reduction and reach only approx. 3 % of the local static pressure. These fluctuations on the rotor are caused by the stator wake <u>and</u> the potential field of the stator. On the stator, they are however caused by the potential effect of the rotor alone.

Therefore, the pressure fluctuations on the stator are less pronounced than on the rotor, amounting to 1–2 % of the local static pressure in the last half of the suction side. They are virtually neglectible on the pressure side.

For today's turbine designers, the knowledge of the time-averaged pressure distribution is no longer sufficient. In order to optimise cooling efficiency, the time-resolved pressure distribution (i.e. the mean value and the amplitude) is of immense interest, for example when determining the maximum pressure at a certain film cooling hole position. For most frozen rotor positions, especially for a shift of 1.125° and 2.25°, the deviations to the time-resolved results are considerable, see Fig. 3. Taking into account the position of the wake at these rotor positions (see Fig. 4), it can be concluded that the closer the stator wake comes to the rotor leading edge, the weaker the frozen rotor model becomes. Due to this fact, the maximum pressure over time on the rotor suction side is underestimated by almost 4 % by the frozen rotor calculations in the first 10 % of the axial chord length.

In the time-resolved calculation, the highest pressure on the near-leading-edge region of the suction side is reached for a rotor shift of 1.6°, on the pressure side for 3.0°. At these positions, the wake becomes attached to the respective position. Neither the absolute value nor the relative rotor position of the maximum pressure is correctly predicted by the frozen rotor calculations because the convection of the wake in the rotor channel cannot be correctly predicted by the frozen rotor calculations.

4.4 Unsteady pressure field

In Fig. 4, the instantaneous pressure and entropy contours are shown at the four positions detailed above. The upper part of the figure shows the effects of the potential interaction between vanes and blades on the static pressure distribution. It causes for example a static pressure increase between the stator and the rotor for t/T between 1/2 and 3/4.

In the design process, a precise knowledge of the static pressure mean value and its amplitude over the seal air injection gap is crucial to determine the seal air pressure needed to keep hot air out of the gap. With a smaller range of uncertainty of the maximum pressure value, the overall efficiency can be raised since the seal air can be taken out of the compressor at a lower pressure level.

For design purposes, the time-resolved pressure amplitude over the seal air injection gap can be derived from the midspan pressure fluctuation at the corresponding axial position provided that the stator twist is small (it is zero in the example). Additionally, the mean pressure value over the hub can be derived from a 3D steady calculation.

In this testcase, it is reasonable to assume the axial location of the seal air injection to be at the location of the grid interface. This makes the monitoring of the fluctuations easier in the specific software environment used but is only applicable if the grid interface may be positioned at the location of the seal air injection gap. With this assumption, the instantaneous pressure and entropy contours over this gap may be derived from Fig. 4 by regarding the variations over the line which represents the grid interface. The time-resolved static pressure distribution at the grid interface is shown in Fig. 5. Plotted on the bottom of this three-dimensional view are the corresponding contour lines.

The pressure fluctuation is mainly caused by the potential effects of the rotor and the stator. Since the static pressure is depicted in the stator system of reference, the diagonal pressure crests correspond to the potential field of the passing rotor. The highest overall pressure appears at the instant t/T=0.67 and a relative pitch length of 0.56. At this time, the rotor has turned 3° away from its original position to a new position where the distance between stator trailing edge and the rotor leading edge is reduced to a minimum, which means that they are aligned. The location with the highest pressure is also situated on this imaginary line. These observations are in excellent agreement with potential theory since the potential interaction between two bodies increases with decreasing distance. The maximum pressure is 5.4 % higher than the pitchwise averaged pressure calculated with the stage interface model at the same axial position and 4.1 % higher than the time- and space-averaged static pressure value derived from the unsteady calculation. As described above, the static pressure fluctuation on the rotor pressure side also reaches its maximum value at this relative position of the rotor.

The calculation shows a time-resolved pressure fluctuation at an arbitrary, fixed position along the grid interface line of 3–3.5 % of the static pressure and a local pressure fluctuation at an arbitrary, fixed instant in time in pitchwise direction at the grid interface of 2–8 %. 8 % are reached for $t/T=0.67$, i.e. the position of the maximum pressure value.

The wake does not appear significantly to influence the pressure distribution over the seal air injection gap in this example. However, the presence of the wake has an influence on the accuracy of the frozen rotor calculations: in the positions where the stator wake does not pass over the seal air injection gap at the location of maximum pressure, the frozen rotor calculation performed at the critical position of 3° has shown its ability to predict the pitchwise pressure distribution and especially the highest pressure obtained over the seal air injection gap as depicted on the left hand side of Fig. 6. For better comparison, the static pressure value for the steady-state mixing plane calculation is also depicted.

On the other hand, the frozen rotor calculations overestimate the pressure fluctuations by more than 100 % when the stator wake becomes attached to the rotor suction side, corresponding to $t/T=0.25$. This is depicted on the right hand side of Fig. 6. The maximum static pressure value is also considerably overestimated. In all time-resolved positions, the pitch-averaged value of the static pressure is captured satisfactorily by the frozen rotor calculation, see Fig. 6.

From these observations, it can be concluded that the pressure required for the injected seal air can be derived from a frozen rotor calculation carried out at the position that corresponds to the minimum distance between stator and rotor. If the stator wakes do become attached to the rotor suction side in the critical rotor position, this approach is less accurate but for the designer it is important to note that it tends to overpredict the maximum pressure thus providing a conservative approximation. The reason for this lack of accuracy is assumed to be a misrepresentation of the wake in the frozen rotor calculation. It is evident that the pressure is overpredicted in the wake region, at least if the wake impinges on the rotor leading edge. In the frozen rotor calculation, the wake's relative position to the rotor remains unchanged. If the wake now enters the boundary layer region of the rotor, it is decelerated and a type of build-up is formed which leads to a pressure increase upstream of the rotor. In reality however, such a build-up is not possible due to the movement of the rotor relative to the wake. Furthermore, the wake is convected faster away from the rotor leading edge due to the blade motion than in the frozen rotor model.

4.5   Unsteady entropy field

The convection of the stator wakes is simulated by the unsteady calculation as depicted in the bottom part of Fig. 4 for four instants in time. Since the chosen grid offers a compromise between a fine resolution of the boundary layer regions and the number of grid points needed, considerable numerical dissipation is found in the mid-channel region. This results in the stator wake almost doubling its thickness when crossing the grid interface at $t/T=3/4$. This phenomenon is much weaker when the wake is entering the rotor grid at a location where the grid cells are both less distorted and finer, for instance at $t/T=1/4$. The overall convection of the wakes is represented sufficiently well. In contrast to the overall pressure field which is captured rather satisfactorily by the frozen rotor calculations, the entropy field in the rotor channel can only be simulated by unsteady calculations since the relative motion of the rotor has to be taken into account to model the convection process.

Zeschky[14] claims that, since the Strouhal number is close to 2, about two wakes are expected to be found in the rotor channel. This is in good agreement with the results.

Fig. 4 shows that the stator wake is chopped by the rotor into segments which then are stretched and bowed due to the lower velocity in the boundary layer regions. In the examined test case, the wake separates from the pressure side shortly after entering the rotor channel so that the stretching effect is not very pronounced. The wake, i.e. the zone of high entropy and high loss then migrates to the suction side to from a kind of sliding "backpack". These results show a similar behaviour as that observed in the measurements of Zeschky[14] and the computations of Walraevens and Jung[13] on the university of Aachens'.1.5 stage test turbine. Depending on Reynolds numbers and blade geometry, this phenomenon can also be less pronounced as can be seen in the calculations of Hodson and Dawes[7] and Michelassi and Martelli[10]. The relative movement of wake fluid versus the rotor suction side can be kinematically explained by the so-called "negative jet"[8]. With the velocity in the wake reduced by approx. 15 % of free-stream velocity while the circumferential velocity component remains constant, the velocity triangles show a deviation of the wake flow towards the suction side. This can be visualized with regard to the relative velocity by depicting the difference between the instantaneous velocity and the time-mean at a certain time step (Fig. 7).

4.6   Further comparison of the different calculation methods

In addition to the study of the pressure and entropy fields, the accuracy of the frozen rotor calculations is analyzed in order to predict other aspects of the flow field that are important for the turbine designer such as the time-resolved mass flow and the time-resolved flow angle distribution.

Whereas the time-averaged mass-flow at the grid interface position is well simulated by the stage interface calculation, the frozen rotor calculation fails to predict its fluctuation over time. Not only is the frozen rotor mass flow oscillation out of phase by 0.95π compared with the unsteady results but it also overpredicts the actual amplitude which equals 0.9 % of the time-mean massflow by 25 %. The frozen rotor calculations also completely fail to predict phase and amplitude of the time-dependent variation of the relative flow angle beta at the inflow and outflow borders of the rotor grid.

The time-resolved analysis predicts a variation over time of the space-averaged rotor inlet flow angle at the grid interface plane of 1.1° (Fig. 2). At a position a short distance upstream the rotor leading edge, it varies by 25°.

5   SUMMARY

The comparison of a stage interface calculation, frozen rotor calculations at several different positions and an unsteady calculation has provided a significant insight into the possibilities and limitations of these three numerical methods of different complexity. Only an unsteady calculation is able exactly to model the time-resolved pressure distribution over the rotor blade regions that are the most critical ones in terms of cooling. However, the precision achieved in the pressure distribution performing steady calculations might be sufficient for a larger number of design tasks. The pressure distribution over the stator and in the downstream section of the rotor is predicted very well by the steady types of calculation.

With regard to the determination of the seal air pressure needed, it can be concluded that a frozen rotor calculation at the critical position (i.e. where the stator trailing edge and the rotor leading edge are aligned) can be used to determine the mean value and amplitude of the pressure field over the seal air injection gap, at least if the stator wake does not pass over this gap at the position of maximum pressure in the critical position. However, the frozen rotor calculation can still be used to provide an upper limit for the pressure since it overestimates the maximum pressure in such a case.

The unsteady flow physics, especially as far as the entropy distribution and the wake convection in the rotor channel are concerned, can only be captured by an unsteady calculation.

References

[1] G. E. Baché, M. E. Thomas, 'Validation of an Efficient and Accurate Turbomachinery CFD Analysis Procedure', AIAA-Paper 90-0681, 1990

[2] C. Casciaro, M. Treiber, M. Sell, A. P. Saxer and G. Gyarmathy, 'A Comparison of Experimental With Computational Results in an Annular Turbine Cascade With Emphasis on Losses', ASME Paper No. 98–GT–146, 1998

[3] M. Deckers, D. Doerwald, 'Steam Turbine Flow Path Optimization for Improved Efficiency', Power-Gen Asia '97, Singapore, 1997

[4] D. J. Dorney and D. L. Sondak, 'Study of Hot Streak Phenomena in Subsonic and Transonic Flows', ASME Paper No. 96–GT–98, 1996

[5] W. Elmendorf, F. Mildner, R. Röper, U. Krüger and M. Kluck, 'Three-Dimensional Analysis of a Multistage Compressor Flow Field', ASME Paper No. 98–GT-249, 1998

[6] P. F. Galpin, R. B. Broberg and B. R. Hutchinson, 'Three-Dimensional Navier Stokes Predictions of Steady State Rotor/Stator Interaction with Pitch Change', 3rd Annual Conference of the CFD Society of Canada, Banff, June 25–27, 1995

[7] H. P. Hodson and W. N. Dawes, 'On the interpretation of measured profile losses in unsteady wake-turbine blade interaction studies', ASME Paper No. 96–GT–494, 1996

[8] H. P. Hodson, 'Bladerow Interactions in Low Pressure Turbines', von Karman Institute for Fluid Dynamics, Lecture Series 1998–02, Rhode Saint Genèse, 1998

[9] B. E. Launder and D. B. Spalding, 'The Numerical Computation of Turbulent Flows', Comp. Meth. Appl. Mech. Eng., vol. 33, pp. 269–289, 1974

[10] V. Michelassi and F. Martelli, 'Blade Row Inteference Effects in Axial Turbomachinery Stages', von Karman Institute for Fluid Dynamics, Lecture Series 1998–02, Rhode Saint Genèse, 1998

[11] M. M. Rai, 'Three-Dimensional Navier-Stokes Simulations of Turbine Rotor-Stator Interaction: Part I – Methodology, ASME Journal of Propulsion, Vol. 5, No.3, pp.305-311, May–June 1989

[12] O. P. Sharma, G. F. Picket and R. H. Ni, 'Assessment of Unsteady Flows in Turbines', ASME Paper No. 90–GT–150, 1990

[13] R. E. Walraevens, H. E. Gallus, A. R. Jung, J. F. Mayer and H. Stetter, 'Experimental and Computational Study of the Unsteady Flow in a 1.5 Stage Axial Turbine with Emphasis on the Secondary Flow in the Second Stator', ASME Paper No. 98–GT–254, 1998

[14] J. Zeschky and H. E. Gallus, 'Effects of Stator Wakes and Spanwise Nonuniform Inlet Conditions on the Rotor Flow of an Axial Turbine Stage', Journal of Turbomachinery, Vol. 115, pp. 128–136, 1993

[15] 'CFX-TurboGrid User Documentation', Version 1.2, AEA Technology, 1997

[16] 'CFX-TASCflow Theory Documentation', Version 2.4, AEA Technology, 1995

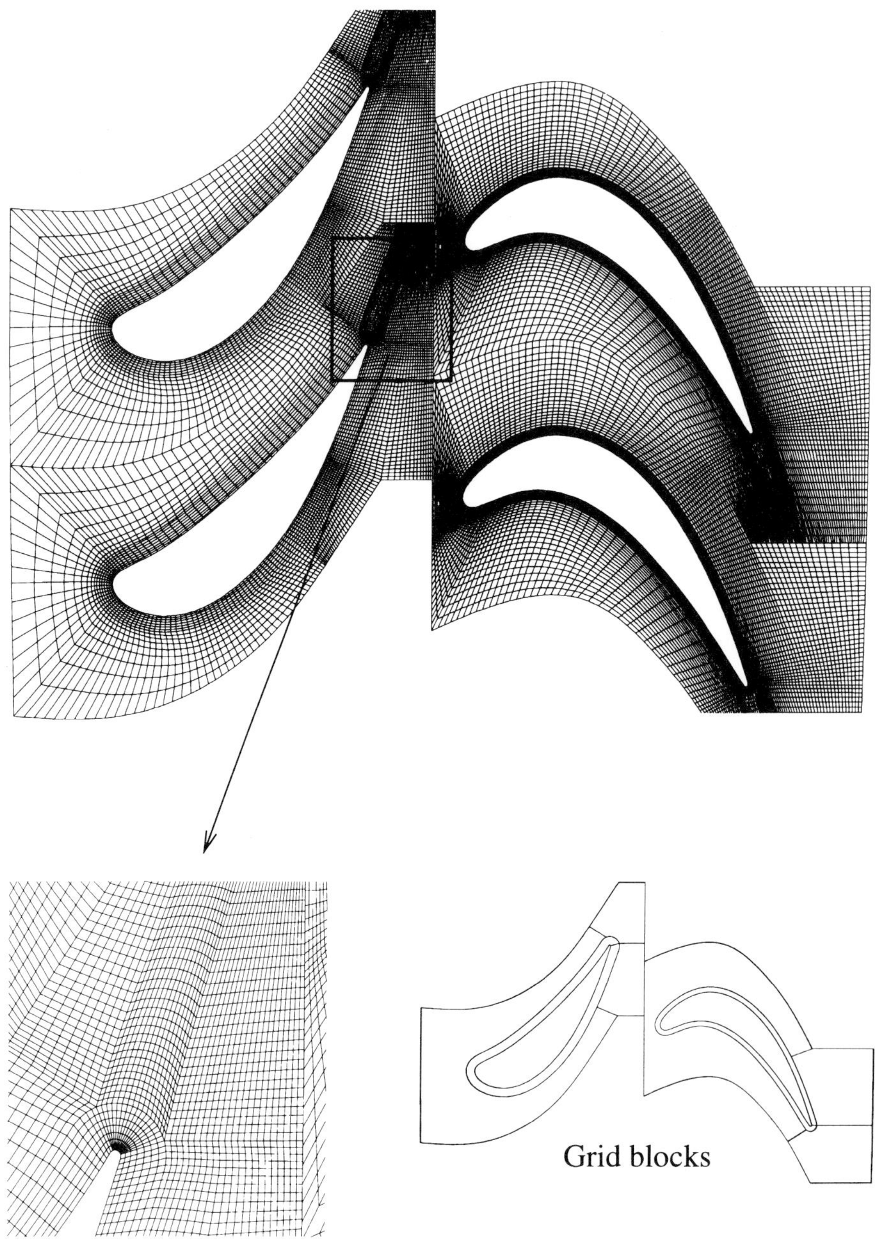

Figure 1: **Geometry and (medium) grid, relative rotor shift: 0°, including close-up view of the stator trailing edge region and sketch of the block borders**

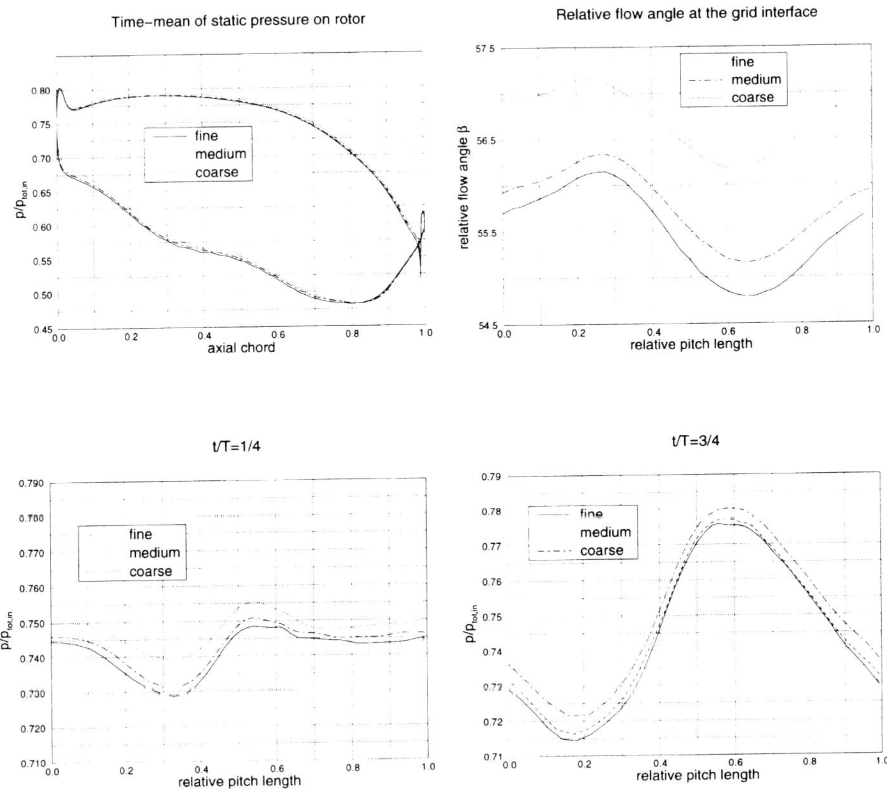

Figure 2: **Grid dependency study for the unsteady calculation. Upper left corner: static pressure distribution over the rotor, upper right corner: relative flow angle at grid interface, below: pressure distribution over the grid interface at two different instants in time**

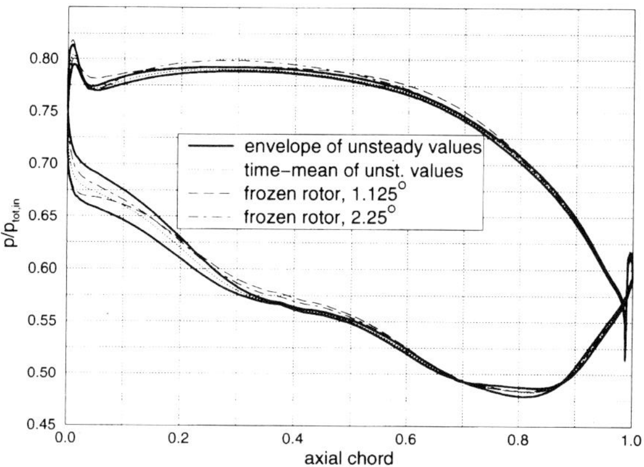

Figure 3: **Pressure distribution over the rotor – close-up picture**

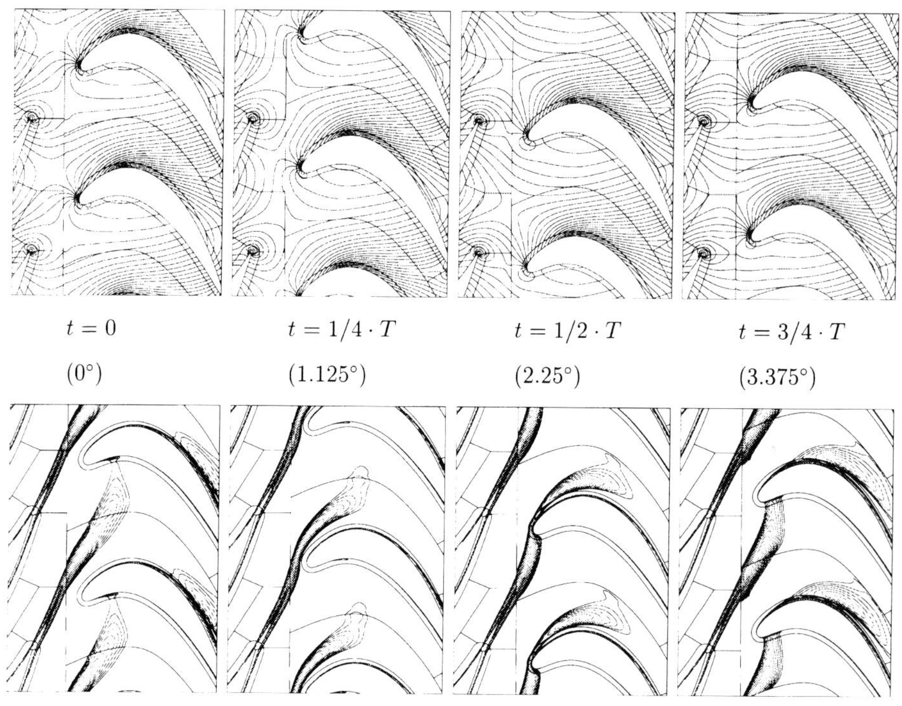

Figure 4: **above:** instantaneous static pressure contours, divided by total pressure at stator inlet (42 contour lines, range: 0.48 – 0.9), **below:** instantaneous entropy contours (42 contour lines, depicted range 5–20 [full range: 0–71])

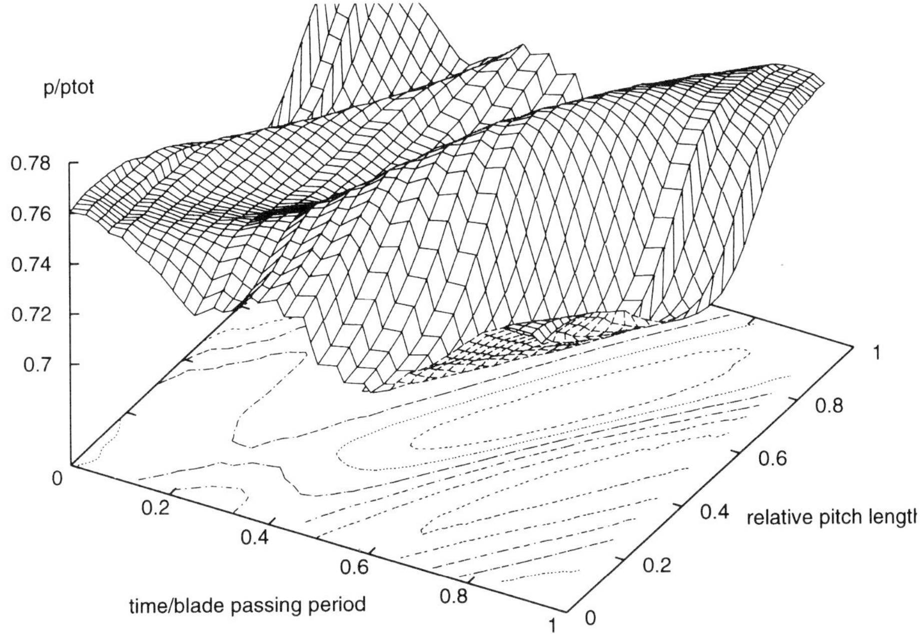

Figure 5: **Pressure distribution over the seal air injection gap, absolute system of reference. The local position is referenced in terms of the relative pitch angle [in °] at the stator outlet increasing in the direction of rotation, 0° being at the lower stator grid boundary**

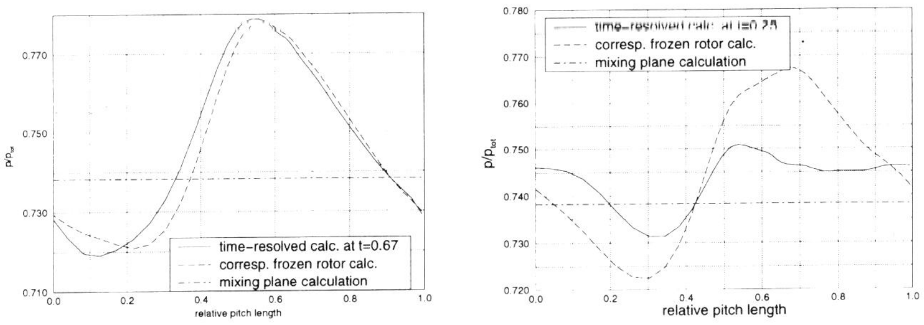

Figure 6: **Pressure distribution over the seal air injection gap at t/T=0.67 and t/T=0.25. Comparison of time-resolved and frozen rotor calculation. The local position is referenced in terms of the relative pitch angle [in °] at the stator outlet increasing in the direction of rotation, 0° being at the lower stator grid boundary**

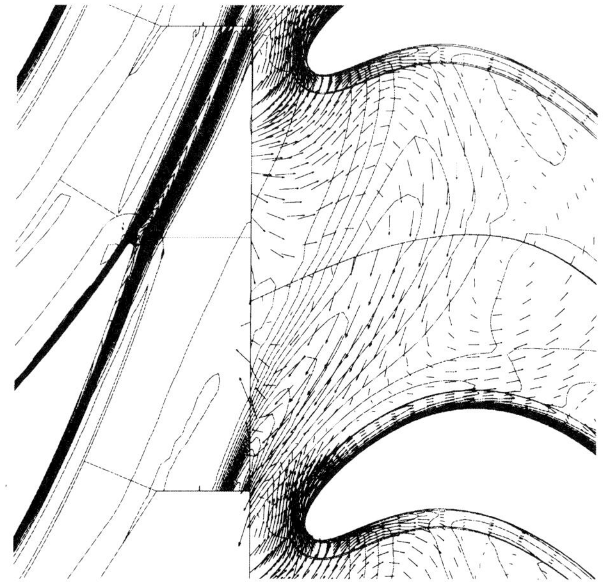

Figure 7: **Negative jet: secondary velocity at $t/T=0$ over entropy contour lines - difference between instantaneous velocity at $t/T=0$ and the time-mean velocity**

C557/017/99

Stator wake clocking effects on three-dimensional unsteady flow in a two-stage low-pressure turbine

J E KRYSINSKI, A SMOLNY, and **J R BLASZCZAK**
Institute for Turbomachinery TU Lodz, Poland
H E GALLUS
IST RWTH Aachen, Germany

Experimental investigations have been conducted of the detailed flow structures in a two-stage low-pressure turbine. The objective of this study is to gain more insight into unsteady flow phenomena affecting flow behaviour due to the stator wake clocking effects in multi-stage machines. A clocking mechanism allows the first stator to be moved relative to the second stage stator. Analysis of the first experimental results includes turbine performance as well as steady and unsteady flow measurements throughout the flow field for different circumferential positions of the first stator.

1. INTRODUCTION

To improve performance and prediction methods for multistage axial turbines, understanding of the unsteady flow is essential (1). A number of experimental studies have been carried out in recent years to investigate these flow phenomena (2) (3) (4) (5) (6).

This experimental work was carried out at the Institute of Turbomachinery (IMP = Instytut Maszyn Przepływowych) of the Technical University of ŁódŸ (Poland) in collaboration with the Institut fuer Strahlantriebe und Turboarbeitsmaschinen of the RWTH Aachen (Germany). The objective of the test program was to experimentally investigate vane clocking effects on the performance of the Model Two-Stage Low-Pressure Turbine (TM-3) situated at IMP TU Lodz. In order to perform these tests the turbine was modified. Modifications included clocking mechanisms to allow the first stage vanes and the second stage vanes to be moved circumferentially independently of the casing. They permitted the clocking positions of the first and the second stator vanes to be changed during the tests without stopping the turbine and dismantling it. In this paper the results in the stator exit planes for six different circumferential positions of the stator vanes (steady flow measurements - every 1/6 of the pitch) and some of the results at the rotor exit planes (unsteady flow measurements, due to the large amount of data for four positions of the second stator - every 1/4 of the pitch) are described.

2. NOMENCLATURE

H   - height of the stator blade (channel)
h   - radial position (along the height H)
T   - pitch of the stator
t   - circumf. position (along the pitch T)
p   - pressure
Tu   - turbulence level (rms)

Greek symbols
α - yaw angle
ζ - local total pressure factor
    $\zeta_1 = (p_{t0} - p_{t1i})/(p_{t0} - p_a)$
    $\zeta_3 = (p_{t0} - p_{t3i})/(p_{t0} - p_a)$
    $\zeta_4 = (p_{t0} - p_{t4i})/(p_{t0} - p_a)$

Subscripts

0 - first stator inlet
1 - first stator exit / first rotor inlet
2 - first rotor exit / second stator inlet
3 - second stator exit / second rotor inlet
4 - second rotor exit

a - ambient (barometric)
i - local
r - radial direction
t - total
u - circumferential direction
z - downstream of the turbine

3. FACILITY

The series of tests were conducted on the two-stage low-pressure turbine. A two-fan set provided a continuous air flow to the test rig. The inlet air parameters were as follows: total pressure - 13.6 +/- 0.2 kPa, total temperature - 318 +/- 3 K, mass flow rate - 3.19 +/- 0.04 kg/s. The rotational speed was 2700 rpm (45 Hz) with a variation of less then 1 percent during all measurements. Figure 1 shows the cross section of the test rig with the measuring planes, and figure 2 shows the turbine geometry. Both stators have 16 constant section blades with the trailing edge inclined to the radial direction at an angle of 22^0. A more detailed description of the test facility geometry and the flow measurements can be found in (7) (8) and (9). Boundary layer conditions were presented in (10).

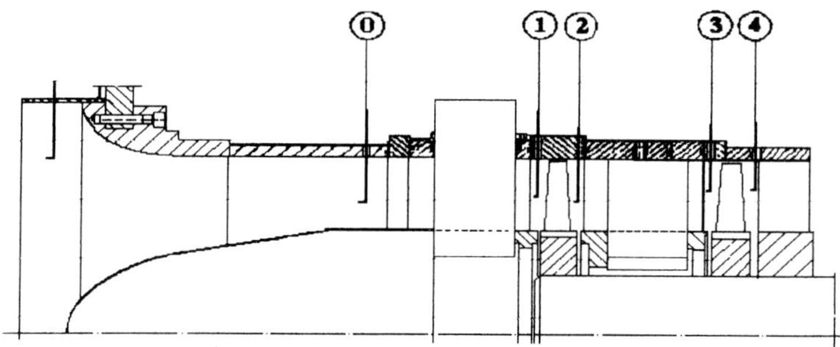

Fig. 1. Turbine cross section

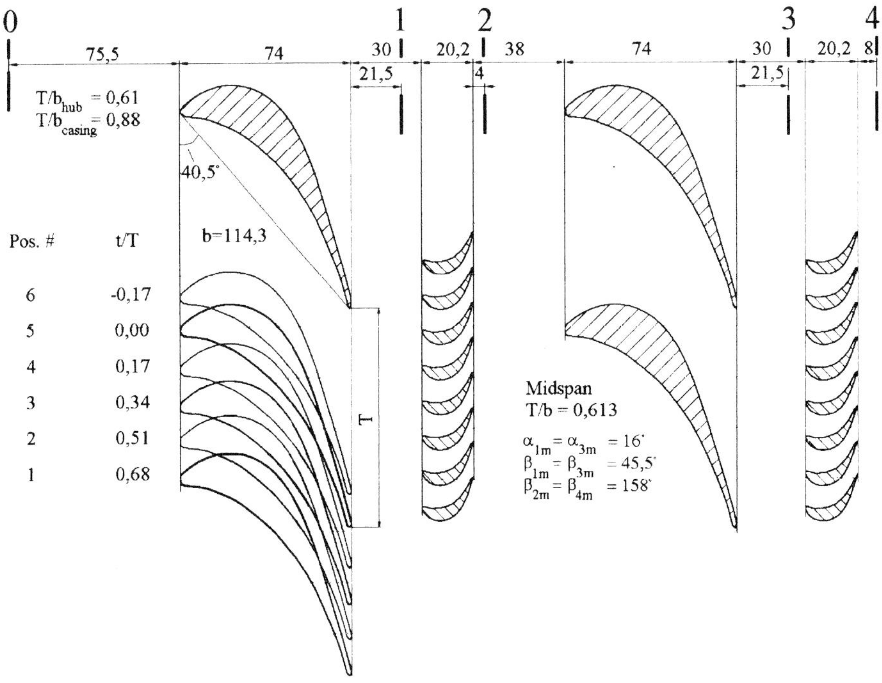

Fig. 2. Turbine geometry

4. INSTRUMENTATION

The flow parameters (total pressure, static pressure, total temperature, velocity vector) in each measuring plane were surveyed with pneumatic five-hole probes with thermocouple mounted in the lower part of each probe. The diameter of the hemispherical head of each probe is 2.5 mm with five holes drilled radially of 0.3 mm diameter. The pneumatic signals were received by a Scanivalve 48-channel system, and recorded simultaneously with a thermocouple signal, by the Keithley 500A data acquisition system controlled by a computer. The probes were traversed by stepper motors from the hub of the turbine to the outer casing. There were 35 locations of each pneumatic probe in the radial direction from $h/H = 0.06$ (near the hub) to $h/H = 0.90$ (near the casing) and 28 circumferential locations over one blade pitch ($t/T = 0.0 \Rightarrow 1.0$). Experimental data have been obtained for 6 different circumferential (clocking) positions of the two stators ($t/T = -0.17, 0.00, 0.17, 0.34, 0.51, 0.67$).

For unsteady flow measurements three-sensor hot-wire probes were used. The sensors have a length of 3 mm with a 9 μm tungsten wire. The data from the hot-wire anemometers were recorded by a digital multimeter (d-c signal) and a four-channel transient recorder (a-c signal). Each channel of the transient recorder has a sampling frequency of up to 1 MHz and a storage buffer of 64 kB. The data acquisition was triggered by a photocell located at the hub of the rotor, so about six wakes were sampled for one-time window with a digital resolution of 256 points at a sampling frequency of 200 kHz. After one rotor revolution, the next time-window was recorded, until 256 of these time-windows were stored in the transient recorder. These real-time data were transferred and stored in a controlling computer. Through on-line control the probe was always aligned with the mean flow direction, which was determined from the pneumatic probe measurements.

5. EXPERIMENTAL RESULTS

Yaw angle α_3 and local total pressure factor ζ_3 distributions at the 2nd stator exit plane, for the 6 clocking positions, are presented on fig. 4 and 5. For comparison yaw angle α_1 and local total pressure factor ζ_1 distribution at the 1st stator exit plane for the 1st clocking position are presented on fig. 3 (for other positions the image is the same but displaced along the pitch).

Local total pressure factor has been chosen for the presentation due to the slight changes of the inlet conditions in different measuring sessions.

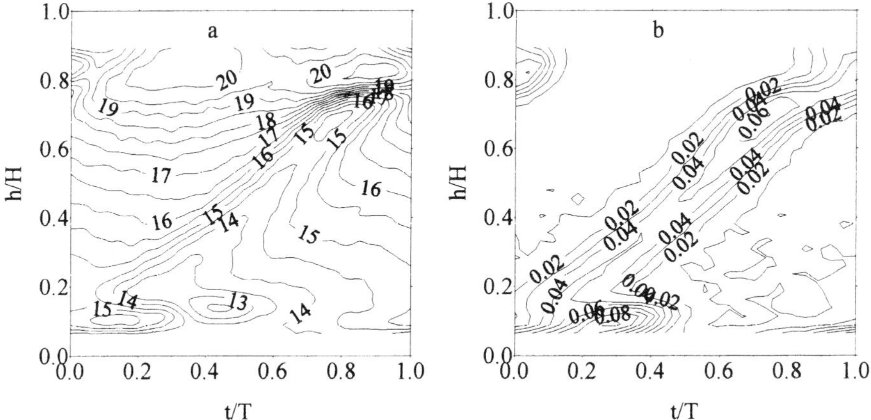

**Fig. 3. Yaw angle α_1 (a) and local total pressure factor ζ_1 (b) distributions
at the 1st stator exit plane for the 1st clocking position**

Fig. 6 shows the pitchwise averaged yaw angles α and total pressure factors ζ at the stator and rotor exit planes for the six clocking positions (absolute frame). For the 1st stator for all clocking positions their values are similar and all the results are within the accuracy range. The total pressure factors behind the 2nd stator are in a wider range than in the case of the 1st stator. The variation of its value may be influenced by the stator to stator clocking position.

Fig. 7 presents the pitchwise averaged values in the absolute frame after the rotors, measured with the three-sensor hot-wire probe. The situation is similar, one can see big differences after the 2nd rotor for each position. Some of the results in the relative frame were presented in (8). A more detailed description of the probe calibration method and the data reduction (decomposition of the signal) is given in (8) (9) and (12).

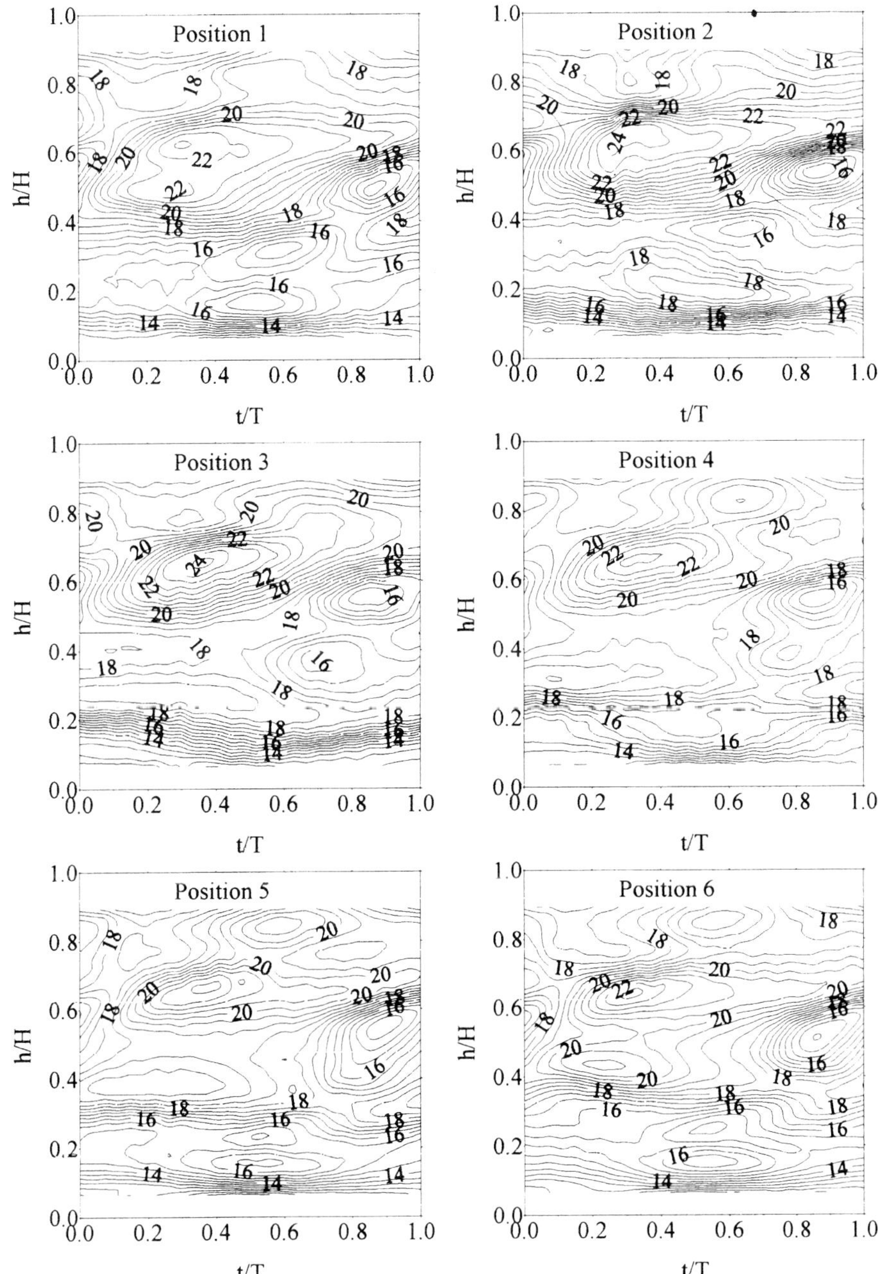

Fig. 4. Flow angle α_3 at the 2nd stator exit plane for the six clocking positions

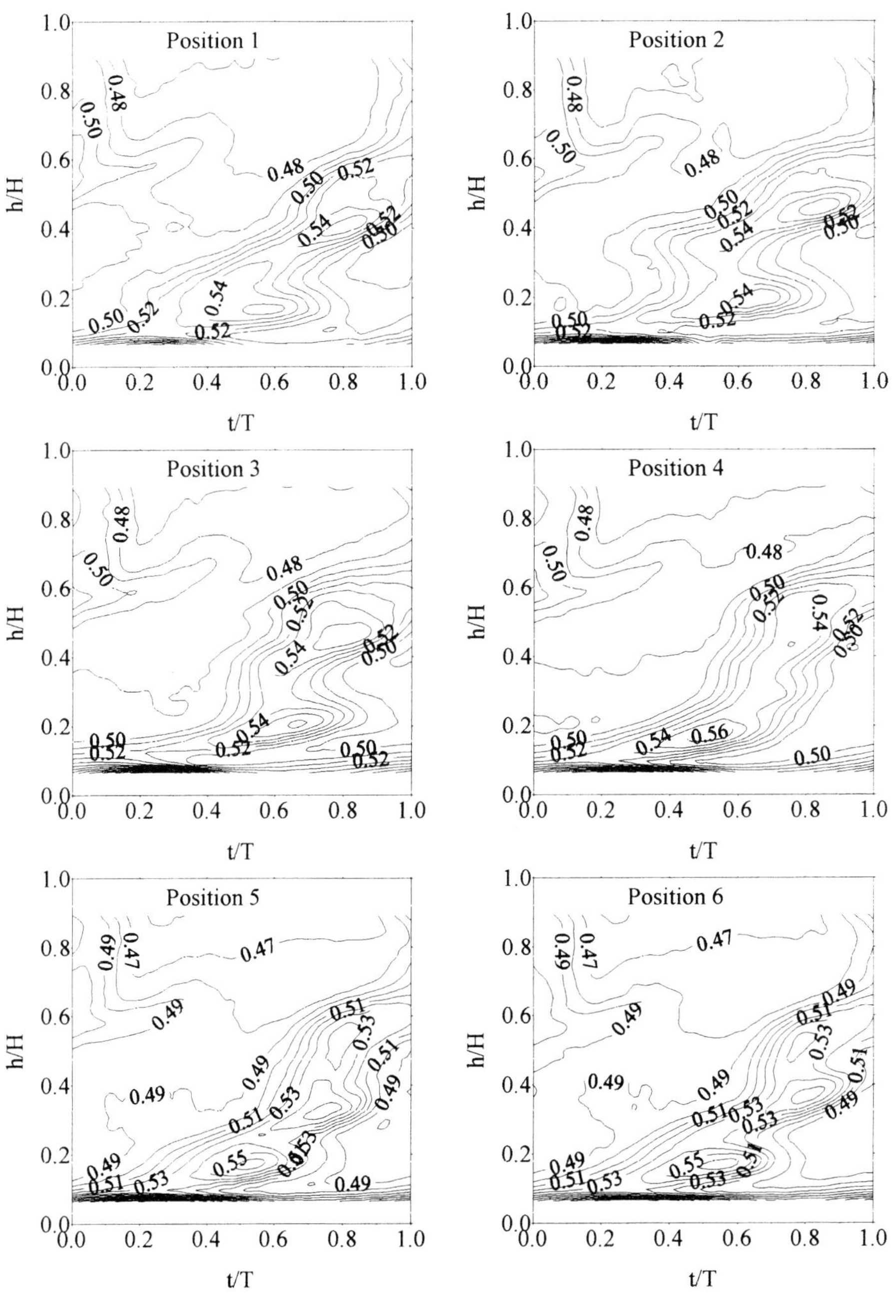

Fig. 5. Local total pressure factor ζ_3 at the 2nd stator exit plane for the six clocking positions

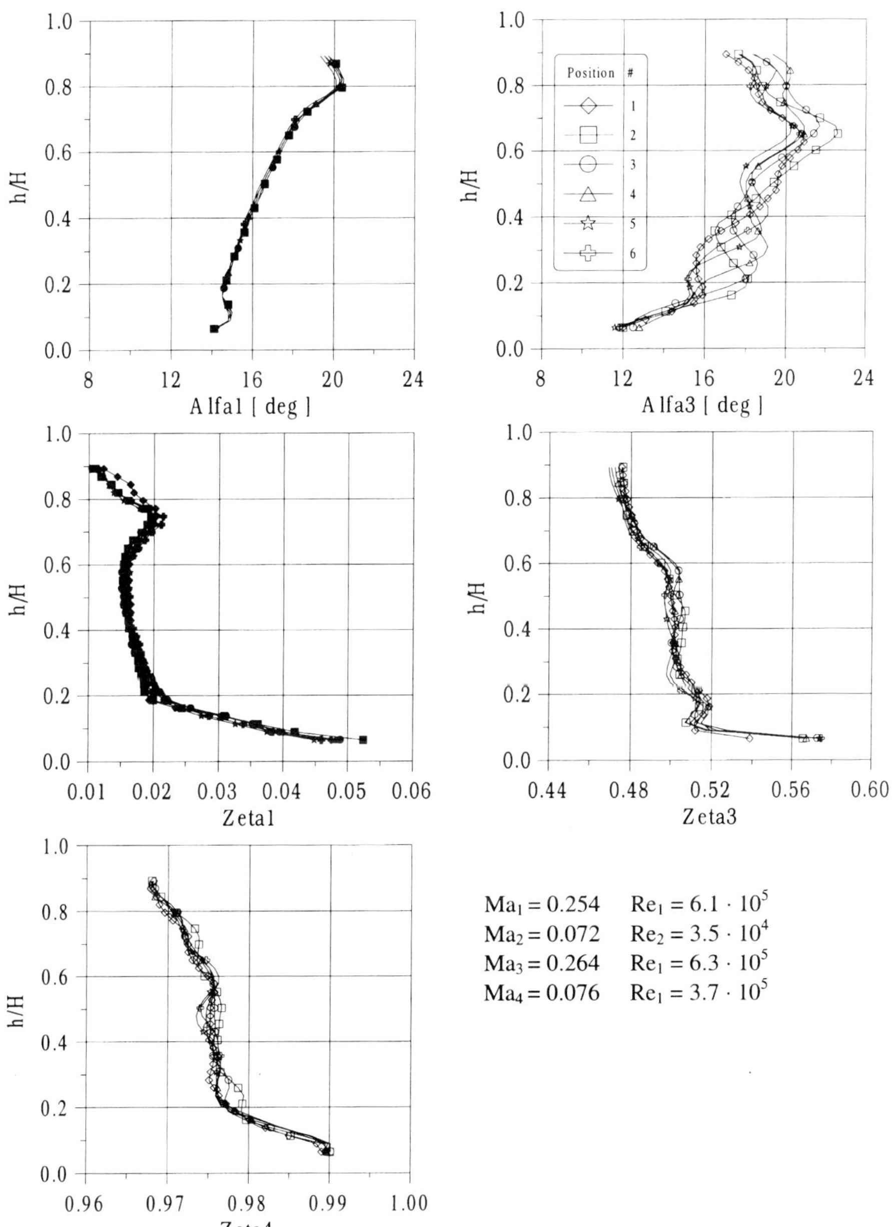

Fig. 6. Pitchwise averaged flow angle and total pressure factor ζ for the six clocking positions

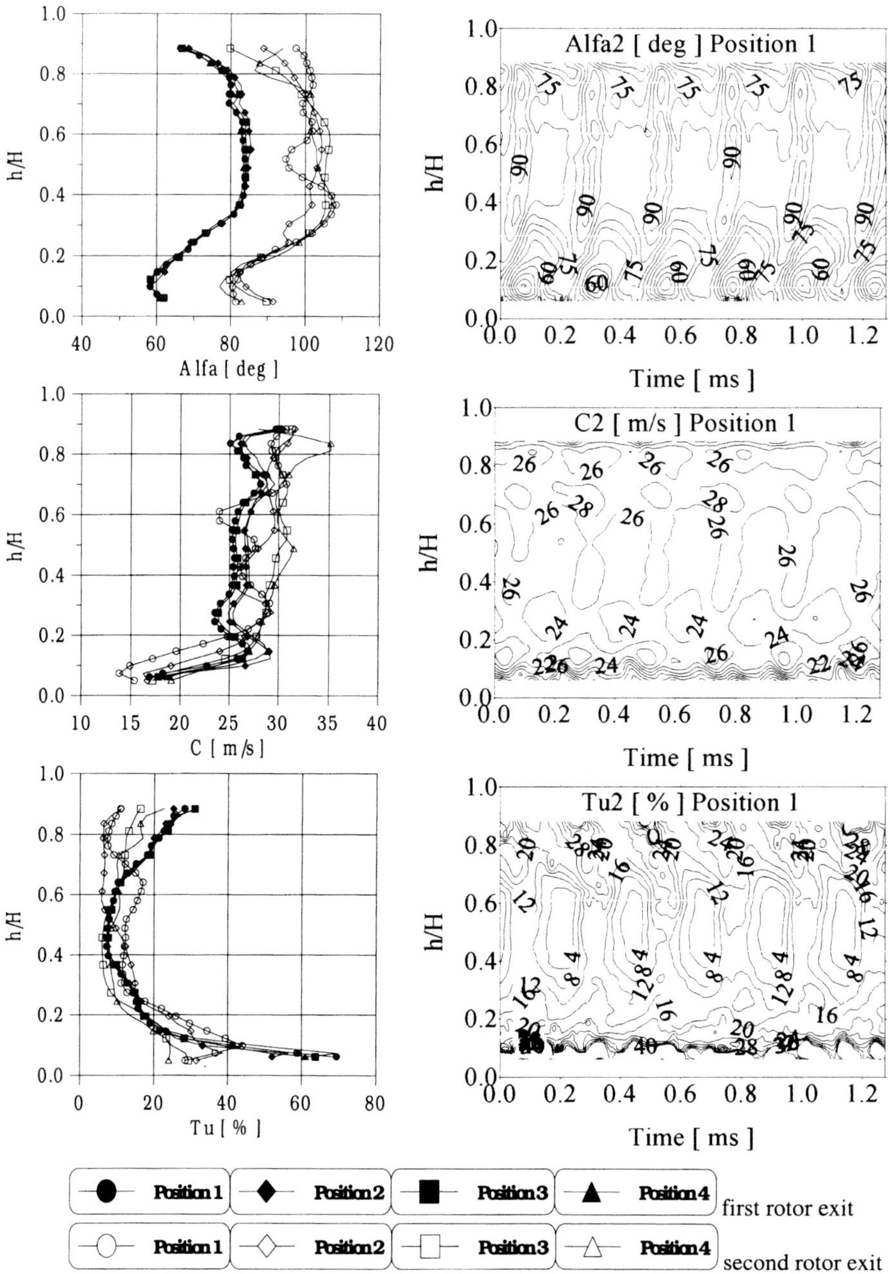

Fig. 7. Pitchwise averaged outlet angle, velocity and total turbulence level after the rotors measured with hot-wire probe for the four 2nd stator positions (the 1st stator was fixed) + examples of parameter distributions after the first rotor

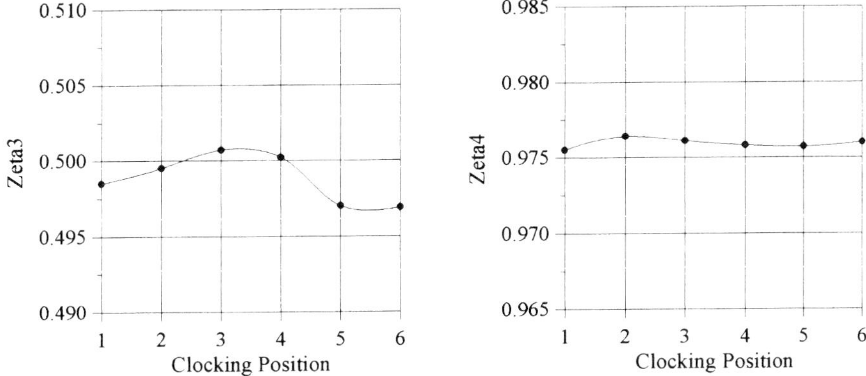

**Fig. 8. Mass averaged total pressure factor ζ_3 and ζ_4
versus clocking position**

The final figure, fig. 8, presents mass averaged total pressure factor versus vane clocking position after the second stator and rotor.

6. CONCLUSIONS

An experimental investigation of stator to stator interaction has been performed in the axial gaps of a two-stage low-pressure turbine. Some of the results acquired with the five-hole pneumatic probes and the three-sensor thermoanemometric probes at different locations have been compared. These give data of the flow inside the turbine, which will provide the base for performing more detailed unsteady flow measurements.

The results lead to the following conclusions:
1. Average yaw angle distributions for the different clocking positions show similarity, but the location of the passage vortices cannot be found behind the second stator. According to (6) the rotor channel vortices dominate over the channel vortices of the second stator (measured after the second stator).
2. Total pressure factors show similar distributions for all clocking positions at each measuring plane.
3. For different stator to stator positions, the exit flow of the second stator is slightly influenced by the clocking position of the first stator (ζ_3). The difference is smaller after the 2nd rotor.
4. The time-averaged values downstream of the second rotor show significant differences not only relative to the first rotor but also at the same plane compared with the other clocking positions. The bigger values of the turbulence level near the hub and the casing are caused by the higher shear stresses and the lower values of velocity in these areas.
5. The next step in testing the turbine will be to focus on the measurements with the three-hole pneumatic and two-sensor hot-wire (type X) probes to determine the boundary layer phenomena (in the region of h/H = 0.00 => 0.05, close to the hub, and 0.90 => 1.00, close

to the casing) and more detailed studies (more clocking positions) of the three-dimensional unsteady flow phenomena in the axial gaps of the turbine. Finally, precise measurement of the power output at the brake will allow the determination of the clocking effect on the two-stage efficiency.

6. When all these results are completed it will be possible to develop a model which helps the understanding of unsteady flow behaviour in multistage turbines and to determine the clocking effect.

REFERENCES

1. Gallus H. E.: *Recent Research Work on Turbomachinery Flow.* Invited Opening Lecture of the Yokohama International Gas Turbine Congress, 95-Yokohama-IGTC-S1, 1995.

2. Boletis E., Sieverding C.H.: *Experimental Study of the Three-Dimensional Flow Field in a Turbine Stator Preceded by a Full Stage.* Trans. of the ASME, Journal of Turbomachinery, vol. 113, 1993.

3. Zeschky J., Gallus H. E.: *Effects of Stator Wakes and Spanwise Nonuniform Inlet Conditions on the Rotor Flow of an Axial Turbine Stage,* ASME J. of Turbomachinery, vol. 115, 1993.

4. Walraevens R. E., Gallus H. E.: *Three-Dimensional Structure of Unsteady Flow Downstream the Rotor in a 1-1/2 Stage Turbine,* Unsteady Aerodynamics and Aeroelasticity of Turbomachines, Elsevier Sience B.V., 1995.

5. Hilditch M.A., Smith G.C, Anderson S.J., Chana K.S., Jones T.V., Ainsworth R.W., Oldfield M.L.G.: *Unsteady Measurements in an Axial Flow Turbine.* CP-571/24, AGARD 1996.

6. Huber F. W., Johnson P. D., Sharma O. P., Staubach J. B., Gaddis S. W.: *Performance Improvement Through Indexing of Turbine Airfoils: Part 1 - Experimental Investigation,* ASME J. of Turbomachinery, Vol. 118, 1996.

7. Krysinski J. E., Gallus H. E., Smolny A., Walraevens R. E., Blaszczak J. R., Mertens B.: *Axial Two-Stage Turbine Test Rig for Unsteady Flow Measurements,* International SYMKOM'95, CMP 108 - IMP TU Lodz, 1995.

8. Smolny A., B³aszczak J. R.: *Experimental Investigations Of Unsteady Flow Fields In A Two-Stage Turbine,* Proceedings of the 2nd European Conference on Turbomachinery - Fluid Dynamics and Thermodynamics, Antwerpen, 1997.

9. Smolny A., B³aszczak J. R., Pawlak R.: *Three-Dimensional Flow Field Measurements In Multistage Turbomachinery Using a Computer Aided System,* Optoelectronic and Electronic Sensors II. Proceedings of SPIE 3054-11, 1997.

10.Smolny A., B³aszczak J. R.: *Boundary Layer and Loss Studies on Highly Loaded Turbine Cascade,* CP-571/4, AGARD 1996.

11.Walraevens R. E., Gallus H. E.: *Stator - Rotor - Stator Interaction In An Axial Flow Turbine And Its Influence On Loss Mechanisms,* CP-571/39, AGARD 1996.

12.Poensgen Ch., Gallus H.E.: *Three-Dimensional Wake Decay Inside of a Compressor Cascade and its Influence on the Downstream Unsteady Flow Field.* ASME J. of Turbomachinery, Vol. 113, 1991.

C557/156/99

Numerical analysis of wakes interactions in a transonic inviscid flow of compressor

G L OLIVEIRA, P FERRAND, and **S AUBERT**
Ecole Centrale de Lyon/Université Claude Bernard, Lyon, France

This paper addresses the interaction problem between wakes and shock waves in ECL axial transonic compressor. The unsteady inviscid 2D compressible equations are solved for the mid-span section of the rotor. An analytical model based on vortical and entropical disturbance is introduced at inlet to simulate passing wakes. Unsteady predictions were carried out for different wake deficits. Correspondent steady computations were also done with the same averaged boundary condition. The main objective of this study is to describe and analyze effects of wake decay due to diffuser and shock waves interaction on compressor performance. Presented results indicates that unsteady flow generate by wakes can modify losses and energy transfer in the compressor, improving its performance.

1 - INTRODUCTION

The physical mechanisms governed by the interactions of rotors and stators in axial transonic turbomachines are not completely understood; in particular there are few publications on transonic aspects. The rise of the load by the blade and the reduction of axial gap increase the flow unsteadiness that modify the time averaged performance characteristics. It is reasonable to expect that temporal changes due to interacting blade rows can change the level of losses and energy transfer. Better understanding of these mechanisms will result in a significant advancement in the turbomachine design.

The unsteady rotor-stator interactions can be decomposed in potential interaction and wake interaction. Potential flow interaction is an inviscid effect due to pressure variation caused by the relative movement of the blades. The potential interaction increases exponentially when the axial gap decreases and is strong if there are shock waves.

A number of publications show that wake interaction can be responsible for the increasing performance of subsonic multistage axial flow compressor. The effect is explained by the ability to recover the total pressure deficit of a wake by a reversible flow process. The goal of this paper is to evaluate the possibility to extend this idea to transonic flow compressors.

2 – WAKE RECOVERY CONCEPT BACKGROUND

Several publications have dealt with wakes transport through turbomachinery blade rows. Kerrebrock and Mikolajczak (1) described the kinematics mechanism of wake convection in a multistage compressor. They associated the wake/blade interaction with pressure and total temperature fluctuations measured on the stator blade walls. Smith (2,3) was the first to verify experimentally that pressure rise and efficiency could be improved reducing axial gap between the blade rows. He suggested the wake recovery concept, referring to the wake decay due to reversible process of wake stretching, occurring in a stator passage. Viscous mixing loss contribution to wake decay are consequently inversely proportional to the inviscid wake recovery.

In a different approach, Adamczyk (4) showed for 2D incompressible invicid flows, that the wake recovery can be expressed in terms of loss of total pressure as a function of the difference of wake kinetic energy flux entering and leaving a compressor row passage. He also indicate that benefits of wake recovery can be quantified by means of a volume integral involving the deterministic stress and the mean strain rate. Valkov and Tan (5) analyzed numerically the wake kinematic convection and its interaction with stator blades producing vortical disturbances and pressure fluctuations. Deregel and Tan (6) used 2D simulations to show that if the rotor wakes of a subsonic compressor are mixed out after (as opposed to before) the stator passage, the time-averaged overall static pressure rise is increased and the mixing loss is reduced. In a recent work, Van Zante et al (7) showed that for typical axial gaps between compressor rows, wake stretching is the dominant wake decay process within the stator with viscous mixing playing a secondary role.

These works have yielded useful and significant results that aid the understanding of wake-induced unsteady flows and its effects on low-subsonic turbomachinery performance. However, there still exists a number of relevant questions which remain to be answered. This work is concerned with unsteady wake phenomena in a transonic high-loading rotor of an axial compressor experimented in ECL. Results from a 2D inviscid computational investigation are presented here. Physical mechanisms regarding wake convection through the mean flow field and its interactions with rotor blades and shock structure are examined. Analyses are carried out for different unsteady conditions of wake deficit and 1:1 wake/blade periodicity.

3 - NUMERICAL APPROACH AND SELECTED TEST CASE

3.1 - Algorithm

The unsteady inviscid flow in a rotor passage is governed by the compressible Euler equations. We make use of a finite volume CFD code, named PROUST (8) to solve numerically these equations. The algorithm has previously been validated and shown to accurately simulate steady and unsteady, inviscid and viscous flows.

The spatial discretization is based on a cell-vertex MUSCL technique over the structured quadrangular meshes. The convective fluxes are estimated using a third order accurate Liou's Advection Upwind Splitting Method. Van Albada limiter, only sensitive to strong discontinuities is employed. The time discretization uses an explicit Runge-Kutta integration with five steps.

Physical boundary conditions are introduced by compatibility relations according to characteristics theory. For steady flows studies, the specified inlet subsonic conditions are relative total pressure (P_{tr}), relative total temperature (T_{tr}) and relative inlet flow angle (β). At the outlet the static pressure is imposed. For unsteady analysis, non reflective conditions must be applied. This rise no problem at the outlet, but at the inlet side, it is impossible to introduce a well established unsteady wake using non reflective conditions. We define in this case a permeability condition resulting from the linear combination of compatibility and non reflective conditions. It assures boundary conditions imposition with minimum waves reflection. The wall surfaces are treated with slip condition. A specific unsteady treatment at periodic boundaries allowing different wake/blade periodicity ratios is introduced. It supposes a constant phase lag between the solution of adjacent passages. This phase lag forms the basis of a cyclic procedure relating the solution on ghost cells along the periodic boundaries of the computational domain to the solution within the domain at an earlier time.

3.2 - Wake Model

The unsteady code uses a time-shifted non uniform upstream boundary condition simulating the relative movement of the incoming wakes. Vortical and entropical distortions associated with wakes are assumed to be Gaussian. Analytical formulations utilized to compute boundary conditions at inlet are:

<u>vortical waves</u>

$$u(y,t) = \overline{u}\left\{1 - \delta U \exp\left[-\alpha\left(\frac{y - U_w t}{L}\right)^2\right]\right\} \qquad [1]$$

$$v(y,t) = \overline{v}\left\{1 - \delta U \exp\left[-\alpha\left(\frac{y - U_w t}{L}\right)^2\right]\right\} + U_w \delta U \exp\left[-\alpha\left(\frac{y - U_w t}{L}\right)^2\right] \qquad [2]$$

<u>entropical waves</u>

$$\rho(y,t) = \overline{\rho}\left\{1 + \delta\rho \exp\left[-\alpha\left(\frac{y - U_w t}{L}\right)^2\right]\right\} \qquad [3]$$

where L is the wake width, $\overline{u}$ and $\overline{v}$ are axial and tangential mean velocity components, U_w is the wake velocity in rotating frame, α is the Gaussian shape coefficient and δU, and $\delta\rho$ are velocity and density fluctuations.

Inlet boundary conditions on P_{tr}, T_{tr} and β are computed from u , v and ρ assuming constant static pressure.

3.3 - Geometry

The transfer from kinetic energy to pressure energy in compressors is obtained by either shock waves or diffuser effect. The present work is focused on analysis of these two kind of energy transfer in an unsteady flow with passing wakes. The ECL4 compressor is a propitious geometry to perform such a study since we can separate quite well the two mechanisms.

The ECL4 configuration is a single-stage with IGV axial-flow transonic compressor designed by SNECMA and experimented by ECL. The rotor geometry was designed for a high rotational speed with a total pressure ratio of 2.25 and a mass flow of 17.5Kg/s. The blade chord at mid-span is 0.052 m. Steady and unsteady simulations were carried out in the rotating frame of a single 2D mid-span passage of the rotor. A structured H-grid with 197 nodes in the axial direction and 51 in the blade-to-blade direction was used. Figure 1a shows the mesh for two consecutive channels with only every other node in both directions.

The steady flow conditions are such that the relative Mach number and the blade-to-blade relative flow angle are respectively 1.15 and 58.5 degrees at inlet and 0.38 and 23 degrees at outlet. The static pressure evolution inside the channel is plotted figure 1b. Values are made non dimensional with respect to inlet dynamic pressure. The shock waves system is composed by two different shocks. The bow shock wave stands in front of the blade and weakens away from the leading edge. It intersects the suction side of adjacent blade and its reflection interacts with the strong passage shock wave. On the suction side the relative Mach number varies from 1.55 to 0.68 through the second shock wave.

The inlet imposed wake has a width to pitch ratio of $L/P = 0.125$. The Gaussian shape coefficient α is set to 0.693. The wake/axial flow velocity ratio is $U_w/u = 1.7$. The reduced frequency based on axial chord and inlet axial velocity $\overline{\omega} = 2\pi f\, u/c$ is 20.5. These high values indicate strong unsteady interactions inside the rotor passage.

4 - RESULTS ANALYSIS

Figure 2 shows the instantaneous wake convection through the inter-blade rotor passage. Entropic waves are presented on upper side of the blade and the associated velocity fluctuations on its lower side. Strong shock waves positions are distinguished by the sonic iso-Mach number line. The wake is stretched moving downstream due to mean flow deceleration and its intensity decays. Through the passage shock wave, the normal velocity component of the mean flow diminishes turning over the wake. Interactions between vorticity waves and blade walls generates two opposite sign vortices that contribute to mass transport inside the wake from suction to pressure side. Mass flow conservation implies inverse mass transport outside the wake.

We define two averaging operators used later in comparative analysis: the area average (applied to velocity components, static pressure and static temperature) and the mass average (applied to stagnation flow variables and *RMS* of fluctuations).

	Unsteady	Steady
Area average	$\overline{A}'(x) = \dfrac{1}{\Delta y . T} \displaystyle\int_0^{\Delta y} \int_0^T A(x,y,t)\, dt\, dy$	$\overline{A}(x) = \dfrac{1}{\Delta y} \displaystyle\int_0^{\Delta y} A(x,y)\, dy$
Mass average	$\tilde{B}'(x) = \dfrac{\displaystyle\int_0^{\Delta y} \int_0^T B(x,y,t)\rho U_x\, dt\, dy}{\displaystyle\int_0^{\Delta y} \int_0^T \rho U_x\, dt\, dy}$	$\tilde{B}(x) = \dfrac{\displaystyle\int_0^{\Delta y} B(x,y)\rho U_x\, dy}{\displaystyle\int_0^{\Delta y} \rho U_x\, dy}$

where A and B are flow variables.

Unsteady computations were done for inlet wakes with 5, 10, 15 and 20 percent of velocity δU and density $\delta\rho$ variations. We designate the corresponding results respectively by A05, A10, A15 and A20. Steady computations were also done with the same inlet/outlet average boundary conditions ($\tilde{T}_{tr} = \tilde{T}'_{tr}$, $\tilde{P}_{tr} = \tilde{P}'_{tr}$ and $\overline{\beta} = a\tan(\overline{v}/\overline{u}')$ at inlet, and $\overline{P}_s = \overline{P}'_s$ at outlet).

Computed steady and unsteady compression ratio and efficiency are presented figures 3a and 3b as a function of mass flow. Results were calculated by means of averaged stagnation flow variables. One can note that although averaged boundary conditions are the same, steady and averaged unsteady compressor performance are not equal. The unsteady computation predicts higher values of averaged mass flow and compression ratio but slightly lower ones for efficiency. Differences are more important as the wake intensity increases. In A20 case, unsteady average results are up to 0.6% higher than steady ones for the mass flow, up to 0.4% higher for compression ratio and up to 0.2% lower for efficiency. Therefore, unsteady operate regime can not be considered quasi-steady. For identical boundary conditions, compression unsteady flow characteristic is shifted up in relation to steady response while efficiency curve is almost the same. Since computations are inviscid, efficiency takes in to account only wake and shock wave losses.

The strong passage shock wave average positions obtained for unsteady and steady simulations are presented figure 4a. The shock moves upstream in a non-linear way approaching leading edge as mass flow decreases (wake intensity increases). As it moves upstream it weakens and the associated losses diminish. For 5 to 15% of wake intensity, the shock is stronger at suction side and steady/unsteady shock wave average positions are almost identical. On the other hand, for 20% of wake intensity, the shock becomes stronger at pressure side and steady/unsteady average positions are quite different. Figure 4b presents the averaged relative total pressure as a function of x-position for A05 and A20 cases. Variations are associated with shock waves and wake losses. In the A05 case, 75% of losses are generated by the passage shock and 25% by the bow shock. Steady and unsteady evolutions are rather similar. In the A20 case, the passage shock wave losses are weaker because of the more upstream shock position (see figure 4a). The steady leading edge shock wave losses represent 60% of total losses. Unsteady average total losses are 50% higher than steady ones. Increase in losses is equally distributed between the bow shock and the passage shock.

Figures 5a and 5b show A05 and A20 axial differences of relative total pressure P_{tr}, static pressure P_s and absolute total pressure P_t, between unsteady and steady results. In the A05 case, more than 50% of P_{tr} unsteady/steady differences are located upstream of the passage shock wave ($x < 0.2\,c$), being related to leading edge shock wave. Moderate variations can also be noticed downstream of the passage shock wave, in the diffuser region $0.4\,c < x < c$. Averaged unsteady static pressure is inferior to steady one at inlet but it is the same at outlet. Relevant averaged unsteady static pressure recuperation is observed between leading edge and the passage shock wave ($0 < x < 0.2\,c$), as well as in the diffuser region. Absolute total pressure recuperation is distributed on 20% between leading edge and passage shock wave, 40% through the passage shock and 40% in the diffuser region. The same trends are observed in the A20 case. Average unsteady passage shock wave position is slightly moved upstream comparing to steady one producing positive static pressures differences and alike losses. About 70% of absolute total pressure rise take place in the diffuser. The remaining 30% is produced through the passage shock.

The axial unsteady/steady difference of relative velocity angle β is presented figure 6. Leading edge incidences differences are not notable up to $x = 0.1\,c$. Downstream of the passage shock ($x > 0.4\,c$) the unsteady flow is overturned. According to Euler equation, we can expect that energy associated with flow deviation is principally recuperated in the diffuser. Moreover, the β outlet-inlet variation is more important for higher wake intensity, what agrees with performance results presented figure 3a.

Figure 7a describe axial evolutions of A20 *RMS* and kinetic energy fluctuations. Results were made non dimensional with respect to inlet values, presented on upper right corner of figures. Unsteady kinetic energy ($UKE = \rho|\tilde{v}|^2/2$) is mainly conditioned by axial velocity fluctuations. It decreases notably in the weak shock and diffuser regions ($x < 0.1\,c$ and $0.3\,c < x < c$ respectively), where wakes are strongly damped. Transversal velocity fluctuations $\tilde{v}'$ increases through the shock waves and are damped after. Although $\tilde{v}'$ is globally amplified between inlet and outlet, it has no significant effect on *UKE* evolution since its inlet level is small compared to the axial one (ten times lower). As showed in figure 7b, all temperature fluctuation evolutions are quite similar. These fluctuations are strongly reduced by the weak shock and the diffuser, but they augment through the passage shock. Figure 7c presents pressure fluctuations. Analysis of static pressure fluctuation evolution indicates that acoustic phenomena plays an important role (short length waves). Total pressure fluctuations seem to be related to P_s fluctuations and *UKE*. Axial evolutions of A05, A10 and A15 unsteady fluctuations (not presented here) are qualitatively similar to A20 ones. Effects of shock waves and diffuser upon wake recovery were found to be nearly proportional to wake deficit.

The averaged mixing loss as function of x-location is shown figures 8a and 8b for A05, A10, A15 and A20 cases. Mixing loss coefficient L_{mix} is defined (6) assuming unsteady energy flow to be complete mixed out for each x position:

$$L_{mix} = \frac{\int_0^{\Delta y} \int_0^T \left[\int_0^{\Delta y} C(x,y,t)\,dy - C(x,y,t) \right] \rho U_x\,dt\,dy}{\left(\tilde{C}'(x_{inlet}) - \overline{D}'(x_{inlet}) \right) \cdot \int_0^{\Delta y} \int_0^T \rho U_x\,dt\,dy} \qquad [4]$$

where C is a relative stagnation flow variable (T_{tr} or P_{tr}), and D is the corresponding static one (T_s or P_s, respectively).

Results shows that mixing loss is significantly reduced due to wake/weak shock wave interaction ($x < 0.1\,c$) and wake stretching in diffuser ($x > 0.4\,c$). Benefit effects are approximately proportional to wakes deficit. The slight loss rise between 20% and 40% of axial cord is associated to wake/strong shock waves interactions.

5 - CONCLUSION

A methodology has been developed to evaluate fluctuations level average effects of unsteady flow produced by wakes. This methodology has been applied to a transonic axial compressor to compare relative effects of bow shock wave, strong passage shock wave and subsonic compression.

Results show, like for subsonic compressor (4,5), that unsteady flow increases performances in terms of compression ratio for all the wake intensities studied. This gain is mainly obtained through the diffuser and eventually through shock waves.

RMS analyses show that unsteady fluctuations decrease through bow shock waves at the opposite of strong passage shock waves. Diffuser deadens always unsteady fluctuations specially velocity fluctuations.

All the results obtained on transonic compressor do not contradict analysis of published literature on subsonic compressor.

REFERENCES

(1)   Kerrebrock, J. L. and Kikolajczak, A. A., "Intra-Stator Transport of Rotor Wakes and Its Effect on Compressor Performance", ASME Journal of Engineering for Power, October 1970, pp. 359-368

(2)   Smith, L. H., "Wake Dispertion in Turbomachines", Journal of Basic Engineering, September 1966, pp. 688-690

(3)   Smith, L. H., "Casing Boundary Layer in Multistage Axial-Flow Compressors", Flow Research in Blading, edited by L. S. Dzung, Elsevier Publishing Company, Amsterdam, 1970

(4)   Adamczyk, J. J., "Wakes Mixing in Axial Flow Compressors", ASME Paper 96-GT-029, 1996

(5)   Vakov, T. and Tan, C. S., "Control of the Unsteady Flow in a Stator Blade Row Interacting With Upstream Moving Wakes", ASME Journal of Turbomachinery, January 1995, Vol. 117, pp. 97-105

(6)   Deregel, P and Tan, C. S., "Inpact of Rotor Wakes on Steady-State Axial Compressor Performance", ASME Paper 96-GT-253, 1996

(7)   Van Zante, D. E., Adamczyk, J. J., Strazisar, A. J. and Okishi, T. H., "Wakes Recovery Performance Benefit in a High-Speed Axial Compressor", ASME Paper 97-GT-535, 1997

(8)   Aubert, S., Ferrand, P., "Study of Unsteady Transonic Flows With a New Mixed Van Leer Flux Splitting Method", 7th International Symposium on Unsteady Aerodynamics and Aerolasticity of Turbomachines, Fukuoka, Japan, 1994

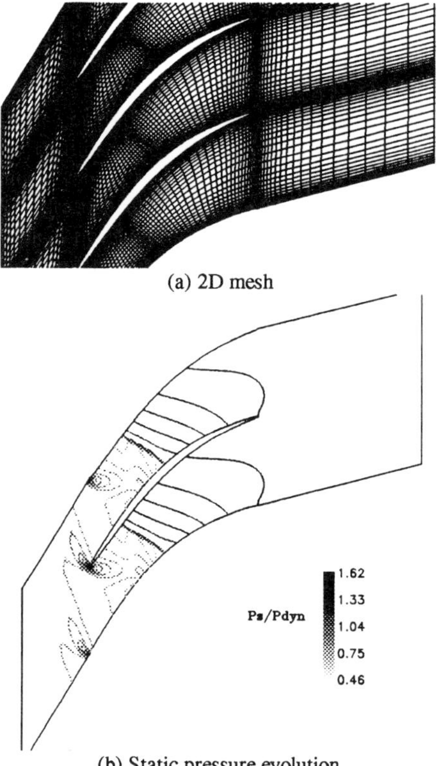

(a) 2D mesh

(b) Static pressure evolution

Figure 1: ECL4 rotor configuration

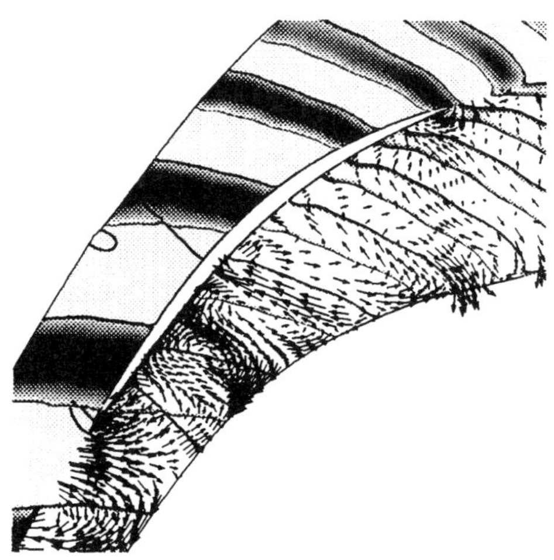

Figure 2: Instantaneous wake convection (top: entropy fluctuation, bottom: velocity fluctuation).

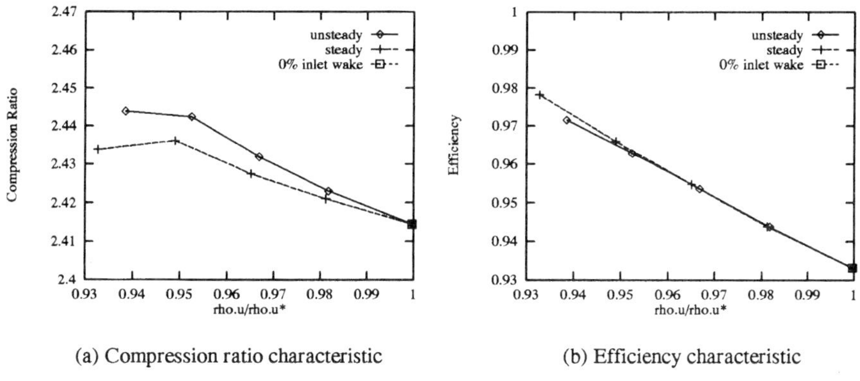

(a) Compression ratio characteristic

(b) Efficiency characteristic

Figure 3: Compressor averaged performance (points correspond to computation results)

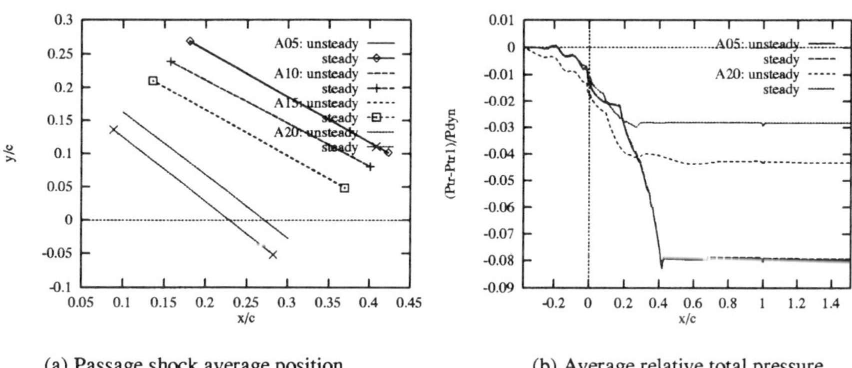

(a) Passage shock average position

(b) Average relative total pressure

Figure 4. Steady and unsteady average shock wave system

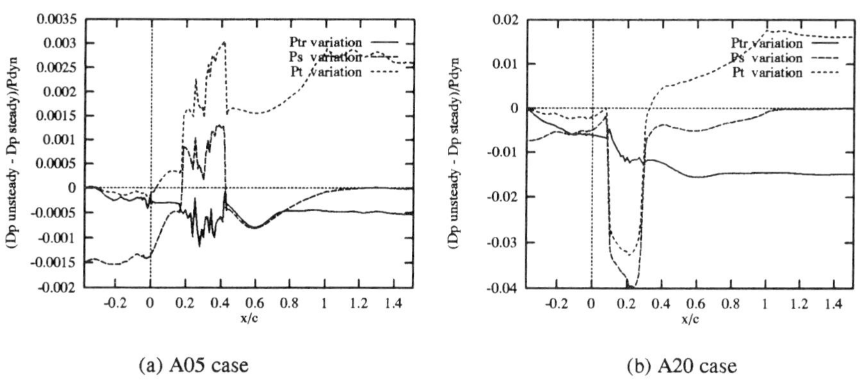

(a) A05 case

(b) A20 case

Figure 5: Average pressure evolution. Unsteady-steady difference.

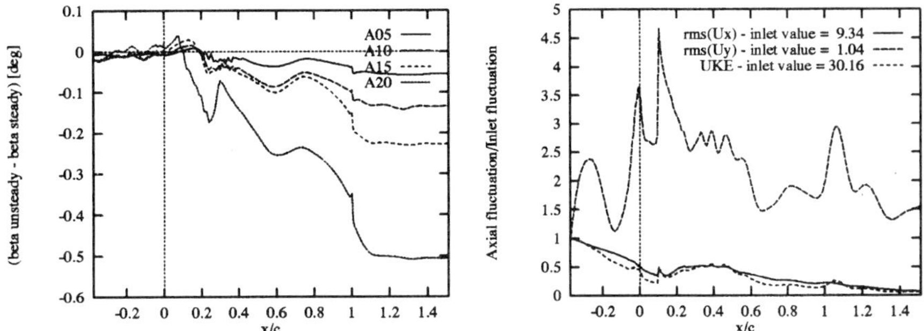

Figure 6: Average relative velocity angle.
Unsteady-Steady difference.

Figure 7(a): Kinematics fluctuations.

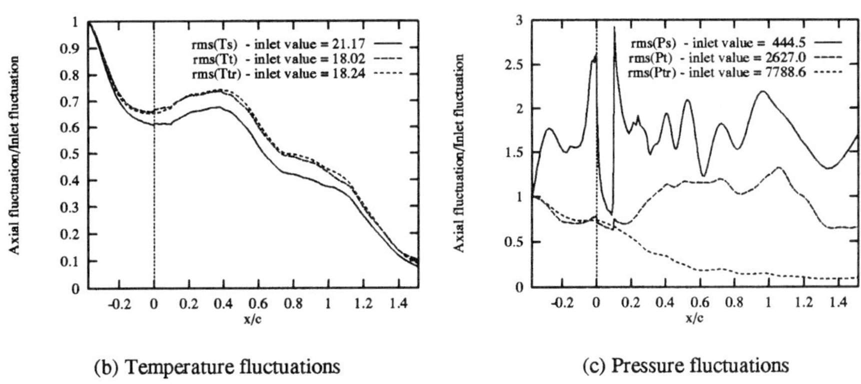

(b) Temperature fluctuations

(c) Pressure fluctuations

Figure 7: A20 Unsteady flow evolution.

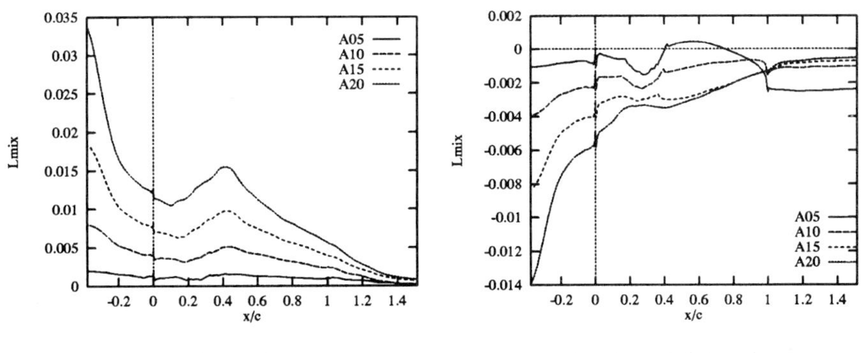

(a) Based on relative total temperature

(b) Based on relative total pressure

Figure 8: Mixing loss coefficient

C557/044/99

Simulation of the unsteady interaction of coupled fans with ventilating systems

B SCHULZE DIECKHOFF and **T H CAROLUS**
Institut für Fluid-und Thermodynamik, Universität-Gesamthochschule Siegen, Germany

A study of the unsteady interaction of fans in ventilating systems was conducted. Emphasis was placed on systems with two fans placed either parallel or in series. The study consisted of the development of a model, its implementation into a computer code and the numerical and experimental investigation of several ventilating systems under various operating conditions.

The flow in the components such as pipes, plenums, fans and valves is assumed to be unsteady, viscous and one-dimensional. The resulting set of partial and ordinary differential and algebraic equations for the entire network is solved numerically by the methods of characteristics.

In this paper two case studies are presented:
- The unstable operation of a system with two parallel fans during a valve closure;
- Surge of a system with two in-series fans upon guide vane adjustment of one fan.

Numerical as well as experimental results are shown and compared. In both cases the model provides comprehensive information about the unsteady flow rate and pressure in the system which agrees well with the experimental results.

NOMENCLATURE
English Symbols

a	m/s	speed of sound	R	J/kg K	gas constant
A	m^2	cross section area	t	s	time
D	m	diameter	T	K	temperature
DR	Pa/(kg/s)^2	resistance coefficient	v	m/s	flow velocity
L	m	length	VOL	m^3	volume
$\dot{m}$	kg/s	mass flow rate	x	m	co-ordinate
p	Pa	pressure			

Greek Symbols

| α | ° | angle of valve or guide vanes | ρ | kg/m³ | density |
| λ | - | friction factor | ζ | - | resistance coefficient |

Subscripts

A	at outlet	U	at point U
E	at inlet	Drall	guide vanes
P	at point P	DR	valve
R	at point R		

1. INTRODUCTION

The flow rate and pressure in complex ventilating systems during various operating conditions are rather difficult to predict, especially if the network contains more than one fan. For example a system with two *parallel* fans may be driven into completely different steady state operating points, depending only on the initial conditions (1). Furthermore, a system with two *in-series* fans may run with large undesired oscillations of flow rate and pressure. The exact interaction of the fans with the network can only be described and subsequently understood when the unsteady flow in the system is taken into account.

Two different, general schemes of networks were developed, one for ventilating systems with two parallel and one with two in-series fans (Fig. 1). Essential components of each network are pipes, valves, plenums and the fans. The mathematical model for the unsteady, viscous one-dimensional flow in the key component 'pipe' uses distributed parameters. That means that inertia, compressibility and frictional losses of the fluid are a function of time *and* the location along the pipe. The flow in all other components is described by lumped parameter models. The resulting set of partial and ordinary differential and algebraic equations for the entire network is solved with the method of characteristics.

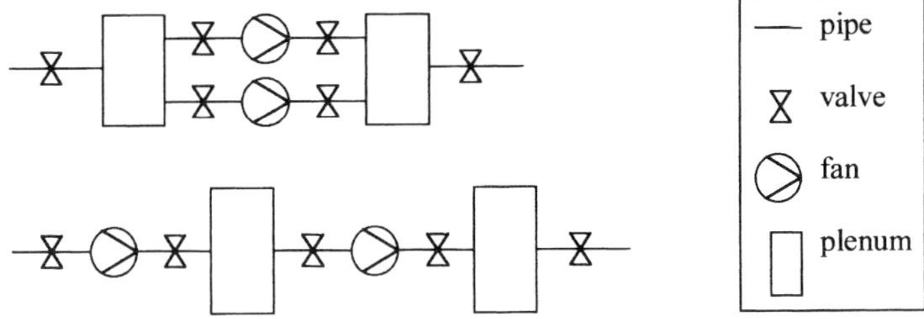

Fig. 1: General schemes of networks. a) Network with parallel fans; b) with in-series fans

To facilitate the quick and easy simulation of the performance of various ventilating systems, the network schemes and the models for the components were implemented into a PC-code. During the simulation the mass flow rate and the static pressure at any location in the system were visualized simultaneously (2). In order to test the model and code, theoretical results were compared to many measurements taken in large heating, ventilating and air conditioning (HVAC) systems as well as in a laboratory test facility.

This paper describes the models used and their implementation into a computer code. Then two case studies are presented: The unstable operation of a system with two parallel fans during a valve closure, and surge of a system with two in-series fans upon guide vane adjustment of one fan. In both cases numerical and experimental results are shown and compared.

2. MODELS FOR THE UNSTEADY FLOW IN THE COMPONENTS

2.1 Pipe
Pipes are key components in any network considered here. Thus the model for the unsteady flow in a pipe is essential. Under the assumptions of
- a one-dimensional flow
- turbulent friction with a constant friction factor λ_R
- no gravity or bouyancy forces
- constant cross section area of the pipe
the equations of continuity and motion, resp., can be written as (references (2), (3), (4), (5))

$$\frac{\partial(\rho A_{Rohr})}{\partial t} + \frac{\partial(\rho A_{Rohr} v)}{\partial x} = 0 \tag{1}$$

$$\rho A_{Rohr} v \frac{\partial}{\partial x} v + \rho A_{Rohr} \frac{\partial v}{\partial t} + A_{Rohr} \frac{\partial p}{\partial x} + \frac{\rho \lambda_R A_{Rohr}}{2D} v |v| = 0 . \tag{2}$$

Here p is the static pressure, v the flow velocity, A_{Rohr} the cross section area of the pipe, D the diameter of the pipe, λ_R the turbulent friction factor and ρ the fluid's density. The independent variables are the co-ordinate x along the pipe axis and the time t.

Further, assuming isothermal change of state

$$p/\rho = \text{const.} = RT, \tag{3}$$

eq.'s (1) and (2) become

$$\frac{\partial p}{\partial t} + \frac{RT}{A_{Rohr}} \frac{\partial \dot{m}}{\partial x} = 0 \tag{4}$$

$$\frac{\partial \dot{m}}{\partial t} + \frac{RT}{A_{Rohr}} \frac{\partial}{\partial x}\left(\frac{\dot{m}^2}{p}\right) + A_{Rohr}\frac{\partial p}{\partial x} + \frac{\lambda RT}{2DA_{Rohr}p} \dot{m} |\dot{m}| = 0 . \tag{5}$$

In references (6) and (3) these two equations are called the 'exact' model of gas flow in a pipe. In ventilating systems usually the flow velocity is small compared to the speed of sound. Thus the convective term in eq. (5) may be neglected and one obtains

$$\frac{\partial \dot{m}}{\partial t} + A_{Rohr} \frac{\partial p}{\partial x} + \frac{\lambda RT}{2DpA_{Rohr}} |\dot{m}|\dot{m} = 0 . \tag{6}$$

These partial differential equations are transformed by the method of characteristics into a set of ordinary differential equations (see references (3) and (6)):

- <u>along a C^+ - characteristic, i.e. dx/dt = +a:</u>

$$\frac{a}{A_{Rohr}} d\dot{m} + dp + \frac{\lambda RTdx}{2DpA_{Rohr}^2} |\dot{m}|\dot{m} = 0 \tag{7}$$

- along a C⁻ - characteristic, i.e. dx/dt = -a:

$$-\frac{a}{A_{Rohr}}\, d\dot{m} + dp - \frac{\lambda R T dx}{2 D p A_{Rohr}^2}\, |\dot{m}|\dot{m} = 0 ,\tag{8}$$

where a is the actual speed of sound (wavespeed) in the pipe. In flexible conduits it can be considerably lower than in the free atmosphere , compare for example reference (7). Eqs. (7) and (8) can be solved numerically. They describe the propagating disturbances along a positive (C^+) and negative (C^-) characteristic as shown in Fig. 2. If the values of $\dot{m}$ and p are known at the points U and R in the $x - t$ plane, one can calculate $\dot{m}$ and p at the point P, etc.

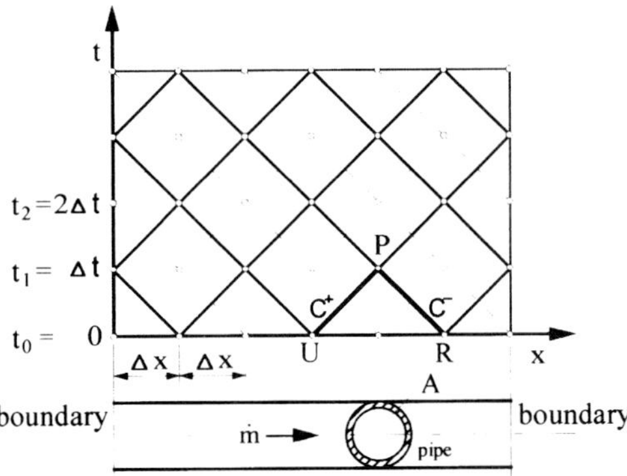

Fig. 2: Grid of the C^+ - and C^- - characteristics in the $x - t$ plane

2.2 Resistances

The essential property of a valve and an in/outlet of a pipe (Fig. 3) is its resistance. The pressure drop due to a resistance is described by a lumped parameter model as

$$p_E - p_A = DR\, \dot{m}|\dot{m}| .\tag{9}$$

It is assumed that the resistance coefficient DR, as defined here, can be calculated from the *steady state* resistance coefficient of the element as found in many handbooks. Of course, if a valve is adjusted with time, DR is a function of time as well. In case of an in/outlet, DR may also be a function of the flow direction.

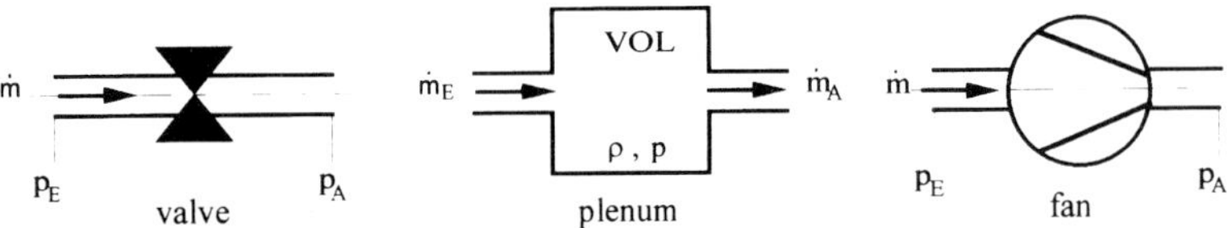

Fig. 3: Some components of a ventilating system

2.3 Plenum

The fluid in a plenum (Fig. 3) is assumed to be compressible but with no kinetic energy. Further assuming a rigid plenum, i.e. VOL = const., and an isothermal change of state, the continuity equation

$$\frac{dm}{dt} = \dot{m}_E - \dot{m}_A \tag{10}$$

yields with $m = Vol \cdot \rho$ for the time dependent pressure in the plenum

$$\frac{dp}{dt} = \frac{RT}{V}\left(\dot{m}_E - \dot{m}_A\right). \tag{11}$$

2.4 Fan

The essential property of the fan (Fig. 3) is a static pressure for a given flow rate. It is given by the steady state mass flow rate - pressure rise characteristic $p_A - p_E = f(\dot{m})$. During a simulation quite often the characteristic curves in regions of negative flow rate and pressure rise are required. This is accomplished by extrapolation, keeping in mind that the fan behaves as a resistance when the flow direction is negative, i.e. from the pressure to the suction side.

Fast variations of flow rate will result in deviations from the steady state characteristic (8,9). Here this is taken into account by adding the inertia of the flow in a short pipe with the length L and the cross section area A, corresponding roughly to the housing, to the steady state pressure rise:

$$p_A - p_E = f(\dot{m}) - \frac{L}{A}\frac{d\dot{m}}{dt} \tag{12}$$

A more precise description of the fan's unsteady behaviour would require a more detailed empirical or theoretical model. In particular this would be necessary if one desired the accurate modeling of the fan in its stall regime.

3. IMPLEMENTATION INTO A COMPUTER CODE

The general network schemes and the models for the components were implemented into a PC-code. At first the code creates a graphical user interface to collect inputs such as
-   the main geometrical dimensions of each component (a component may also be left out by assigning a negliable small length or a very large volume, etc. to it)
-   friction factors and resistance coefficients of the pipe, valves, in/outlets
-   the characteristic curves of the fans
-   time functions of the fans' speed and valve and guide vane settings
-   start and stop signals for the fans
-   density, speed of sound.

Then the scheme of the complete ventilating system and the pressure rise – flow rate characteristic of each fan are displayed on the screen. Once the simulation is started, transient pressures and mass flow rates at any location of the system are visualized in a bar graph manner. Also the fans' operating points and their traces in the pressure rise – flow rate plane are shown. In addition, valve and guide vane settings and the fans' speeds may be altered within an ongoing simulation by key stroke. Thus a variety of operations can be examined. Finally, the time functions of many variables are stored and thus are available for further processing.

4. CASE STUDIES

It is the goal of the two following case studies to (i) demonstrate the expected or unexpected operation of simple ventilating systems with either parallel or in-series fans, and (ii) to show a comparison of theoretical results to measurements. The experimental results shown here are obtained with a laboratory test facility, which emulates exactly the systems investigated, performs precisely the desired operations and allows to measure all time-dependent quantities of interest. Although many measurements were carried out in actual environment as well (e.g. in a HVAC system of a large hospital and a big lecture hall), they are not reported here as these systems did not allow systematic investigations.

4.1 Case study A: Ventilating system with two parallel fans

The operation of systems with two parallel fans may lead to unexpected results. If the characteristics of the fans have a positive slope in a certain region, unstable operation of the whole network may occur. This is even evident from a steady state point of view, since more than one value of the flow rate corresponds to certain values of each fan's pressure rise. Further, the start-up of such systems can be problematic. Usually a fan, parallel to another one already in operation, will be started up against a closed valve to avoid operation in the stall regime, reference (10).

In this case study a simple system with two almost identical parallel fans is investigated. The fans work with the same rotational speed and discharge into the same downstream plenum (Fig. 4). The upstream plenum is set to a very large volume, thus the intake of the upper and lower branch is decoupled.

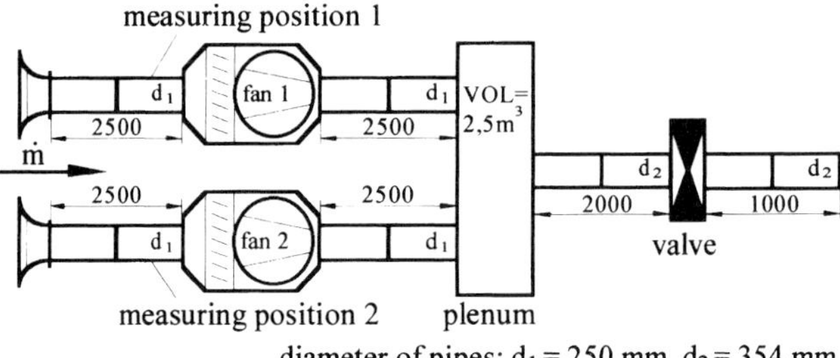

Fig. 4: Case study A: System with two parallel fans

At first, for a clear understanding of the phenomenon, the *calculated* time functions of flow rates and pressures are discussed. As depicted in Fig.5 and 6 both fans work at a stable operating point A (i.e. in a region of negative slope of the characteristic) at the beginning. When the valve is slowly closed, the operating points of the fans jointly pass the maximum of their characteristics and move into the region with positive slope, i.e. to B. Now the operating points start to drift apart. The flow rate of fan 1 decreases rapidly until it reaches C_1 in a region of negative slope while the flow rate of fan 2 increases corresponding to C_2. Then, as the valve continues to close, the flow rates of both fans are decreasing more or less steadily until both fans reach their final operating point D (i.e. D_1 and D_2).

Fig. 6 shows that the calculated results agree quite well with the measurments. The differences in the instability inception point might be due to the fact that as compared to the simulation the fans in the experiments have slightly different characteristics due to installation effects. Also, the simple one-dimensional model of the fans' unsteady behaviour has obvious limitations to represent the complex flow involved in the fan stalling mechanism. In addition, many similar experiments have shown that the onset of the instability may be triggered by the stochastic fluctuations of pressure and flow rate usually encountered in this region of most fans' characteristics.

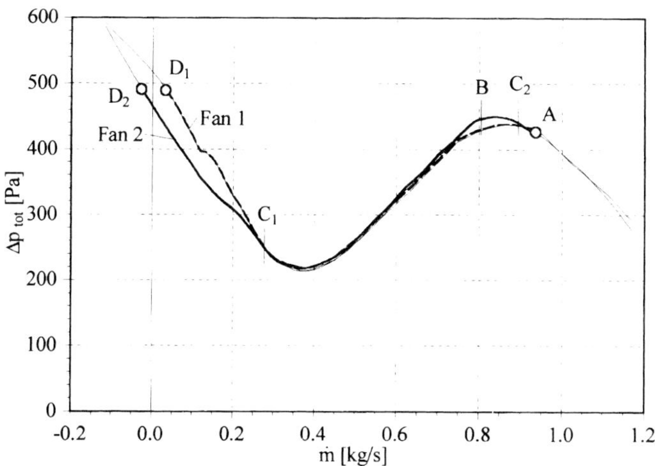

Fig. 5: Calculated operating points of the fans in the pressure rise – mass flow rate plane during valve closure (closure time = 25.0 s); the solid lines represent the (slightly different) steady state characteristics of both fans

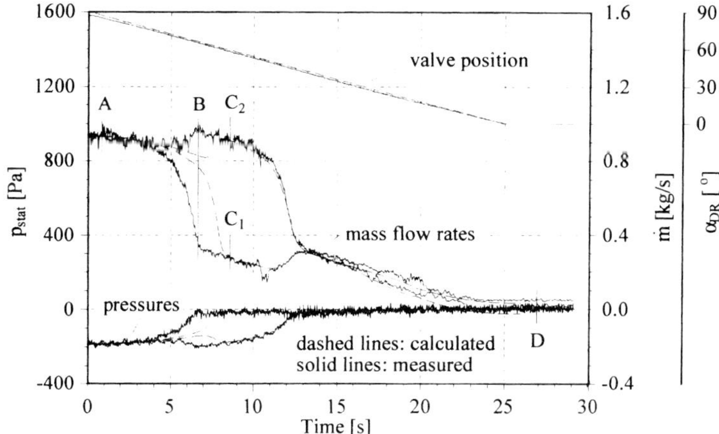

Fig. 6: Valve closure function and time functions of pressure and mass flow rate at the measuring positions 1 and 2; valve closure time = 25.0 s; (calculated vs. measured)

It is remarkable that the phenomenon strongly depends on the valve closure time. If the valve closure time is much shorter, the operating points may not drift apart but move jointly along their nearly identical characteristics from A to D (Fig. 7).

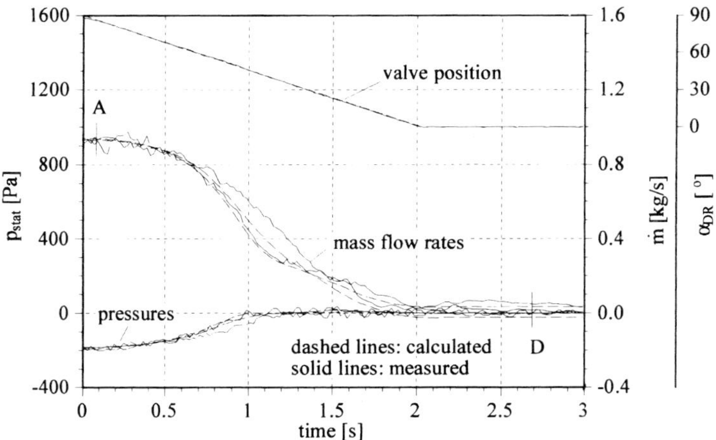

Fig. 7: Valve closure function and time functions of pressure and mass flow rate at the measuring positions 1 and 2; valve closure time = 2.0 s; (calculated vs. measured)

4.2 Case study B: Ventilating system with two fans in series

Banzhaf (11) describes a failure in the exhaust system of a power plant with two fans in series due to excessive fluctuating pressures. Erraneously the adjustable rotor blades of one of the two fans suddenly were closed and, as a result, the second fan was driven into its stall regime. This led to surge in the whole system and subsequently caused damages at the steam generator and filters. This phenomenon is investigated in the following case study with a simplified system (Fig. 8) under slightly different operating conditions.

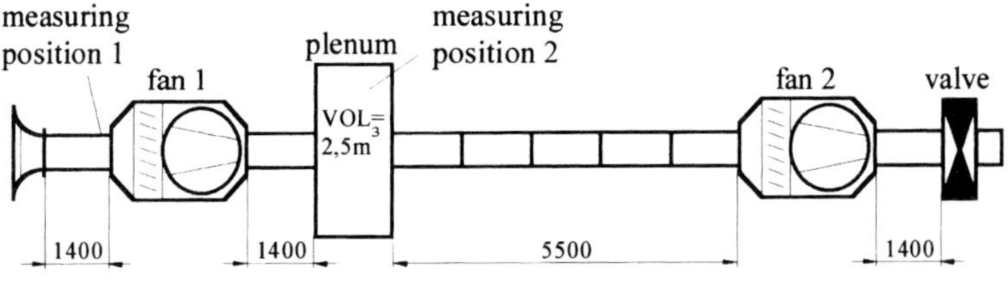

Fig. 8: Case study: System with two in-series fans

At the beginning of the operation both fans work in the stable region of their characteristics with open guide vanes (operating points A in Fig. 9). As the guide vanes of fan 2 are closed, its operating point shifts to the region C on its characterstic for a guide vane angle α_{Drall} = 45° (between the characteristics for α_{Drall} = 40° and α_{Drall} = 50°). Simultaneously the operating point of fan 1 moves to the region B, a region with positive slope. This causes large scale self-excited oscillations (surge) of the flow with excessive pressure peaks. In Fig. 9 the steady cycles of the oscillations are denoted by the arrows at B and C. In Fig. 10 the development of the oscillations as a function of time can be seen more clearly.

In Fig. 10 again experimental results are compared to the calculations. The onset of surge, the frequency and the amplitudes of flow rate and pressure agree very well. However, one has to keep in mind that the exact onset may be triggered by stochastic disturbances in the system (12). The slight differences at the beginning of the time record are most probably caused by an inaccuracy in the flow rate sensor reading.

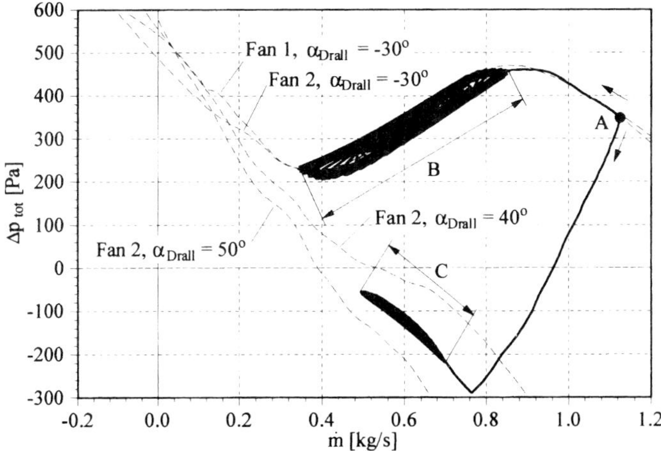

Fig. 9: Calculated traces of the fans' operating points in the pressure rise – mass flow rate plane during guide vane adjustment of fan 2; the thin dashed lines represent the steady state characteristics of both fans for different guide vane angles

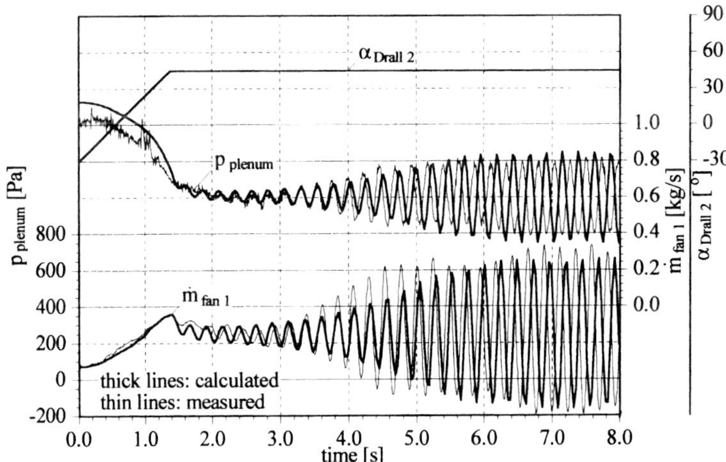

Fig. 10: Guide vane closure function and time functions of pressure in the plenum and mass flow rate through fan 1; (calculated vs. measured)

It is remarkable that the oscillations strongly depend on the size of the plenum and the pipes. These components essentially determine the elasticity, the mass and the damping forces in the whole system (8).

5. CONCLUSIONS

The model presented allows quite accurate simulation of the transient flow in ventilating systems with two parallel or in-series fans under various operating conditions. This is useful in the design stage of such systems as well as in understanding and correcting existing systems that have encountered undesirable operating conditions or even damages. Presently the model is being further developed to include components such as controllers for constant flow rate or pressure so that the function and stability of such systems can be investigated.

ACKNOWLEDGEMENT
This work is supported by the German Bundeswirtschaftsministerium through contract AiF 9840 and AiF 11120 N / 1 under supervision of the Forschungsvereinigung für Luft- und Trocknungstechnik (FLT) e.V. We gratefully acknowledge this support.

REFERENCES
/1/ Eck, B.: Ventilatoren. Springer Verlag 1972
/2/ Schulze Dieckhoff, B.; Carolus, Th.: Simulation des Betriebsverhaltens lufttechnischer Anlagen mit parallelgeschalteten Axialventilatoren, VDI-Verlag 1996, VDI Berichte 1249, pp. 335-350
/3/ Horlacher, H.B.; Lüdecke, H.J.: Strömungsberechnung für Rohrsysteme, expert verlag, (1992)
/4/ Bender, E.; Konrad, B.: Zur Simulation dynamischer Vorgänge in Gasnetzen, Regelungstechnik rt, 27. Jahrgang (1997), Heft 4, pp. 105-140
/5/ Kolnsberg, A.; Wachenberg, B.: Simulation instationärer Strömungsvorgänge in Gas-Rohrleitungen, gwf-gas/erdgas 124(1983) H. 6
/6/ Wylie, E.B.; Streeter, V.L.: Fluid Transients. Mc Graw Hill-Verlag 1978
/7/ Maier, E.; Untersuchung strömungsmechanischer Schwingungen sowie der Wellenausbreitungsgeschwindigkeit in lufttechnischen Anlagen, VDI-Verlag, 1994, VDI Fortschritt-Berichte Reihe 7, 250
/8/ Carolus, Th.: Theoretische und experimentelle Untersuchungen des Pumpens von lufttechnischen Anlagen mit Radialventilatoren. Dr.-Ing. Diss. Universität (TH) Karlsruhe, 1984
/9/ Kullmann, L., Carolus, Th.: Pumpschwingungen in einer lufttechnischen Anlage: Vergleich zweier Rechenmodelle anhand experimenteller Ergebnisse. Periodica Polytechnika, Vol. 29 Nr. 2-4, 1985
/10/ Banzhaf, H.-U.: Anlagenspezifische Fragen bei Ventilatoren in Parallelschaltung, BWK 41 (1989), Nr. 1/2
/11/ Banzhaf, H.-U.: Anlagenspezifische Fragen bei Ventilatoren in Reihenschaltung, BWK 41 (1989), Nr. 3
/12/ Schulze Dieckhoff, B., Carolus, Th.: Betriebsverhalten lufttechnischer Anlagen mit parallelgeschalteten Axialventilatoren, FLT-Abschlußbericht zum Forschungsvorhaben AiF 9840, Frankfurt/Main, Oktober 1996

C557/114/99

Viscous flutter calculations for a fan assembly using hybrid grids

M VAHDATI, A I SAYMA, and **M IMREGUN**
Imperial College, London, UK

Abstract

This paper presents an integrated non-linear numerical model for aeroelasticity predictions of fan assemblies. The compressible turbulent flow, represented by the Reynolds-averaged Navier-Stokes equations together with a one-equation turbulence model, is coupled to a modal model of the structure. The flow domain is discretised using unstructured grids which can contain a mixture of tetrahedral, prismatic and hexahedral cells. The flow equations are discretised in space using a cell-vertex finite volume methodology represented by an edge based data structure. The equations are advanced in time using a point-implicit procedure. The aeroelastic coupling is via an exchange of boundary conditions at the interface between the fluid and the structure at every time step. The flutter analysis of a complete fan assembly with an intake duct was undertaken using such a numerical model. The results were found to be in good agreement with available experimental data in the sense that sharp reductions in stability were observed for very narrow speed ranges.

1 INTRODUCTION

The aeroelasticity analysis of turbomachinery blades requires the study of a fluid-structure system exposed to unsteady dynamic loading. The nature of the flow in turbomachines is complicated due to the co-existence of subsonic, supersonic and transonic regions in addition to shock waves and shock-boundary layer interactions. The situation is further complicated by the presence of acoustic waves which may cause acoustic resonances. The blade vibration and its interaction with the fluid flow add another dimension to the problem. Due to such factors, a realistic simulation of aeroelasticity phenomena requires a time-accurate viscous representation of the unsteady compressible flow and the inclusion of blade flexibility.

In recent years, the rapid development of numerical methods for the solution of the flow equations and the availability of powerful computers led to the emergence of various systems for the prediction of complex turbomachinery flows [1, 2]. Most of the prediction methods for turbomachinery flows use structured grids [3, 4]. On the other hand, unstructured grids received a great deal of research and development for external compressible flows [5, 6]. It is only in recent years that unstructured tetrahedral grids found their way into turbomachinery applications[7].

A literature survey on flutter prediction methods is beyond the scope of this paper and the interested reader should consult [8]. However, a brief overview will be given here for the sake of completeness. Most flutter computations consider a typical sector vibrating in some given assembly mode (or interblade phase angle) for which flutter is expected to occur. In other words, the flutter mode must be known before the analysis, though it is also possible to consider the individual stability of each mode in turn. In such, usually linear, analyses, the interblade phase angle must be prescribed at the periodic boundaries. Although such an approach has relatively modest CPU and in-core memory requirements, it also has the following serious shortcomings: (i) the flutter frequency of the aeroelastic system must be assumed to be identical to the natural frequency of the structural system, (ii) no interaction can take place between the various assembly modes, (iii) the approach is valid only when the aeroelastic system has reached a periodic steady-state condition. In most cases, extra computations are needed to impose the required interblade phase angle pattern over several cycles, and (iv) if the flutter mode is not known in advance, all possible interblade phase angles need to be tried as separate computations. In any case, if the geometric and acoustic properties of asymmetric intake ducts are to be modelled, the only possibility is a full assembly analysis.

Another consideration is whether the aerodynamic analysis needs to be conducted in a linear or non-linear fashion, since the structural model is almost always linear. For investigative studies, one of the essential features of fan flutter is the limit cycle behaviour which cannot be predicted with a linear aerodynamic model. Also, a linear model has some of the disadvantages above, such as the requirement for a-priori knowledge of the flutter frequency and of the critical interblade phase angle. Therefore, a non-linear aerodynamic model was adopted in this study.

Since the geometry of the fan assembly is complex, the natural way to go for spatial discretization is to use unstructured grids. The most widely available mesh generators fill volumes with tetrahedral cells. Although relatively easy to generate, they do not provide the most efficient discretization in terms of computer time. Furthermore, boundary layers are more accurately and efficiently represented by structured, body-fitted hexahedral grids. Due to the nature of the application, where intake ducts or bypass ducts need to be modelled together with the blade-rows, it is possible to use hybrid meshes which can contain a mixture of hexahedral and prismatic cells around the blades and tetrahedral cells elsewhere[9]. Together with an edge-based data structure, such an approach provides a good compromise between compactness and a flexible spatial discretization.

2 THE FLOW MODEL

The flow governing equations are cast in their conservation form in a Cartesian co-ordinate system which is fixed to a rotating blade:

$$\frac{d}{dt}\int_{\Omega}\mathbf{U}\,d\Omega + \oint_{\partial\Omega}\mathbf{F}.\mathbf{n}\,d\Gamma + \oint_{\partial\Omega}\mathbf{G}.\mathbf{n}\,d\Gamma = \int_{\Omega}\mathbf{I}\,d\Omega \tag{1}$$

where Ω refers to any piecewise smoothly-bounded fixed volume with boundary Γ, $\mathbf{U}$ is the vector of conservative variables, $\mathbf{F}$ and $\mathbf{G}$ are the inviscid and viscous flux vectors respectively, and $\mathbf{I}$ contains the terms due to the rotation of the co-ordinate system. These terms can explicitly be written as:

$$\mathbf{U} = \begin{bmatrix} \rho \\ \rho u_i \\ \rho\epsilon \end{bmatrix}; \quad \mathbf{F} = \begin{bmatrix} \rho(u_j - w_j) \\ \rho u_i(u_j - w_j) + p\delta_{ij} \\ \rho E(u_j - w_j) + Pu_j \end{bmatrix}; \quad \mathbf{G} = \begin{bmatrix} 0 \\ \sigma_{ij} \\ u_k\sigma_{ik} + \kappa\dfrac{\partial T}{\partial x_i} \end{bmatrix} \tag{2}$$

where ρ is the density, ϵ is the specific total energy, u_i is the absolute velocity vector, w_i is the grid velocity relative to the rotating frame of reference, κ is the summation of the molecular and eddy diffusivity. The viscous stress tensor σ_{ij} is given by:

$$\sigma_{ij} = \mu\left(\frac{\partial u_i}{\partial x_j} + \frac{\partial u_j}{\partial x_i}\right) - \frac{2}{3}\delta_{ij}\mu\frac{\partial u_k}{\partial x_k} \tag{3}$$

where μ is the summation of the molecular and eddy viscosities.

The term $\mathbf{I}$ of equation (1) is given by:

$$\mathbf{I} = \begin{bmatrix} 0 \\ 0 \\ \rho\omega u_2 \\ \rho\omega u_3 \\ 0 \end{bmatrix} \tag{4}$$

where ω is the rotational speed. The above system of equations is complemented by the perfect gas equation :

$$p = (\gamma - 1)\rho\left(\epsilon - \frac{u^2}{2}\right) \tag{5}$$

where γ denotes the ratio of the fluid specific heats.

The Eddy viscosity is calculated using the one-equation turbulence model of Baldwin and Barth [10].

2.1 NUMERICAL METHODOLOGY

The three dimensional spatial domain is discretised using unstructured grids which, in principle, can contain cells with any number of boundary faces. The solution vector is stored at the vertices of the cells. Furthermore, the use of mixed cells allows more flexibility for both mesh generation and solution accuracy.

For clarity, the numerical discretization of the flow equations is illustrated on a 2D mesh. However, the resulting formulation is equally applicable to 3D cells. Using an edge-based scheme the typical 2D mesh of figure 1 can be discretised by connecting the median dual of the cells surrounding an internal node. For an internal node, (1) can then be discretised as:

$$\Omega_j \frac{d\mathbf{U_i}}{dt} + \sum_i \frac{1}{2} \left(\mathbf{F}_i^h - \mathbf{F}_0^h \right) w_{0i} - \sum_i \frac{1}{2} \left(\mathbf{G}_i^h - \mathbf{G}_0^h \right) w_{0i} = \Omega_j I_j \tag{6}$$

where Ω_j is the area of the control volume which is represented by the shaded area in fig 1. The side weight is given by the summation of the two dual median lengths around the side times their normals. For example, the weight of the side connecting nodes 0 and 1 is given by:

$$w_{0i} = -w_{i0} = \vec{AB} + \vec{BC} \tag{7}$$

In 3D, the resulting weights are the summation of the areas times their normal over the cell faces resulting from connecting the centroids of the cells and the middle points of the sides. The four sided areas resulting in hexahedral and prismatic cells are calculated by dividing them to triangular faces in a consistent manner for the neighbouring cells such that conservation is assured.

The resulting numerical scheme is second-order accurate in space for tetrahedral meshes. For prismatic and hexahedral cells, the scheme is still second order accurate for regular cells with right angles. In the worst case of a highly skewed cell, the scheme will reduce to first order accuracy. However, hexahedral meshes are usually generated in boundary layer where orthogonality results in regular cells. Similarly, prismatic cells are usually generated in a structured manner by projecting triangular meshes on radial layers and then connecting them. Highly skewed meshes are unlikely to occur in these situations.

The resulting central difference scheme for the advection terms is stabilized using a mixture of second and fourth order matrix artificial dissipation with a pressure switch which guarantees that the scheme is TVD and reverts to first order Roe scheme in the vicinity of discontinuities [10].

There are mainly four different types of boundary conditions: inlet, outlet, solid walls and periodicity. At the solid walls, the viscous computations are performed in conjunction with the wall function, thus a slip velocity boundary condition is applied, with the velocity normal to the wall set to zero in a weak sense. For the single passage steady calculations, the points at the periodic boundary are treated as mirror image of their corresponding points at the other boundary, so that periodicity is enforced by considering such points as internal to the solution domain. Concerning inflow and outflow boundary conditions, different treatments are used for steady and unsteady calculations. For steady flows, the pseudo 3D non-reflecting boundary conditions of [11] are applied. These are obtained from the linear, harmonic unsteady non-reflecting boundary conditions by considering the limit case of zero-frequency unsteadiness and they are exact for a linear solution at the far-field boundary. The formulation of suitable boundary conditions for non-linear unsteady aerodynamics is somehow more difficult because of the extra degree of freedom arising from the unsteady variations and the zero-frequency assumption is clearly no longer valid in this case. For this reason, the standard 1D unsteady boundary conditions [12] were

used in this work. Such a treatment guarantees good accuracy if the unsteady waves are
mainly in the axial direction (normal to the boundary) and/or if they are relatively small
at the boundary. In the present work, the far-stream inflow boundary extends to about 10
times the intake duct length. However, it is not realistic to extend the outflow boundary
to a similar length and this is fixed at the leading edge of the splitter.

The semi-discrete system of (1) is advanced in time using an implicit relaxation algorithm, the details of which can be found in reference [7].

3   THE STRUCTURAL MODEL

The structural model is based on a linear modal model, the mode shapes and natural
frequencies being obtained via standard FE analysis techniques. The mode shapes are
interpolated from the structural mesh onto the aerodynamic mesh as the two discretization
levels are unlikely to be coincident.

The structural part of the aeroelastic equations of motion are uncoupled by using the
mode shape matrix ϕ.

$$\frac{d^2\eta_r}{dt^2} + \omega_r^2\eta_r = \sum_{i=1}^{N} \phi_{i,r}.F_i \tag{8}$$

where r is the mode index, i is the node number, η is the modal deflection, N is the
number of surface nodes and ω is the natural frequency. F is the aerodynamic force
which is obtained using the relation: $F_i = p_i\delta A_i\mathbf{n}_i$, where p denotes the static pressure, δA
denotes the elemental area and $\mathbf{n}$ denotes the unit normal vector, all at node i. Equation
8 is advanced in time using the Newmark-β method. Boundary conditions from the
structural and aerodynamic domains are exchanged at each time step and the aeroelastic
mesh is moved to follow the structural motion using a spring analogy algorithm.

4   NUMERICAL RESULTS

The above described system was used for the flutter analysis of a full fan assembly with
an intake duct. The fan assembly has 26 blades and a chord to span ratio of 0.35. The
mesh was generated in a way to exploit the advantages of the present mixed-cell, semi-
structured scheme. The single passage mesh, shown in fig 2, consists of three types of cells.
Hexahedral cells are generated in the boundary layer and at the inner part of the intake
duct; triangular prisms are used to fill the inside of the blade passage; and tetrahedral
cells are used in the outer part surrounding the intake duct. The outer part of the mesh
surrounding the duct is taken to about 10 times the duct radius in a hemispherical shape.
The total number of points in the blade passage is about 120,000. The intake duct in this
case was symmetric, although a non-symmetric duct can be handled in a straightforward
manner. The interface between the fan domain, which is treated in a rotating frame of
reference, and the outer domain, which treated in a stationary frame of reference, was
represented as a sliding boundary and flow information were exchanged every time step
between the two domains. The grid used is relatively coarse for viscous computations.
However, it was necessary to use the coarsest possible mesh to enable the full assembly
unsteady computations to be performed in a relatively short time. The steady solutions
obtained on such a grid were in acceptable agreement with those obtained on finer grids.

The inflow boundary conditions applied at the far-stream were total pressure, total temperature and flow angle, corresponding to stagnation conditions at sea level. A radial static pressure distribution was applied at the outflow. The distribution was obtained from through-flow calculations at the relevant operating conditions, each speed considered being on an elevated operating line close to the expected flutter boundary.

A steady-state flow solution, to be used as a starting condition for the unsteady computations, was obtained for a single blade passage mesh. The 60- 80% speed range was covered in steps of 2% along an elevated working line. The Mach number contours at the tip section for 60%, 70% and 80% speeds are shown in fig 3. Figure 4 shows the pressure profile at the suction side of the blade for the same three speeds while fig 5 shows the Mach number contours at the periodic boundary for the 80% speed. It should be noted that the left boundary in fig 4 is the at the sliding plane between the rotating fan domain mesh and the stationary intake domain mesh. Figure 6 shows the pressure profile around the blade at the tip section for the range of speeds analysed. From the steady-state results, it can be noticed that, as the rotational speed goes down, the suction-side shock moves further upstream and the interaction of this shock with the leading edge expansion fan becomes stronger, a feature which results on the weakening of the shock away from the blade.

The computed mass flow rates against pressure ratios were plotted on the compressor map and good agreement with the measured quantities was obtained. The full assembly mesh, needed for the flutter calculations, was built by expanding the single sector mesh. Two configurations, one with the intake duct and one without, were considered. The full assembly viscous mesh, shown in fig 7, contains about 2.8 million points. different structural analyses were performed for all speeds of interest to obtain the mode shapes required for the computations. The mode shapes were then interpolated onto the aerodynamic mesh. All assembly modes corresponding to the first three blade vibration modes were included in the aeroelasticity analyses. The computations were started by exciting all modes that were included in the computation. Aeroelastic motion is started by giving each mode an initial displacement, velocity or acceleration. Given the non-linear nature of the model, it is theoretically possible that different initial conditions will yield different time histories. However, numerical experience shows that the amplitude of the limit cycle will remain the same with varying initial conditions, though the time history to reach limit cycle may be different for each case. Also, the initial disturbance must not lead to excessive mesh distortion or structural motion. In the present case, an initial velocity of $0.1m/(\sqrt{Kg}S)$ was applied to each mode.

The modal time histories were then monitored for about 20 cycles, with 100 time steps each. The in-core computer memory requirement for such a viscous computation with a 2.8 million point mesh is about 1.3 GBytes in single precision arithmetic. The computing time to obtain 20 vibration cycles is about 500 hours on a 633 MHz 21164 CPU DEC Alpha workstation. The case without the intake duct, which has about 2 million grid points, required an in-core memory of 0.9 GBytes and 350 CPU hours for an equivalent computation. The damping, which is a measure of aeroelastic stability, is estimated by calculating log-dec values from each modal time history. Two typical time histories, obtained with and without intake duct, are shown in figs 8 and 9.

The stability of the first flap mode for 2, 3 and 4 nodal diameters as a function of speed are shown in figs 10, 11 and 12 for with and without intake duct configurations. The least stable mode for each speed is shown in fig 13 for both configurations. Three main points can be observed from these figures. First, flutter stability becomes worse with the intake duct in general. Second, with and without intake duct configurations yield different least stable nodal diameter modes at different speeds. Third, there are several dips in the stability curve for each nodal diameter. This behaviour is believed to be due to intake duct acoustics, when the unsteady excitation matches the acoustic modes of the duct. For both configurations, the stability predictions were found to be in good agreement with experimental data. The sharp stability drops for very narrow speed ranges are routinely observed during rig tests of development engines.

5 CONCLUDING REMARKS

A hybrid semi-structured mesh was used for the non-linear aeroelasticity analysis of a full fan assembly with an intake duct. The edge-based data structure and the use of mixed cells provide a compact way for spatial discretization, which enables unsteady viscous computations of full fan assemblies and intake ducts with currently available computer power. It is believed that such an analysis is reported for the first time. It was observed that flutter stability exhibited sharp drops for very narrow speed ranges, the so-called *flutter bite*. The numerical findings were found to be in good agreement with experimental observations which showed very similar behaviour. A following paper will report similar computations for fans with non-symmetric intake ducts and will focus on the post-processing of the unsteady flow data for better physical understanding of the flutter phenomenon.

Acknowledgements

The authors would like to thank Rolls-Royce plc for both sponsoring this work and allowing it publication. They also particularly acknowledge many useful discussions with Drs J.G. Marshall, J.W. Chew and Messrs C. Freeman and D. Halliwell.

References

[1] O. O. Bendiksen(1986), Role of shocks in transonic/supersonic compressor rotor flutter, *AIAA Journal* **24**, 1179-1186.

[2] W. N. Dawes(1988), Development of a three-dimensional Navier-Stokes solver for application to all types of turbomachinery, ASME paper 88-GT-70.

[3] Rai M.M. and Madavan N.K, 1990 "Multi-Airfoil Navier Stokes Simulation of Turbine Rotor-Stator Interaction" *J. of Turbomachinery* , **112**, 377-385.

[4] Giles M. B., 1987 "Calculation of Unsteady Wake/Rotor Interactions" AIAA 87-006.

[5] A. Jameson, T. J. Baker and N. P. Weatherill(1986), Calculation of inviscid transonic flow over a complete aircraft, AIAA Paper 85-0103.

[6] M. Vahdati, K. Morgan and J. Peraire(1993), The computations of viscous compressible flows using an upwind algorithm and unstructured meshes, *Computational Non-linear Mechanics in Aerospace engineering*, AIAA Progress in Aeronautics and Astronautics series.

[7] M. Vahdati and M. Imregun(1995), Non-linear aeroelasticity analyses using unstructured dynamic meshes, in Y. Tanida and M. Namba (eds.), *Unsteady Aerodynamics and Aeroelasticity of turbomachines*, Elsevier Publishers, pp. 177-190.

[8] J. G. Marshall & M. Imregun(1996), A survey of aeroelasticity methods with emphasis on turbomachinery, *J. of Fluids and Structures*, **10**, 237-267.

[9] L.Sbardella, A.I. Sayma and M. Imregun (1997), Semi-unstructured Meshes for Turbomachinery Blades, in Unsteady Aerodynamics and Aeroelasticity of Turbomachines, T. Fransson (Editor), Kluwer Academic Publishers.

[10] B. S. Baldwin and . J. Barth(1991), A one equation turbulence transport model for high Reynolds number wall-bounded flows, *AIAA Paper* 91-0610.

[11] Saxer, A., P. and Giles, M. B. Quasi-Three-Dimensional Non-reflecting Boundary Conditions for Euler Equations Calculations. *AIAA journal of propulsion and power*, 9:263-271, 1993.

[12] Thompson, K. W. Time Dependent Boundary Conditions for Hyperbolic Systems, II. *Journal of Computational Physics*, 89:439-461, 1990.

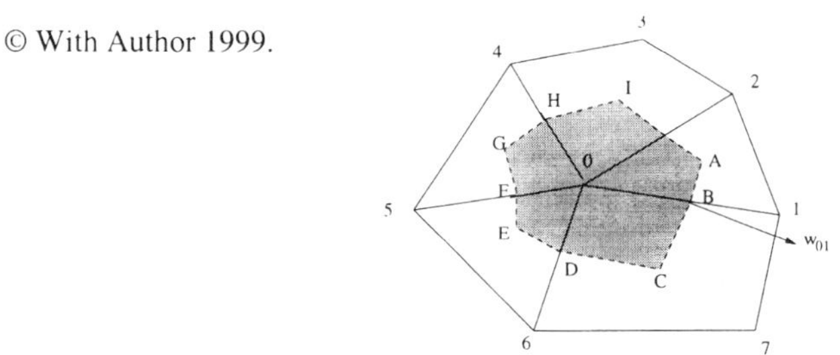

Figure 1: **Typical 2D mixed-cell mesh**

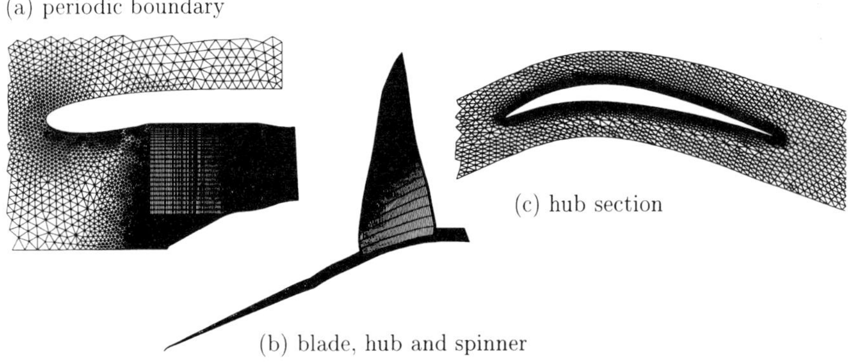

Figure 2: **Mesh for a sector containing one blade passage**

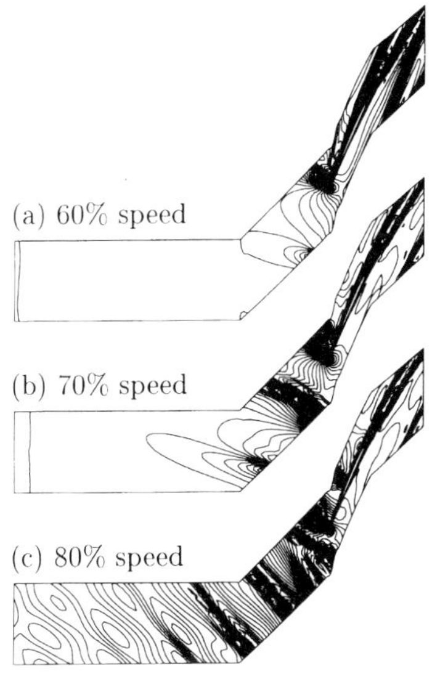

Figure 3: **Steady-State Mach number contours at tip section**

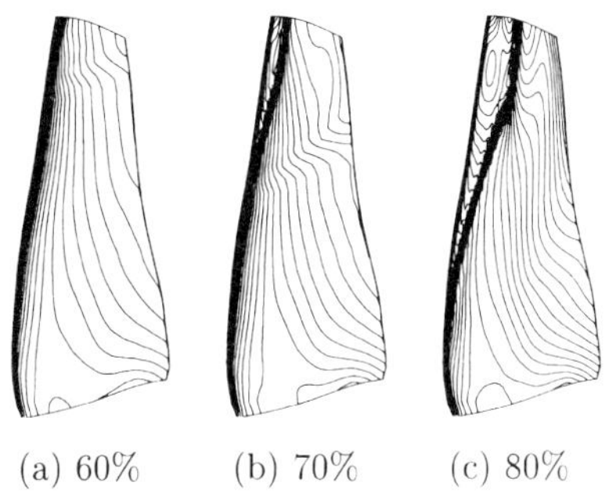

Figure 4: **Steady-State pressure contours at the suction side**

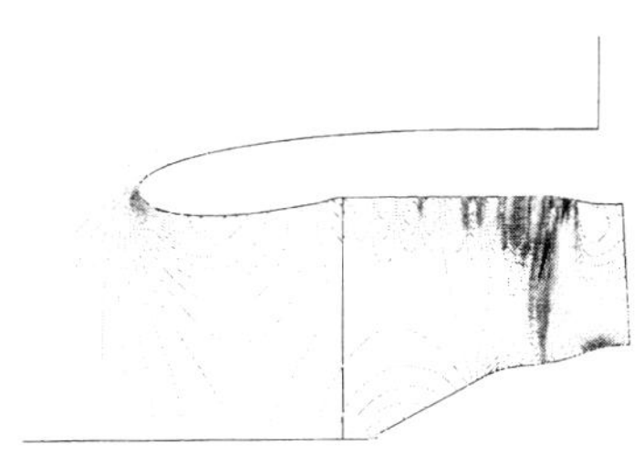

Figure 5: **Steady pressure contours at periodic boundary, 8% speed**

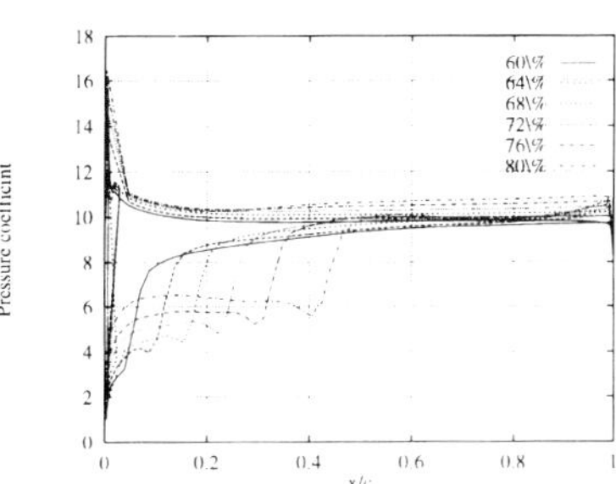

Figure 6: **Pressure profile at blades tip as a function of speed**

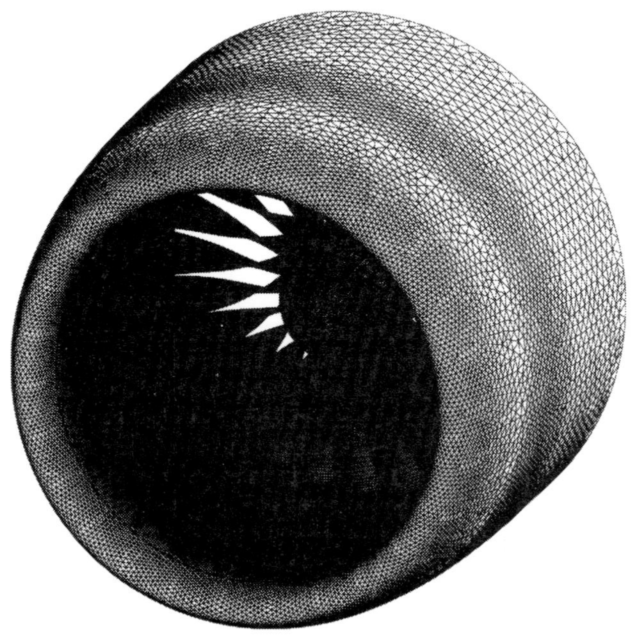

Figure 7: **A view of the full fan assembly mesh with the intake duct**

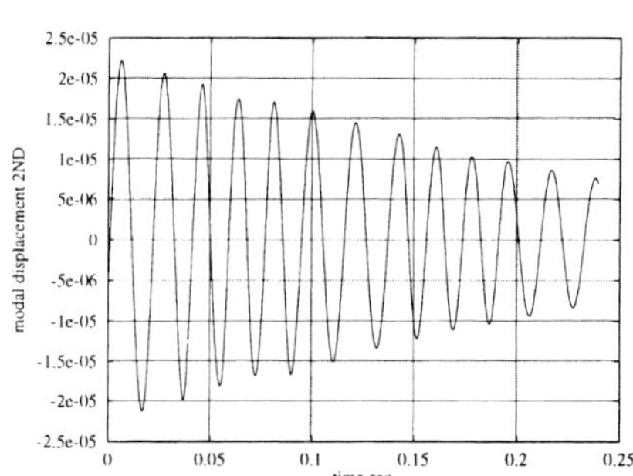

Figure 8: **Modal displacement time history for 2ND mode at 62% speed, no intake**

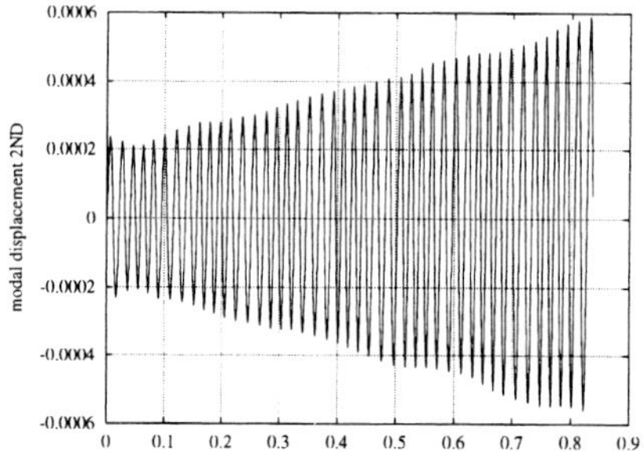

Figure 9: **Modal displacement time history for 2ND mode at 62% speed, with intake**

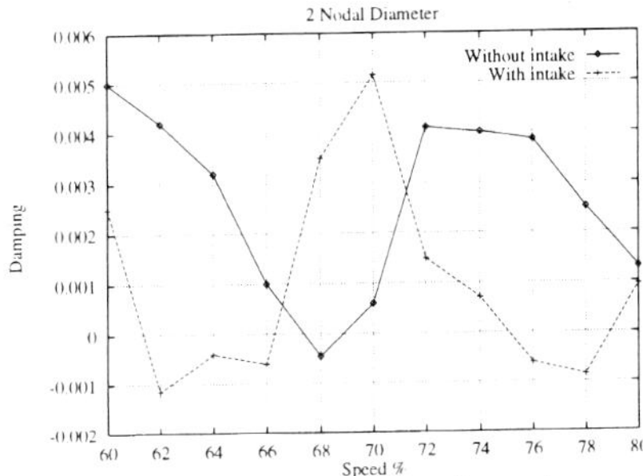

Figure 10: **Damping as a function of speed for 2ND mode**

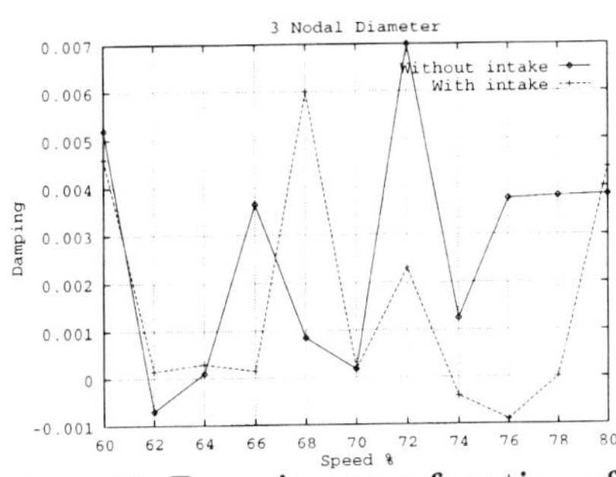

Figure 11: **Damping as a function of speed for 3ND mode**

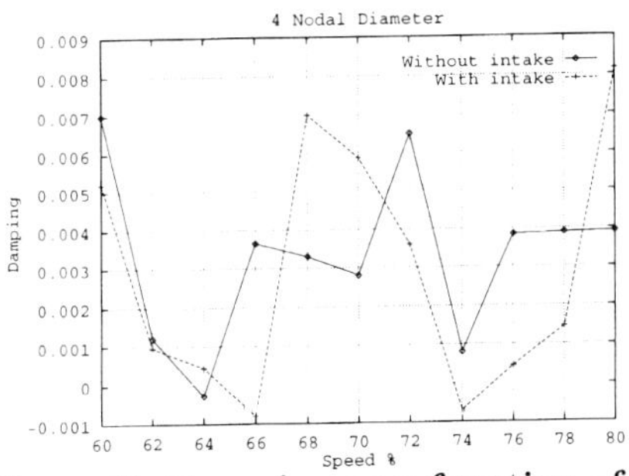

Figure 12: **Damping as a function of speed for 3ND mode**

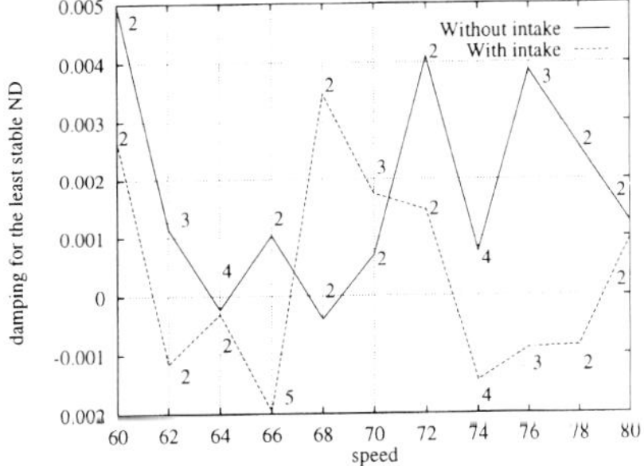

Figure 13: **Damping as a function of the least stable nodal diameter**

C557/116/99

Flutter stability analysis of a bird-damaged fan assembly

E FERRARI, M VAHDATI, and **M IMREGUN**
Imperial College, London, UK

Bird strike is a major consideration when designing fans for civil aero engines. In addition to bird impact tests and structural optimisation, it is now possible to investigate the aeroelastic behaviour of fan assemblies after actual or predicted damage due to ingested birds. This paper reports such a study which is believed to be the first of its kind. It is found that such investigations require the use of whole assembly models, as the cyclic symmetry is lost by one or more blades undergoing plastic deformation under the effect of the bird impact. Here, it was assumed that two consecutive blades had suffered unequal amount of bird damage, the so-called heavy- and medium-damaged blades. A viscous steady state solution of the bird-damaged assembly was computed first. The results indicated the formation of a strong wake from the heavy-damaged blade onto the downstream medium-damaged blade. It was also found that the mass flow had reduced considerably due to the blockage effect of the damaged region. Viscous unsteady flow calculations with and without blade motion were performed for the whole assembly and the results indicated the possibility of rotating stall, due to flow separation behind the heavy-damaged blade. The findings are in good agreement with available experimental data as both predictions and observations indicated a torsional instability for the medium damaged blade.

1 INTRODUCTION

This paper presents a set of aeroelasticity analysis results for a civil aero engine fan assembly which contains two damaged blades. The "raison d'être" of such a study stems from the vulnerability of fan blades to foreign object impacts. The possibility of severe bird damage is a significant consideration in fan blade design as the engine is expected to remain functional for some time after the event. Such a requirement involves the assessment of its flutter stability, in addition to the study of other factors such as blade strength, tip clearance, etc. Given the practical difficulties of experimentally determining the flutter stability for a large number of possible damage configurations, a proven computational route becomes very attractive. The aim of this work is to assess the feasibility of such a route and to gain further physical insight into the flutter of bird-damaged fan assemblies. The blockage due to the damage usually causes a reduction of the overall mass flow and the efficiency drops. This leads to increased steady and unsteady forcing on some blades and reduced flutter stability, which might reduce the fatigue life.

Flutter is defined as an unstable and self-excited vibration of a body in an airstream and it results from a continuous interaction between the fluid and the structure, either or both of which may be non-linear in nature. In turbomachinery blade rows, the structure-to-fluid mass ratio tends to be high while the same parameter is much lower for wings. Thus, whereas the wing flutter usually occurs as a result of coupling between the (bending and torsional) modes, turbomachinery blade flutter tends to be a single mode phenomenon as the aerodynamic forces, which remain much smaller than the inertial and stiffness ones, cannot usually cause modal coupling. In the former case, the aeroelastic mode can be significantly different from the structural one, in both frequency and mode shape, while this is usually not the situation in the latter case, though there may still be discrepancies between the aeroelastic mode and the corresponding structural mode.

The development of aeroelasticity methods for specific turbomachinery applications started in the early 1950s but the scale of the problem is such that after some 50 years of focused and innovative research, there are still no established numerical methods that can routinely predict the aeroelasticity behaviour of representative geometries. Reviews of advances in the field of aeroelasticity have been made by Sisto [1], Collar [2], Fleeter [3], Garrick & Reed [4], Platzer [5], Kaza [6], Bendiksen [7, 8] and Marshall and Imregun [9]. The main topics of turbomachinery aeroelasticity are also discussed in some detail in the AGARD Manual [10] which is one of the most comprehensive contributions to the subject, though significant advances have been made after its publication. The particular area of stall flutter has been surveyed by Chi & Srinivasan [11].

One particular regime of flutter can occur when the fan operates at part-speed, a phenomenon which is sometimes called supersonic stall flutter [12]. In the main, this type of flutter is observed in 1 to 6 forward travelling nodal diameter assembly modes, the blades usually vibrating in their first flap mode. The worst condition, i.e. that closest to the working line, typically occurs between 60% to 80% speed range. This terminology is based upon the fact that it is observed close to stall conditions and it has often been assumed that flow separation is an essential part of the flutter mechanism. Recent results, reported by Chew et al. [13], indicate that this is not necessarily the case. The part-speed computations for a 26-bladed-fan assembly showed that inviscid mechanisms could also produce flutter and the findings are in good agreement with the experimental data obtained during engine tests. The results were obtained for unsteady inviscid flows computed on a relatively coarse whole-assembly mesh, an approach that allows a general treatment of the interblade phase angle. The viscous losses were approximated via a loss model that is described by Sayma et al. [14].

2 THE FLOW MODEL

A realistic simulation of the aeroelasticity phenomena requires a time accurate viscous representation of unsteady compressible flows. For this reason the Reynolds-Averaged Navier-Stokes equations and the one equation Baldwin-Barth turbulence model have been adopted here [15]. The governing equations are written in their conservation form in a three dimensional Cartesian co-ordinate system, which is fixed to the rotating fan and the x-axis coincides with the engine axis [16].

$$\frac{d}{dt}\int_\Omega \boldsymbol{U}\,d\Omega + \oint_{\partial\Omega} \boldsymbol{F}\cdot\boldsymbol{n}\,d\Gamma + \oint_{\partial\Omega} \boldsymbol{G}\cdot\boldsymbol{n}\,d\Gamma = \int_\Omega \boldsymbol{I}\,d\Omega \tag{1}$$

where the computational domain Ω is a fixed volume with boundary Γ, $\mathbf{U}$ is the vector of convervative variables, $\mathbf{F}$ and $\mathbf{G}$ are the inviscid and viscous fluxes respectively, and $\mathbf{I}$ contains the rotational terms.

These terms can be written explicitly as:

$$\boldsymbol{U} = \begin{bmatrix} \rho \\ \rho u_i \\ \rho\epsilon \end{bmatrix}; \boldsymbol{F} = \begin{bmatrix} \rho(u_i - w_i) \\ \rho u_i(u_i - w_i) + p\delta_{ij} \\ \rho E(u_i - w_i) + pu_j \end{bmatrix}; \boldsymbol{G} = \begin{bmatrix} 0 \\ \sigma_{ij} \\ u_k\sigma_{ik} + \kappa\frac{\partial T}{\partial x_i} \end{bmatrix}; \tag{2}$$

where ρ is the density, ϵ is the specific total energy, u_i is the absolute velocity vector, w_i is the contra variant velocity vector, κ is the summation of the molecular and eddy diffusivities, and $(i, j = 1, 3)$.

The viscous stress tensor σ_{ij} is given by:

$$\sigma_{ij} = \mu\left(\frac{\partial u_i}{\partial x_j} + \frac{\partial u_j}{\partial x_i}\right) - \frac{2}{3}\delta_{ij}\mu\frac{\partial u_k}{\partial x_k} \tag{3}$$

where the effective viscosity $\mu = \mu_l + \mu_t$ is the summation of the laminar and turbulent viscosities. The thermal conductivity has a similar expression:

$$\kappa = \frac{\mu_l c_p}{Pr_l} + \frac{\mu_t c_p}{Pr_t}$$

For air, $Pr_l = 0.72$ and $Pr_t \simeq 0.9$ approximately. The laminar viscosity is modelled accordingly to Sutherland's law while the turbulent viscosity μ_t is calculated using the one-equation turbulence model of Baldwin and Barth [17]. The choice of the turbulence model is dictated by several factors, including numerical robustness and previous experience with the particular model. In [18], it is stated that "...the Baldwin-Barth turbulence model shares a few of the failures as well as most of the successes of the mixing-length model. It shows improved predictive capability and has achieved closer agreement with measurements for a limited number of separated flows than is possible with algebraic models.." In the authors' experience, Baldwin-Barth model is well suited to the modelling of unsteady fan blade flows, though it would be optimistic to expect an accurate prediction of flow separation and re-attachment.

The rotational term $\boldsymbol{I}$ is given by:

$$\boldsymbol{I} = \begin{bmatrix} 0 \\ 0 \\ \rho\omega u_2 \\ \rho\omega u_3 \\ 0 \end{bmatrix}; \tag{4}$$

where ω is the rotational speed.

The static pressure p can be determined by the perfect-gas equation:

$$p = (\gamma - 1)\rho \left(\epsilon - \frac{u^2}{2} \right) \tag{5}$$

where $\gamma = c_p/c_v$ denotes the ratio of the fluid specific heats.

A velocity vector, obtained with a wall function approximation, and a zero-turbulence Reynolds number are imposed at the solid walls. Although such conditions are perhaps not directly applicable to separated flows with large structural displacements, their use is dictated by computational considerations. The boundary conditions are imposed in a weak form by solving unidimensional Riemann problems at the boundary surfaces. These are both conservative and independent of blade motion.

The flow domain is discretized using an unstructured grid which contains a mixture of tetrahedral, prismatic and hexahedral elements. The flow equations are discretized in space using a cell-vertex finite volume approach, and an edge-data structure is used to minimize both the memory requirements and the computational time. The equations are advanced in time using a point implicit procedure. The structural modes are computed using a finite element code and the mode shapes are interpolated onto the fluid mesh. The equations of structural motion are advanced at each time step according to the instantaneous pressure distribution acting on the fan blades, subsequently the flow field is recomputed accordingly to the new blades geometry. The flutter analysis of the bird damaged fan assembly was undertaken using such a numerical model.

3 THE AEROELASTICITY MODEL

The structural model is based on a linear modal model. Such an approach is valid provided the vibration amplitude remains within the bounds of linear behaviour. The mode shapes and natural frequencies are obtained via finite element analysis techniques. The global equations of motion will the have the following form in generalized coordinates:

$$\mathbf{M\ddot{q} + C\dot{q} + Kq = P}(t) \tag{6}$$

where $\mathbf{q}$ is the generalized coordinates vector, $\mathbf{M}$, $\mathbf{C}$ and $\mathbf{K}$ are respectively the mass, damping and stiffness matrices, and $\mathbf{P}(t)$ is the generalized loads vector.

A transformation into principal coordinates via $\mathbf{q} = \mathbf{\Phi}\,\eta$ yields:

$$\ddot{\eta}_i + (2\zeta_i\omega_i)\dot{\eta}_i + (\omega_i^2)\eta_i = \phi_i^T \mathbf{P}(t) = \pi_i(t) \qquad i = 1, N \tag{7}$$

where i is the mode index, η_i is the modal deflection, ϕ_i is the mode shape vector, ω_i is the natural frequency, ζ_i is the non-dimensional damping coefficient ($\zeta_i = 0$ for the rest of the analysis), N is the number of mode shapes retained in the analysis, and $\pi_i(t)$ is the modal projection of the aerodynamic load vector $\mathbf{P}(t)$ onto the mode ϕ_i. Having interpolated the mode shapes onto the aerodynamic mesh, the i-modal force can be obtained via the relationship:

$$\pi_i(t) = \sum_{j=1}^{nodes} \phi_{i,j}(\delta A_j \mathbf{n}_j) p_j(t)$$

where $p_j(t)$ is the static pressure, $\mathbf{n}_j$ is the inward unit normal vector and δA_j is the surface element area all considered at the surface boundary node j.

The structural motion is computed via modal summation over the modes which are of interest in flutter stability. It is important to note that the system of second-order coupled differential equations (6) has been transformed into a much smaller number of uncoupled equations (7). Each equation is advanced in time using the Newmark-β method [19].

Boundary conditions from the structural and aerodynamic domains are exchanged at each time step. The aerodynamic load vector $\mathbf{P}(t)$ is obtained at every time step from the flow field solution, including in such a way non-linear aerodynamic effects. Once the modal displacements have been calculated, the aeroedynamic mesh is moved to follow the structural motion using a spring analogy algorithm [20] and then a new flow solution can be computed.

4 TIME ACCURATE VISCOUS FLOW ANALYSIS

A time-accurate viscous flow analysis of the damaged fan assembly without blade motion will be presented here. The mesh used for this calculation contains approximately $1,400,000$ nodes and $2,000,000$ elements. The adoption of an unstructured grid with mixed elements has been a natural choice, because it can easily cope with the distorted geometry and it allows a local mesh refinement around the blade surface and near the damaged region. A front view of the mesh is shown in Fig. 1, in which the heavy-damaged (HD) blade is clearly visible. The medium-damaged (MD) blade is located just downstream the HD one. Mesh details at tip section of the HD blade are shown in Fig. 2. A semi-unstructured mesh generator was used to cope with the complex fan geometry [21].

To avoid spurious reflections which can occur at the mesh boundaries, the inlet domain was extended by approximately 2 chords. This approach also allows to deal with the reverse flow generated by the damaged blades. A fixed time step of 5.0×10^{-5} s was adopted in the calculation, so that the CFL number in the smallest computational cell does not exceed the value of 80. The "steady-state" solution required 3 weeks computing time on a 500 MHz 21164 CPU on a Dec Alpha workstation with 1 Gb memory. Even though the residuals of Fig. 2 decreased by three orders of magnitude, the solution did not converge to a steady-state value as it kept on changing smoothly, a feature which is a good indicator of the unsteady nature of the flow. The largest flowfield fluctuations are believed to have taken place in the 'wake region' past the HD blade.

The Mach number and density contours of the last computed solution near the tip section of the damaged blades are shown in Fig. 3. From these pictures it can be noted that there is no shock on the HD blade's suction surface, and that the flow is completely separated. A strong recirculating area, that will be referred as 'wake', is released from the leading edge of the HD blade and hits the medium-damaged blade. Important viscous effects take place in the wake and the flow appears to be strongly unsteady. Reverse flow can be observed from the velocity vectors in Fig. 4. Another consequence of the wake and of the flow blockage is the change in the shock's position on the suction surface of the blades, expecially on those closer to the damaged ones (see Fig. 4). Where the flow incidence is locally increased, the shock gets stronger and viceversa. The HD blade causes a considerable blockage of the flow field. At this stage of the computation the overall isoentropic efficiency has decreased by -1.4% and, compared to the undamaged fan assembly [22], the mass flow has dropped as much as 8%.

A key question is to determine the extent of the region where significant unsteady effects take place: is this area limited to the wake region past the HD blade or does it affect several passages of the fan assembly? The unsteady viscous flow analysis without blade motion shows very little time dependency in the domain away from the damaged blades, but this feature will be discussed later when describing the unsteady flow with blade motion.

5 STRUCTURAL MISTUNING

The mistuned whole-assembly mode shapes were computed using the finite element code SC03 [23]. With such a geometry, there is no cyclic symmetry and the disc flexibility has to be taken into account. A number of preliminary studies indicated that it was better to use mistuned assembly mode shapes rather than using an approximate representation based on single blade mode shapes. Considerable structural mistuning, caused by the bird strike, changes significantly the mode shapes patterns. The nodal lines have preferred orientations since they either tend to go through or to be orthogonal to the damaged blades. A frequency drop for the first modes of each family has also been detected. The shape of the most significant modes are shown in Fig. 5. It is important to observe that the assembly modes corresponding to the 2F exhibit a significant deviation from the standard nodal displacement pattern. The disc stiffness couples the individual blade modes and the resulting assembly modes are markedly different from the single blade modes. However, some modes resemble closely the single blade modes and can be interpreted as the 1T modes of the HD and MD blades respectively (modes 53 and 54 in Fig. 5). On the other hand, modes 55 and 56 can be considered as the orthogonal 1T modes of the rest of the blades.

6 AEROELASTICITY ANALYSIS

A time accurate unsteady viscous flow analysis of the damaged fan assembly with blade motion will be presented here. The same mesh and the same time step of Section (4) were used in the calculations which took another three weeks on the same DEC Alpha workstation.

The unbalanced pressure distribution on the blades (Fig. 4), due to severe aerodynamic mistuning, causes a large blade displacement at time = 0 when the mesh movement is switched on. Therefore, there was no need to apply an initial disturbance to the fan assembly.

The modal displacements for modes 53 and 54 are shown in Fig. 6 (left). These modes, as explained before, correspond approximately to the 1T mode shapes of the HD and MD blades (see Fig. 5). The modal displacements for modes 55 and 56 are shown in Fig. 6 (right). The maximum peak-to-peak displacements are in the order of millimetres. While the 1T modes of the HD blade and of the undamaged ones are stable, the 1T modal displacement of the MD blade shows a growing time history which indicates instability, and this is in good agreement with the experimental results.

The iso-contour plots of the negative axial velocity component at tip section are shown in Fig. 7. These plots have been taken after 0, 32, 160 and 320 time steps after the mesh movement was switched on. The corresponding real times are 0, 0.0016, 0.008 and 0.016 s, representing 0, 0.06, 0.29 and 0.58 rotations of the fan. The flow is truly and globally unsteady, and the main cause of unsteadiness is the strong wake released from the HD blade. From these plots it is also possible to see a rotating stall phenomenon. The stall cell, identified by recirculating and reversed flow, moves downsteam and the flow tends to re-attach after a short while. The phenomenon is closely related to the viscous nature of the flow. It is also affected by the blade vibration as the unsteady viscous effects become larger. The time-accurate unsteady viscous flow analysis using fixed blades only resulted in a somewhat different rotating stall event which had a lower frequency and a smaller area.

For stability reasons that are related to mesh movement and to the size of the small boundary cells, the time step is kept very small. Such a viscous unsteady calculation is computationally very expensive, especially if compared with other approaches like the inviscid+loss model [16]. Moreover the use of the full-assembly mode shapes increases the memory storage requirements as well as the computational time.

However, such a complete model is probably the only way to capture the physics of the problem and simplifications are likely to introduce significant errors. The vortex generated in the tip region of the suction surface and casing viscous effects interact with the boundary layer and affect the aeroelastic behaviour. Therefore, it is highly desirable to include such effects by having additional mesh layers at the blade's tip. However, no such attempt was made here because of restrictions on available computational resources.

7 CONCLUSIONS

The results for flow and flutter analyses on a damaged fan assembly have been presented. Unsteady viscous effects and mistuned mode shapes are found to be essential factors in order to model representatively this kind of aeroelastic phenomenon.

- A separation zone past the damaged blades and a decreased mass-flow have been predicted for the damaged configuration. In the separation region the flow is strongly unsteady and produces an unsteady wake excitation on the medium-damaged blade.

- The assembly mode corresponding to the 1T mode of the medium-damaged blade is unstable while the 1T assembly modes for the rest of the blades are stable. This is in good agreement with experimental results.

- The time accurate viscous analysis with blade motion has shown the appearance of a possible rotating stall, which is periodically generated from the damaged area. This global instability is likely to affect the whole fan assembly.

- A viscous flutter analysis with mesh movement is computationally very expensive. Not only a viscous mesh is much finer than its inviscid counterpart, but also the maximum allowable time step is dictated by the smallest cell. The next version of the current code will be running in parallel mode on distributed memory machines with an expected efficiency of 90%. Hence the run times are expected to reduce significantly on relatively inexpensive hardware such as multi-CPU PCs.

8 ACKNOWLEDGMENTS

The authors would like to thank Rolls-Royce plc for both sponsoring this work and allowing it publication. They also acknowledge many useful discussions with and significant contributions from their colleagues at Rolls-Royce: Messrs Chris Freeman, D. Halliwell and Drs. J.G. Marshall and J.W. Chew.

References

[1] F. Sisto. Stall flutter in cascades. *Journal of the Aeronautical Sciences*, 20:598–604, 1953.

[2] A. R. Collar. The first fifty years of aeroelasticity. *Aerospace 5*, 1978.

[3] S. Fleeter and W. H. Reed. Aeroelasticity research for turbomachine applications. *Journal of Aircraft*, 16:320–326, 1979.

[4] I. E. Garrick and W. H. Reed. Historical development of aircraft flutter. *Journal of Aircraft*, 18:897–912, 1981.

[5] M. F. Platzer. Transonic blade flutter: A survey of new developments. *Shock and Vibration Digest*, 14:3–8, 1982.

[6] K. R. V. Kaza. *Development of aeroelastic analysis methods for turborotors and propfans - including mistuning*, volume 1. NASA Lewis Research Centre, 1990. Lewis Structures Technology.

[7] O. O. Bendiksen. Recent developments in flutter suppression techniques for turbomachinery rotors. *Journal of Propulsion*, 4:164–171, 1988.

[8] O. O. Bendiksen. A new approach to computational aeroelasticity. In *31st Structural Dymnamics and Materials Conference*, Baltimore, MD, USA, 1991. AIAA Paper 91-0939.

[9] J.G. Marshall and M. Imregun. A survey of aeroelasticity methods with emphasis on turbomachinery applications. *Journal of Fluid and Structures*, 10:237–267, 1996.

[10] M. F. Platzer and F. O. Carta. *Aeroelasticity in axial-flow turbomachines*, volume 1 & 2. Agard Manual, 1987. AGARD AG-298.

[11] R. M. Chi and A. V. Srinivasan. Some recent advances in the understanding and prediction of turbomachine subsonic stall flutter. *ASME Journal of Engineering for Gas Turbines and Power*, 107:408–417, 1985.

[12] N. A. Cumsty. *Compressor Aerdynamics*. Longman Scientific & Technical, London, 1989.

[13] M. Vahdati J. W. Chew and M. Imregun. Part-speed flutter analysis of a wide-chord fan blade. In *8th International Symposium on Unsteady Aerodynamics and Aeroelasticity of Turbomachines*, Stockholm, September 1997.

[14] J. S. Green A. I. Sayma, M. Vahdati and M. Imregun. Whole assembly flutter analysis of a low pressure turbine blade. *The Aeronautical Journal*, to appear.

[15] C. Hirsh. *Numerical Computation of Internal and Externals Flows*, volume 2. John Wiley & Sons, 1990.

[16] M. Vahdati. Non-linear aeroelasticity analyses using unstructured dynamic meshes. In Y. Tanida and M. Namba, editor, *Unsteady Aerodynamics and Aeroelasticity of Turbomachines*, pages 177–190. Elsevier, 1995.

[17] B. Baldwin and T. Barth. A One-Equation Turbulence Transport Model for High Reynolds Number Wall-Bounded Flows. *AIAA paper*, 91-0610, 1991.

[18] David C. Wilcox. *Turbulence Modelling for CFD*. DCW Industries, Inc., 1993.

[19] J. G. Marshall. Mathematical formulation of partially-integrated non-linear methods of aeroelastic analysis. Technical Report VUTC/C/93001, IC Vibration UTC, 1993.

[20] J. Peraire K. Morgan and J. Peiró. Unstructured grid methods for compressible flow. Technical report, AGARD 787, 1993. Special Course on Unstructured Grid Methods for Advection Dominated Flows.

[21] A. I. Sayma L. Sbardella and M. Imregun. Semi-unstructured mesh generator for flow calculations in axial turbomachinery blading. In *8th International Symposium on Unsteady Aerodynamics and Aeroelasticity of Turbomachines*, Stockholm, September 1997.

[22] E. Ferrari. Flow and Flutter Calculations for the Trent 800 Bird-Damaged Fan. Technical Report VUTC/CB7/98008, IC Vibration UTC, 1998.

[23] T. M. Edmunds. SC03 Finite Element Methods. Technical Report DNS14442, Rolls-Royce plc, 1995.

Figure 1: 3D Mesh used for unsteady viscous flutter analysis

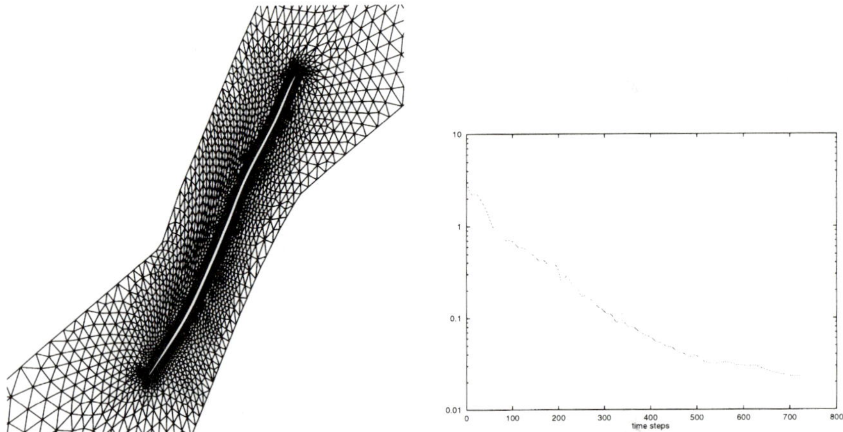

Figure 2: HD blade tip mesh (left) and residuals convergence (right)

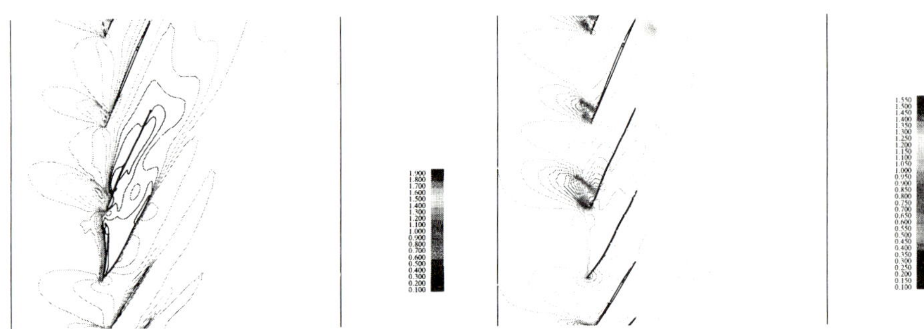

Figure 3: Mach numbers (left) and density contours (right) at tip section

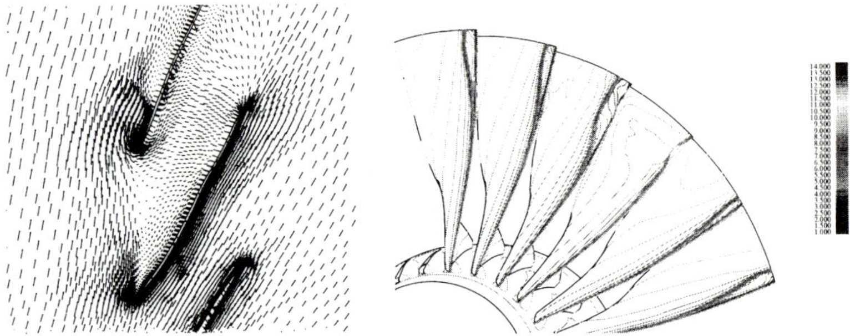

Figure 4: Velocity vectors (left) and front pressure distribution (right)

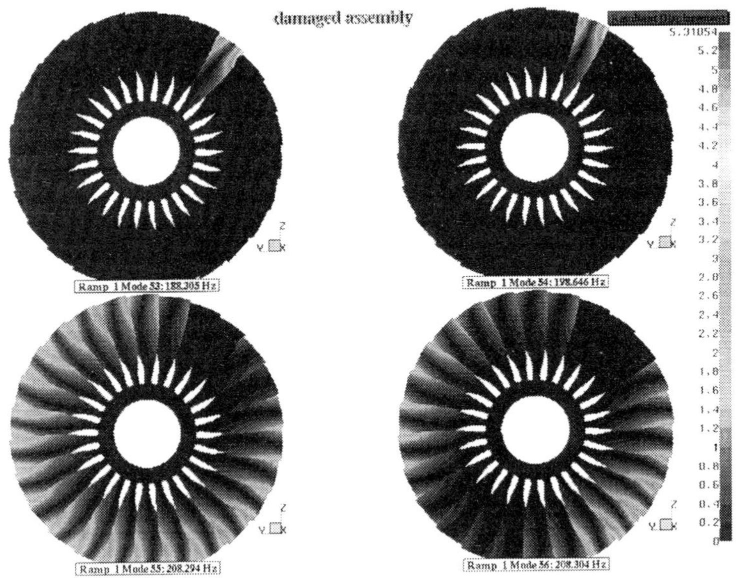

Figure 5: 53-56 Structural modes (1T)

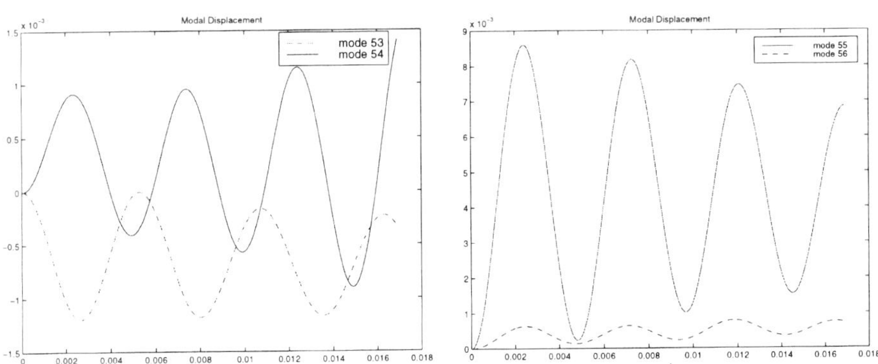

Figure 6: Modal displacements for modes 53 and 54 (left) and 55 and 56 (right)

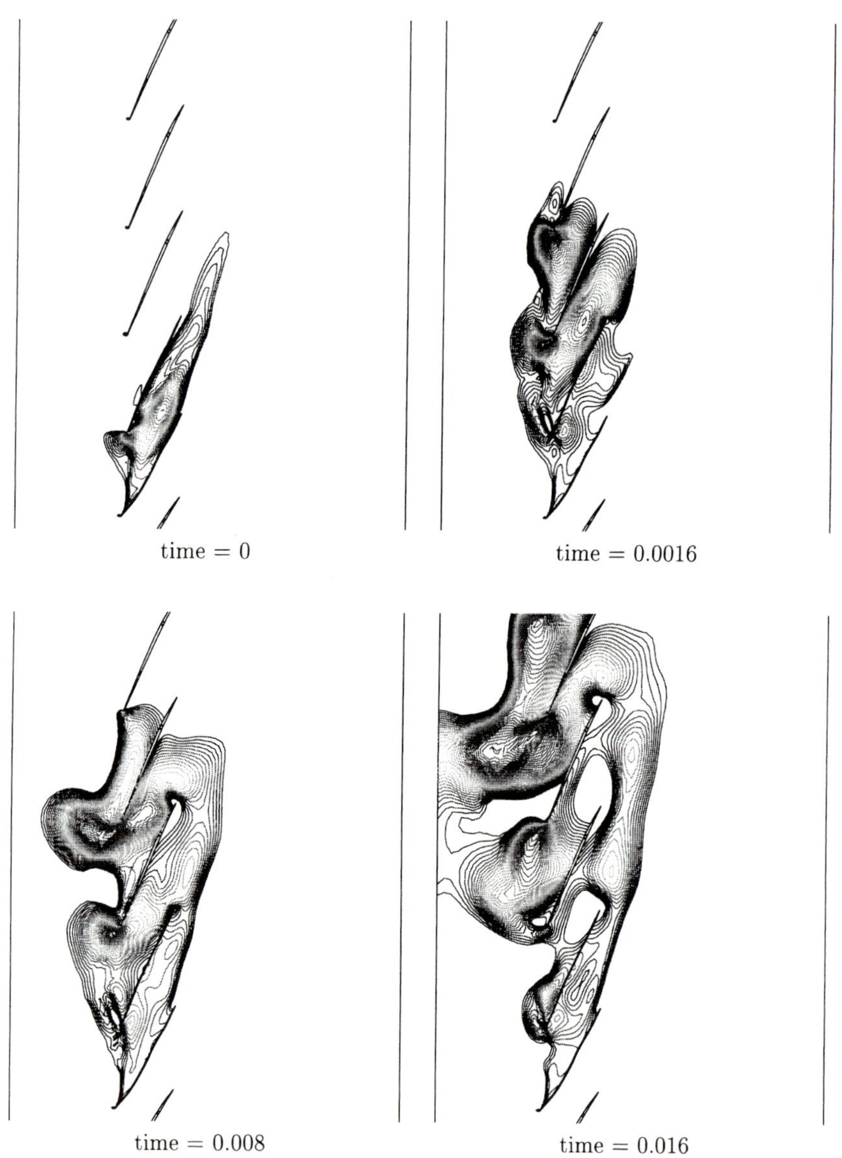

time = 0

time = 0.0016

time = 0.008

time = 0.016

Figure 7: Negative axial velocity component (time evolution)

Forced response prediction within the design process

J S GREEN and **J G MARSHALL**
Rolls-Royce plc, Derby, UK

SYNOPSIS

High cycle fatigue caused by vibration is a serious problem for turbomachinary blading. The traditional method for assessing vibration acceptability is to rely on past experience and engine strain gauge testing. This approach obviously has its problems, as unacceptable designs are only revealed late in the development programme.

This paper presents the application of a numerical method for the prediction of forced response in turbomachinary blade rows. The unsteady aerodynamic model uses the three-dimensional linearised Euler equations. Unsteady pressures are passed to a separate transient dynamics program to predict the blade resonant responses, including non-linear mechanical effects. Computational results are presented for a typical turbine stage and compared to in-engine strain gauge measurements of resonant frequency and amplitude.

The emphasis here is not the detail of sophisticated unsteady CFD tools, but instead the focus is on the application and key elements of a system which can influence component design at an early stage in the process. The method models an acceleration of the rotor using scaling and interpolation rather than a single fixed point. Good agreement was found with experimenal results in terms of amplitude and frequency shift.

1. NOMENCLATURE

ζ	critical damping ratio	k	damper element contact stiffness
Q	dynamic magnification factor	μ	friction coefficient

2. INTRODUCTION

Excessive vibration can cause a component to fail due to high cycle fatigue. In aero-engines this must be avoided because it could lead to sudden engine failure, risking the safety of the aircraft. Airworthiness regulations therefore require in-situ engine tests to prove the vibration integrity. Testing obviously takes place late in the development program and vibration problems which come to light at this stage are very costly.

The traditional approach for avoidance of vibration problems during the design process is to calculate the modes of vibration using a Finite Element approach and to plot these on a Campbell diagram. This shows the engine shaft speeds at which resonances occur but it is not clear whether vibrations will be acceptable. The designer relies on previous experience and may put unnecessary constraints on the design due to vibration concerns. A method for predicting the vibration response is required if engine development costs and time scales are to be reduced. Indeed it may also give the designer freedom to study unconventional airfoils which may improve efficiency, life or weight.

A forced response method must include models of both the aerodynamic and mechanical properties of the structure. Reviews by Verdon (1) and Marshall (2) show a large range of possible approaches, highlighting advantages and limitations. Fully coupled 3D non-linear viscous methods have been developed (3), but these models are computationally expensive. Linearised CFD methods are extremely attractive due to the speed advantage and there is a growing body of evidence that linear models can be used successfully for unsteady flow in turbomachinary bladerows (4) & (5).

The majority of the forced response approaches reported calculate the vibration response at a specific aerodynamic condition. However, damping devices such as under platform dampers can increase the resonant frequency by as much as 20% . The resonance will then occur at a different operating point i.e. the shaft speed and aerodynamic boundary condition will both change.

The purpose of the work reported here was to investigate the requirements of a system which could be used to predict the vibration characteristics of a turbine blade with an under platform damper. The method is required for use within the design process and must therefore be fast enough to have an impact on the design. One method for achieving this is presented and results are compared against experimental data.

3. METHOD OVERVIEW

The method adopted uses existing methods for the aerodynamic and mechanical parts of the calculation, but combines them in a new way. A 3D linearised Euler method developed by Giles (4), known as SLiQ was used for the unsteady aerodynamic parts of the calculation. The blade surface forcing function produced by SLiQ is applied to a reduced finite element model of the structure. The vibration response is then calculated using a transient dynamics method of Thomas & Gladwell (6) and includes both linear damping and a non-linear model of the under platform damper, developed by Sanliturk (7). The novel aspect is the forcing function, mechanical properties and damping all vary with time using a simple scaling approach so that an engine acceleration can be modelled. A flow chart shown in figure 1 indicates how each of these elements interact.

Aerodynamic Method

SLiQ performs the unsteady aerodynamic calculation in two stages. In stage 1 the program calculates the steady flow in a single passage using non-linear Euler equations with a standard cell vertex finite volume scheme on a structured hexahedral mesh. The solution is time marched using a 4-stage Runge-Kutta integrator, with standard second and fourth order smoothing.

In stage 2 a linearised unsteady flow is calculated assuming that it is a harmonic perturbation about the steady flow. The sources of unsteadiness that be analysed are: prescribed upstream wakes, up/downstream potential fields and prescribed blade motion. The unsteady perturbation equations are formulated in a similar form to the non-linear steady equations but at one particular frequency. They can, therefore, be solved by the standard numerical scheme used for the steady flow above using psuedo time-marching. The details of this procedure are recorded in (4).

In the approach presented here, two sources of unsteadiness are treated separately and superposed later in the analysis. Firstly the aerodynamic damping is calculated for each vibration mode in turn. This is done by interpolating the modeshape amplitude and phase from a cyclic finite element solution to the grid points at the solid boundary of the SLiQ mesh. Since SLiQ calculates the flow in a passage between two blades, a phase lag is applied between the solid boundaries assuming a rearward travelling wave with the given inter-blade phase angle. The flow perturbations are calculated in response to this prescribed motion, and the aerodynamic damping is then the work done on the fluid, integrated over the solid boundary for one vibration cycle. Thus a value of damping may be calculated for each mode.

The second calculation assumes a sinusoidal disturbance at the inflow boundary as the unsteadiness and again the perturbation to the steady flow is calculated. The upstream flow disturbance is caused by the blade rotating through a non-uniform flow field produced by the upstream nozzle guide vanes. The steady flow in the nozzle is calculated using a standard Navier-Stokes CFD code, from which both the potential and vortical components may be extracted. The perturbation at the inlet is calculated by a Fourier decomposition of the nozzle steady exit flow to give magnitude and phase of each harmonic. The unsteady flow for each harmonic must be calculated separately, but it is usually only the fundamental that is required since higher harmonics would be at a different frequency i.e. off resonance. The result of the calculation is a flow perturbation expressed as real and imaginary parts at a given frequency, from which the forcing on the blade surface is determined.

Mechanical Model

A standard Finite Element method is used to model the structure (8). The model contains a full description of the geometry, temperature, material properties etc. and is first used to calculate the undamped frequencies and modeshapes of the structure. When plotted on a

Campbell diagram this defines the shaft speed and hence boundary conditions for the CFD calculation.

The model is reduced using a method first proposed by Guyan (9) to significantly reduce the size of the mechanical model. The pressure perturbations caused by upstream flow disturbances are interpolated onto the structural model and converted into equivalent reduced forces for each of the reduced structural variables.

A proprietary transient dynamics program (SJ18) calculates the response of a structure to applied forces using the implicit time marching method of Thomas and Gladwell (6), and can include non-linear elements. Figure 1 shows that many inputs are required to SJ18: linear aerodynamic and mechanical damping, non-linear mechanical elements, forcing function and the definition of the model.

The linear mechanical damping represents the inherent damping in the structure such as material damping and damping due to friction between the blade root and the disk etc. Damping values can be measured in the laboratory or extrapolated from previous experience, but can differ widely for different components and different modes on the same component. The non-linear damping elements are based an under platform damper element developed by Sanliturk (7). The model requires geometric and surface contact parameters, such as friction coefficient and surface stiffness, which must be measured for the material combination in question, at representative operating conditions (temperature, roational speed etc.)

Transient Modelling Approach

The under platform dampers can cause a change in the resonant frequency, so to ensure that the resonant amplitude and frequency is predicted, the system is used to simulate an engine acceleration through the resonance. All input parameters vary with time: force frequency and amplitude, damper weight, material properties etc. to mimic a true engine acceleration as would be typical for an engine strain gauge test. The stresses can be recovered and used in a blade life assessment.

The most difficult parameters to scale are the unsteady pressures. In general, the flow around an aerofoil is highly dependant on the operating point and scaling should only be used with extreme care. The method allows the use of unsteady pressures at several operating points which may be used for interpolation.

In an aero-engine the turbine is designed to stay at a constant non-dimensional operating point due to the way it interacts with the compressor. The high pressure rotor in particular is stationed between two choked nozzles and therefore the pressure ratio remains constant and Mach number distribution changes very little between idle and maximum thrust. Figure 2 shows a graph of HP stage inlet pressure against shaft speed for a selection of ground (i.e. zero altitude) running conditions. The scatter is mainly due to variation in ambient atmospheric conditions and taking this into consideration a single best fit curve can be calculated. This curve is then used as the basis for scaling the unsteady pressures on the blade surface.

4. CASE STUDY

A High Pressure Turbine blade, typical of modern design, was chosen to demonstrate essential features of the approach and highlight implication to any forced response prediction method.

The HP stage consists of 40 nozzle guide vanes (NGVs) and 92 development standard blades. The rotor has transonic exit conditions and its aerodynamic behaviour changes very little across its useful working speed range. The rotor is shrouded (but not interlocked), and also has an under-platform damper which introduces non-linear friction damping. The resonances of interest are 1st Torsion (1T) and second flap (2F) vibration modes which are excited by wake passing and occur at low speeds. The results are presented with and without friction damping and the displacements are compared with gauge results from development engines.

Normal Vibration Modes

A 3D finite element model of a blade and disc sector was used, (See figure 3). Cyclic symmetry was assumed at the disc sector boundaries for vibration with 40 nodal diameters (ND). The first five modes were calculated and plotted on the Campbell diagram (figure 4). First torsion (1T) at 51.0% and second flap (2F) at 67.4% shaft speed were identified for further investigation

Aero Forcing

Based on the Campbell diagram above, an appropriate aerodynamic operating point was identified and a through flow calculation was performed at 69.2% speed to provide boundary conditions. The flow through the upstream nozzle guide vane was calculated using a 3D Navier-Stokes steady code and this provides the definition of the inlet disturbance for SLiQ. Figure 5 shows the flow distortion that is produced by the nozzle at the inlet plane to the SLiQ mesh. The nozzle guide vane is designed to produce a very curved wake so that the phase relationship between different radial positions tend to minimise vibration.

A single operating point was used because previous experience had shown that for a similar case, the error in scaling from an operating point at 52% speed to 84% was less than 10% in terms of response. This error is considerably smaller than the scatter in some of the other variables, such as the damping and friction parameters. A larger database of results could have been used, with interpolation, but it was not thought necessary for the purpose of this study.

The unsteady linearised part of SLiQ was used to determine the unsteady flow in the rotor due to the 1st harmonic of non-uniform flow in the nozzle. The incoming 'wake' was evaluated due to the combined potential and vortical field. The unsteady flow results may be shown in the form of raw real and imaginary parts - but these results can be hard to interpret so the results are plotted over a number of blade passages, animated in time. The unsteady results for 4 time points at mid- height are plotted in figure 6a. Here the static pressure is plotted - as this is what ultimately forces the structure. Figure 6b shows the steady static pressure and the envelope of unsteadiness around the aerofoil at mid height. It can be seen that there is a significant level of unsteadiness in the rotor row (equivalent to about 10% of the steady lift).

Aero Damping

The aerodynamic damping for the rotor has been calculated for the same steady flow condition. This was done by interpolating the vibration modes from the cyclic FE model of one of the blades onto the aerodynamic mesh. It was found that the aerodynamic damping in all modes is very small: a critical damping ratio, ζ, typically of 0.03%, which is equivalent to a dynamic magnification factor, Q, of 1500. These levels of damping are dwarfed by the mechanical 'inherent' damping which ranges from $\zeta=0.25\%$ to $\zeta=1.0\%$. These values of mechanical damping are based on laboratory measurements of the blade, held in a block whose firtree profile matches that of the blade firtree. This experimental set-up is not identical to the engine configuration, but experience has shown that they are in broad agreement with measurements obtained by engine running.

Model Reduction

The full FE model was reduced using Guyan reduction. The full model has approximately 50000 degrees of freedom and the reduced model has just 44, but even with this huge reduction the frequency error is less than 0.25% for the first 5 modes. Guyan master nodes were chosen to be in areas of mechanical non-linearity and additional degrees of freedom were included to improve frequency and modeshape match. The modeshape match was compared visually and showed no discernable difference.

The static pressure is mapped onto the structure and reduced to apply to the generalised variables. This model now consists of a mass and stiffness matrix with 44 degrees of freedom and 44 forces each with independent magnitude and phase.

Transient Dynamic Simulation

The under platform dampers were modelled as 3 elements on each side of the platform with the mass realistically distributed. Each damper applies forces to the structure based upon geometric properties, instantaneous shaft speed and the relative displacements at its points of application. The relationship between displacements and force are based on a macroslip model which required knowledge of the friction coefficient, μ, and surface stiffness, k, as seen in figure 7. The hysteresis loop shown can be considered as a stiffness part (linear relation between force and displacement) , which causes the rise in frequency; and a damping part (the area within the loop) which causes the reduction in amplitude. The damper elements are connected between two points one of which is under the blade platform, at the true point of application, and the other is connected to earth. In the real configuration the damper would be connected to the adjacent blade, but its amplitude and phase are unknown. Mistuning effects in these high nodal diameter modes are significant and it is likely that the adjacent blade has a slightly different resonant frequency and is therefore not at resonance at the same time. This situation is more akin to the earth attachment condition than the assuming the adjacent blade has the same amplitude as the blade of interest but at different phase.

A graph of stage inlet pressure was plotted for a variety of engine manoeuvres (see figure 2) and is then normalised to the operating point of the CFD calculations. This was then used as a factor to scale the reduced forces to other conditions. This is equivalent to scaling the unsteady pressure amplitude on the blade surface based on the inlet total pressure, whilst keeping the phase constant.

A rotor acceleration from 50% speed to 100% speed is assumed to occur linearly over a period of 4 seconds. The excitation frequency is therefore ramped accordingly to produce the 40 engine order excitation line and other parameters are scaled based on the instantaneous shaft speed.

5. RESPONSE RESULTS

An example of the damped results in the form of a 'z-shot' for the friction damped case is shown in figure 8. This form of post-processing looks rather complicated but it is basically a Campbell diagram and is very similar to the form used for engine strain gauge result interpretation. The central plot, labelled Envelope, shows the envelope of the entire time signal. This time signal is broken into small blocks and converted to the frequency domain using Fourier analysis. The main part of the figure (top-left) uses the results of the Fourier analysis and plots the amplitude at each frequency and time as a logarithmic grey scale. The other two views, labelled Peak-hold and Component are different views on the same data. The peak-hold graph shows the maximum amplitude along a horizontal line (constant frequency), and the component graph show the maximum for a vertical line (constant time).

The 40 engine order line can be clearly seen. The modes of vibration can also be seen as (almost) horizontal lines, and where these cross the 40 engine order excitation line a peak can be clearly seen in the Component and Peak-hold views.

The predicted amplitudes from the transient dynamic analysis are shown in table 1.

	Mode	Shroud Corner Amplitude [mm]	Speed [%]	Freq [Hz]
Undamped	1T	0.012	51.0%	3607.8
	2F	0.017	67.4%	4766.1
	2E	0.306	111.9%	7915.8

	Mode	Shroud Corner Amplitude [mm]	Speed [%]	Freq [Hz]
Damped	1T	0.010	52.1%	3686.0
	2F	0.008	78.3%	5536.3
	2E	0.033	115.7%	8182.0

Table 1: Results from forced response analysis.

Amplitude Correlation.

Strain gauge engines tests on Test_A and Test_B gave measurements of vibration amplitude for First Torsion(1T) & Second Flap (2F) modes. These 2 tests are nominally identical in all major respects. Data from both engine tests is compared against the prediction in table 2, and this data is shown graphically in figure 9.

		Shroud Amplitude [mm]	
Mode		1T/40EO	2F/40EO
Test_A			
	Max	0.0064	0.0184
	Mean	0.0056	0.0099
Test_B			
	Max	0.012	0.0185
	Mean	0.0099	0.0106
Predicted			
	undamped	0.012	0.017
	damped	0.010	0.008

Table 2: Measured Amplitudes compared to Predictions

When the damped prediction is compared with the mean measured response a good agreement can be seen . The ratio between the undamped and damped predictions shows the effectiveness of the damper for each mode. In this case the damper is much more effective for the second flap (2F) mode than for the first torsion mode (1T) and this agrees with previous experience from development engines. The scatter in the experimental results is not typical but it is certainly not unusual and highlights the difficulties of method validation.

The prediction for the undamped case appears to agree well with the measured maximum. It is possible that the worst blade in an engine set has the highest amplitude because the damper is not working effectively but this conclusion is flawed because it neglects the amplitude scatter caused by aerodynamic and mechanical mistuning.

Frequency Shift Correlation

The effect of the under-platform damper on frequency is also of key importance. The predicted frequency shift due to the damper is taken from the above tables and compared against experimental results.

Mode	Frequency Shift	
	Predicted	Experimental
1T	2.2%	0.13%
2F	16.2%	18.37%
2E	3.4%	no data

Prediction - undamped vs. damped resonant freq.
Experimental - damped resonant freq (in engine) vs cold/static (lab)

Table 3: Measured Amplitudes compared to Predictions

The frequency shift correlation is good for modes 1T & 2F, especially considering the slight difference in calculation method between prediction and experimental results (see footnote to table 3).

6. DISCUSSION

For the system to be used as a design tool it must predict vibration amplitudes equivalent to existing methods for vibration assessment i.e. an engine strain gauge test. This allows a valid comparison between the prediction and measurement methods and also allows the designer to use the same assessment criteria. It is therefore very important to match the resonant shaft speed to a particular flight point and use these parameters as the aerodynamic boundary conditions. The method can then be used to optimise damper mass, assess changes to the mechanical definition and to study the effect of NGV wake shaping.

The scaling approach used has limited range of validity. For this case the validity range is quite broad because the blade operates between choked nozzles and the pressure ratio and therefore flow distribution are fairly constant. However for other stages where this is not the case it is more appropriate to calculate a unsteady pressure at a selection of shaft speeds and interpolate.

In the design of a real component it is not sufficient to predict only the mean, because it is the first blade to fail within each assembly that is important. The first blade to fail is most likely to be the highest amplitude but it will also be affected by scatter in material fatigue strength. In this case the measured results give an indication of the amplitude scatter due to mistuning for each mode by comparing the maximum to mean response for each mode.

The response of a blade to an unsteady pressure field depends upon its modeshape. When the under-platform dampers are incorporated, the modeshape of the blade will change due to the additional stiffness at the blade platform. The approach described in this paper predicts the damped modeshape and therefore the change in modal force. The change in modeshape is also important to the blade life prediction due to the change in stress distribution.

7. CONCLUSIONS

This approach can be used within design time scales to assess vibration frequency and amplitudes on Turbine Blades. It has also been shown to give good agreement with experimental results.

Accurate prediction of the resonant frequency can be as important as the vibration amplitude, since a mode could be pushed into or out of the running range by the action of the damper.

Frequency changes can cause significant changes to aero boundary conditions and this must be taken into account in the prediction method.

All codes which predict forced response of systems with mechanical non-linearity need some form of boundary condition correction and some form of resonance tracking algorithm to correctly predict vibration amplitude and frequency. For a linear system this is not required because the frequency and amplitude are not inter-related, i.e. then can be calculated independantly.

This method relies upon scaling of the aerodynamic results. This is a reasonable but not perfect assumption for this case but if this method were to be used for a fan or compressor then more operating points would be required. The characteristic of fans and compressors are quite different to turbines and also depend on the operating line as well as speed.

8. ACKNOWLEDGEMENT

The authors would like to thank Rolls-Royce plc. for their support and allowing the publication of this paper and Prof. M.B. Giles for support with the SLiQ code.

9. REFERENCES

(1)   Verdon, J.M. 'Review of Unsteady Aerodynamic Methods for Turbomachinary Aeroelastic and Aeroacoustic Applications.', AIAA Journal, Vol 31 No 2. Feb 1993.

(2)   Marshal, J.M. & Imregun, M. 'A review of Aeroelasticity Methods with Emphasis on Turbomachinary Applications', Journal of Fluids and Structures (1996) 10, 237-267

(3)   Vahdati, M. and Imregun, M. 1994, Non-Linear Aeroelastic Analyses Using Unstructured Dynamic Meshes, Proc of the 7th ISUAAT, Fukuoka Armstrong, I. &

(4)   Marshall, J.G. & Giles, M.B., 1997, 'Some Applications of a Time-Linearized Euler Method to Flutter and Forced Response in Turbomachinery', 8th ISUAAT, Stockholm

(5)   Suddhoo, A., Giles, M.B. & Stow, P., 1991, 'Simulation of Inviscid Blade Row Interaction Using a Linear and a Non-Linear Method', ISABE Conference Giles, M.B. 1992, 'An Approach for Multi-Stage Calculations Incorporating Unsteadiness', ASME Paper 92-GT-282

(6)   Thomas R.M. & Gladwell, I., 1988, Variable-Order Variable-Step Algorithms for Second-Order Systems Parts I&II, International Journal for Numerical Methods in Engineering 26 39-80

(7)   Sanliturk, K.Y., Imregun, M. & Ewins, D.J., Harmonic Balance Vibration Analysis of Turbine Blades with Friction Dampers. ASME J. Vibration and Acoustics, Vol 119, 1997, pp 96-103.

(8)   Edmunds. T.M., 1989, 'Fully Automatic Analysis in an Industrial Environment', 2nd NAFEMS International Conference, Stratford

(9)   Guyan, R. J. 1965, Reduction of Mass and Stiffness Matrices, AIAA Journal 3(2) 380

                                                          C557/067/99

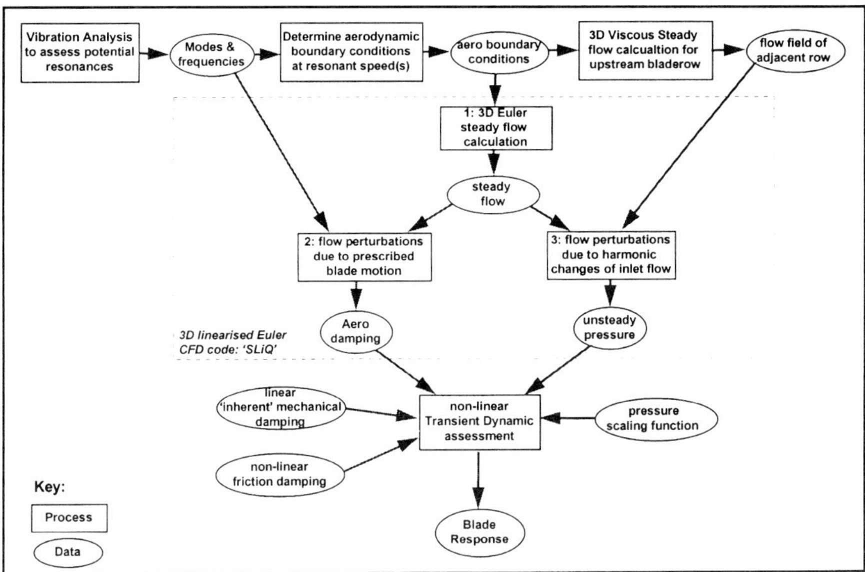

Figure 1 - Process Flow Chart

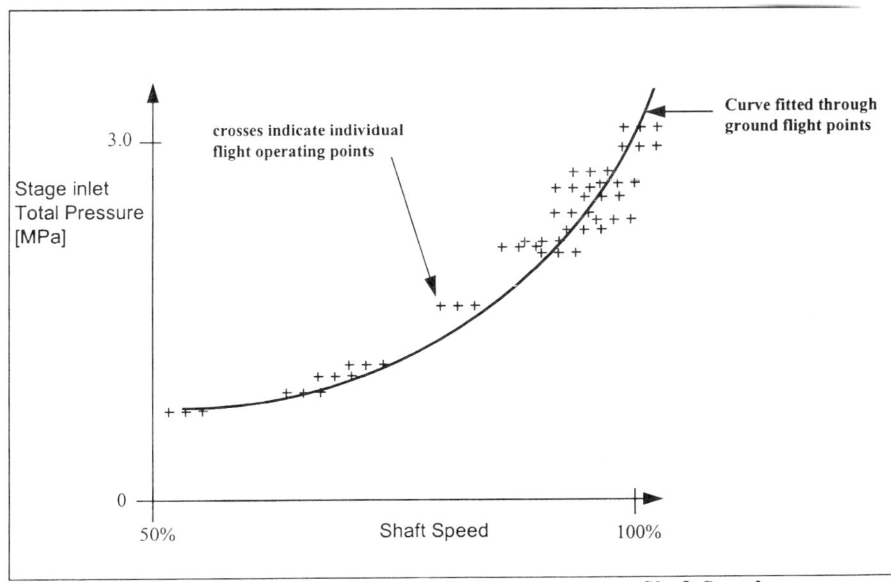

Figure 2- Variation of Stage Inlet Pressure versus Shaft Speed

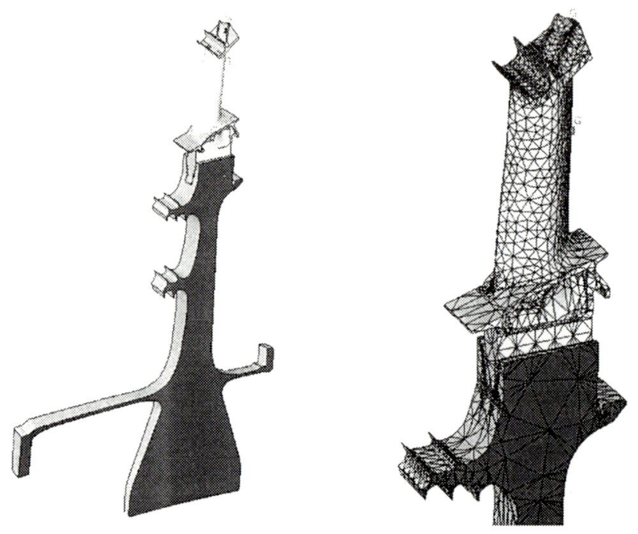

Figure 3 - Structural Model

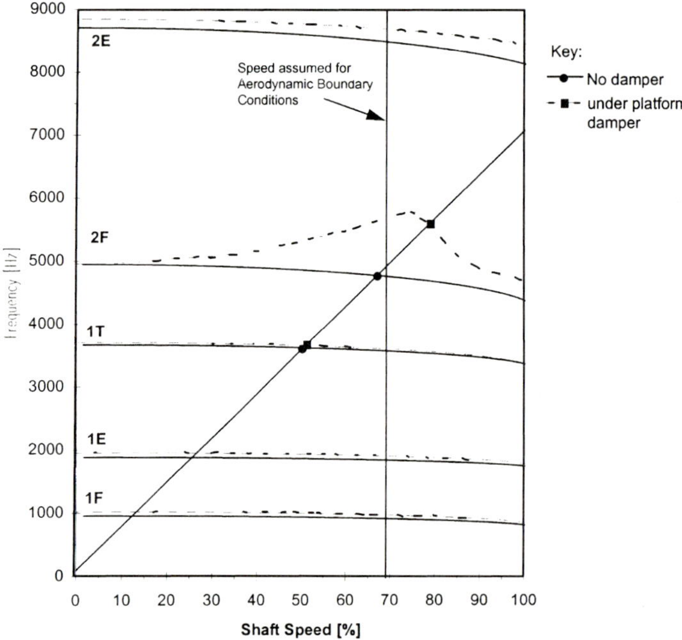

Figure 4 - Campbell Diagram

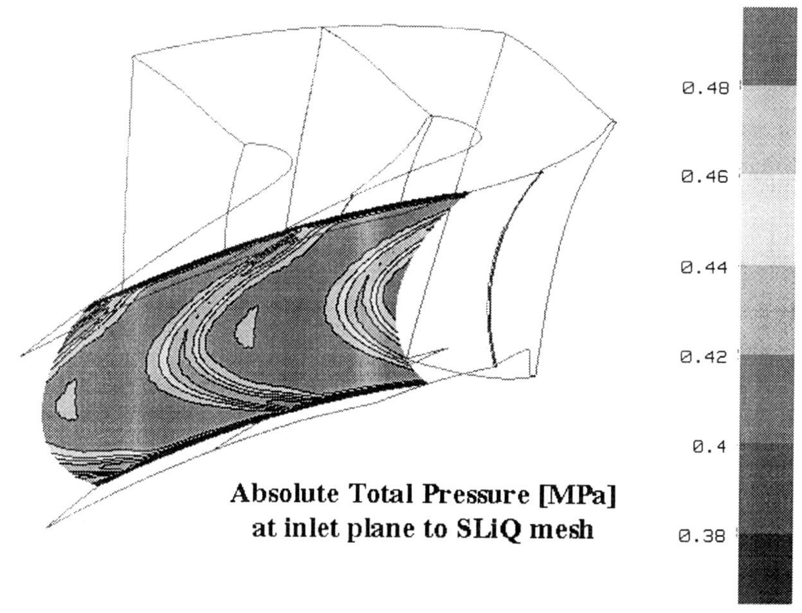

Figure 5 - Flow Distortion Caused by Nozzle Guide Vane at SLiQ Inlet Boundary

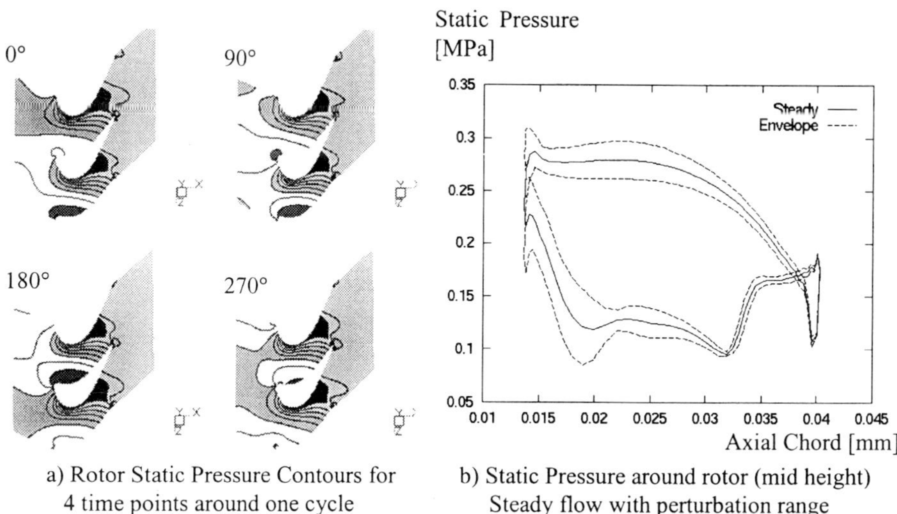

a) Rotor Static Pressure Contours for 4 time points around one cycle

b) Static Pressure around rotor (mid height) Steady flow with perturbation range

Figure 6 - Rotor Static Pressure at Mid-height Section using SLiQ

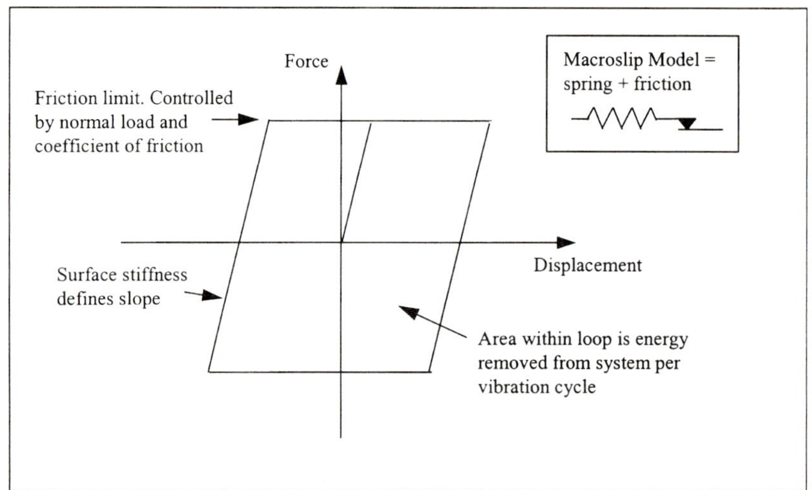

Figure 7 - Unsteady Pressure on Structure

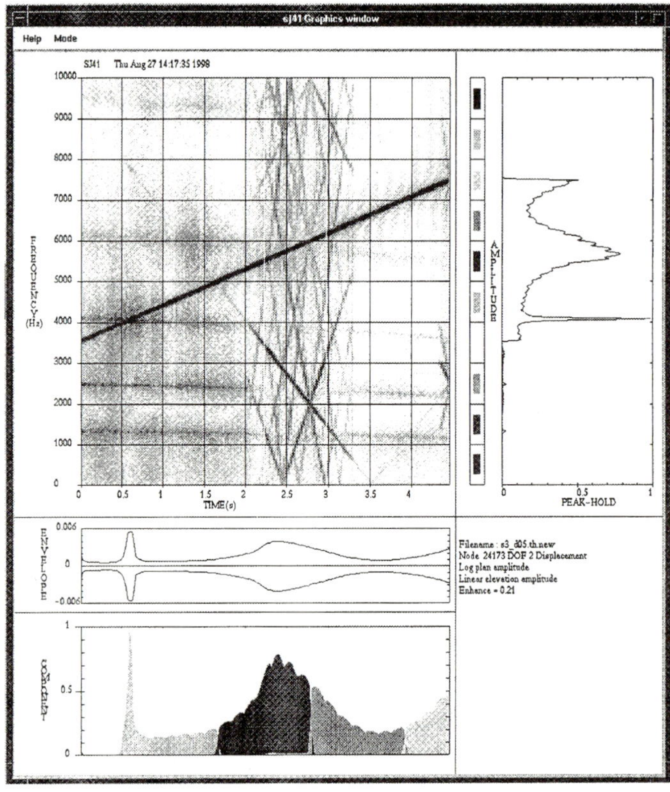

Figure 8 - 'Z-Shot' - damped

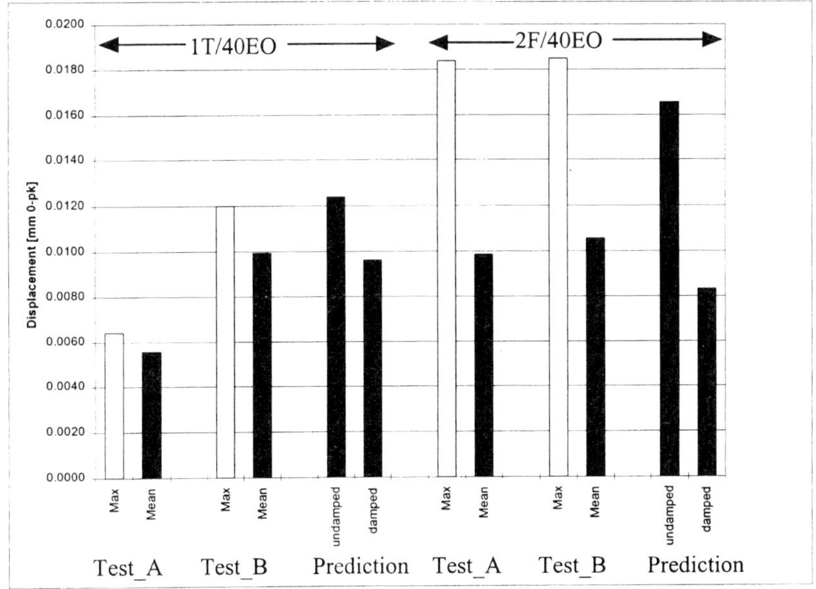

Figure 9 - Amplitude Comparison (Prediction Versus Experiment)

Erosion

Novel design for a gas expander for highly particle loaded gases

E ZETTL and **H JERICHA**
University of Technology Graz, Austria

ABSTRACT

An expansion machine suitable for highly particle loaded gases is presented here. Particle load at the entrance of the machine and the maximum size of the particles as well may be higher than are usually accepted for common gas turbines. No or at least fewer cyclones are needed here as dust separation takes place in the intake of the expander itself. In the intake of the machine the gas flow is accelerated into a vortex around the turbine axis an in addition the flow is bent in axial direction. Due to the centrifugal forces particles cannot follow the gas stream to the inner part of the machine, but move in circumferential but also in axial direction. this movement leads them towards the channel wall, where they tend to deposit and from where they can be sucked off through slits in this wall.

NOMENCLATURE

F_D	drag force	v_{rel}	particle relative velocity
F_L	centrifugal force	v_u, v_m	particle velocity components
F_S	Saffman force	$m, \dot{m}$	mass, mass flow
F_M	Magnus force	$V_p,$	particle volume
I	moment of inertia	d_p	particle diameter
c_D, c_{LS}, c_{LM}	coefficients for drag, Saffman and Magnus force	A_{proj}	projected sectional area
		A	area , sectional area
c_T	coefficient (momentum)	b, h	width, height
η	fluid dynamic viscosity	δ	boundary layer thickness
ρ_f, ρ_p	fluid density, particle density	δ_l	displacement thickness
$\dot{V}$	volume flow	R, R_c	radius, radius of curvature
c	fluid velocity	D, D_m	blade diameter
c_r, c_{ax}, c_u, c_m	components of fluid velocity	l	blade length

INTRODUCTION

Gas turbines many times have to be operated with particle loaded flue gases – e.g. at gasification of coal or at bio-mass combustion. These particles cause serious damage due to erosion, which leads to reduced performance and all in all to a shorter lifespan. All methods of conventional high temperature gas cleaning have certain disadvantages. Ceramic filters may clogg, crack and cause high pressure losses in any case. Cyclones, even in several stage arrangement are limited in cleaning capacity (regarding minimum size of separated dust grains) and cyclones require high investment costs for high their high temperature structures.

The authors propose a new and in several ways improved solution. For the inverted gas turbine cycle a maximum pressure ratio of 2.5 is reasonable, so that such a turbine can be built as a fast one stage turbine. According to our proposal the intake to the turbine can be combined with an effective dust separation device. The investment costs for large high temperature cyclones are saved as well as their pressure loss and the cooling requirements for the first vanes, since cooling can be affected on the side walls of the inlet volute. So the cooling system is limited to the gas turbine rotor blades.

1 DESIGN AND FUNCTION OF THE MACHINE

The main parts of the single stage machine are the **intake**, the **rotor blading** and the **exhaust**. A special feature is that no guide vanes a needed.

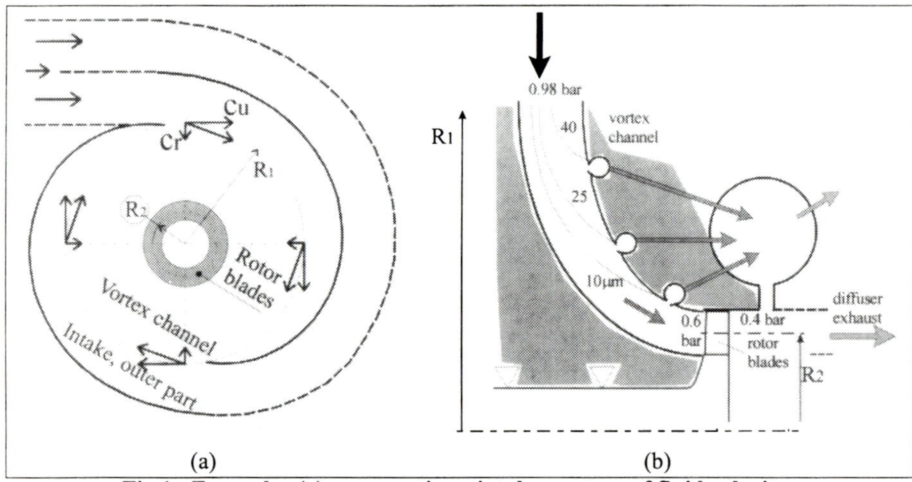

Fig.1   Expander (a) cross section - intake, vectors of fluid velocity;
(b) longitudinal section – vortex channel, rotor blades, particle separation

1.1 Intake

The Intake can be divided into two sections, a first one with an in plain flow and a second one with spatial (three-dimensional) flow. In the first section a vortex around the machines axis is generated, in the second one (entitled as *"vortex-channel"* from now on) the flow is accelerated in a free vortex and in addition bent into the axial direction. Velocity and angle of the flow will fit for the following rotor blades. Stream effects in this vortex channel force most of the particles to move towards the outer channel wall, where they accumulate in the boundary layer area. The highly particle loaded boundary layer flow can be sucked off through slits in this wall, e.g. by using the reduced pressure behind the rotor blades (Fig.1(b)).

1.2 Rotor blades

At the end of the *vortex-channel* axial gas velocity as a result of the bending of the channel varies over its height (blades length) in the manner shown in Fig.2. The *axial velocity* is higher a the tip of the blade and lower at the root section of the blade. Contrary to that, the distribution of the *circumferential velocity* will decrease with increasing radius (according to the free vortex law (2)). Due to this fact it is necessary to use twisted rotor blades. Prescribing irrotational flow behind the rotor blades (no outlet swirl) and a constant pressure distribution over the channels height, we can calculate the inlet angle to the blade row and assuming constant stage work for all radius sections we arrive at the velocity triangles and the blade section profiles also shown in Fig.2.

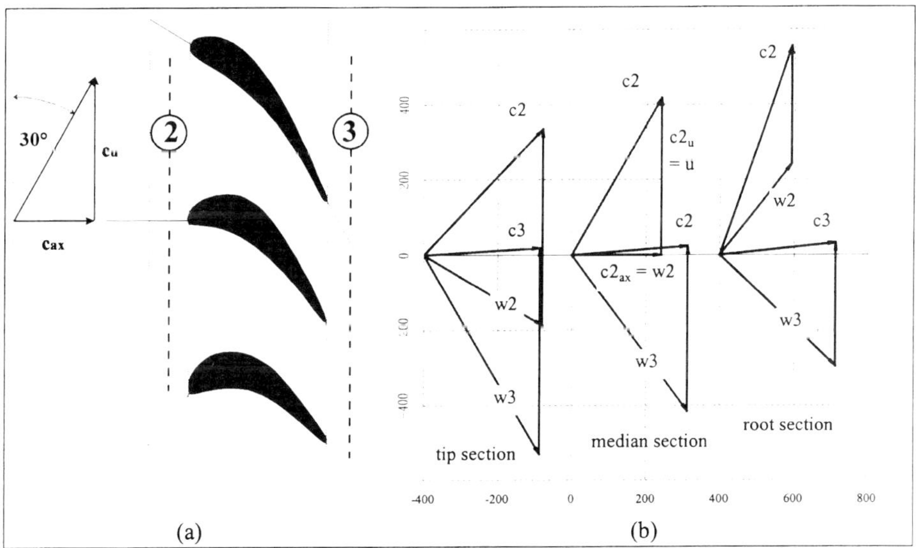

Fig.2   Rotor blade sections (a) - corresponding velocity triangles (b)

1.3 Diffuser and exhaust

For the flow is still at a high velocity after exiting the stage, it has to be retarded in a diffuser and the succeeding exhaust.

2 CALCULATIONS FOR A REALISTIC MACHINE

The waste heat of a process with atmospheric combustion (industrial or biomass) can be utilised in a inverted gas turbine cycle [1, 2]. The flue gas, which may have a temperature of 700°C to 900°C, is first expanded into vacuum and again recompressed to ambient pressure after having passed through a heat exchanger.

2.1 Specifications, data

A pressure ratio of 2.5, which is reasonable for such a cycle, can be realised in a fast one stage turbine. In the case of a reaction turbine this will mean a pressure of about 0.6 bar before and 0.4 bar behind the rotor blades. As a transmission gearbox to the generator drive is necessary any way, any applicable rotor speed can be selected. Since this design will have to cope with a high flow volume, it is proposed to design an expansion machine with high speed of revolution in order to keep the size of the expander as small as possible. For this calculation here a maximum speed of 10 000 rpm is chosen.

Table 1

mass flow	35	[kg/s]
turbine inlet temperature	700	[°C]
turbine inlet pressure	0.98	[bar]
velocity at the entrance	50	[m/s]

2.2 Rotor blading

Depending on an estimation (rating) of the rotor blade dimensions, gained by continuity and Euler equations the size of the whole machine can be determined. The volume flow $\dot{V}$ of the gas is given by equation (1) for a linear distribution of the axial gas velocity c_{ax}, where D_m is the median blade-diameter and l the length of the blade.

$$\dot{V} = \int_{R_i}^{R_a} c_{ax}(R) \cdot 2\pi \cdot R \cdot dR \tag{1}$$

simplified $\qquad \dot{V} = c_{ax} \cdot D_m \cdot \pi \cdot l$

Regarding blade stress, stage clearance loss and cooling requirements a maximum ratio of $D_m/l = 4$ is realistic, which means relatively long blades. A stream-angle of $\alpha = 30°$ (measured from the circumferential direction) at the entrance of the rotor blade channel is common. With the relative gas flow just in axial direction at the median section, a first estimation for the rotor blades data can be done.

Table 2

rotor speed	n	9500	[rpm]
median blade diameter	D_m	0.84	[m]
blade length	l	0.21	[m]
stream angle	α_m	30	[°]

The rotor blades length may vary in the following (more exact) calculation, but the values for the revolution speed n and the median blade diameter D_m are held constant and thus will directly influence the size of the entire machine and the dimensions of the intake as well.

The median stream line of the *vortex-channel* ends at a radius R_2, which is $D_m/2$ at the same time (Fig.3). As supposed the circumferential velocity of the flow equals that of the rotor ($c_{u,2} = u$) at this certain radius, R_2 and $c_{u,2}$ are the reference values for determination of the free vortex flow in the *vortex channel*.

2.3 Flow field (intake flow)

The objective of the outer part of the intake is to generate a vortex around the machine axis. The incoming gas flow has to be accelerated from a low speed in the feed pipe to an inwards bent flow with already high circumferential velocity at the transition to the *vortex-channel*. As one possible solution the authors propose a construction with two spirals as it is shown in Fig.1(a). The main point is to guarantee a steady vortex with a certain stream velocity and stream angle at radius R_1.

The *vortex channel* is the subject part of the presented machine, where dust separation shall take place. The first task is to determine the flow parameters. Pressure and temperature can be deduced from table 1 and the total gas inlet condition. By choosing the pressures ahead and behind the blade row, the law of isentropic expansion can be applied. Radius R_1 and stream angle α_1 define the velocity components at the starting point of the median stream line. The circumferential velocity follows from the free vortex law

$$R \cdot c_u = const , \qquad (2)$$

which documents the acceleration with decreasing radius. The meridional velocity component c_m should increase as well along the (projected) streamline between radius R_1 and radius R_2.

$$c_m(R) = c_{m,0} \cdot R^{exp} \qquad (3)$$

In this arbitrary function $c_{m,0}$ is a reference velocity and *exp* is an exponent, both are adapted to fit the curve to the calculated values of $c_{m,1}$ and $c_{m,2}$ at radius R_1 and R_2. The absolute gas velocity c then is given by

$$\vec{c} = \vec{c}_u + \vec{c}_r + \vec{c}_{ax} \qquad (4)$$

with c_m separated into a radial (c_r) and an axial velocity component (c_{ax}):

with	$R = R_1$	$c_{ax} = 0$	$c_{m,1} = c_{r,1}$
	$R = R_2$	$c_r = 0$	$c_{m,2} = c_{ax,2}$

Table 4

radius	stream angle	velocity	components
$R_1 = 1.55$ m	$\alpha_1 = 20°$	$c_{u,1} = 113$ m/s	$c_{r,1} = 41$ m/s
$R_2 = 0.42$ m	$\alpha_2 = 30°$	$c_{u,2} = 418$ m/s	$c_{ax,2} = 241$ m/s

To describe the curvature of the median stream line (which does a 90 degrees bending), the function of an ellipse is taken. The major axis is given by $R_1 - R_2$ and the minor axis is chosen to *0.7(R_1-R_2)*. (This provides a proper velocity distribution for c_m over the *vortex-channels* height). This velocity distribution is deduced from the equilibrium of forces in the perpendicular direction of the meridional curve. Thus the velocity distribution normal to the curve is a function of the radius of curvature R_c and Radius R according to the radial equilibrium equation. For an isentropic flow temperature, pressure and density can be easily determined for each point in the flow field. The channels height h is deduced in a iterative loop (with h as variable) by numerical integration of the mass flow $\dot{m}$ through the sectional area A normal to the median stream line.

$$\dot{m} = \int_{-H/2}^{H/2} \rho_f \cdot c \cdot 2\pi \cdot R \cdot dh \qquad (5)$$

Here ρ_f is the fluid density, c is the stream velocity normal to the flow area $2\pi \cdot dh$ is a function of ψ and R_c (see Fig.3).

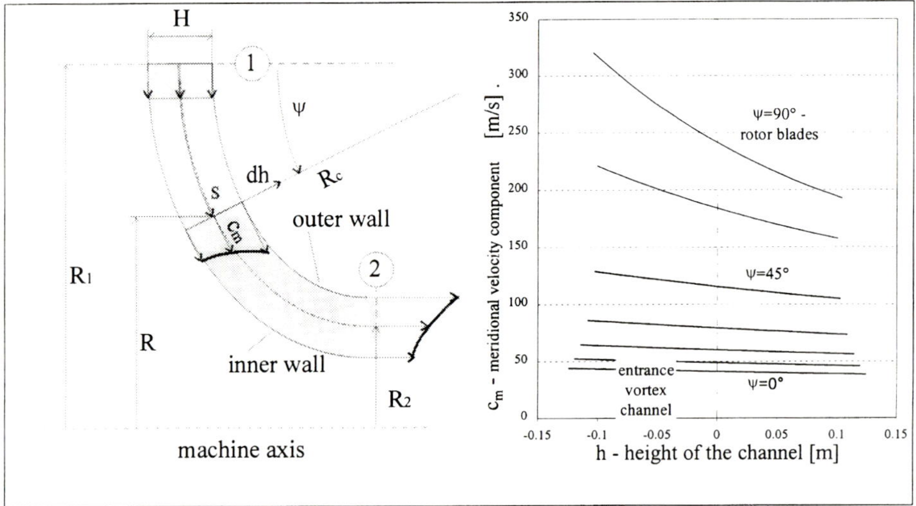

Fig.3   Velocity distribution along the vortex channel and over the channels height (c$_m$ meridional velocity component applied to the angle Ψ)

The boundary layer in the inner/outer wall of the vortex channel. The boundary layer thickness δ is determined from a comparable calculation for the flat plate. The streamline s in the very near of the inner/outer vortex-channel wall is rectified and is considered as a stream line s' in the (accelerated) main flow near a flat plate. The velocity distribution along the length of this stream line is derived from Fig.3. The transition point is varied between the beginning and once at end of the observed length. The boundary layer thickness $\delta(s)$ will grow for the turbulent case up to 15 mm and the laminar case to 4 mm after 1.5 meters flow length. The flow areas will be iterated according to the values of the displacement thickness.

2.4   Particle trajectories

Data of particle load and particle size distribution were taken over from former investigations concerning gas turbine erosion [3] as a realistic example, where the initial particle load of 234 mg/Nm^3 was reduced in a cyclone to 65 mg/Nm^3. No cyclone is planed here, so particles with diameters up to 50µm (according to table 3) may enter into the expansion machine. The chemical structure of the sandy particles would be aluminium oxide and quartz (Al$_2$O$_3$, SiO$_2$) with an average density of 3500 kg/m^3.

Table 3 Cumulative particle size distribution

diameter [μm]	[weight %]
5	7.5
10	18.5
20	58
30	89
40	97.5
50	100

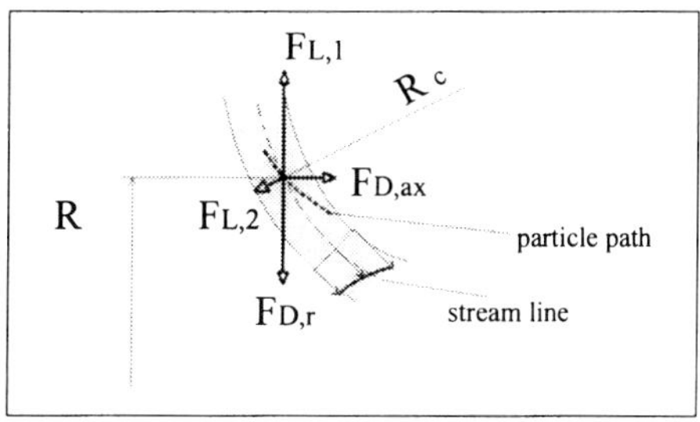

Fig.4 Forces on a single particle (in the main stream of the vortex channel)

The following assumptions were made: A *uniform particle contribution* is presupposed for the entrance of the *vortex-channel* as well a uniform particle velocity as high as 95% of the fluid velocity. Particle-particle interaction same as reactions on the fluid due to high particle load are neglected. Still particle-wall interactions are taken into account.

Forces on a single particle. In the **main flow** the drag force F_D and centrifugal forces F_L influence the particle path, whereas the acceleration due to gravity is negligible for these small particles of such a size and at such high velocities. F_D can be separated into its component in radial ($F_{D,r}$), axial ($F_{D,ax}$) and circumferential direction ($F_{D,u}$). The drag coefficient c_D for spherical particles is given by [4].

$$F_D = c_D \cdot \frac{1}{2} \cdot \rho_f \cdot v_{rel}^2 \cdot A_{proj} \tag{6}$$

$$F_{L,1} = (\rho_p \cdot V_p) \cdot \frac{v_u^2}{R} \tag{7}$$

$$F_{L,2} = (\rho_p \cdot V_p) \cdot \frac{v_m^2}{R_c} \tag{8}$$

$F_{L,1}$ is the centrifugal force as a result of the particle movement around the machines axis at the instantaneous radius R with the particle velocity component v_u. $F_{L,2}$ is the centrifugal force in the meridional section with the bending radius R_c and the particles velocity component v_m as (multiplying) factors. In total all these forces cause a deflection of the particle path from the fluids stream line. The particle moves towards the (outer) wall of the channel. In the **boundary layer flow** additional forces come into effect: The *Saffman force F_S* is a transverse force due to the fluid shear [5]. As the shear lift coefficient c_{LS} is only defined for very small Reynolds numbers, the influence of this lift force can only be analysed qualitatively.

$$F_S = c_{LS} \cdot \frac{1}{2} \cdot \rho_f \cdot v_{rel}^2 \cdot A_{proj} \tag{9}$$

$$F_M = c_{LM} \cdot \frac{1}{2} \cdot \rho_f \cdot v_{rel}^2 \cdot A_{proj} \tag{10}$$

The *Magnus force F_M* is a transverse force as well and is acting on a rotating particle in relative motion to the surrounding fluid. The lift coefficient for the Magnus force c_{LM} is a function of particle rotation and Reynolds number [6,7]. Particle rotation (with a particle angular velocity ω_p) mainly results from wall collisions and varies with time according to

$$I \cdot \dot{\omega}_p = c_T \cdot \eta \cdot \left(\frac{d_p}{2} \right)^3 \cdot \omega_p, \tag{11}$$

where I is the moment of inertia of the particle, d_p is the particle's diameter and η is the fluid's dynamic viscosity. For spherical particles and low Reynolds Numbers the authors of [5] found $c_T = 8\pi$. The initial particle angular velocity (in the moment just after wall contact) is obtained from the following estimation: The kinetic energy after reflection shall be as much as 90% of that before, half of the difference shall be converted into rotational energy and the rest shall be lost due to friction, damping and deformation.

Reflection from the wall. The velocity after reflection is given through arbitrary definition of the ratio kinetic energy before and after the particle contacts the wall. A reflection angle varies according to the combination of materials (particle, wall). Examples are given e.g. by [8,9], which can not be generalised. Additional uncertainty results from wall roughness and particle shape. So for this calculation it was decided to define the angle of reflection to be just equal to the impingement angle. The bouncing back and forth of particle from main flow to boundary layer, casing wall and back is thus described. This results will facilitate the layout of the dust suction slits (Fig.1).

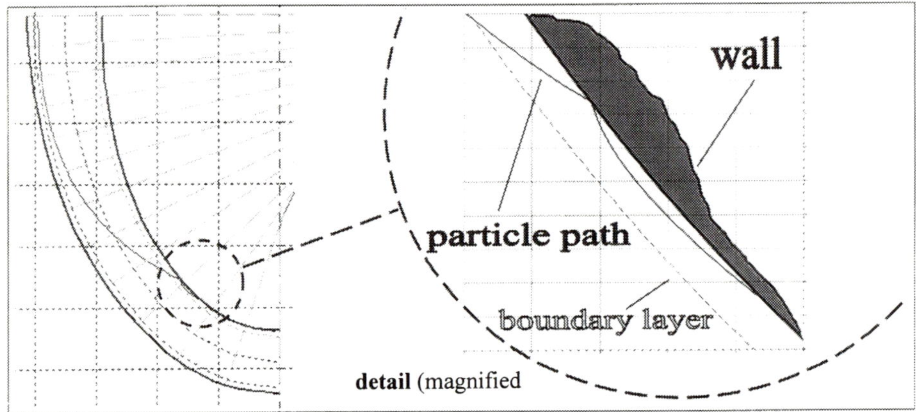

Fig.5   Particle path (projection into the meridional section)
particle-wall interaction with multiple wall contact and reflection

Particle trajectory calculation has been carried out using a **Four-Step-Runge-Kutta-Integrator** with a constant time step in the main flow.

3 RESULTS AND DISCUSSION

The particle trajectory calculation shows: The *vortex-channel* is an appropriate solution for a reliable separation of particles from a gas flow. This separation will be complete for particles in a wide range of their sizes. According to the concept of the machine (dimensions, velocities) particles with diameters from 15 µm – 40 µm will be separated completely. For larger particles it will be necessary to have a separator before the expander. Otherwise these particles - once inside the intake of the expansion machine - will not pass through the machine, nor will deposit, but will accumulate in an other part of the machines intake.

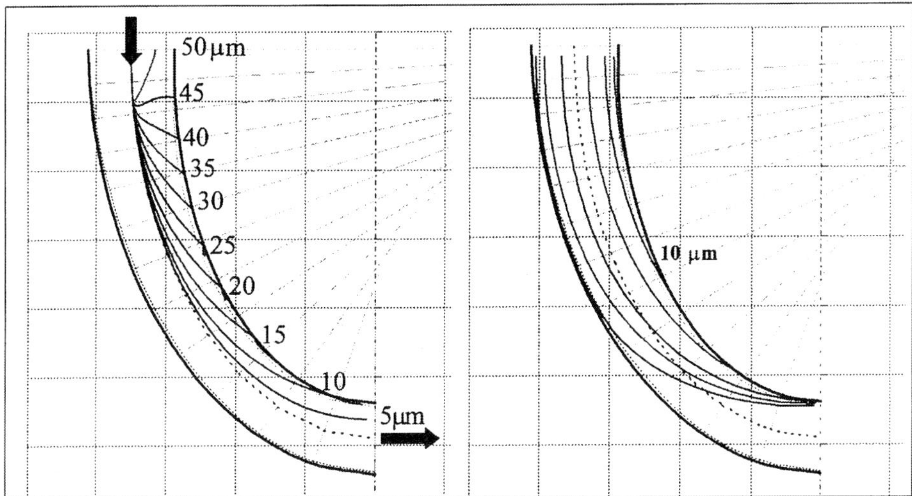

Fig.6   Particle trajectories – (projection into the meridian section)

Some (very) small particles can be trapped in the inner channel wall boundary layer and will be transported towards the root section of the rotor blades, where they can be sucked off through a slit arranged ahead.

Smaller particles of a size of less than 10 μm may not deposit on the outer channel wall (from where they can be sucked off), but will reach the rotor blades. These particles may reach the rotor blades channel, they nearly have the same velocity and angle as the flow and follow the stream so well that they will hardly hit the blades surfaces. Only a small portion will hit the blade leading edge at high impact angles.

Since there is a limited number of blades provided (there are no guide vanes in this machine) high quality alloy may be chosen for the rotor blading, so the blades can withstand the erosive attack of the few particles left in the working gas. Thus a long blade life span can be expected. For the materials used in the vortex channel ceramics is supposed to be the best choice. The calculation showed that particle impact on the wall always happened with relatively high velocities but under very small angles (of not more than 6° measured from the wall surface). The authors of [10] found the erosion loss to be most at 90° impingement angle and less for angles below 10° for ceramic surfaces, while metallic surfaces have its maximum of erosion loss at an impingement angle of about 30 degrees.

The need of a cooling system should be limited to the turbine rotor blades, so higher gas entry temperatures (even up to 900 °C) can be put to use. Rising gas temperatures would mean increased machine dimensions according to continuity.

The output given by the total enthalpy difference (computed for the median section) multiplied with the mass flow would be around 5.75 MW for this machine shown here. Taking into account that this mass flow is reduced by a boundary layer suction (in order to separate the particles accumulated in the most near wall boundary layer flow from the gas flow) the power output will decrease to a certain amount. An assumption of the loss of working gas is given here: The highly particle loaded gas in the very near of the outer channel wall is sucked off through three slits (over the whole circumference) located in the outer wall of the *vortex-channel* (Fig.1(b)). The mass flow through this slits is approximated by

$\dot{m}_{slit} = \overline{\rho}_f \cdot \overline{c}_m \cdot \overline{A}$, where $\overline{A}$ is defined as a section perpendicular to the outer wall with a height as much as 1/3 of the maximum boundary layer thickness δ_{max}. The stream velocity normal to this section is given by the local meridional velocity $\overline{c}_m$ (at $\delta_{max}/3$), and $\overline{\rho}_f$ is the local fluid density. In total this will lead to a reduction of output by 5% to 7.5%.

CONCLUSION

The authors believe to have demonstrated, that it seems possible to achieve an effective dust separation in the intake of a gas expander with a novel design as presented here. The disadvantages of large machine dimensions and the complicated shape of the intake is compensated by the fact that for a specific plant layout no (or at least fewer) preceding cyclones are needed.

ACKNOWLEDGEMENT

This work was sponsored by the **FWF**, Austrian Science Foundation grant Nr: 56807-TEC

REFERENCES

[1] **Fesharaki, M., Halozan, H., Jericha, H., Moser, S.,** "Biomass fired atmospheric gasturbine plant", ASME, Birmingham, UK, paper 96-GT-246

[2] **Fesharaki, M., Halozan, H., Jericha, H., Kulanek, G.,** "Biomass fired atmospheric gasturbine plant-I", ASME, Orlando, Florida, paper 97-GT-290

[3] **Sanz, W.,** "Schaufelerosion in einer Industriegasturbine", Diplomarbeit, Technische Universität Graz, 1989

[4] **Rudinger, G.,** "Fundamentals of gas-particle flow",Handbook of Powder Technology. Vol.(J.C.Williams and T.Allen.Eds.).Elsevier Scientific. Amsterdam, 1980

[5] **Saffman, P.G.,** "The lift on a small sphere in slow shear flow", J.Fluid Mech., 1965 (See also: **Saffman, P.G.,** "Correction", J.Fluid Mech., 1968)

[6] **Rubinow, S. I. and Keller, J. B.,** "The transverse force on a spinning sphere moving in a viscous fluid", J. Fluid Mech., 1961

[7] **Humphrey, J.A.C.,** "Fundamentals of fluid motion in erosion by solid particle impact", Int. J. Heat and Fluid Flow, 1990

[8] **Tabakoff, W.; Malak, M.F.; Hamed, A.,** 1987, "Laser measurements of solid-particle rebound parameters", AIAA Journal, Vol. 25, 721-726

[9] **Elfeki, S., Tabakoff, W.,** "Erosion study of radial flow compressors with Splitters", ASME Paper 86-GT-240, 1986

[10] **Watanabe, S., Sorrel, G.,** "Erosion studies with fluid catalytic cracking catalysts" Tokyo Japan, Exxon research, USA

C557/091/99

Erosion due to particle impact in supersonic flow

R PÖSCHL, J WOISETSCHLÄGER, and H JERICHA
Institute for Thermal Turbomachinery and Machine Dynamics, University of Technology, Graz, Austria

1. ABSTRACT

The paper has the intention to clear up the mechanics of erosion at high velocity impact at flat impact angles. Experimental methods are presented which should give improved insight into the mechanism of mechanical erosion by particles harder than the metal surface.

At the institute for Thermal Turbomachinery and Machine Dynamics at the University of Technology in Graz, an *erosion- test rig* was designed for investigating particle impacts for particle sizes of up to 63μm diameter in supersonic velocities up to Mach number 2.2. The Technique of 4- Colour-Schlieren-Visualisation was applied to visualise the flow and the shock pattern. Flow investigation were done using *laser doppler anemometry* (LDA). The measurements were compared to numerical results from an *viscous unsteady flow simulation* performed with the software package FIRE (AVL-Graz, Austria). To examine the impact and restitution angles and particle velocities *laser sheet measurements* as particle image velocimetry (PIV) were used. The *erosion scratches* on the polished specimens were investigated using a scanning electron microscope (SEM). So it was possible to define important parameters of the erosion mechanism and to allow better prediction of erosion rates.

On the bases of theoretical investigations as published before (1) a suggestion of how to correlate impact velocitiy, particle mass and loss of specimen material is presented here. From this more detailed survey (SEM) of specimen volume loss a better correlation of the plowing friction factor with the gas turbine blade metal reality comes into reach. This investigations were done for different particle sizes (1 to 63μm) in the region of small impact angles (around 15°) where severe damage is done to ductile materials.

2. INTRODUCTION

Often in industrial process heat is available in the form of hot gases at **atmospheric pressure level**. A typical example is biomass combustion where the heat is available at ambient pressure level. A novel solution for this case is given in the **inverse gasturbine process (2)** where at first biomass is burned at atmospheric pressure and the hot exhaust gases are conducted to a gas turbine for expansion into vacuum. At a vacuum pressure of about 0.4bar the gases are cooled in heat exchangers and the moisture content is made to condense. Later on the uncondensable gases are compressed to ambient pressure thus closing the cycle. In those plants we have to handle gases highly loaded with dust particles. The gas turbines which are needed for proper expansion have to be designed as one stage turbines for cost reasons. In that harsh environment of operation with transonic and supersonic flow information about erosion behaviour of blade material is especially important. Erosion can alter blade profiles severely and decrease flow efficiency drastically **(3)**.

Particles elutriated from a fluid bed passing through a gasturbine blading are **continuously reduced in size** due to cracking as consequence of repeated impacts on blade surfaces this has been observed in FCC expanders **(4)**, where the size distribution shows a shift to smaller particle size from inlet towards the exit of the turbine.

It is stated in literature **(5)** that the value of **material loss** increases with the impact velocity of particles to the power of higher than 2, sometimes even 5 in special cases ($e \sim v^n$ $2 < n < 5$)! Ductile materials like heat- resistant steels show considerable sensibility to **small impact angles (6)** which occur mainly on the blade pressure surfaces.

We hope that the results presented here will form one step closer to the solution of the problem described. The authors will further try to analyse their measurement data in several respects to increase the benefit for the designer.

A further theme of research will be the erosion resistance of **coatings**, as corrosion protecting and also thermal barrier coatings. Coatings of this kind are indispensable for the protection of the high temperature gas turbine blades as described by **(7)** and **(8)**.

3. GAS FLOW AND SHOCK WAVE PATTERN IN THE EROSION TEST RIG WITHIN A SUPERSONIC CASCADE

3.1. Supersonic Erosion Test Stand

The erosion test stands is designed as a **replaceable cylindrical insert** into the structure of the Transonic- Cascade (**Fig. 3.1-1**). The main parts of the erosion test stand are the wedge-shaped specimen which is attached to an specimen-holder beeing adjustable within the two contour parts. The latter give the channel the needed shape of a naval nozzle. For observing particles in the half depth plane of the 2D-Channel, a **light sheet window** made of glass has to be included in the contour part. This design feature requires a linear part in the continuously curved contour at the position of the leading edge of the specimen. To generate an appropriate mathematically defined contour a B- Spline function was used to guarantee continuity of shape and curvature.

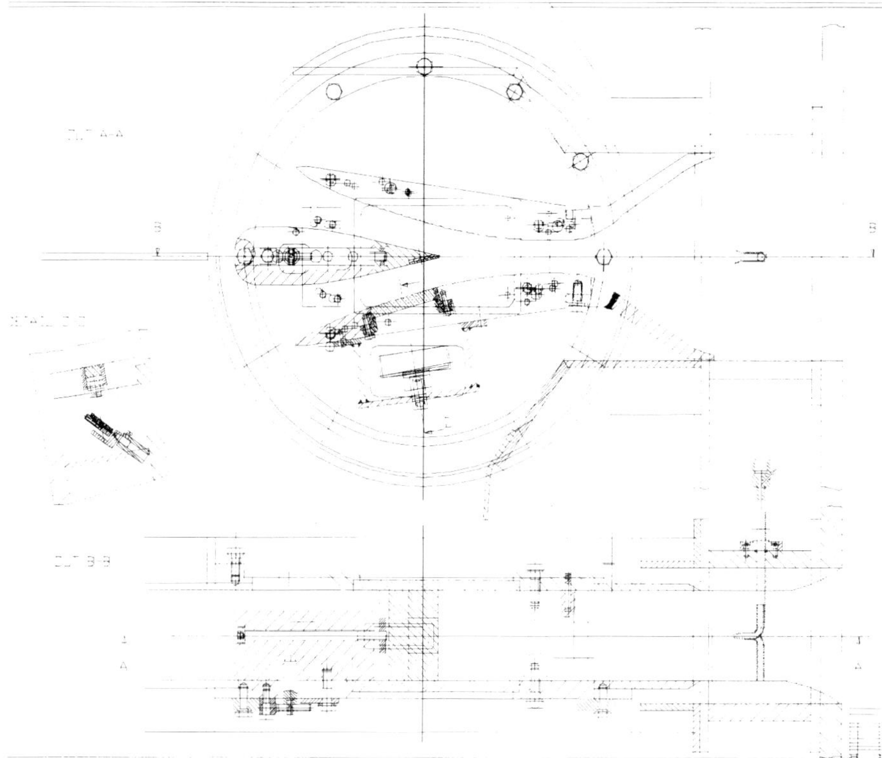

Fig. 3.1-1:   Design Drawing of the Erosion Test Rig

3.2.   4-Colour- Schlieren- Technique

Parallel light waves passing through a flow field with density gradients are deflected towards the larger value of density. Those **deviated light beams** can be focused afterwards and filtered out at a knife mask, which results in an grey scale image of the gradients in the direction perpendicular to the knife edge. Settles **(9)** improved this technique by colouring the light waves in an diaphragm and used according to the 4 colours 4 knife edges to enable the filtering. With that method **location and gradient orientation** of shock waves could be visualised easily. Unfortunately there is no direct relationship between the brightness and the magnitude of the density gradient.

The **flow pattern around a blunt wedge** (test specimen) in the erosion test stand is characterised by two oblique shock waves which appear nearly parallel on both sides of the specimen (**Fig. 3.2-1**). Using the 4-Colour- Schlieren- Technique both of them can be seen. The detached shock is located a few mm in front of the leading edge and another one starting from the wedge surface. Thus incoming *oblique shock wave* initiates a local very small separation zone in the boundary layer of the channel contour which causes another oblique shock wave (i.e. reflected one) and an expansion wave to turn the supersonic flow around this tiny obstacle **(10)** or **(11)**. After a short expansion zone another oblique shock wave crosses the channel and raises the pressure level to approximately ambient condition.

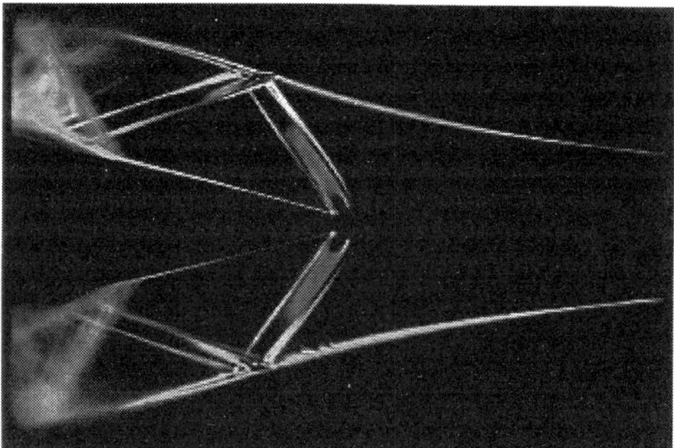

Fig. 3.2-1:   Shock Wave Distribution At Blunt Wedge

3.3.   Numerical Simulation With FIRE

The **two dimensional unsteady flow simulation** was done by a finite volume based, pressure correction code for complex flow structures.

The calculated *density contours* can directly be compared to the Schlieren – pictures, keeping in mind that the latter one visualises gradients. The position of the detached shock wave and the shock reflection on the contour walls are presented.

Downstream to the wedge a considerable unstable *recirculation area* is present. Its extensions can be appraised on a plot of velocity- vectors. The tiny separation zone and reattachment zone at the position of the reflected shock wave on the lave contour could hardly be seen.

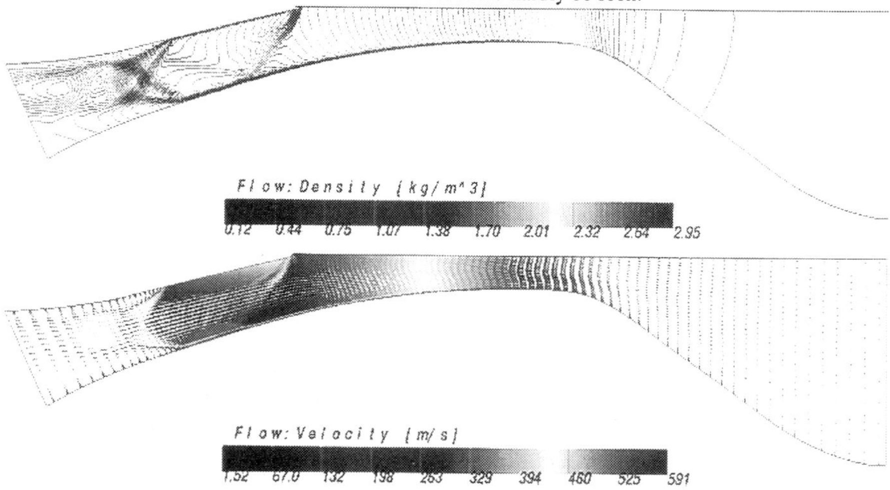

Fig. 3.3-1:   Flow Calculation Results (FIRE)

3.4. LDA Measurements

The used DANTEC laser equipment consists of a **twin-ray argon-ion laser** and a Burst Spectrum analyser which examine the signal from the photo-multiplier for evaluating the calculation.

The *velocity magnitudes* 1.5mm above the surface of the wedge starting at the leading edge as a comparison between LDA and calculation with FIRE show **Fig. 3.4-1.**

For some location along the flow direction two values have been analysed from the LDA- signal according to *two distinct peaks* which correspond to the fluctuating position of the unsteady second shock wave. The signal upstream to the shock wave position at the leading edge could not be analysed properly. The *starting condensation* of humidity generates droplets in the range of the applied seeding material of the LDA- system an effect we sought to avoid by raising air inlet temperature.

The extensions of the *recirculation zone* on the laval cotour pulsates according to the unsteady change of the second shock wave position, which in some extreme situation reaches nearly the location of the first shock wave at the leading edge.

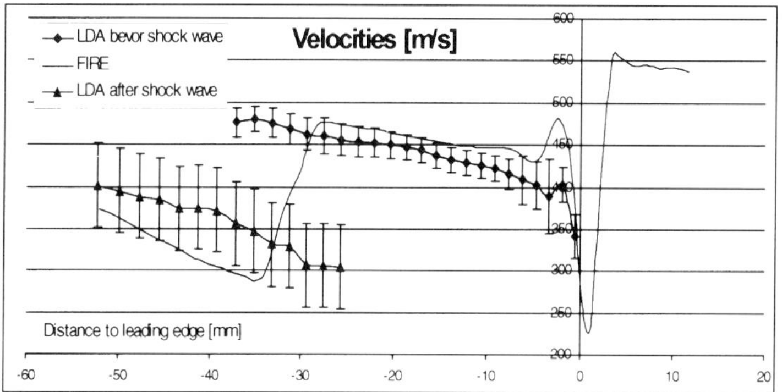

Fig. 3.4-1: Comparison of LDA Measurement and CFD Calculation close to the wedge surface

4. INVESTIGATION OF PARTICLE MOTION AND IMPACT PARAMETERS

4.1. Particle Ingestion

The different experiments conducted required two types of **solid particle injection.** In the case of *continuous* introduction for the laser measurements a cylindrical tank of approximately 1lit capacity was filled with particles. Compressed air of about 4bar flows through a perforated tube in the centre of this tank and sucks particles into the injection stream. The particles reach the main flow of the test stand through a 90° pipe bent.

To provide an *intermitting* stream of a distinct amount of particles, the quartz grains are loaded via a funnel directly into the sealed pipe bend, and are afterwards ingested as few by few particles by short duration jet of pressurised air into the main stream.

In both cases the main difficulty is the **agglomeration** effect of particles smaller than 20µm diameter. The particle distribution observed near the leading edge is considerably influenced by the starting conditions of the introduced particles. The **Laval- nozzle focuses** the particles towards its axis, therefore most of them would hit the flat leading edge, and afterwards impact several times on the specimen surface. To force the particles to hit the surface **directly on the specimen** in its inclined position behind the leading edge the ingesting pipe bend was inclined -15° to the axial direction. Another severe effect for the radial deviation of the particles in the main stream is the different inertia force due to the **acceleration within the ingesting pipe** bend which tended to deviate the particle sidewise.

Nevertheless, the direction of the first impact on the surface *do not differ very much* from the axial direction.

4.2. PIV Measurements

For evaluating the erosion phenomena the **impact velocities** and **angles** are of special interest. Therefore, the very advanced PIV-method has be applied using a *double pulse ruby laser*. The laser is optically widened to a horizontal *light sheet* which is deflected via a mirror and a window in the lower LAVAL- contour part in the mid plane of the 2D channel. Particles moving along the light sheet deflect light in perpendicular direction where a camera is located to expose a photographic film. They are displayed twice, according to the pulse delay (2 to 5µs). The photo- camera was equipped with an exchangeable CCD-chip to enable optimal adjustment very close to the surface.

The PIV-images are enlarged 5 times during the photographic developing procedure and digitised afterwards for being used in an **automatic analysis (Fig. 4.2-1)**. An *auto-correlation procedure* computes the correlation values for every interrogation area subsequently for the whole PIV image. The program select the three most probable correlation direction out of the correlation field. Finally, an validation procedure figure out the most realistic velocity vector by comparing adjacent vectors and applying other selecting criteria.

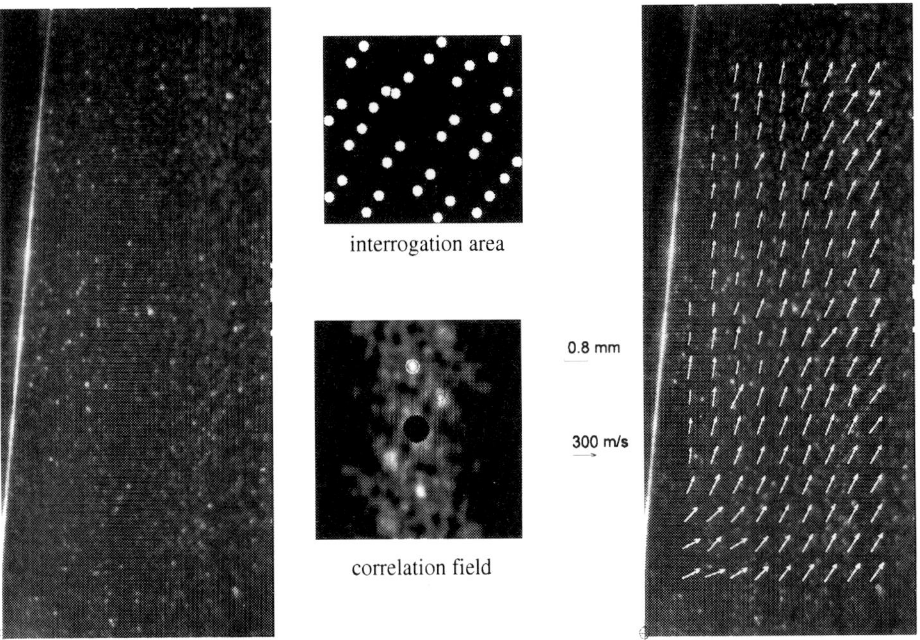

Fig. 4.2-1:   **Steps in PIV Analysis Procedure**

After a statistical analysis of all different calculations and averaging the significant values of all CFD calculations they could be compared to the PIV measurements as shown on **Table 1.**

particle size	v_min	v_average	v_max	α_min	α_average	α_max
		PIV measurement				
μm	m/s	m/s	m/s	°	°	°
10-25	352	**388.7**	430	8	**10.0**	12
25-40	267	**292.8**	321	8	**11.0**	13
40-63	182	**222.5**	265	6	**9.0**	12
		Calculated Results (FIRE)				
5.0	441	**448.5**	454	6	**7.7**	9
10.0	415	**418.1**	421	11	**11.3**	12
17.5	368	**369.2**	371	12	**12.7**	13
25.0	333	**335.3**	338	12	**13.2**	14
32.5	306	**307.0**	309	12	**12.9**	14
40.0	284	**285.4**	287	11	**12.4**	14
51.5	264	**267.2**	270	10	**10.5**	12
63.0	240	**240.2**	240	11	**11.7**	12

Table 1: Comparison Of Calculated And Measured Impact Parameters

From the results of the PIV measurements *rebound coefficients* for different particle sizes as shown in **Table 2** could be derived easily as ratio of according rebound to impact value.

particle size	v_min	v_average	v_max	α_min	α_average	α_max
	m/s	m/s	m/s	°	°	°
10-25	0.71	**0.76**	0.79	0.73	**0.81**	0.85
25-40	0.79	**0.88**	0.90	0.63	**0.79**	0.92
40-63	0.91	**0.98**	1.01	0.71	**0.97**	1.09

Table 2: Rebound Coefficients for different particle size

4.3. Calculated Particle Trajectories

The influence of the **introduction parameters** on the impact position at the specimen has been investigated on trajectories for 5 different *starting locations* and 5 different *starting angles*. (As a means to simulate the variation in particle starting conditions in reality.) There are different starting velocities for every particle size, considering the distinct particle *acceleration in the ingestion duct* itself.

The *deviation* in the particle trajectories due to different *starting location* are shown on **Fig. 4.3-1** for particles with diameters of 5μm. Larger particles show smaller differences between trajectories from different starting position.

The influence of *starting angles* (+/- 10°) on impact position on the specimen is very small. The trajectories form smaller particles could not be separated. The **focusing effect** of the Laval nozzle is obviously strong due to low starting velocities especially for bigger particles.

9 particle sizes (1,5,10,17.5,25,32.5,40,51.5,63μm) have been selected to compare the **inertia effects** on the different particle trajectories. On the according *particle velocities* more details about the differences in acceleration after ingestion and at impact position on the wedge are presented. Negative Velocities indicate the flow from right to left.

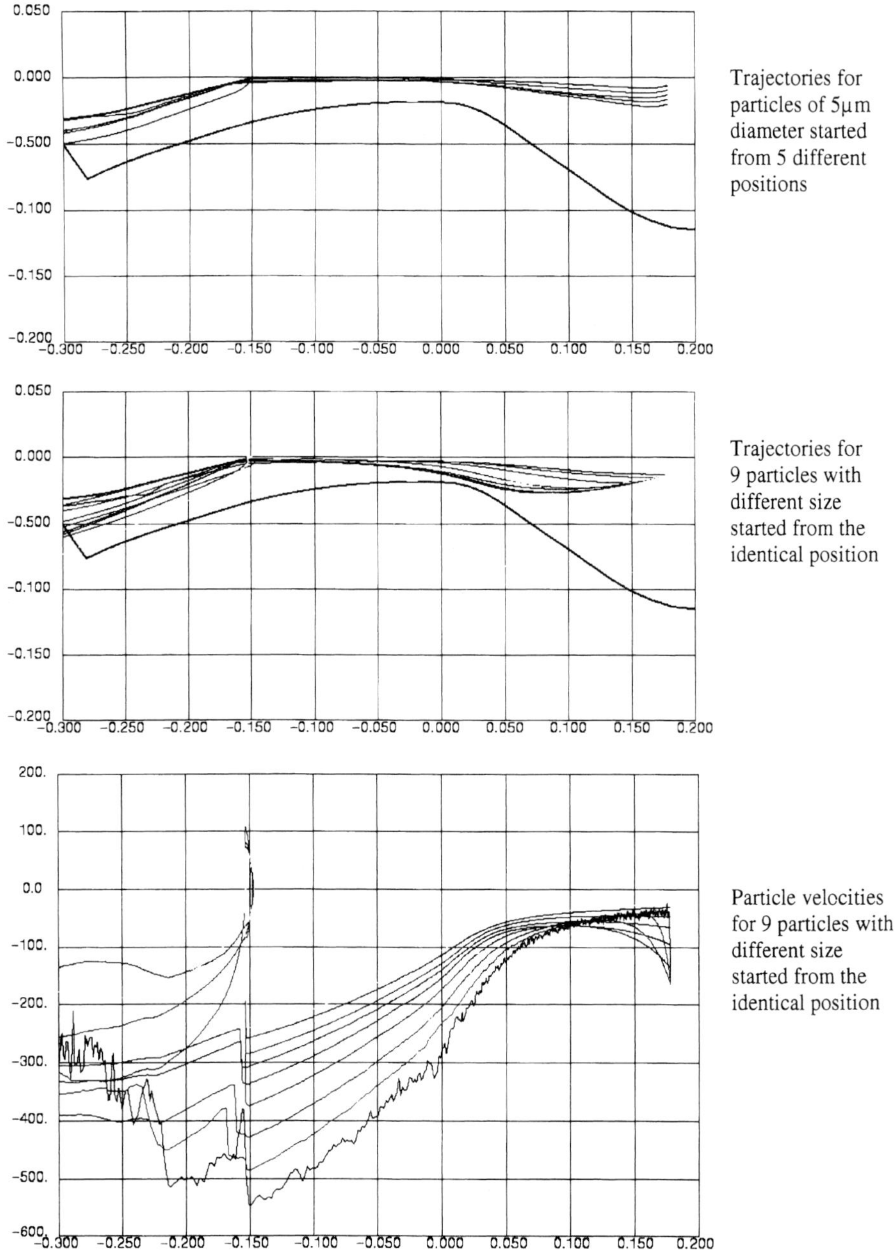

Fig. 4.3-1: **Calculated Particle Trajectories and Particle Velocities**

5. ELECTRON-MICROSCOPICAL ANALYSIS OF EROSION PATTERNS ON THE SUR-FACES OF THE SPECIMEN

5.1. Scanning Electron Microscope

A beam of electrons is shot on a distinct point on the specimen and set electrons free from the metal surface. The number of the detected secondary electrons define the brightness of that particular point on the image. As more electrons are released from edges and corners than from a plain or an concave region a perfect criteria for generating best spatial pictures with optimum depth of focus is given.

Thus, the SEM images provide us the possibility to understand the microscopic erosion mechanism of particles with supersonic speed.

5.2. Theory Of Abrasive Material Loss Due To Single Particle Impact

Due to differences in hardness of the erodent and the **ductile** target material the particle edges usually penetrate the metal surface. Severe erosion could be observed at angles around 15 to 30 degrees by scratching and cutting mechanisms. Plowing of particle edges is often observed together with breaking of edges. It is stated in literature **(12)** that the criteria for big material loss from smallest particles is *stress fatigue* by repeated particle impacts. Due to elastic properties they are very resistant against perpendicular impacts.

Whereas **Brittle** materials suffer most from that perpendicular impacts. These generating *cracks* below the metal surface which leads to material loss in form of flat plates. The dimension of those flat plates could exceed the diameter of the impacting particles **(13)**.

5.3. Erodent Material

Dried **Quartz-powder** ZE 36 1600 from *Quarzwerke Zelking* (density 2650kg/m^3) was used as eroding material. A tendency to **agglomeration** due to intermolecular forces are readily apparent at particle diameters lower than 20µm.

On each specimen **1g of quartz sand** with defined range of size has been shot.

For the majority of the particles the break-up criteria (tensile strength due to bending moment at impact) has been reached.

5.4. Erosion Specimen

The specimens used for analysing the scratches after being exposed to the supersonic particle impacts, are flat plates in dimensions of 35x40x2mm of blade material (X20CrMoV12-1 [DIN 17175; 1.4922]). They were fixed on the lower leg of the Erosion wedge in the middle axis of the nozzle.

On every specimen **4 impact regions** have been investigated with different densities of particle impacts. The closer the region to the leading edge and the closer to the middle axis, the more particle impacts per mm^2 could be observed. Therefore two positions close to the leading edge of the wedge have been chosen with different distances to the middle axis (2mm and 18mm). Two according locations at a distance of about 25mm to the leading edge have been added.

To figure out the influence of the particle size **3 different sizes of particles** have been introduced.

5.5. Erosion Pattern

Differences in the **location** of scratches on the probe surface is mainly due to the number of impacts which is of course variable in the length and width of the specimen. The scratches seem to indicate that impact velocities and angles do not differ significantly, although towards the leading edge the *density of impacts* increases. Since the particles are ingested into the symmetry plane of the channel particle impacts probability towards the side walls decrease (Gaussian distribution).

Some **typical shape** of erosion scratches could be divided. *Long scratches* (above one diameter in length) indicate direct material loss by *cutting* metal chips (estimated as one tenth of an particle diameter in height). In some impact conditions the particles could not directly remove material but *plough* through the metal surface. Repeated impacts of this kind cause stress fatigue and therefore material loss on long-term.

Circular small dots with small penetration depths indicate the contact of single particle edges and lead not directly to erosion but deform the metal surface. In most cases there remain a part of the attacking particle sticking in the surface.

Curved scratches may arise due to single sharp edges penetrating the metal surface being in *eccentric* location at particle outer surface, thus turning the particles sidewise as well as rolling it in the direction of impact.

According to the **size of the particles** different dimensions and a slightly change in the shape of the erosion pattern could be observed. **Fig. 5.5-1** present a comparison of SEM pictures of specimen suffer on different particle size impacts (10-25μm respectively 40-63μm). The different magnification is listed in the bottom left corner of the pictures. To illustrate the dimensions a size bar is added. The main flow direction is from bottom to the top of the image.

For larger particles the penetration depth and width of the scratches seems to be relatively larger than those of smaller particles. More broken edges of particles remain in the specimen shot with bigger particles. The number of small particle impacts decrease with particle size.

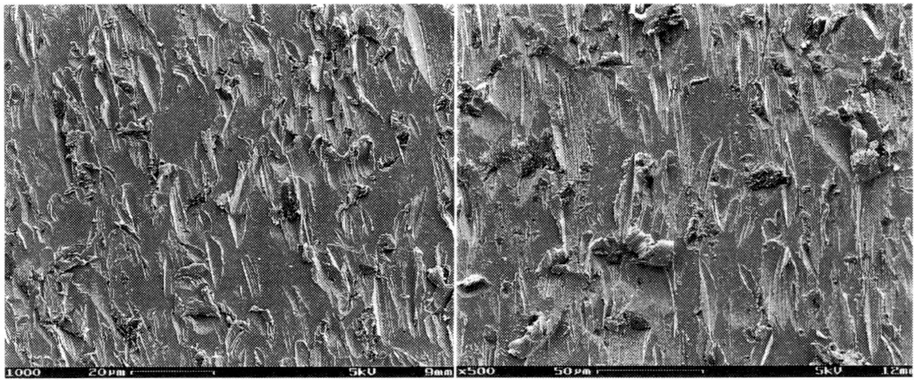

10-25μm   40-63μm

Fig. 5.5-1:   SEM Picture Of Different Erosion Scratches

5.6.   Material Loss

The electron-microscopic analysis of erosion patterns enabled the authors to roughly estimate a average volume loss at single impact. This was done baring in mind the measured impact velocities and impact angles of the particles and providing thus the necessary correlation for the theory already put forward. As a first result the **mean values** are shown in the following **Table 3**. The correlation of average values of impact velocity to average value of metal loss at single impact has thus be established.

particle size μm	v_average m/s	α_average °	kin Energy 1E6kgm^2/s^2	scraped Volume 1E-18 m^3
10-25	388.7	10.0	0.35	16.1
25-40	292.8	11.0	3.90	39.8
40-63	222.5	9.0	8.96	96.4

Table 3: Average Material Loss Due to Single Particle Impacts

6. CONCLUSIONS

- A quite good agreement between measurements and CFD calculation could be achieved for the case of flat angle impact on ductile metal specimens.
- Single impact of particles leads to scratches which make the impression of ploughing the hard particle edges through the more soft metal surface. Thus tilted up side walls of ductile materials are generated which have to be counted as erosion loss since impacting smaller particles will remove them easily by fatigue bending.
- It can be sated that at these high velocities particles are prone to cause extreme damage.
- Severe erosion damage occurs due to stress fatigue on repeated impacts of the higher number of small particles.
- Most of the particles break apart at impact or lose broken edges which remain stuck in the metal even at small impact angles.

7. ACKNOWLEDGEMENT

The author thank the *Austrian Foundation for Scientific Research* (FWF) which has granted our institute fund for this research project S6809.

The author would like to thank the management of *AVL LIST GmbH*, author of applied software package FIRE, for permission in the application of their code.

Respectful thanks are offered to *Prof. El-Sayed* (Zagazig University, Cairo) for the intense discussions and interesting suggestions during his stay in Graz in Nov. 97.

8. REFERENCES

(1) Jericha, H., Göttfried, T., Sanz, W., 1989, "Industrial gas turbine blade erosion", ASME Cogen Turbo, IGTI 4, 193-200

(2) Fesharaki, M, Halozan, H., Jericha, H., Kulhanek, G., 1997, "Biomass-Fired Atmospheric Gas Turbine Plant-I", ASME Turbo Expo 97, Orlando, Florida, ASME paper 97-GT-290

(3) El-Sayed, A. F. A. A., Brown, A. 1987, "Computer Predictions of Erosion Damage in Gas Turbines", ASME 87 GT 127

(4) Bauer, E., and Jericha, H., 1987, "Refinery FCC Plant Energy Recovery 75000h of cyclone and expanding gas turbine", CIMAC Warsaw, T-1

(5) El-Sayed, A. F. A. A., 1995, "Prediction of Erosion in Pipe Bends Due to Particulate Flow", Engineering Research Journal Vol. 4 of University of Helwan, ISSN 1110-2330

(6) Kotwal, R., Tabakoff, W., 1981, "A New Approach for Erosion Prediction Due to Fly Ash", Journal of Engineering for Power Vol. 103

(7) Elsner, W., 1998, " Schäden im Heißgaspfad von Gasturbinen und ihre Ursachen", Allianz Report 2/1998

(8) Tabakoff, W, Metawally, M., Hamed, A., 1993, "High Temperature Coatings For Protection Against Turbine Deterioration", ASME Cogen-Turbo, IGTI-Vol.6

(9) Settles, G., 1970, "A Four-Colour Schlieren Technique With Sensitivity In All Directions", Tennessee Engineer, pp. 4.5. 7-9

(10) Shapiro, A., " The Dynamics and Thermodynamics of COMPRESSIBLE FLUID FLOW", The Ronald Press Company , New York

(11) Sieverding, C. H., 1976, "Transsonic Flows in Axial Turbomachinery", Von Karman Institute for Fluid Dynamics

(12) Zum Gahr, K.H., et. al., 1983, "Reibung und Verschleiß", Deutsche Ges. für Metallkunde

(13) Watanabe, S., 1985, "Erosion Studies With Fluid Catalytic Cracking Catalysts" ASME

C557/118/99

A pump separator of mechanical impurities

A WILK and **S WILK**
Mining Mechanization Center, Komag, Gliwice, Poland

Synopsis
The paper deals with the results of investigations concerning the design of a pump separator of mechanical impurities contained in the pumped liquid. A mathematical model of the movement of solid bodies has been developed, which leave the centrifugal pump together with the liquid. The applied method of calculations makes it possible to determine the path and parameters of the moving solid. Assuming the given shape of the discharge channel, the results of calculations have been analyzed.

Test-stand investigations were carried out on a multi-stage centrifugal pump, in which instead of one of the stages a separator of mechanical impurities had been installed. The impurities which were precipitated in the pump were removed from the casing of the pump through a sink. It has been found the relation of degree of purification (the mass fraction of precipitated impurities) to volume fraction of the liquid discharged through the sink. Such investigation have been run at various deliveries of the pump.

Notation schedule

c_{m3} -   radius component of liquid velocity at the impeller outlet, m/s
c_{u3} -   circumference component of liquid velocity at the impeller outlet, m/s
F_A -   buoyant force, N
F_C -   pressure force, N
F_G -   gravity force, N
F_N -   medium resistance force, N
F_O -   centrifugal force, N
g -   acceleration of gravity, m/s^2
k -   velocityi coefficient
m -   total mass of separated sand, kg
m_t -   mass of the sand from the pump outlet collected on the sieve, kg
m_u -   mass of the sand from the pump sink collected on the sieve, kg
Q_s -   volumetric flow rate at the pump inlet, dm^3/min

Q_t -   volumetric flow rate at the pump outlet, dm^3/min
Q_u -   volumetric flow rate at the pump sink, dm^3/min
r -   radius, m
R -   radius of grain position, m
r_z -   grain (sphere) radius, m
u_t -   mass fraction of the sand from the pump outlet collected on the sieve
u_u -   mass fraction of the sand from the pump sink collected on the sieve
u_{wt} -   volume fraction of flow rate on the pump outlet
u_{wu} -   volume fraction of flow rate on the pump sink
v_c -   liquid velocity, m/s
v_{cr} -   radius component of liquid velocity, m/s
v_{cu} -   circumference component of liquid velocity, m/s
v_o -   axis component of liquid velocity, m/s
v_z -   grain velocity, m/s
w -   relative velocity of grain and liquid, m/s
η_{sk} -   effectiveness measure of separated solid grains from liquid
ψ -   resistance coefficient
ρ_c -   liquid density, kg/m^3
ρ_s -   solid density, kg/m^3
ω_c -   angular velocity of liquid, 1/s
ω_z -   angular velocity of grain, 1/s

1. INTRODUCTION

In some technological circuits or installations mechanical impurities free water is required. Mechanical impurities present in liquid can cause damage to equipment or diminish its reliability. For these reasons, when installation is supplied with water containing mechanical impurities, water purification is required (e.g. filtration). Liquid supplying such installations is usually under pressure.

Such situations make it necessary to employ two separate processes - liquid purification and increasing the liquid pressure. Commonly used liquid purification methods, using filters or hydrocyclons, create many operational problems. These problems lower the installation reliability and increase operational costs (changing filter packs). Also, there is a pressure drop on filter units.

The new proposed solution combines these two separate processes - liquid purification and increasing the liquid pressure - into one process. The essence of this solution lies in encasing separator of mechanical impurities directly in the pump. The separated mechanical impurities can be carried away periodically or continuously and directed into a settling tank. The advantage of this solution lies in the fact that the whole water stream is being purified at once, the mechanical impurities are carried outside, and the pressure drop is minor.

A mathematical model of solid bodies movement inside of discharge channel of centrifugal pump, a technical solution enabling the mechanical impurities separation and tests with their results have been presented in the paper.

2. MATHEMATICAL MODEL OF GRAIN MOVEMENT FOR A GRAIN COMING OUT OF A CENTRIFUGAL PUMP IMPELLER

Separator of mechanical impurities has been installed inside the multi-stage centrifugal pump, instead of one of the stages. The incoming mechanical impurities are separated on leaving the impeller. In order to design so positioned a separator, it has been necessary to develop a mathematical model of a solid body grain movement in water stream, coming out of a centrifugal pump impeller.

When developing the model the following assumptions have been made:

1) the radius component of the liquid velocity is equal to

$$v_{cr} = c_{m3} \tag{1}$$

and the circumference component of the liquid velocity

$$v_{cu} = c_{u3} \tag{2}$$

2) the angular momentum is constant

$$r v_{cu} = const \tag{3}$$

3) in the moment of leaving the impeller, the velocity of the solid body grain differs from the velocity of the liquid

$$v_z = k \cdot v_c \tag{4}$$

4) the grain contained in the liquid stream coming out of the impeller, is subject to the action of the following forces:
   - medium resistance (resulting from different velocities of the grain and the liquid)

$$F_N = 4 \cdot \psi \cdot \rho_c \cdot w^2 \cdot r_z^2 \tag{5}$$

   - centrifugal force

$$F_O = \frac{4}{3} \cdot \Pi \cdot r_z^3 \cdot \rho_s \cdot R \cdot \omega_z^2 \tag{6}$$

- gravity

$$F_G = \frac{4}{3} \cdot \Pi \cdot r_z^3 \cdot \rho_s \cdot g \tag{7}$$

- buoyant force

$$F_A = \frac{4}{3} \cdot \Pi \cdot r_z^3 \cdot \rho_c \cdot g \tag{8}$$

- pressure force (resulting from parabolic distribution of pressure in spinning liquid)

$$F_C = \frac{4}{3} \Pi \cdot r_z^3 \cdot \rho_c \cdot R \cdot \omega_c^2 \tag{9}$$

5) separator of mechanical impurities has a horizontal orientation (the pump axis is horizontal),
6) the impurities grain are of a spherical shape,
7) the grains of impurities come out along the whole circumference of the pump impeller,
8) in the moment of leaving the impeller the axis component of the grain velocity

$$v_o = 0 \tag{10}$$

In order to calculate the grain movement parameters after leaving the impeller with the water stream, the liquid velocity components are needed: the circumference component c_u and the radius component c_r (c_{u3} i c_{m2}). For nominal parameters of pump operation these values are calculated when designing the impeller.

Taking the above assumptions into consideration, an algorithm and a program for determining the grain movement trajectory have been developed. The calculations have been done for various parameters resulting from flow-through system dimensions, density, grain diameters and kinematic parameters of the grain and the liquid.

An analysis of the calculations shows that it is possible to chose such dimensions of an mechanical impurities separator encased in the pump, that the grain coming out of the impeller will cross the outer surface of the liquid discharge channel; so it is also possible to construct such a device. In practice the main problem is to separate and hold back the grains in such a way, that they do not re-enter the liquid stream flowing to the next pump stage.

3. DESCRIPTION OF THE PUMP SEPARATOR OF MECHANICAL IMPURITIES

The pump separator of mechanical impurities has been constructed using elements of a centrifugal 3-stage pump with single-suction impellers, enclosed in a horizontal system with centrifugal and centripetal stators. The pump casing id divided in planes perpendicular to the shaft axis.

For the purpose of testing, various types (5 structural versions) of separator of mechanical impurities have been installed in the multistage pump at the casing of the second stage. The separator of impurities which gave the best results has been shown in Fig. 1.

The liquid coming out of the impeller flows into the centripetal stator inside a cylindrical housing. Between the outer surface of the housing and the pump casing there is a chamber from which the mechanical impurities are carried away from the pump. There are dividers and spacer rings on the inner surface of the housing. The spacer rings have holes for taking the mechanical impurities away. The inside edges of the dividers are sharpened, which prevents the impurities grains from bouncing off the inside surface of the dividers.

4. PURPOSE AND METHODOLOGY OF TESTS

The purpose of the testing was to determine the effectiveness of mechanical impurities separation and to establish the flow-through system characteristics of pump with a separator of impurities. Testing has been conducted for all structural types of developed separator solutions.

4.1. Effectiveness of impurities separation testing

The separation effectiveness has been tested for pump operation with different rates of delivery and with different volumetric flow rates of the liquid taken away with mechanical impurities through the sink. The purpose was to determine the range of separation effectiveness. In place of mechanical impurities, quartz sand of specific granularity was used, dosed at the pump inlet. The sand coming out from the pressure conduit and from the sink was collected on sieves, then dried and weight.

4.2. Pump operation parameters measurement

The pump operation parameters were taken for a "normal" pomp i.e. without the separator of impurities and for pump with separator of impurities. The purpose of the testing was to determine the pump operation characteristics of a pump with a separator in comparison to a "normal" pump.

Testing and measuring pump operation parameters was done at a test site for pumps in an open system. Parameters were taken in accordance with the requirements of Polish Standards for measuring impeller pump characteristics (PN-85/M-44005), in close class of fit. The testing was done across the whole range of delivery rates, for a pump with a closed valve on the pumping side, through to a fully open one. The measurements were taken for clean water.

5. TESTS RESULTS

Based on the testing, test results have been calculated.

5.1. Separator of impurities measurement results

Based on the testing, the following values have been calculated:
- flow rate at the pump inlet

$$Q_s = Q_t + Q_u \tag{11}$$

- volume fraction of flow rate at the pump sink

$$u_{wu} = \frac{Q_u}{Q_s} \tag{12}$$

- volume fraction of flow rate at the pump outlet

$$u_{wt} = \frac{Q_t}{Q_s} \tag{13}$$

- total mass of collected sand

$$m = m_t + m_u \tag{14}$$

- mass fraction of sand from the pump sink collected on the sieve

$$u_u = \frac{m_u}{m} \tag{15}$$

- mass fraction of sand from the pump outlet collected on the sieve

$$u_t = \frac{m_t}{m} \tag{16}$$

As the effectiveness measure of separating solid body grains from a stream of liquid the following were taken

$$\eta_{sk} = \frac{m_u}{m_u + m_t} \tag{17}$$

Test results for sand granularity of 0.2 to 0.8 mm in the $u_u = f(u_{wu})$ system, shown in figure 2. Fig. 3 shows the impurities separator characteristics in the $\eta_{sk} = f(Q_t, Q_u)$ system.

5.2. Pump operation parameters measurement results

In the pump with separator of impurities the flow-through system has been changed. One of the pump stages contains the separator of impurities. The proposed structure is in a sense one of the pump stages. It has an impeller and centripetal stator. It has no centrifugal stator; the flow-through channel between the impeller outlet and the centripetal stator has built-in elements which impede the flow. For these reasons the pump operation characteristics of a pump with the separator of impurities may differ from the operation characteristics of a pump without the separator.

Based on the tests and calculations, graphs (fig. 4) of pump operation characteristics of the pump with the separator of impurities and "normal" pump have been constructed. These characteristics show relations between the pump delivery head, power input and the pump efficiency as a function of delivery rate.

Comparing the two systems - one with and one without the separator of impurities, one can see that a pump with the separator has a lower delivery head as well as efficiency. Differences in power input on the shafts are minor.

Of all the tested solutions, the one we discussing has shown the smallest drop in parameters.

6. CONCLUSIONS

Analyzing the tests results one can see that:

1) using the separator of mechanical impurities build-in the pump one can separate mechanical impurities from the liquid,

2) the separation effectiveness increases when the flow rate at the pump sink and decreases when the flow rate on the pumping outlet increases,

3) building into the pump the separator of mechanical impurities changes pumps operation parameters and characteristics, i.e. lowers delivery head as well as efficiency.

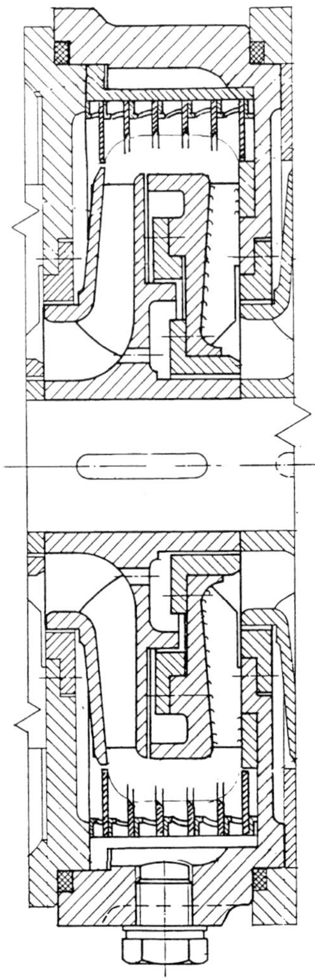

Fig.1. The separator of impurities

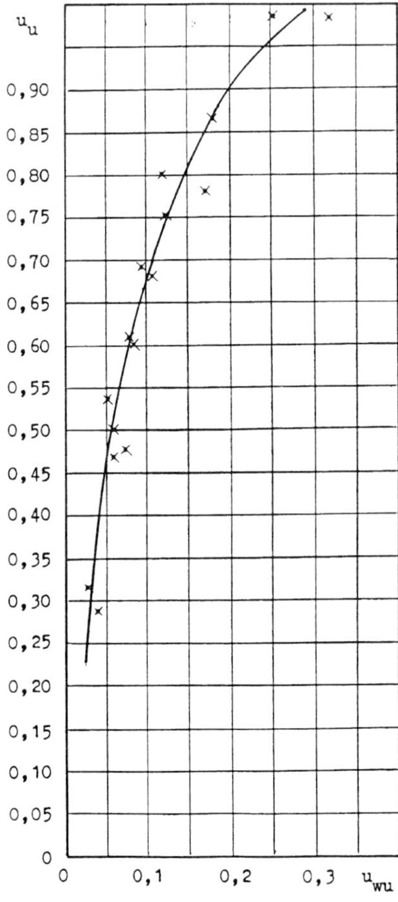

Fig.2. Relation of mass fraction of the impurities at the sink
to liquid flow rate friction through the sink

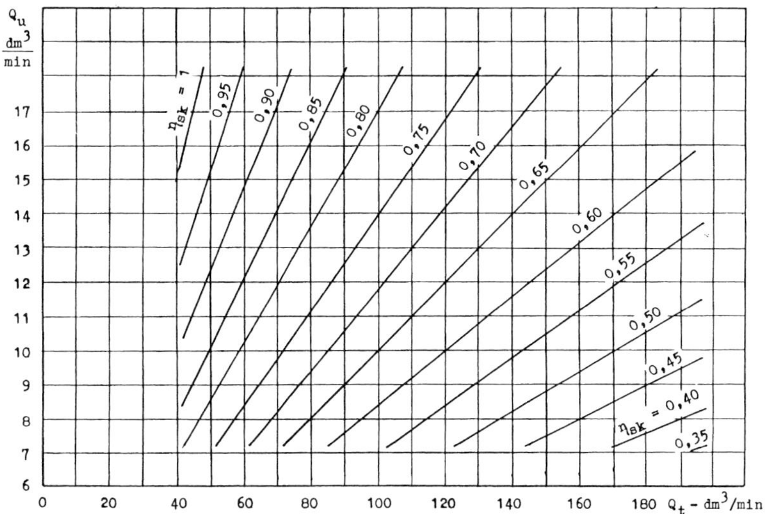

Fig. 3. Separator of mechanical impurities characteristic

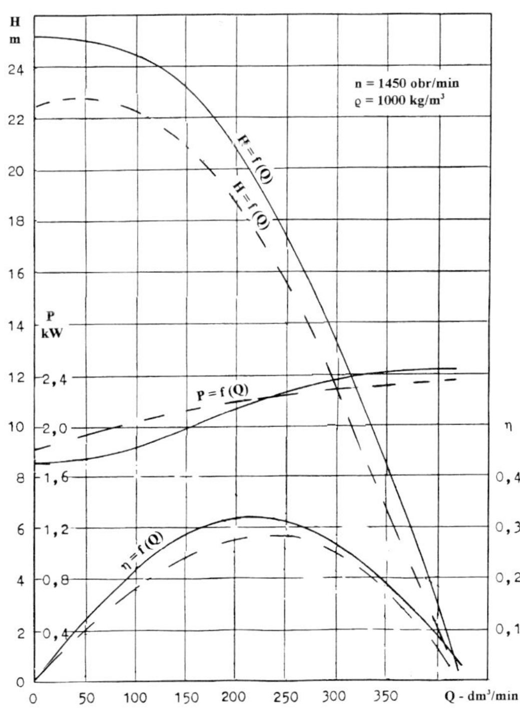

Fig. 4. "Normal" pump operation characteristics (solid line) and pump with separator of impurities operation characteristics (broken line)

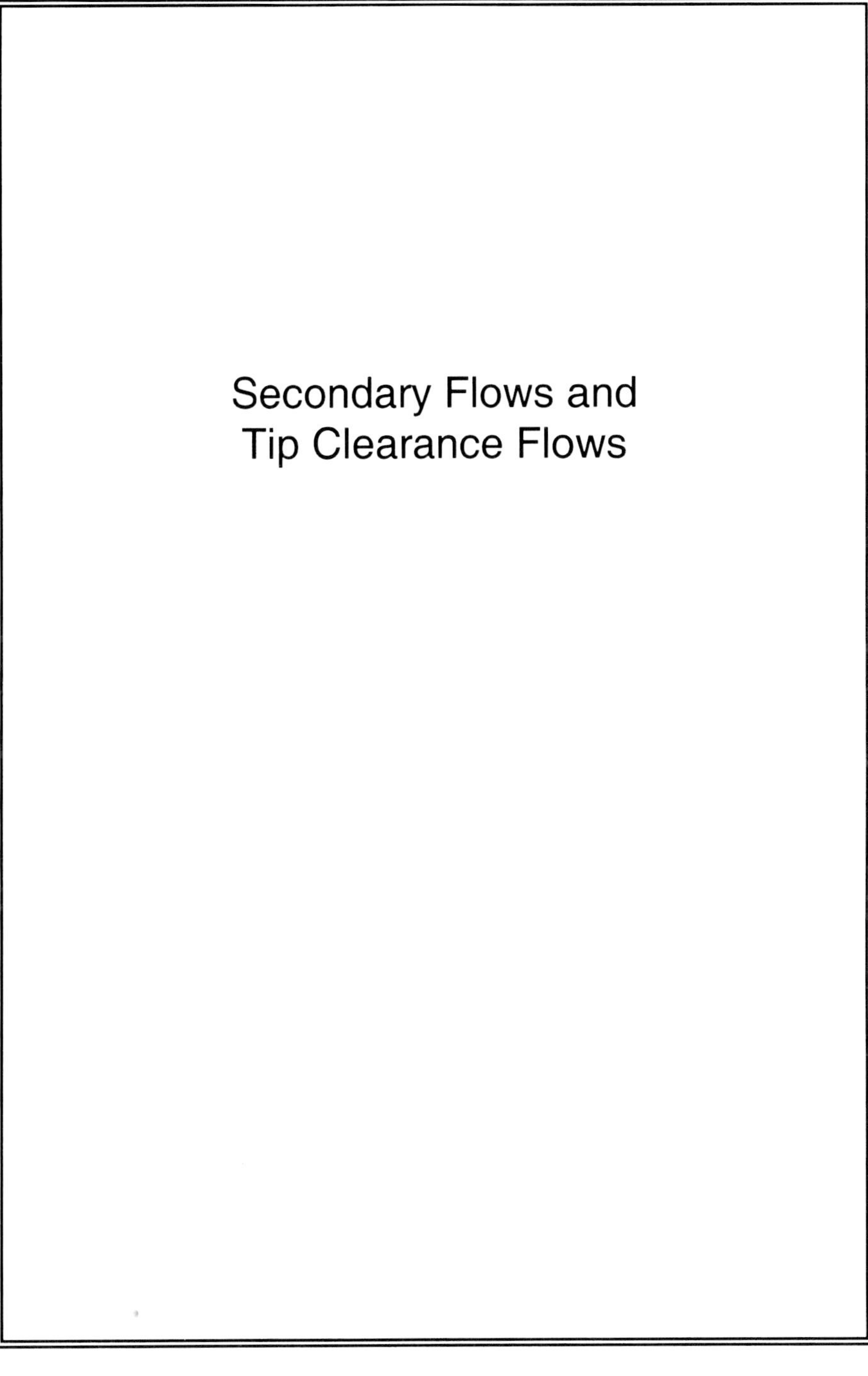

Secondary Flows and Tip Clearance Flows

Horseshoe vortex control by suction through a slot in the wall cylinder junction

D P GEORGIOU and **V A PAPAVASSILIPOULOS**
Department of Mechanical Engineering and Aeronautics, University of Patras, Rion-Patras, Greece

SYNOPSIS

The present study investigates experimentally the evolution of the secondary flow field in a bluff body (formed by a rectangle with a half-cylinder leading edge) flat endwall junction when light suction of the air is implemented through a slot. Three slot configurations were employed in the present study, one on the plane of symmetry and the two others on the junction corner covering arcs of $60°$ and $180°$ respectively.

The study investigates the merits of light suction in eliminating the horseshoe vortex since the no-suction data indicated that this vortex originates in the plane of symmetry and then it simply dissipates exponentially in strength as it bends around the junction.

The data (velocity vector, total and static pressure) were taken by transversing a miniature (1.2mm OD) 5-hole Pitot tube probe in 4 radial planes (with respect to the half-cylinder centerline), inclined at $30°$, $50°$, $70°$ and $90°$ from the symmetry plane.

The results indicate that although the light suction eliminates the horseshoe vortex, after the $60°$ plane a new streamwise vortex is formed by a mechanism similar to that leading to the "passage" vortex in bend ducts. This leads to the conclusion that the elimination of the horseshoe vortex requires a complex (possibly two step) procedure.

NOMENCLATURE

C_{PT}	Total pressure coefficient	z_c	Vortex core distance from bluff body wall
D	Diameter of the bluff body leading edge	y_c	Vortex core distance from the endwall
u	Streamwise velocity component		

v	Velocity component along the y direction	$\overline{C_{PT}}$	Mass average pressure loss coefficient $= \dfrac{\int \rho u C_{PT} dA}{\int \rho u dA}$
w	Velocity component along the z direction	v^*	Dimensionless resultant secondary flow field vector $= \sqrt{v^2 + w^2}/U_0$
x	Local streamwise direction	**Greek**	
y	Direction normal to the endwall	β	Radial plane angle relative to the plane of symmetry
z	Direction parallel to the endwall	Γ	Circulation $= \int \omega_x dA$
U_o	Free stream velocity	δ	endwall boundary layer thickness
S	Vortex trajectory arc length from the plane of symmetry	ω_x	Streamwise vorticity component
L	Width of the slot	$\omega_{x\,max}$	Maximum streamwise vorticity at vortex core

1. INTRODUCTION

The horseshoe vortex is encountered in many engineering applications, such as the wing-fuselage junction of the aircraft, the blade-hub junction of the turbomachinery, the sail-hub junction of submarines, the bridge-piers junction in the river bridges etc. As many previous studies point out, the effects of this vortex on the flowfield around and downstream of the bluff bodies that generate it are usually undesirable.

Eckerle and Langston (1) have shown experimentally that the horseshoe vortex formation around a circular cylinder is associated with high total pressure losses and a large streamwise vorticity component in the core of the vortex. Pierce and Sin (2) have studied the evolution of the horseshoe vortex around a wing-like bluff body. They showed that the vortex persists even up to a distance of two wing chords downstream of the wing. Devenport and Simpson (3) performed LDV measurements on the symmetry plane upstream of the bluff body (a symmetric airfoil wing) and have found that the vortex was unsteady. In a blade this will lead to vibration and dynamic stresses. Baker (4) found that there can be two types of vortex unsteadiness, i.e. either oscillations of the entire vortex system or instabilities in the vortex core only. Besides the total pressure losses and the velocity non-uniformity, the horseshoe vortex has been shown by Fisher and Eibeck (5) to modify significantly the convective heat transfer coefficient distribution on the endwall. The measurements by Papavassilopoulos and Georgiou (12) on the evolution of the secondary flow field in a horseshoe vortex indicated that:

(i) The vortex is created on the plane of symmetry. The vortex maximum streamwise vorticity and corresponding circulation dissipate exponentially as the vortex bends around the junction and then nearly parallel the bluff body side.

(ii) Although the vortex dissipates in a continuous way, as it turns around the junction the distance of the vortex core from the endwall and the bluff body increases suddenly after the $60°$ radial plane, indicative of another mechanism in the evolution of the vortex.

In the turbomachinery blades channels the two horseshoe vortex legs are part of the total secondary flow phenomena. Their contribution to the total pressure losses is rather small, but they modify the flow field in the early part of the channel significantly (Georgiou et al.(11)).

This modification influences not only the velocity field (transition, separation etc.) but it also increases the endwall heat flux considerably, which in turn leads to an increased consumption of cooling air.

The apparent need to reduce (or better eliminate) the horseshoe vortex has driven a number of experimental efforts to study the merits of various ideas. Mehta (6) compared experimentally the effect various leading edge shapes have on the horseshoe vortex characteristics and found that a smaller and weaker vortex was created when a less blunt leading edge was used. Kubendran et al. (7) reported a reduction of the streamwise vorticity with the use of a leading edge swept fairing. Devenport et al (8) examined the effects of a simple fillet on the unsteadiness of the horseshoe vortex. They also examined the effects of the angle of attack and the boundary layer thickness. Phillips et al. (9) reported on a novel approach, where the incoming boundary layer was removed by strong suction through a slot on the endwall, immediately upstream of the wing. They found that the suction reduces the size and circulation of the horseshoe vortex significantly. When the suction volumetric flow rate was twice the boundary layer flow rate ahead of the bluff body width, the vortex was no longer visible.

These data by Papavassilopoulos and Georgiou (12) lead to the speculation that the horseshoe vortex could be eliminated by eliminating the recirculation zone in the plane of symmetry. The present study investigates experimentally this by introducing suction through a slot on the plane of symmetry and of a length comparable to the distance between the center of the recirculation and the junction point. The results indicated that although the suction did eliminate the horseshoe vortex up to the 50° plane, downstream of this plane another vortex was created by a mechanism similar to that of the "passage vortex" in bend ducts. The suction study was then extended to suction through thin slots (L/d=0.05) forming 60° and 180° arcs on the junction corner. The data produced similar results. This leads to the apparent conclusion that, for the elimination of the horseshoe vortex two mechanisms must be applied in parallel, one for the recirculation zone in the plane of symmetry and another for the "passage" vortex.

In turbomachinery applications (i.e. compressor or turbine stages) the light suction method has to be balanced against the mass flux leak and the associated fluid enthalpy losses. In general however, this loss is smaller than the gains, the lower total pressure losses and much reduced heat flux and velocity non-uniformity inside the blade channel.

2. EXPERIMENTAL FACILITIES AND PROCEDURE

2.1 Experimental Facilities

The measurements were performed in the subsonic wind tunnel facility of the Thermal Engines Laboratory, University of Patras. The rectangular test section was 0.30m wide x 0.20m height. It had a length of 2m. The operating velocity range was 0-22 m/sec and the freestream turbulence intensity measured by a HWA was found to be below 0.1%. The mean velocity in the test section was measured at the nozzle exit by a Prandtl pitot tube.

The vortex was generated at the junction between the test section endwall and a bluff body formed by a 0.21m x 0.06m rectangle with a D=0.06 m diameter half-cylinder in front of it. The height of the body was 0.20m, thus extending to the entire test section height. The body presented a blockage of 20% to the test section cross-sectional area. Its centerline was aligned with the test section centerline, while its leading edge was positioned 1.5m downstream of the test section inlet. A schematic view of the body coordinates is shown in figure (1a).

The five hole Pitot tube used for the measurements had an overall tip diameter of 1.2 mm and it was made in the Thermal Engines Laboratory by hypodermic tubes. Each hole of the five hole probe was connected to a differential pressure transducer with an output voltage range of ±5V over a pressure range of ±1000Pa. The five voltage signals of the pressure transducers were acquired simultaneously by a data acquisition card with a 12 bit resolution. Data were stored in a PC connected with the data acquisition card.

The five hole probe was mounted on a two-dimensional transversing system allowing measurements in y-z planes with a positioning accuracy of 0.01mm.

Three slot configurations were employed. The first two had a slot width of 3mm and formed arcs of $180°$ and $60°$ in the body wall junction, while the third was located along the plane of symmetry of the body with a cross section of 0.005m x 0.02m just upstream the leading edge. The slot configurations are shown in figure (1b).

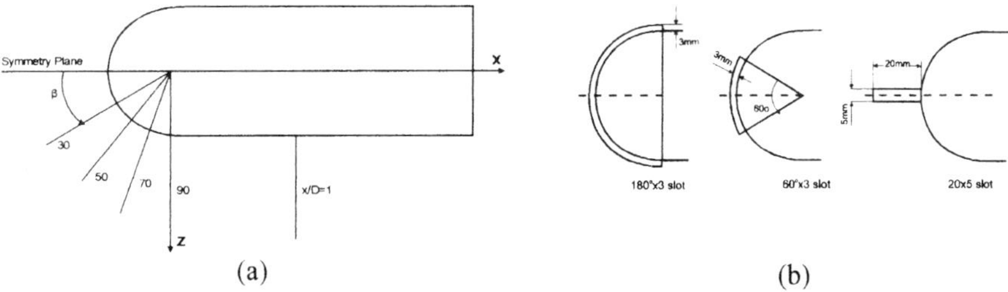

Figure 1. The test conditions, (a) Measurement planes and body coordinates. (b) The three slot configurations.

These slots allowed for the elimination of the initial horseshoe vortex at suction rates that were less than 5% the suction rate Phillips et al (9) found necessary in order to eliminate the entire horseshoe vortex. In addition the slot widths were of order 10% to 20% the boundary layer thickness. Both values indicate a minimum disturbance.

2.2 Experimental Procedure

All experiments were performed with a freestream velocity of U_o=13.5 m/sec on the test section inlet. On the position of the bluff body leading edge the undisturbed turbulent boundary layer thickness was δ =0.03m, while the Re number based on the bluff body width (D) was 5.36×10^4.

Measurements were taken in four radial planes (β =30°,50°,70°,90°) and in a plane located at one body width downstream the β =90° radial plane, (fig (1a)). Data were acquired from y=2mm up to y=50mm from the endwall. Grid points were spaced at intervals of 1mm in the range $2\text{mm} \le y \le 10\text{mm}$ and at intervals of 2mm in the rest of the height. In the β =30°, 50°, and 70° radial planes measurements were taken form z=2mm to z=30mm away from the body leading edge, while in the β =90° and x/D=1 planes from z=6mm to z=60mm from the body wall. In all measurement planes the grid spacing in the z direction was 2mm. The grid density was about one order of magnitude larger when compared to those of other similar studies.

The Pitot probe was no closer than at least 1.5 times the tip diameter to the endwall in order to reduce the wall interference effects. Near the bluff body the measurements indicated yaw angles greater than $\pm40°$ (i.e. outside the calibration limits). When this yaw angle was exceeded the data were considered unreliable and rejected.

The five hole probe was calibrated at a velocity of U_o=13.5m/sec so as to minimize the Reynolds number effects and it was used in the non-nulling mode. It was calibrated in a jet facility following the method recommended by C. D. Wickens and C. D. Williams (10) . The calibration was performed for yaw angles between -45° and +45° and for roll angles between 0 and 360°. During the calibration and the measurements, each pressure signal was sampled at 500Hz over a period of 10 seconds.

The vorticity data were generated by differentiation from the velocity data. The average of the forward and the backward difference approximations (which is equivalent to the central differencing scheme) was used for the required partial derivatives. The circulation was determined from the vorticity data by a area integration. The area of integration was defined from the vortex center up to where the vorticity decreased to 20 percent of the peak value at the vortex center. The same area was employed for the C_{PT} integration.

The error estimates for the data acquisition and analysis process for this configuration have been discussed in an earlier paper by Georgiou et al. (12).

The suction was done by exhausting through the slot into the atmosphere below through a ventouri that was employed to measure the flow rate. Based on the structure of a turbulent boundary layer with a thickness of 30mm and a free stream velocity of 13.5m/sec, the ratio of the suction flow rate to the flow rate of the incoming boundary layer over a width equal to that of the projected slot width was: (a) 0.037 for the 180° x 3mm slot, (b) 0.0616 for the 60° x 3mm slot, (c) 0.44 for the 20mm x 5 mm slot. When compared to the ratio of 2 reported by Phillips et al (9) (suction rate over the boundary layer rate for the entire width of the bluff body) these suction rates can be classified as "light"

In a turbomachinery application, where the typical blade leading edge diameter to blade spacing is of order less than 5%, the blade spacing to height ratio is of order 1 and the endwall boundary layer thickness is less than ¼ the blade height, the symmetry plane slot suction leads to an airflow leakage ratio of less than 0.1% per stage, i.e a very small amount.

3.RESULTS AND DISCUSSION

The data in figure 2. give the secondary velocity vector distribution on the planes where measurements took place. The solid wall (no suction) data, in all measurement planes, present a well defined vortex structure (the horseshoe vortex leg) that gradually grows in size while its center moves away from the two walls. These data have been analyzed in Ref. (12). In all velocity vector figures only the lower part of the measurement grid is shown (y/D=0.45) for reasons of clarity. The data indicated that near y/D=0.6 the flow was two-dimensional.

In the 180° x 3mm slot (figure 3), a vortex structure appears only after the $\beta=70°$ plane. In the earlier planes ($\beta=30°$ and $\beta=50°$) there is no vortex, the flow moving into the slot. At the $\beta=70°$ plane the secondary flow near the slot (and the corner) does not exhibit any tendency to flow into the slot, probably due to the small pressure difference between the mainstream and the air below. In all planes however the secondary velocity vectors are smaller when compared with the solid wall case. In the 60°x3mm slot the $\beta=30°$ data are similar to the 180°x3mm slot ones. The vortex is clearly much weaker than the vortex for the solid wall case at the corresponding planes.

The data for the 20mm x 5mm slot, presented in figure 4, show that a small very weak vortex may exist in a radial position near the upstream edge of the slot. After the $\beta=70°$ plane however, a significant vortex appears. Apparently something is generating a new vortex between the planes $\beta=50°$ and $\beta=70°$. The lifting of the vortex core from the endwall surface is quite apparent.

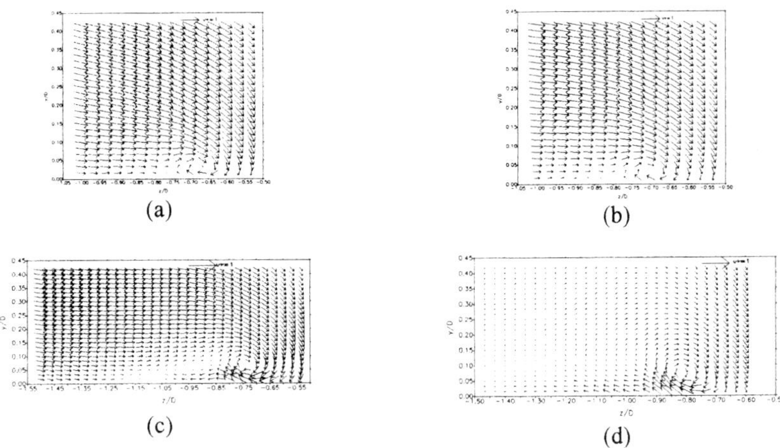

(a)   (b)

(c)   (d)

Figure 2. Secondary velocity flowfields for the solid wall case, (a) 30° plane, (b) 50° plane, (c) 70° plane, (d) 90° plane.

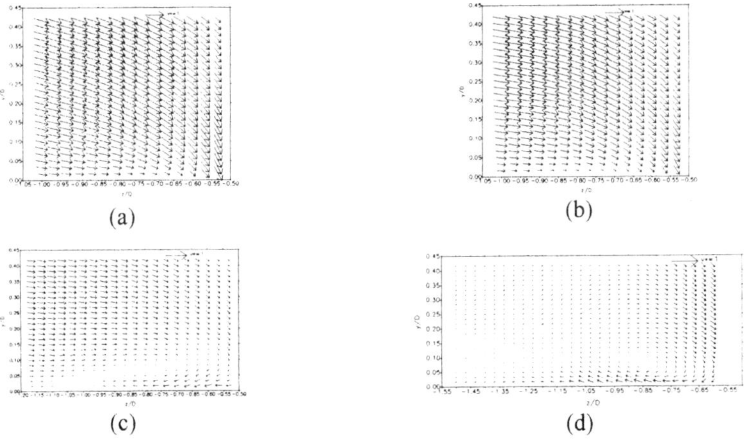

(a)   (b)

(c)   (d)

Figure 3. Secondary velocity flowfields for the 180° x 3mm suction slot case, (a) 30° plane, (b) 50° plane, (c) 70° plane, (d) 90° plane.

Figure 5 gives the corresponding streamwise flowfields (i.e. the u / U_0 distribution) for the solid wall case. All planes $30° \leq \beta \leq 90°$ exhibit the well known corner distribution, where the ISO-(u / U_0) contours form radial lines originating near the corner. The presence of the horseshoe vortex leg modifies this distribution accordingly. An important indicator of the much reduced size and strength of the vortex leg was the much reduced influence of the vortices on the ISO -(u / U_0) lines in all the three suction slot configurations. In most cases

C557/135/99

the presence of the vortex can not be detected from these distributions. The 20mm x 5mm slot (presented in figure 6) exhibits the smallest modification on the ISO-(u / U_0) lines.

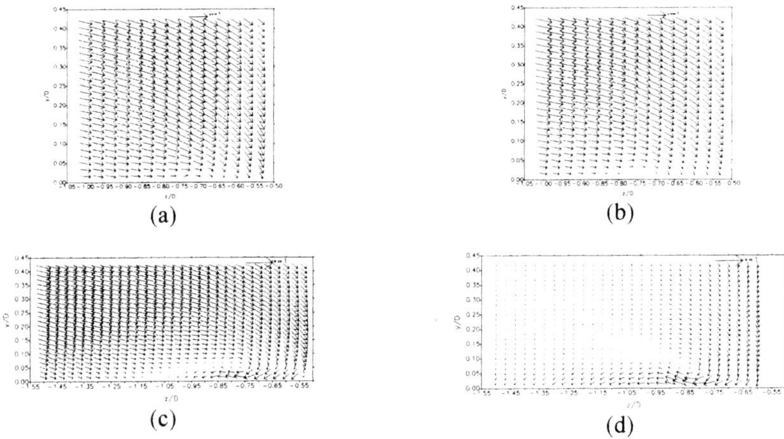

Figure 4. Secondary velocity flowfields for the 20mm x 5mm suction slot case, (a) 30^0 plane, (b) 50^0 plane, (c) 70^0 plane, (d) 90^0 plane.

The wall jet effect observed in the solid endwall study (Papavassilopoulos and Georgiou (12)), remains in the presence of light suction. This supports the assumption that the flow acceleration near the bluff body, (appearing as a wall jet) is due to the sweeping of the endwall boundary layer, as it accumulates on the body leading edge.

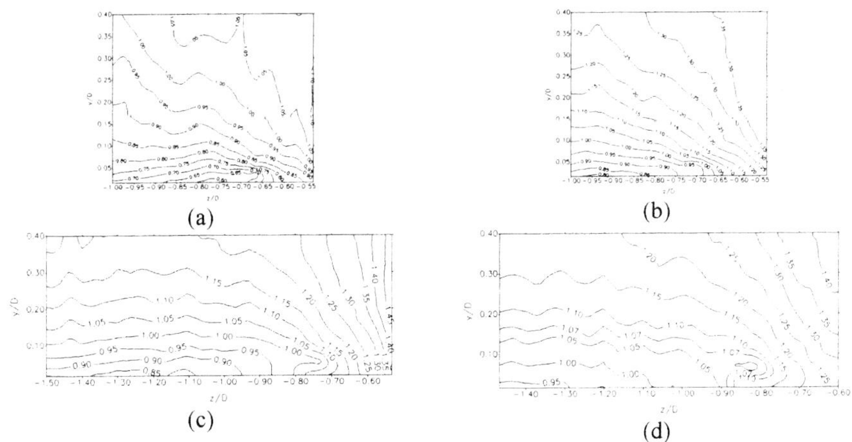

Figure 5. Streamwise velocity contours (u / U_0) for the solid wall case, (a) 30^0 plane, (b) 50^0 plane, (c) 70^0 plane, (d) 90^0 plane.

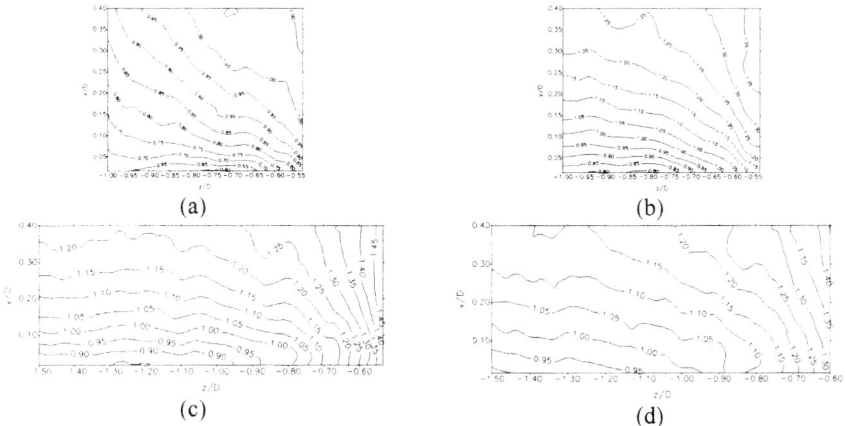

(a)

(b)

(c)

(d)

Figure 6. Streamwise velocity contours (u / U$_0$) for the 20mm x 5mm slot case, (a) 30° plane, (b) 50° plane, (c) 70° plane, (d) 90° plane.

Figure 7 gives the distribution of the streamwise vorticity. All data show a sharp reduction in the strength of the vortex. If we employ the x/D=1 plane as a basis, all three suction configurations lead to a vortex of nearly half the strength when compared to that of the solid wall case. The evolution of the maximum streamwise vorticity ($\omega_{x\,max}$)and vortex circulation (Γ) are given in the figures 8 and 9.

The data in figure 8 show that there is a significant difference between the evolution of the max. ω_x in the solid wall and those in the three suction slot cases. In the solid wall case

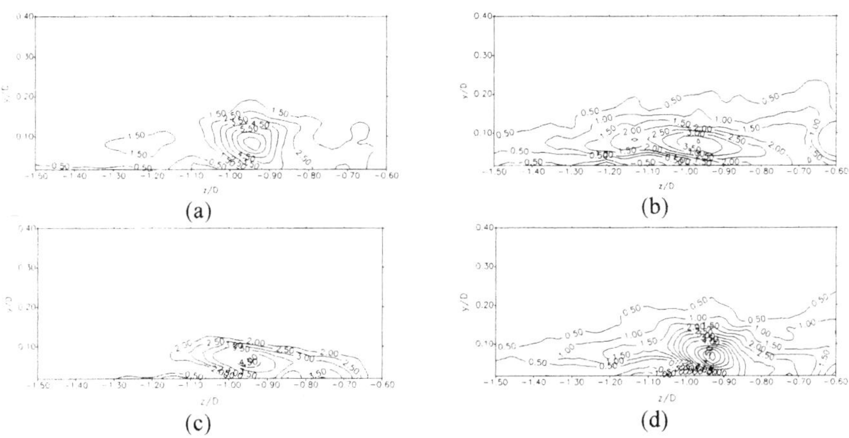

(a)

(b)

(c)

(d)

Figure 7. Streamwise vorticity ($\omega_x\, D/U_o$) contours in the x/D=1 plane, (a) solid wall case, (b) 180° x 3mm slot case, (c) 60° x 3mm slot case, (d) 20mm x 5mm slot case.

C557/135/99

the vortex is formed earlier than the $\beta=30^{\circ}$ plane and dissipates exponentially after its creation, while in the suction slot cases the vorticity increases continuously up to $\beta=70^{\circ}$ and then dissipates.

The circulation (Γ) data in figure 9 give a better picture. In the solid wall data, the circulation dissipates exponentially, as in any swirling flow. Apparently the vortex originates on the recirculation zone of the symmetry plane. The suction slot data suggest a different phenomenon. No significant vortex is observed earlier than the $\beta=70^{\circ}$ plane. The circulation increases up to 70° and then dissipates. This is exactly the behavior observed for the streamwise circulation in bend ducts (e.g. Georgiou and Papavassilopoulos (13)). Hence, the vortex observed in the suction data is nothing but the equivalent of the "passage vortex" of the ducts and is due to the sweeping of the incoming boundary layer away from the symmetry plane. In the solid wall data this second mechanism has been preceded by the one that creates the vortex on the plane of symmetry and can not be identified. At the $\beta=70^{\circ}$ plane the circulation for the 20x 5mm slot is only 2/3 that of the corresponding solid wall value, while for the other slots it is even smaller. Apparently, in the solid wall case the initial vortex is much stronger than the "passage vortex" component and dominates the secondary flowfield.

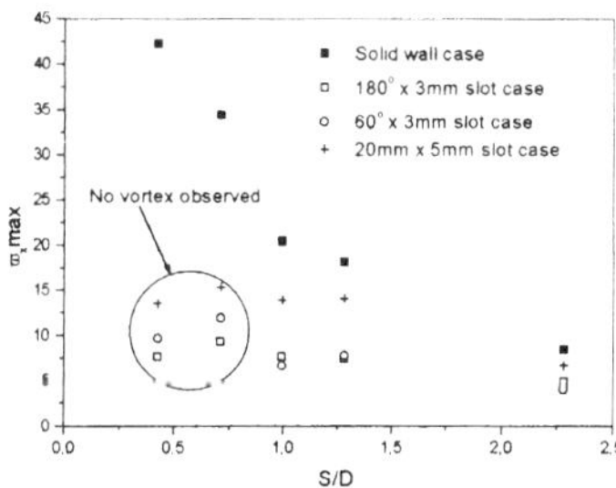

Figure 8. Evolution of the maximum value of streamwise vorticity $\omega_{x\ \mathbf{max}}$

Figure 9. Non dimensional circulation in the leg of the horseshoe vortex.

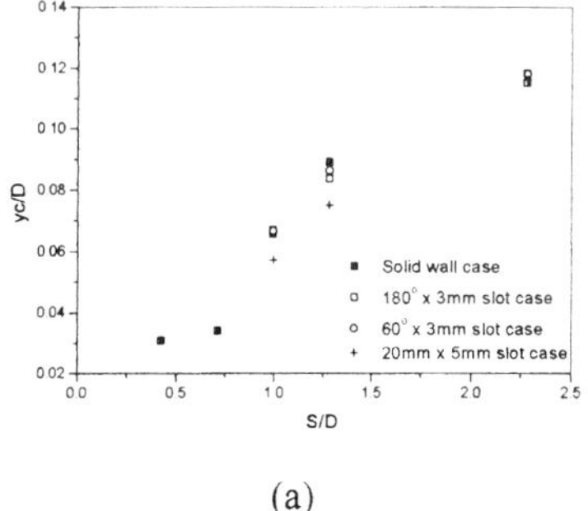

(a)

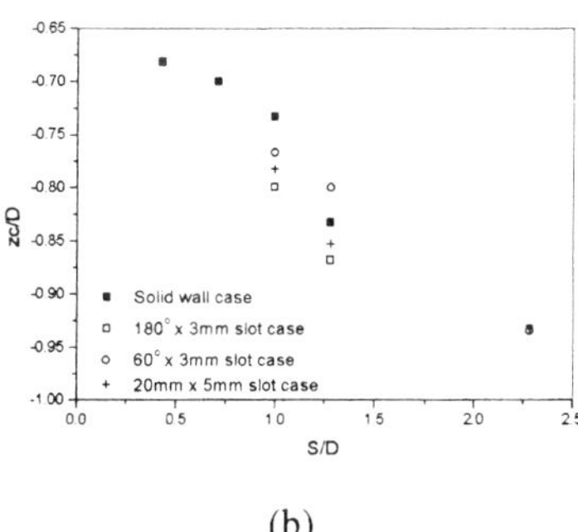

(b)

Figure 10. Trajectories of the maximum streamwise vorticity, (a) vertical, (b) lateral.

The sweeping however of the incoming boundary layer influences the position of the vortex core after the $\beta=70°$ plane in all cases. Here, the horseshoe vortex leg is assumed to be a vortex line originating on the symmetry plane so that its trajectory length (S) can be calculated from the vortex core (y, z) position on the various measurement planes. Since the measurements started from the $\beta=30°$ plane, the origin is assumed to be on the $\beta=0°$ plane the same radial position as for the $\beta=30°$ plane. The vortex core center points for the suction cases in any plane (when they are observed) are positioned against the corresponding (S/D) distance of the solid wall configuration for the same plane. Figure 10 shows that both the vertical and the lateral positions of the maximum streamwise vorticity point are not affected significantly by the suction. In addition the data for the solid wall (Papavassilopoulos and Georgiou (12)) showed a sudden change in the (y, z) position of this point after the $\beta=50°$ plane, apparently due the action of the "passage" vortex mechanism after that plane.

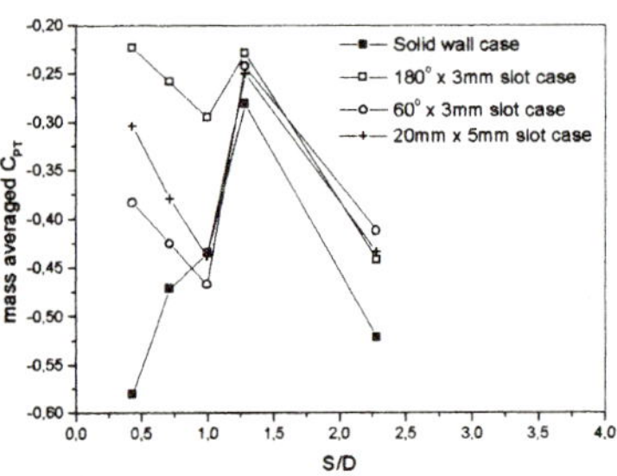

Figure 11. The evolution of the mass averaged pressure loss coefficient $\overline{C_{PT}}$.

The evolution of the total pressure loss coefficient $\overline{C_{PT}}$ (integrated over the same region with the streamwise circulation of the solid wall case) is presented in figure 11. The sharp difference between the suction and solid wall distributions can be explained as follows: (i) In the solid wall case the initial horseshoe vortex core is a high loss region but as the vortex bends around the junction the very strong entrainment of low loss fluid, leads to a $\overline{C_{PT}}$ reduction up to the $\beta=90°$ plane. Downstream of this plane the dissipation process overtakes the entrainment and $\overline{C_{PT}}$ increases again. (ii) In the suction cases the lack of the vortex entraining action leads to a continuous $\overline{C_{PT}}$ rise in the same region that the solid wall data gave the vortex core. After the $\beta=70°$ plane, however, the "passage vortex" action leads again to a reduction of $\overline{C_{PT}}$ up to the $\beta=90°$ plane and a subsequent rise after that plane.

4. CONCLUSIONS

Weak suction was applied in the junction of a body with its supporting endwall. The body was of a rectangular cross-section with a half cylindrical leading edge. Measurements of the mean velocity vector on a number of planes lead to the determination of the secondary flowfields associated with the endwall boundary layer near the junction. The no suction case produced the well known horseshoe vortex structure. Three suction slot geometries were employed one on the symmetry plane and the other two in the corner of the junction, the first with a $\pm90°$ arc and the other with a $\pm30°$ arc around the symmetry plane.

All three suction cases produced similar data. No horseshoe vortex leg was observed up to the plane $\beta=50°$, but downstream of it a weaker vortex was observed. This vortex appeared to possess the character of the passage vortex observed in the bend ducts.

The results lead to the conclusion that any effort to eliminate the horseshoe vortex must incorporate solutions for both the recirculation zone and the "passage vortex" mechanisms, but the observed reduction in the vortex circulation is sufficient to overcome the leakage losses in turbomachinery applications.

5. REFERENCES

1. W. A. Eckerle and L. S. Langston, «Horseshoe Vortex Formation Around a Cylinder», ASME Journal of Turbomachinery, Vol. 109, pp. 278-284, April 1987.
2. F. J. Pierse and J. Shin, «The Development of a Turbulent Junction Vortex System», ASME Journal of Fluids Engineering, Vol. 114, pp. 559-565, December 1992.
3. W. J. Devenport and R. L. Simpson, «Time-Dependent and Time-Averaged Turbulent Structure Near the Nose of a Wing-Body Junction», Journal of Fluid Mechanics, Vol. 210, pp. 23-55, 1990.
4. C. J. Baker, "The Oscillation of Horseshoe Vortex Systems", ASME Journal of Fluids Engineering, Vol. 113, pp. 489-495, 1991.
5. E. M. Fisher and P. A. Eibeck, «The Influence of a Horseshoe Vortex on Local Convective Heat Transfer», ASME Journal of Heat Transfer, Vol. 112, pp. 329-334, May 1990.
6. R. D. Mehta, «Effect of Wing Nose Shape on the Flow in a Wing Body Junction», Aeronautical Journal, Vol. 88, pp. 456-460, December 1984.
7. L. R. Kubendran, A. Bar-Sevar, W. D. Harvey, «Flow Control in a Wing-Fuselage Type Juncture», AIAA Paper 88-0614, 26th Aerospace Sciences Meeting, Reno NV, Jan 1988.
8. W. J. Devenport, M. B. Dewitz, N. K. Agarwal, R. L Simpson and K. Poddar, «Effects of a Fillet on the Flow Past a Wing-Body Junction», AIAA Journal, Vol. 28, No 12, pp. 2017-2024, Dec 1990.
9. D. B. Philips, J. M. Cimbala and A. L. Treaster, «Suppression of the Wing-Body Junction Vortex by Body Surface Suction», Journal Of Aircraft, Vol. 29, No 1, pp. 118-122, Jan-Feb 1992.
10. C. D. Wickens, C. D. Williams, «Calibration and Use of Five Hole Flow Directional Probes for Low Speed Wind Tunnel Application», NAE Report, NAE-AN-29, 1985.
11. D. P. Georgiou, V. A. Papavasilopoulos and M. Alevisos, «Experimental Contribution on the Significance and the Control by Transverse Injection of the Horseshoe Vortex», ASME Paper 96-GT-255, International Gas Turbine and Aeroengine Congress & Exhibition, Birmingham, UK , June 10-13 1996.
12. Papavassilopoulos V. A., Georgiou D. P. "The evolution of the secondary flow in the horseshoe vortex. Part 1: The mean flow", Journal Of Fluid Mechanics (submitted).
13. Georgiou D. P., Papavassilopoulos V.A, "The evolution of the mean secondary flow and the total pressure losses in a $90°$ bend duct", Journal Of Fluid Mechanics (submitted).

An experimental investigation of secondary flow in a low aspect ratio impulse cascade at different inlet flow angles

V MOLNÁR, F RIDZON, and **T SPORINA**
Slovak Technical University, Bratislava, Slovak Republic

The paper describes the effect of incidence angle on the three-dimensional flow in a linear large scale impulse cascade. The blade chord length in the experiment was 500 mm. The results are presented for a blade aspect ratio of 0.8 and incidence angles of $-14.4°$, $-0.6°$ and $+8.6$. The details of the 3D flow were examined at five planes within the blade passage and at one plane behind the trailing edge. The 3D flow field has been obtained by traversing a 4 mm spherical fivehole probe. The flow details are presented in terms of loss coefficient, axial vorticity, secondary velocity and flow angles. The mass averaged results are presented in terms of spanwise distribution of loss coefficient, secondary flow angle deviation and growth of the loss through the cascade. The test results includes static pressure measurements over the whole blade surface. These show the 3D flow development on the blade under various incidence angles.

Nomenclature

Symbols		Superscripts	
a.r.	aspect ratio (span/chord)	MS	mid span
Cp	static pressure coefficient	-	pitchwise mass averaged value
i	incidence angle (β-blade inlet angle)	=	area mass averaged value
p, p$_0$	static pressure, total pressure	Subscripts	
R,U,Z	radial/spanwise, pitchwise, axial direction	0,1,,6	measuring planes
w	flow velocity	in	measuring plane 0
α	flow angle measured from endwall	out	measuring plane 6
β	flow angle with respect to pitchwise direction	R,U,Z	radial/spanwise, pitchwise, axial direction
ξ'	total pressure loss coefficient	s	secondary velocity
Abbreviations			
LE,TE	blade leading, trailing edge		
PS,SS	blade pressure, suction side		

1. INTRODUCTION

The operation conditions of a turbine blade row may extend over quite a wide range. Changes in rotational   speed and in flow rate are the main cause of incidence angle variations in turbine stages. In low aspect ratio stages the secondary effects are often dominant. Secondary flow is generally understood as the difference between the actual 3D flow and a primary flow, usually the flow at the mid span in linear cascades. Difficulties occur in this understanding if the blade aspect ratio is low and the flows originated from both ends of the blade interact at mid span. In the past a large number of experimental investigations on secondary flow in turbine cascades have been carried out in order to understand the cascade internal flows and the loss generation mechanisms in more detail [1-5]. Most of the experimental work was focused on low speed linear or annual cascades with nozzle type blades, fixed geometry and on design conditions. Detailed studies aimed at the effects of incidence on turbine internal flows seem to be few in number. Langston et al. [2] presented ink-trace flow visualisation in a low speed linear cascade for two incidences (about 0° and +11.8°) and showed that a saddle point of the cascade endwall inlet flow separation moved to the suction side of the cascade passage as the incidence was increased. In [6] the internal flow mechanism of a high speed linear cascade for two incidences -20° and +8.6° is discussed. The work of Yamamoto and Nouse [7] refers to a low speed turbine nozzle linear cascade at moderately positive (+7.2°) and high negative (-53.3°) incidence. The flow was surveyed in 15 planes. The blade chord length was only 73.5 mm long. In [8] the results of an experimental investigation of 3D flow downstream of a linear turbine nozzle cascade with the chord length 55.2 mm for five incidence angles between -60° and +35° are described. The present work aims to complete the data obtained earlier at variable aspect ratio [9] for typical off design conditions of the SKODA B1 linear impulse cascade.

2. EXPERIMENTAL FACILITY AND METHODS

At the Slovak Technical University in Bratislava a low speed wind tunnel was constructed to enable detailed measurements within a family of large scale linear cascades. A linear cascade with four or five blades and a chord length of up to 550 mm can be placed in the test section. The test cross section has dimensions 1000 x 800 mm. The blades are constructed of laminated, reinforced polystyrene. The tunnel is an open cycle configuration (Fig 1).

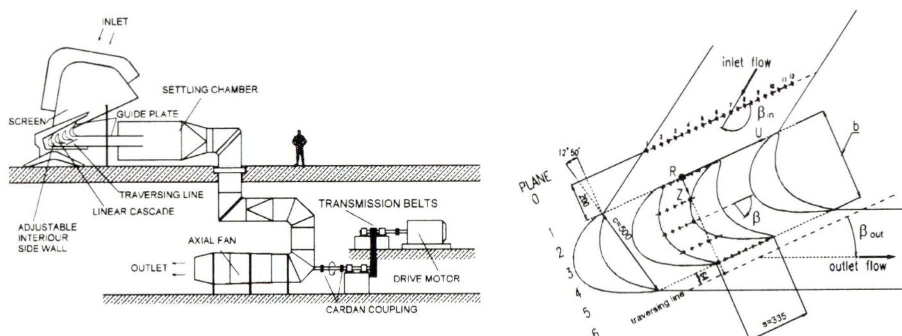

Fig.1. Wind tunnel for large scale cascade tests at STU and the detail of the cascade test section

The facility is designed for low speed tests at an inlet velocity of about 65 m/s. The inlet boundary layer on the endwall can be adjusted by screens with adjustable spacing. The flow periodicity is adjusted by varying the position of the tailboards and bleeds at exit from the test section. The blade aspect ratio is adjusted using a movable internal plexiglas wall. The cascade selected for this experiment is a SKODA B1 high turning steam turbine high pressure impulse cascade. General characteristics are given in Fig.1 and Table.1. Rows of static pressure tappings are located on the two central blades and at the distances of 10, 25 50, 125, 200 and 300 mm from the fixed endwall. The experiments were performed with angles of incidences of -14.4°, -6.8°, -0.3° and +8.6°.

Table.1 General data of the SKODA B1 cascade (Fig.1)

Profile geometry	SKODA B1
Blade chord length	500 mm
Blade axial chord length (b)	483 mm
Blade height	800 mm
Blade aspect ratio	0.8
Blade pitch to chord ratio	0.670
Stagger angle	12°50'
Blade camber angle	125°30'
Blade inlet angle	140°
Blade outlet angle	14°30'

The probe used was a fivehole 4 mm spherical probe. It was positioned spanwise and pitchwise and rotated by the probe traversing mechanism mounted on the endwall. In the blade to blade direction the spanwise traversing took place at 6 to 8 defined pitchwise positions. At exit the data were obtained from traversing across two pitches to test the periodicity and in the spanwise direction. The traversing within the cascade was carried out for the half blade span. The nearest position of the probe to the solid wall was 10 mm. The flow data were measured with the probe automatically equalised in the blade to blade direction. In this way at each measuring point two flow angles (α, β) and two pressures (p, p_0) were obtained. The inlet 2D flow details were measured with a cylindrical threehole probe (Table.2). Potential measuring uncertainties are discussed in [9].

Table.2 Endwall boundary layer characteristics in plane #0

	-14.4°	-0.3°	+8.6°
Incidence angle			
Inlet flow velocity at mid span	22.9 m/s	28.3 m/s	35.5 m/s
Turbulence intensity	~1.5%	~1.5%	~1.5%
Displacement thickness / blade height	0.0270	0.0264	0.0194
Momentum thickness / blade height	0.0186	0.0181	0.0114
Energy thickness / blade height	0.0354	0.0344	0.0213
Shape factor H12	1.45	1.46	1.70
Shape factor H32	1.90	1.90	1.87
Displacement thickness Reynolds No.	15180	18830	17310
Momentum thickness Reynolds No.	10470	12890	10180

The test outlet Reynolds number based on the mass averaged outlet velocity at plane 6 and the blade chord was 1.36×10^6. In this case as the incidence increases the inlet velocity increases in experiment. The streamline length between the plane #0 and #1 increases. This is the reason for bigger boundary layer in the plane #1 for the positive incidence angle.

3. EXPERIMENTAL RESULTS

From the measured values of α, β, p, p_0 the flow velocity w, its components w_R, w_U, w_Z and a set of local flow items were calculated. The local loss coefficient was calculated from the equation

$$\xi' = \frac{\overline{p}_{0in} - p_{0local}}{\overline{p}_{0in}^{MS} - \overline{p}_{out}^{MS}} \qquad (1)$$

This definition of loss coefficient includes only additional losses downstream of the inlet plane. The local dimensionless axial vorticity in the plane k=1,2,3,4,5,6 was calculated from:

$$\Omega_Z = \frac{\mathrm{rot}_Z \vec{w}}{\left| \overline{W}_{Uk}^{MS} - \overline{W}_{Uin}^{MS} \right| / b} = \frac{\dfrac{\partial w_U}{\partial R} - \dfrac{\partial w_R}{\partial U}}{\left| \overline{W}_{Uk}^{MS} - \overline{W}_{Uin}^{MS} \right| / b} \qquad (2)$$

The definition of the local secondary vorticity includes the ratio of the actual axial component of the vorticity and the absolute value of the "primary" vorticity in the direction R introduced in the 2D flow by the blade passage curvature between the plane 0 and k.

The local secondary velocity in this study was considered as the difference between the local flow velocity and the mass averaged flow velocity in the measuring plane:

$$\vec{w}_{sec} = \vec{w} - \overline{\overline{w}} \qquad (3)$$

To obtain the secondary flow vectors for visualisation, the secondary velocity was first projected onto a plane perpendicular to the mass averaged flow velocity vector

$$\vec{w}_{sec\,p} = \vec{w} - \left(\vec{w}.\vec{n}_0 \right)\vec{n}_0, \qquad (4)$$

with the normal vector $\vec{n}_0 = \overline{\overline{w}} / \left| \overline{\overline{w}} \right|$, and then decomposed in R, U direction:

$$w_{R\,sec}\,\vec{i}_R + w_{U\,sec}\,\vec{i}_U + w_{Z\,sec}\,\vec{i}_Z = \vec{w} - \left(\vec{w}.\overline{\overline{w}} / \left| \overline{\overline{w}} \right| \right)\, \overline{\overline{w}} / \left| \overline{\overline{w}} \right| \qquad (5)$$

The magnitude of the local secondary velocity was calculated from the R and U components:

$$w_{RU\,sec} = \sqrt{w_{R\,sec}^2 + w_{U\,sec}^2} \qquad (6)$$

The total pressure gradient in axial direction Z was calculated from the data obtained in the given measuring plane from the Z component of the incompressible Helmholz equation:

$$w_R\left(\frac{\partial w_R}{\partial Z} - \frac{\partial w_Z}{\partial R} \right) - w_U\left(\frac{\partial w_Z}{\partial U} - \frac{\partial w_U}{\partial Z} \right) = \frac{1}{\rho}\frac{\partial p_0}{\partial Z} \qquad (7)$$

Required velocity gradients in axial direction are calculated from the remaining two components of the Helmholz equation using $\dfrac{1}{\rho}\dfrac{\partial p_0}{\partial R}$ and $\dfrac{1}{\rho}\dfrac{\partial p_0}{\partial U}$.

3.1 Incidence effects on 3D flow in the measuring planes

The results obtained show details of the development of the secondary flows and losses under various incidence conditions. Results are presented here for traverse half-planes #1,2,3, and 5.

At cascade inlet plane #1 (Fig.2) the pressure side leg (Hp) and the suction side leg (Hs) of the horseshoe vortex can be seen under all incidence conditions. A good resolution of the horseshoe vortex shows the contour plot of the total pressure gradient with its negative values. The axial vorticity contours and the secondary velocity charts indicate Hp and Hs too. The contour plots show, that the loss and the vorticity are introduced mainly by the inlet boundary layer. Hs reduces the axial vorticity near the blade SS, because it is rotating in the opposite direction. The magnitude of the secondary velocity is larger at blade PS for negative incidences and larger at blade SS for positive incidence.

Figure 3 shows the results from plane #2. For the incidence -14.4° Hs is still visible in the secondary velocity chart and Hp moved towards the blade SS. The contours of the flow angle $\beta > 180°$ indicate the backward flow as a result of the flow separation between the plane 1 and 2 caused by a strong pressure gradient at blade LE. Consequently increased loss and additional secondary flow are present near the blade PS. For the near zero incidence the presence of Hs is weak and Hp is already merged into the passage vortex. The loss is located mainly near the endwall and near the blade SS. For the positive incidence at plane #2 the

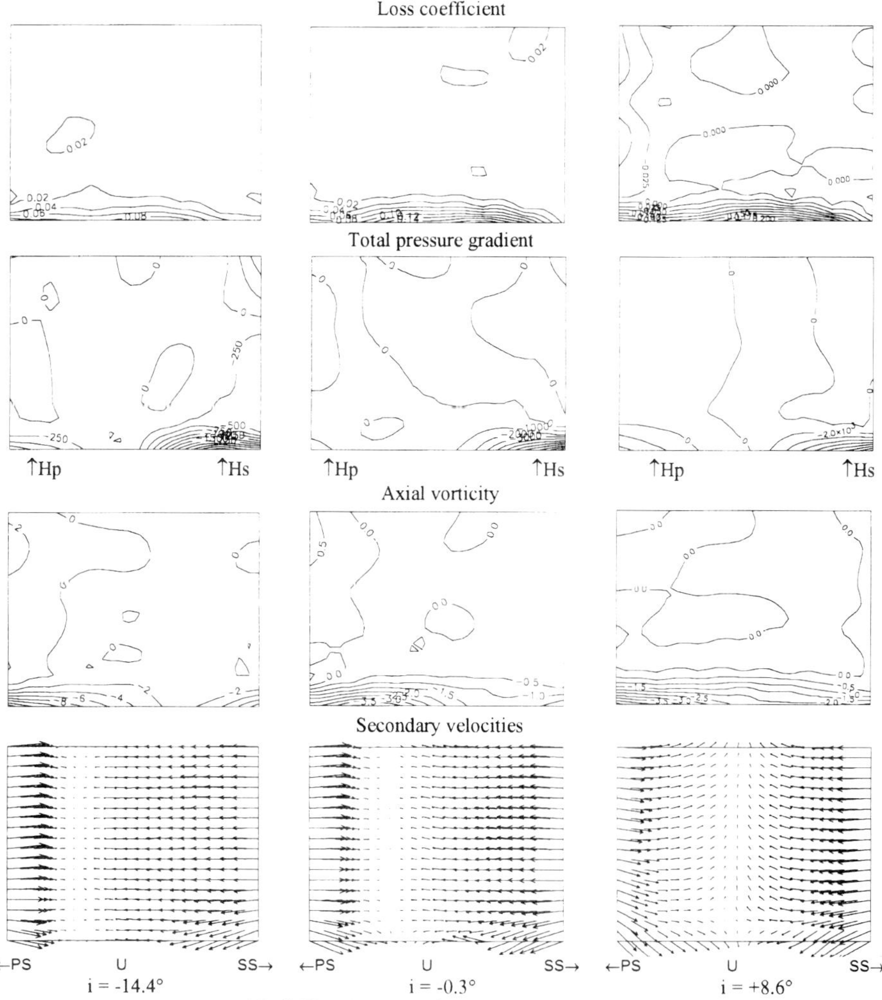

Fig.2 Flow details in the measuring plane #1

most of the loss remains near the endwall. Near the blade PS negative loss occurs due to the secondary mixing process.

A similar flow situation is found in the plane #3 at the end of the accelerating part of the cascade passage (Fig.4). Due to the flow separation at $i = -14.4°$ the axial vorticity and the secondary velocity indicate additional secondary positive vorticity near the midspan and in the blade PS corner. The central area between these vortices is integrated to form a passage vortex with negative vorticity. In the cases of near zero incidence and positive incidence the passage vortex is well developed and merged with Hp.

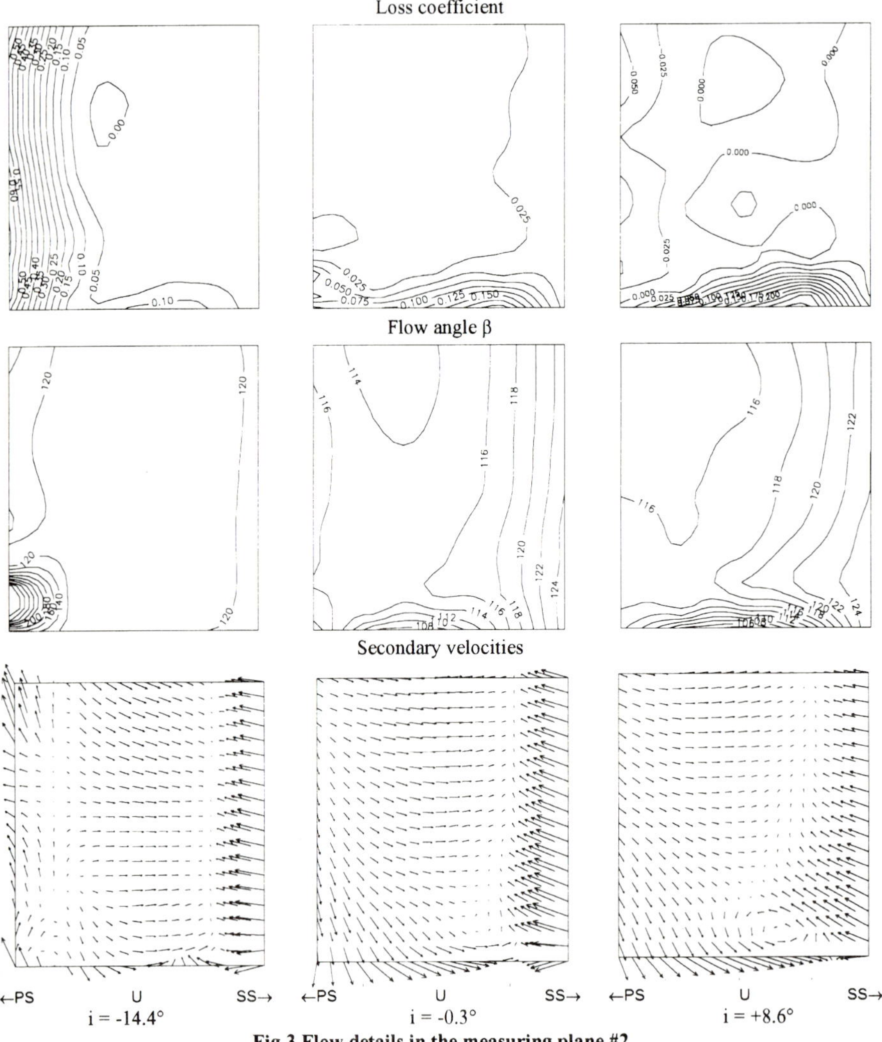

Fig.3 Flow details in the measuring plane #2

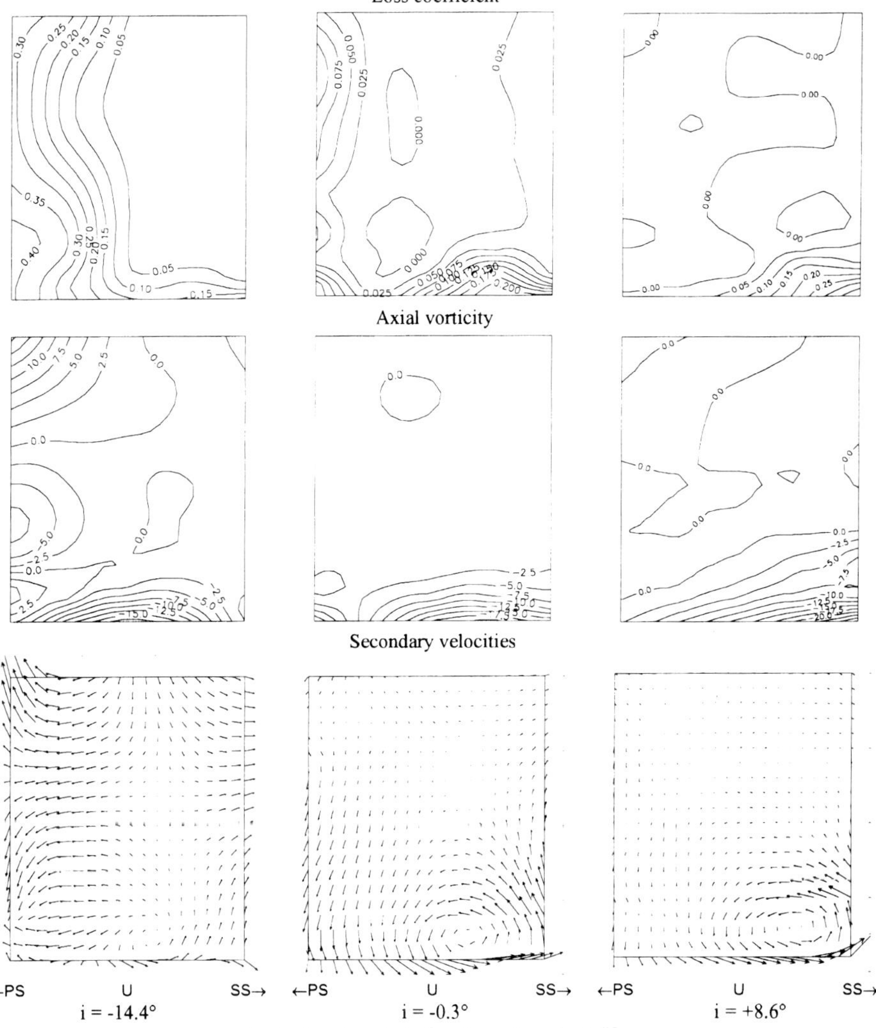

Fig.4 Flow details in the measuring plane #3

The results from plane #5 which is located near the blade T.E. at the end of the decelerating part of the passage are shown on figure 5. Similar structure of axial vorticity, secondary velocity, and loss coefficient are found under all incidence conditions as the result of the flow deceleration. Typical are the closed contours of the loss coefficient near the blade SS in the region of the passage vortex centre. High loss originating in the flow separation behind the blade L.E ($i = -14.4°$) is distributed over the whole measuring plane. The peak loss is more than 40%. A similar maximum level of loss occurs in the case of the positive

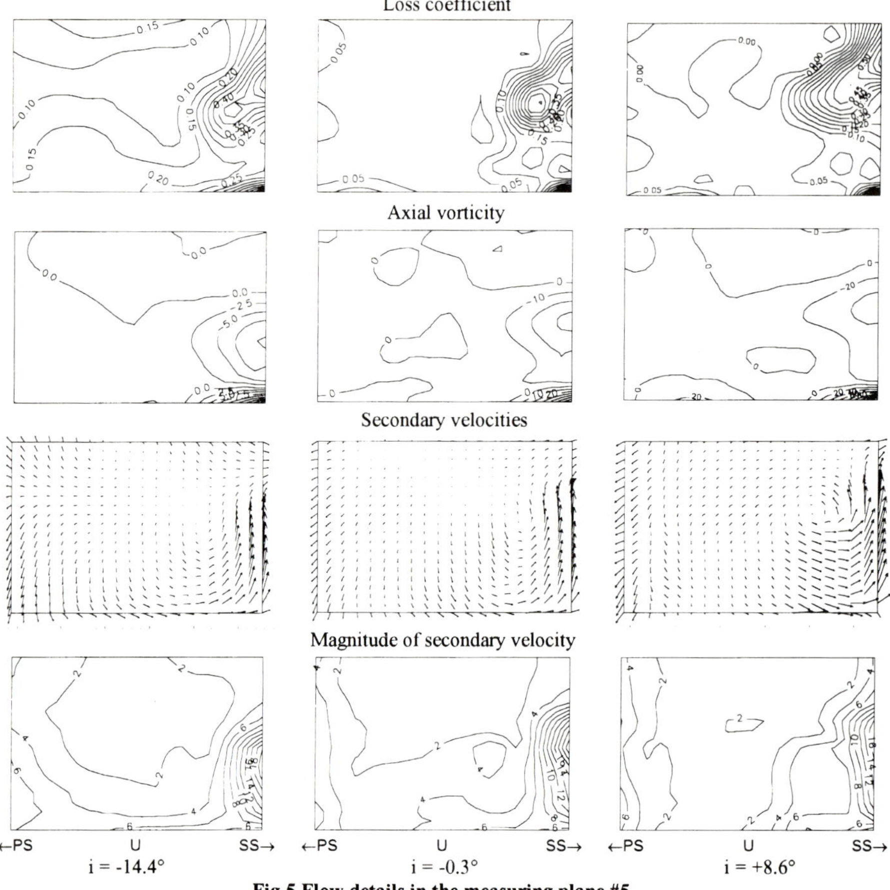

Fig.5 Flow details in the measuring plane #5

incidence +8.6°. The peak absolute value of the local dimensionsless vorticity increases with increasing incidence angle, while the magnitudes of the secondary velocity remain approximately constant. Near the blade SS positive corner vorticity is developed.

3.2 Incidence effects on mass averaged results

Figure 6 shows the incidence effects on the pitchwise mass averaged results. The spanwise distribution of the flow angle deviation is shown as the difference $\overline{\beta} - \overline{\beta}^{\mathrm{MS}}$ between the local flow direction and the flow direction at midspan. The deviation distribution at negative incidence shows generally smaller underturning (+) and overturning (-) in planes #1, 2, 3 in comparison with the planes in the deceleration part of the cascade passage. Increasing the incidence, maximal secondary deviation moves downstream from measuring plane #4 to #5. The spanwise distribution of the total pressure loss coefficient was calculated from

$$\overline{\xi}' = \frac{\overline{\overline{p}}_{0in} - \overline{p}_0}{{}^{-MS} \quad {}^{-MS}} \qquad (8)$$

In the accelerating part of the blade passage the maximum loss is located near the endwall. At plane #4 the structure changes for all incidences. Near the wall the loss is reduced in plane #4 and #5 (for i= -0.3°,+8.6°) but rises a way from the wall. This is due to the secondary mixing process removing low energy fluid from the endwall. For all incidences in the outlet plane #6 the maximum loss occurs near the endwall. In the case of i=-14.7° loss at the midspan increases significantly.

Figure 7 shows the growth of the overall pressure loss coefficient through the cascade

$$\overline{\overline{\xi}}' = \frac{\overline{\overline{p}}_{0in} - \overline{\overline{p}}_0}{\overline{\overline{p}}_{0in} - \overline{p}_{out}}{}^{-MS} \qquad (9),$$

and the overall cascade characteristic $\overline{\overline{\xi}}'$=f(i). The overall loss coefficient as it is defined, accounts only for loss apparent downstream of the inlet plane and starts with zero value. The measuring planes are not orthogonal to the cascade passage mean flow (with exception of plane #3) and due to the intensive secondary mixing at positive incidence the loss seems to decrease in plane #4 due to the measuring error. This phenomenon has been observed also for larger blade aspect ratio [9]. The secondary loss $\overline{\overline{\xi}}'$ - $\overline{\xi}'^{MS}$ = f(i) plotted on figure 7 is only an estimation, because the flow is not fully 2D at midspan.

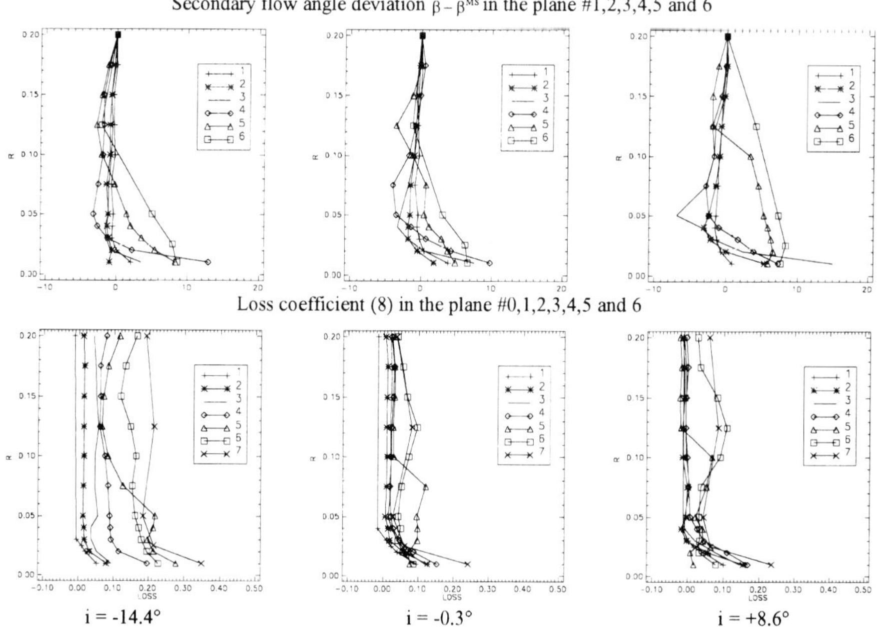

Fig. 6 Incidence effects on pitchwise mass averaged results

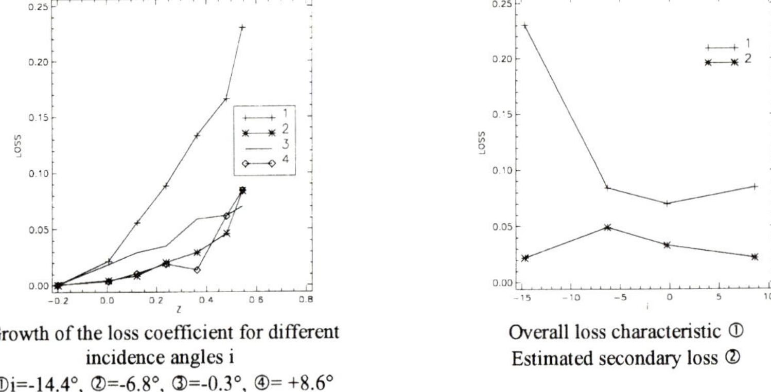

Growth of the loss coefficient for different incidence angles i

①i=-14.4°, ②=-6.8°, ③=-0.3°, ④= +8.6°

Overall loss characteristic ①

Estimated secondary loss ②

Fig.7 Incidence effects on area mass averaged results

i | Pressure side | Suction side

-14.7°

-0.3°

+8.6°

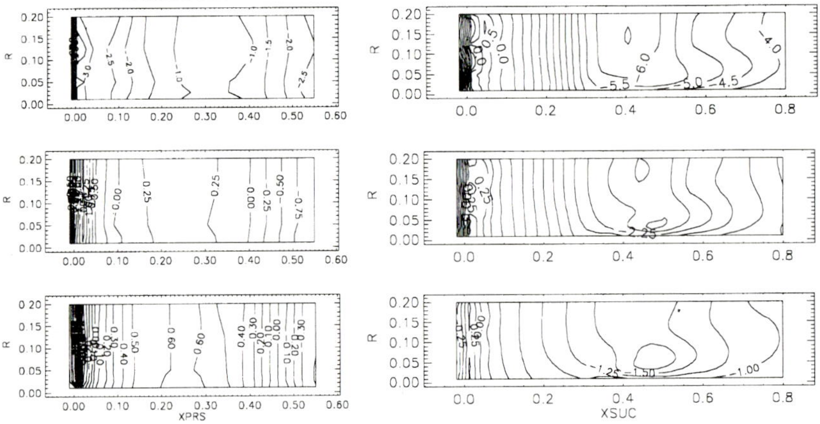

① R=10mm, ② R=25mm, ③ R=50mm, ④ R=125mm, ⑤R=200mm from the endwall

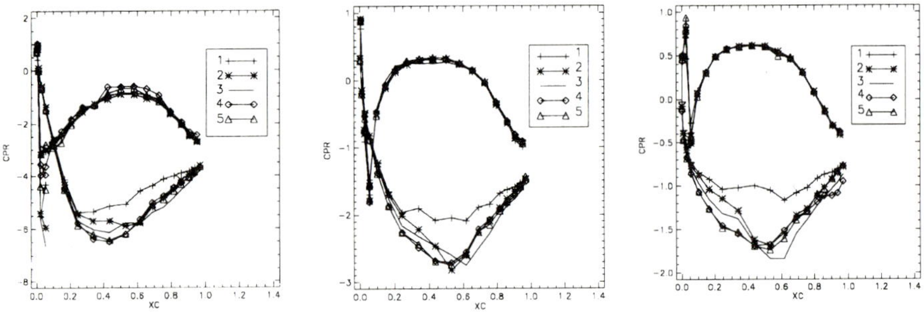

i = -14.4° i = -0.3° i = +8.6°

Fig.8 Incidence effects on 3D flow at blade surface

3.3 Incidence effects on 3D flow at blade surface

The blade surface pressure coefficient distribution on the blade surface shows figure 8. The pressure coefficient is

$$C_p = \frac{p - \overline{p}_{in}^{MS}}{\overline{p}_{0in}^{MS} - \overline{p}_{in}^{MS}} .$$

The contour plots are presented at half height blade surface. On the blade pressure side the flow is influenced by the separation (i =-14.7°) and is three dimensional. The separation originates from the blade LE and reattaches downstream onto the blade pressure surface. As the incidence angle increases, the extension of the 3D effects on the blade surface is reduced. On the suction side the flow is accelerated due to the blockage effect of the LE separation (i =-14.4°) and the 3D effects are reduced. The distributions of C_p over the blade chord coordinate x_c shows at the blade SS significant differences in C_p between the section near the endwall and midspan. The minimum C_p moves from midspan closer to the endwall with increasing the incidence.

4. CONCLUSIONS

The performance of a linear impulse turbine cascade at low aspect ratio has been investigated for a range of off-design conditions in addition to the previous study of aspect ratio effects on the secondary flows [9]. The limited extend of the paper allows only limited presentation of the results obtained. It was not possible for instance to present the flow details of the corner vortices or near the the walls.

Several conclusions can be drawn from the current experimental work:

- A relationship between the blade loading in the acceleration part of the blade passage and the intensity of secondary flows has been observed. For all incidences Hp generates less total pressure drop as Hs. As the incidence angle is increased the region covered by 3D effects at the blade PS decreases and increases at blade SS. Spanwise deviation angle distribution varies in the deceleration part in a wide range of max. 15 to 25°. The study shows details of the formation of the secondary flow after the LE separation at negative incidence.
- The test with a ratio of the fivehole probe diameter to the blade chord of 0.8% allows one to analyse flow details and prepare the data for assessment of a 3D computational method.
- The evaluation of the secondary velocities (eq. 5) was introduced using only flow velocity components to avoid errors when using flow angles.
- The total pressure gradient calculated from the data obtained in the measuring plane was introduced as an important tool to analyse the flow details /horseshoe vortex/.
- The growth of loss through the impulse cascade determined in parallel measuring planes is not smooth and the "loss decrease" is influenced not only by the incidence angle but also by the blade aspect ratio.
- The secondary flow in impulse cascades is mainly determined by the accelerating part of the flow passage. Downstream in the deceleration part of the passage the differences in the flow details caused by the incidence angle tend to equalise.
- Classical secondary flow theory would indicate that the strength of the secondary flows depends on the turning and inlet vorticity. Incidence will only effect the secondary flow through its effect on turning. The separation is a main cause of the cascade loss generation at negative incidence. Additional secondary flow diffuses the separation loss downstream.

ACKNOWLEDGMENT
The work described in this paper was supported by the SKODA Turbines Ltd. Plzen, Czech Republic and by the Slovak Scientific Grant Agency. The authors thank SKODA for permission to publish the results presented in this paper.

REFERENCES
[1] Marchal, Ph., Sieverding, C.H.- Secondary Flows Within Turbomachinery Bladings. AGARD Conf.Proc. No. 214 Secondary Flows in Turbomachines, 1977
[2] Langston, L.S., Nice, M.L., Hooper, R.M.- Three-Dimensional Flow Within a Turbine Cascade Passage. ASME J.Eng.Power, Jan. 1977
[3] Sieverding, C.H.- Recent Progress in the Understanding of Basic Aspects of Secondary Flows in Turbine Blade Passages. ASME J. Eng. Gas Turbines and Power, Vol 107, 1985.
[4] Jilek, J.- An Experimental Investigation of the Three-Dimensional Flow Within Large Scale Turbine Cascades. ASME Paper 86-GT-170
[5] Gregory-Smith, D.G. et al.: Growth of Secondary Losses and Vorticity in an Axial Turbine Cascade. ASME Paper No. 87-GT-114, May 1987.
[6] Hodson, H.P., Dominy,R.G.: The Off-Design Performance of a Low Pressure Turbine Cascade Passage. Trans. ASME, J. of Turbomachinery, Vol. 109, April 1987
[7] Yamamoto,A., Nouse,H.: Effects of Incidence on Three-Dimensional Flows in a Linear Turbine Cascade.. Trans. ASME , J. of Turbomachinery, Vol. 110, October 1988
[8] Perdichizzi,A., Dossena,V.: Incidence Angle and Pitch-Chord Effects on Secondary Flows Downstream of a Turbine Cascade. Trans. ASME , J. of Turbomachinery, Vol. 115, July 1993
[9] Molnar,V.,Ridzon,F., Nyíri, J.: Secondary Flow Studies in a Large Scale Turbine Cascade. Proc. 2nd European Conference on Turbomachinery, Antwerpen, 1997

Profiled end-wall design for a turbine nozzle row

J YAN and **D G GREGORY-SMITH**
School of Engineering, University of Durham, UK
N Z INCE
Steam Turbine Group, ALSTOM Energy Limited, Rugby, UK

SYNOPSIS

The design of a profiled end-wall for a steam turbine HP nozzle guide vane is described. The design tool was a CFD code developed by ALSTOM Energy Ltd., and adapted for use with profiled end-walls which are not axisymmetric. The code has been developed and validated against a standard rotor blade test case and a cascade of the nozzle guide vanes. The experimental and CFD results have been compared showing that the CFD gives good predictions.

In order to reduce the secondary loss, various non-axisymmetric end-wall profiles were designed with the aim of reducing the cross passage pressure gradient on the wall. These were evaluated by the CFD code, and an optimum profiled end-wall chosen to give reduced secondary flow and total pressure loss. A reduction of 6.6% of the loss is predicted.

1. INTRODUCTION

When a non-uniform flow is turned within a blade row, the flow becomes three-dimensional and there is a variation of flow angle along the span of the blade. This effect is known as secondary flow. In a low aspect ratio highly loaded turbine, the effects of secondary flow will be evident throughout most of the flow field, and the secondary loss may account for about half of the total loss. Profiling of the endwall(s) of the nozzle row holds promise for achieving improved performance through the reduction of secondary loss. Deich et al [1] reported much work investigating the effects of the end-wall profiling in a linear cascade and annular cascade. The selected profiles reduced the velocity in the region of highest turning and provided increased acceleration of the flow later in the blade passage. Increases in efficiency of up to 1.5% were reported with blades of aspect ratio of 0.5. Morris and Hoare [2] described experiments in a linear cascade with one end-wall modified with a profile similar to that used by Deich. A reduction of the total loss was found in the half span next to the flat wall, and no change in the half span near the profiled wall. They also investigated a non-axisymmetric profile in order to move the velocity maximum nearer the leading edge and give a more rapid deceleration toward the trailing edge. A large region of high total pressure loss adjacent to the profiled wall indicated that the wall boundary layer had been greatly thickened and may have separated. This illustrates the point that it is very important to consider carefully the effects of a proposed profile on the blade and end-wall pressure distributions. Ewen [3] and Boletis [4] both report achieving an increase in efficiency in rotating rigs using axisymmetric end-wall profiling, probably due to more uniform flow into the downstream blade row rather than a direct reduction in loss in the row with end-wall profiling.

With the development of prediction techniques using computational fluid dynamics (CFD) some workers have used CFD to design end-wall profiles. Atkins [5] suggested two important points for investigations of the end-wall profiling. Firstly, the investigations should not be restricted to a small number of end-wall profiles because this leads to inconclusive results. Secondly, the profiles should be chosen in a systematic way, on the basis of their influence upon the pressure distribution in the cascade. Atkins used an inviscid CFD program to select profiles for a linear cascade. Some of the profiles were tested, but no clear conclusions were drawn regarding the reduction of loss. Rose [6] used a CFD method to design a non-axisymmetric end-wall profile to reduce the non-uniformity of static pressure distribution at the platform edge downstream of the trailing edge of nozzle guide vanes. The object was to reduce the cooling air flow requirements to avoid hot gas ingestion. Hartland et al [7] used Rose's idea to design a non-axisymmetric end-wall profile for a row of rotor blades, and showed that the method worked well when tested in linear cascade. They also showed that there was a significant affect on the secondary flow. The profile had convex curvature on the end wall near the pressure surface, reducing the pressure there, and concave curvature near the suction surface, raising the pressure there. Thus the cross passage pressure gradient was reduced, reducing the secondary flow.

The aim of this work was to design an end-wall profile using CFD by carefully considering the effect of the profile on the pressure distributions on the end-wall and blade surfaces. A systematic study was to be conducted and an assessment based on the effects on secondary flow and loss produced in the cascade. Firstly a CFD method was required, followed by validation against experimental data. This paper describes the method and its validation, and then details the design process and the assessment of the chosen non-axisymmetric end-wall profile.

2.0 COMPUTATIONAL FRAMEWORK

The calculations reported in this paper have been performed with a non-orthogonal 3D finite volume code developed within the Aerodynamics group of ALSTOM Energy Ltd. The code solves the Reynolds and continuity equations in an iterative manner over a collocated grid arrangement.

A pressure correction scheme, SIMPLE [8], is used to determine the pressure field. It is well known that the fully-collocated grid arrangement creates a numerical decoupling between the collocated velocity and pressure values and results in chequerboard oscillation. The non-linear interpolation scheme proposed by Rhie & Chow [9] is used to achieve numerical stability within the collocated storage arrangement. The convective fluxes are approximated with the quadratic QUICK scheme of Leonard [10]. This scheme is third-order accurate for uniform grids. The unbounded nature of QUICK scheme prevents its use for turbulence quantities. Several researchers have formulated bounded forms of the QUICK scheme. The code used in this study employs the MUSCLE approach proposed by Lien & Leschziner [11]. The calculations have been started by using the UPWIND scheme to enhance stability and the QUICK scheme is activated after a number of iterations.

The code used in this study offers a number of eddy-viscosity models (Baldwin-Lomax, k-ε, k-ω and a two-layer model standard k-ε model away from the wall and one-equation model (Norris and Reynolds [12]) in the near wall region).

The production term for kinetic energy equation in linear eddy-viscosity models is proportional to the square of strain invariant (S). In irrotational strain fields such as a stagnation region near the leading edge of a blade, the strain invariant has a very large value, and excessive levels of turbulence are generated. This high level of turbulence is convected into the flow passage and causes poor flow prediction. The vorticity invariant (Ω) is very small in near irrotational strain while it has the same value as the strain invariant in simple shear flow. Several researchers have therefore used the vorticity invariant in place of the strain in the production rate of kinetic energy to solve the so-called '*stagnation point anomaly*'. The S-Ω version of this technique, proposed by Kato & Launder [13], is used here.

For the inlet boundary condition, the total pressure, total temperature, inlet flow angle and turbulent levels are prescribed. Uniform static pressure is applied at the exit plane. A non-slip boundary condition is applied on the blade and end wall surfaces. The calculation domain is limited to one passage by using periodic boundary conditions. For most of the computations, in order to reduce computer resource requirements only the half span was calculated, with a symmetrical boundary at mid-span.

3.0 VALIDATION OF CFD CODE

3.1 Durham Cascade (ERCOFTAC Test Case)

The code was first validated against the Durham linear turbine cascade which is one of the ERCOFTAC (European Research Community On Flow Turbulence And Combustion) test cases for their Turbomachinery Workshops (Gregory-Smith [14]). It is a high pressure gas turbine rotor blade with design inlet angle of 42.75° and exit angle -68°, giving just over 110° of turning. The flow is low speed (exit Mach Number ~0.1), and the chord is 230mm. The Reynolds number (based on chord and exit velocity) is 4×10^5 so the blade boundary layers are transitional. The grid used was a 'H' Type grid with a size of 25×99×48 (Pitch, Axial, Span), and it is shown in Figure 1. Only the half span was computed.

The area plots of total pressure loss and secondary velocity vectors are shown in Fig. 2 for the traverse slot 10, which is 28% axial chord downstream of the trailing edge. The contours of total pressure loss are expressed as a loss coefficient with respect to *inlet* velocity. The general features of the flow are well represented, with the wake distortion and loss core in the contour plot and the passage vortex in the vector plot. However the vortex centre is predicted a little closer to the end-wall, about 45mm compared to 65mm. This is an indication that the secondary velocities are under-predicted giving less vortex migration. Consequently the wake distortion and definition of the loss core are not so clear. The counter vortex on the end-wall is seen in the vector plot, although its pitchwise position is not quite correct. The associated loss on the end-wall is somewhat masked by the end-wall loss in the predictions, which were not measured in the experiment. This is partly due to the nearest measurement position being 5mm from the wall, but may also be due to the assumption of turbulent flow everywhere. In the experiment, the flow on the end-wall shows laminar transitional characteristics (Moore & Gregory-Smith [15]). Similarly the wake shows too much loss, probably for the same reason.

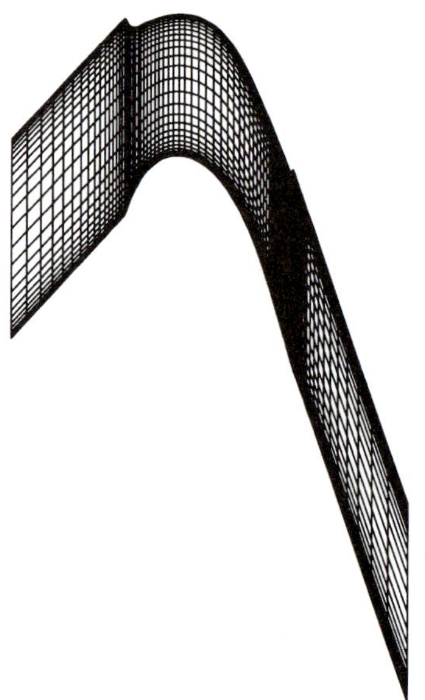

Figure 1: Grid for Durham Cascade

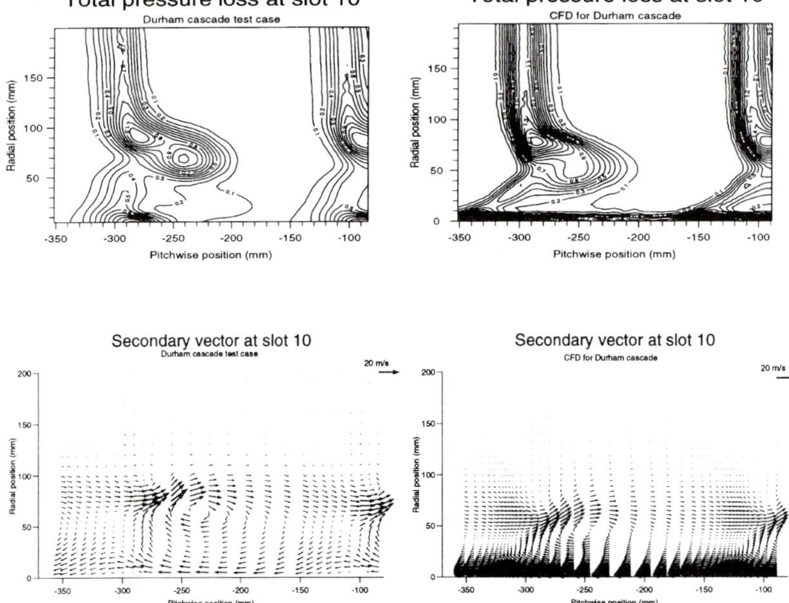

Figure 2: Area Plots at Slot 10 for Durham Cascade

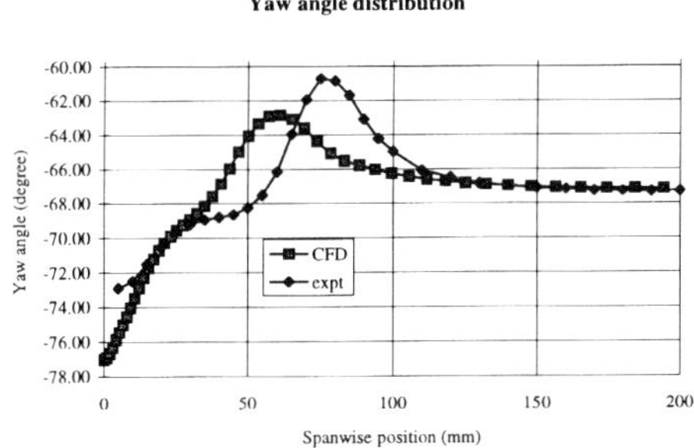

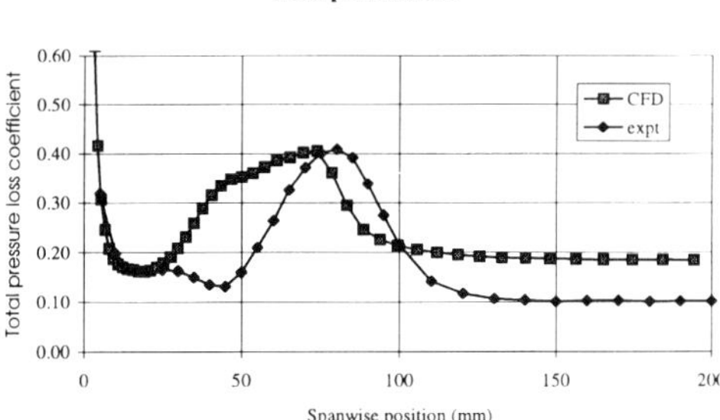

Figure 3: Pitch Averaged Results for Durham Cascade

Quantitative results are shown for pitch averaged yaw angle and loss in Fig. 3. At mid-span, the angle is predicted well, but nearer the end-wall, the lower secondary velocities give less under-turning and the peak is nearer the end-wall. Close to the end wall the overturning is higher because the vortex is too close to the wall. The pitch averaged loss at mid-span is too high, as expected from the loss contour plot. The loss peak is again closer to the end-wall than in the experiment, but near the end-wall, the loss looks satisfactory. Table 1 shows the area averaged results at slot 10. For the CFD, the mid-span loss causes the high overall loss, and the secondary kinetic energy indicates the smaller secondary velocities.

Table 1: Area Averaged Results for Durham Cascade

(Coefficients based on Inlet Velocity)

	Total Pressure Loss	Mid-span Loss	Secondary Loss	Secondary Kinetic Energy
CFD	0.250	0.185	0.065	0.012
Experiment	0.179	0.095	0.084	0.025

Overall these results are typical of those achieved by CFD codes with a k-ε model assuming turbulent flow over all surfaces (Gregory-Smith [16]). The test case is a severe one, particularly with the regions of laminar transitional flow, so that the turbulence model may give too high mid-span loss and dampen the secondary velocities, giving low vortex migration.

3.2 ALSTOM Nozzle Cascade

The end-wall profiling design was to be applied to a nozzle row of a high-pressure steam turbine. The design inlet angle was 0° and exit angle 77°. So compared to the Durham Cascade, the turning was lower and the acceleration much higher. It was therefore necessary to validate the code for these rather different conditions.

A linear cascade of the blades was constructed with a chord of 225mm. A trip was placed on the suction surface to ensure that the boundary layer became turbulent, because the Reynolds number was an order of magnitude smaller than in a typical turbine stage. The flow was traversed using 5-hole pressure probes upstream and downstream and within the blade passage. Blade surface static pressures and end-wall static pressures were also recorded. Yan [17] gives details of the apparatus and experiment.

A 'H' grid was constructed with same number of grid points as for the Durham cascade. Yan [17] gives a detailed comparison of the flow results through the blade passage, but here only the slot 10 (26% axial chord downstream) traverse results are shown. The loss contours and secondary vectors in Fig. 4 agree qualitatively as they did for the Durham cascade. However secondary flow is confined to a much smaller spanwise extent because of the lower turning and higher acceleration. For this reason the vector plots have been expanded to show only the first 50mm from the end-wall. At the mid-span, the loss values in the wake are again higher than those of the experiments. (The uneven appearance of the wake is due to the plotting program.) The loss peak near the endwall also shows a much higher value than that of experiments. The positions of the loss core and counter vortex loss are well predicted by the CFD. The secondary flow is very similar to the experimental results, although the vortex appears a little larger and the secondary velocities higher, unlike the predictions for the Durham Cascade.

The pitch averaged total pressure loss and yaw angle at slot 10 are compared with the experimental results in Fig. 5. The yaw angle shows a small error at mid-span. Near the endwall, the predicted yaw angle is almost the same as the experimental results, although it appears the under-turning is slightly higher. This would indicate stronger secondary flow, as

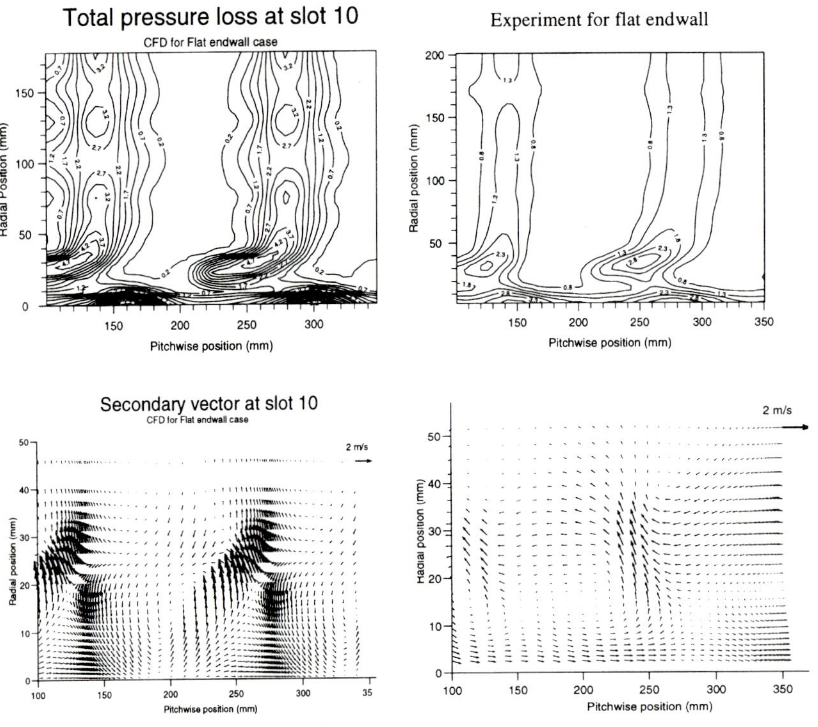

Figure 4: Area Plots at Slot 10 for Nozzle Cascade

seen in the vector plot. The total pressure loss at mid-span is higher than the experimental results, although by not so much as for the Durham Cascade. This may be due to the use of the turbulence trip causing the suction surface boundary layer to be largely turbulent. Near the endwall, the CFD gives a thinner boundary layer and a much higher loss peak compared with the experimental results. This would be consistent with higher secondary velocities, which would sweep more low energy fluid into the loss core. Area averaged data at slot 10 gives an overall view of the cascade flow, and these are compared in Table 2. The mid-span loss is a little high, giving high overall loss. The secondary kinetic energy is predicted more than the experiment, confirming the impression of stronger secondary flow seen in the vector plot.

Table 2: Area Averaged Results for Nozzle Cascade

(Coefficients based on Inlet Velocity)

	Total Pressure Loss	Mid-span Loss	Secondary Loss	Secondary Kinetic Energy
CFD	1.14	0.98	0.16	0.0088
Experiment	1.03	0.83	0.20	0.0042

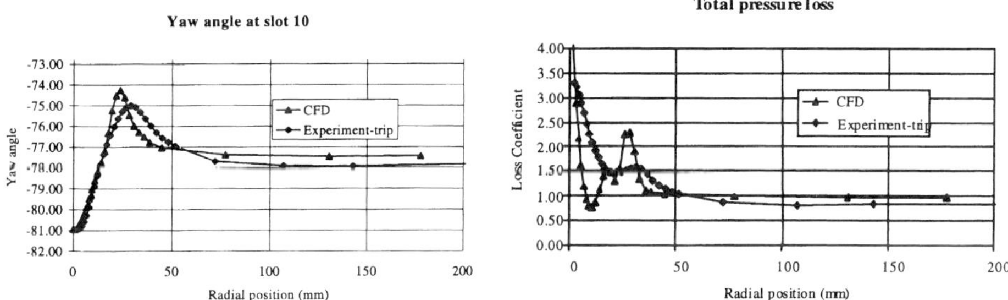

Figure 5: Pitch Averaged Results for Nozzle Cascade

3.3 Overall Assessment

The CFD code has been validated against experimental data for a high turning rotor blade (Durham Cascade) and a lower turning but high acceleration Nozzle Cascade. The code gives higher total pressure loss for the Durham case where the flow is transitional but the predictions for the lower turning nozzle cascade are more accurate. This is probably due to the use of the trip in the nozzle case forcing turbulent flow. Hence, the assumption in the turbulence model of turbulent flow everywhere appears more reasonable for the nozzle cascade. A contrast appears between the two cascades for the strength of the secondary flows, the Durham cascade being predicted too low and the nozzle cascade too high. In the vortex region the turbulence structure is very complex, with non-isotropic eddy viscosity, and even negative values being observed (Gregory-Smith & Biesinger [18]). Thus it may be that the secondary kinetic energy is not being damped quickly enough for the nozzle cascade, but that the fully turbulent assumption causes too much damping for the Durham cascade. Overall it may be concluded that the CFD code gives results sufficiently accurate to predict trends in the design study, even if absolute values (particularly of loss) are not precise.

4.0 END-WALL PROFILE DESIGN

4.1 End-wall Profile Definition

The idea of the non-axisymmetric endwall profile design was to influence the local pressure field. The end-wall profile was designed based on the assumption that applying concave curvature to the endwall could locally raise the endwall static pressure. Likewise, the static pressure could be lowered by convex curvature. The highest loading on the blade is in the later part of the blade passage (see Fig. 9), and so a profile in the axial direction was designed with curvature towards the trailing edge. This profile is shown in Fig. 6, as it appears near the pressure surface. The change starts 15% upstream from the leading edge and finishes at the trailing edge. The upstream and outlet shapes of the blade passage are the same as for the flat endwall. This makes it possible to apply the shape to a real turbine. The profile is divided into three parts in the axial direction. The first part is an arc that makes the change smooth with the flat upstream end-wall and the linear second part. The third part is a sinusoid that connects the second part and the trailing edge platform as shown in Fig. 6. In the pitchwise direction the change is two sine waves which have a smooth connection between them as shown in Fig. 6. The final non-axisymmetric profile is the combination of the axial and the pitchwise variation.

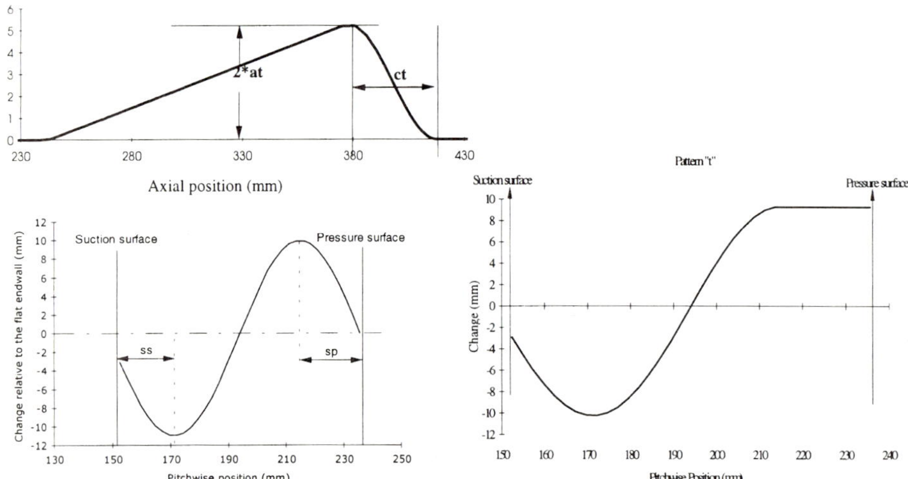

Figure 6: End-wall Profile Definition   **Figure 7: Type 't' Pitchwise Profile**

The important parameters to define the endwall shape are shown in Fig. 6. *ct* is the distance between the trailing edge and the peak of axial profile and *at* is the amplitude of the peak. *sp* and *ss* define the peaks of the pitchwise sine waves near the pressure and suction surfaces. They are expressed as fraction of the local pitch (distance between the blade surfaces), with the mid-pitch end-wall height being fixed at zero. The names of end-wall profiles are defined in a simple way. n is for nozzle; the first two digits is the value of *ct* in mm; the third digit is half the value of *at*; the last digit or letter is for types of pitchwise profile (values of *ss* and *sp*). For example, the name of n4042 profile means *ct* = 40mm, *at* = 2x4 = 8mm, and the pitchwise shape is type 2 (*ss* = 0.25, *sp* = 0.225). A special type 't' was also used which is the same as type 2, but the re-entrant corner on the pressure surface is removed, as illustrated in Fig. 7. In Fig. 8 contours of the final end wall are shown as the combination of the axial and pitchwise profiles (the blade profile has been distorted to preserve confidentiality). To change the endwall design, a program was written which generated a file to include the information of the geometry and the computational grid of the cascade for input into the CFD code.

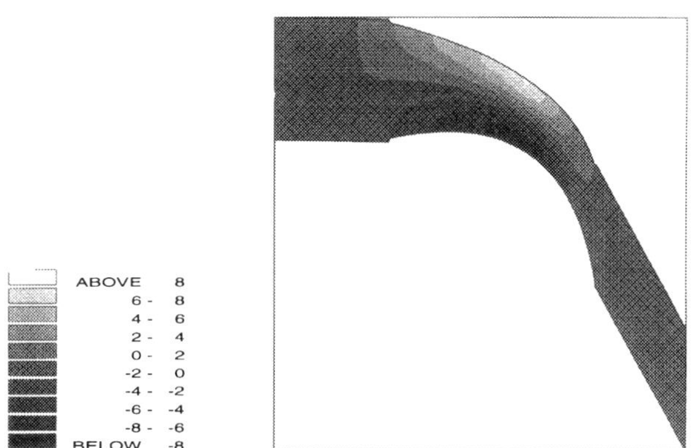

Figure 8: Contour Map of Final Profile
(Heights in mm)

4.2 CFD Investigation of Profiles

After their definition, the different profiles were calculated systematically by running the CFD code. The total pressure loss and secondary kinetic energy were compared as a basis for the design choice. In order to make the CFD tests more computationally efficient the grid size was reduced in the spanwise direction. The effect of the grid change was investigated before the grid reduction was used to ensure no significant change in loss or secondary kinetic energy. The final grid used was 25×99×35, and required 70% of the time of the original grid.

From some preliminary tests, the type 2 was selected to be tested systematically. The profiles with *at* = 8 mm, 10 mm and 12 mm were tested with different *ct*. In order to see the effect of different *ct*'s, the comparisons are made for fixed *at* values. The mass averaged values of total pressure loss, yaw angle and secondary kinetic energy were calculated and compared. The yaw angle varied very little which indicates an almost constant mass flow rate. The results are shown in Fig. 9, along with some results for type "t". It can be seen that the secondary flow is reduced as *ct* increases, with a minimum point around *ct* = 50mm. The loss is reduced compared to the flat end-wall, without a very clear trend, but some indication of a minimum around *ct* = 40-55mm. For *at* = 8, the reductions are generally less than for the higher values. Pitch averaged results were also studied, and these indicate that the high loss for *at* = 10 at low *ct* is probably due to separation of the end wall boundary layer. The low value of *ct* means their peak is near the trailing edge, and so there is a rapid change between the peak and the exit. This was also seen on the static pressure distributions around the blade near the end wall, the low value of *ct* giving a higher diffusion on the suction surface.

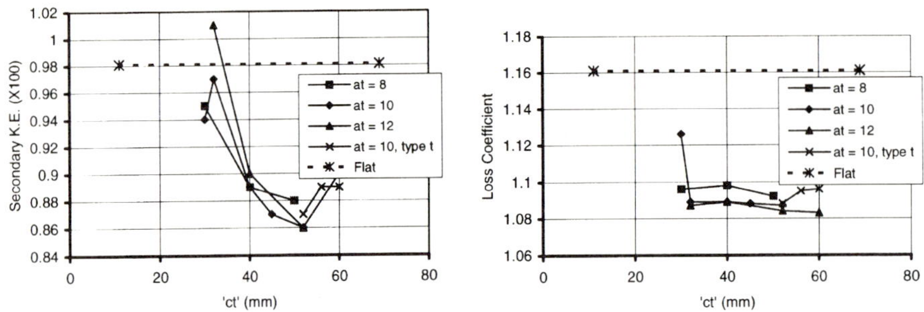

Figure 9: Design Study Results for Type 2

The final choice was the type "t" profile with at=10mm and ct=52mm (code n525t). The optimum condition as indicated by the secondary K.E. appeared around ct=52mm, so that the peak was 100mm from the leading edge (axial chord=152mm). The loss also appeared low, but the loss changes are probably not significant in view of likely errors due to the turbulence model and numerical scheme. A value of 10mm rather than 12mm was chosen for at as there seemed no advantage with the higher value, and it would be easier to manufacture. The type "t" was chosen for the same reason, and also because it would reduce the 'wetted area' of the end wall and pressure surface.

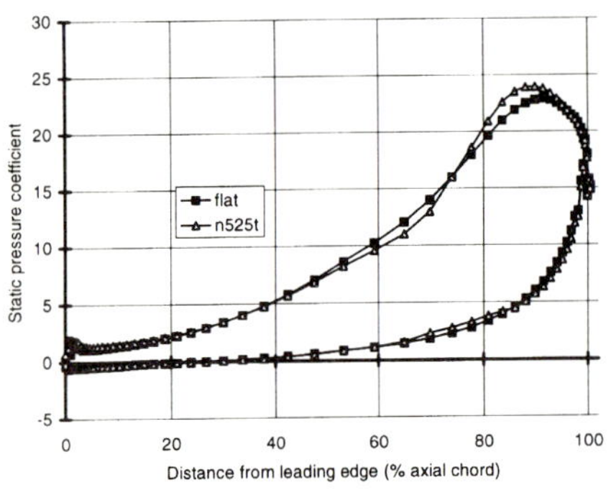

Figure 10: Static Pressure Profile close to End-wall

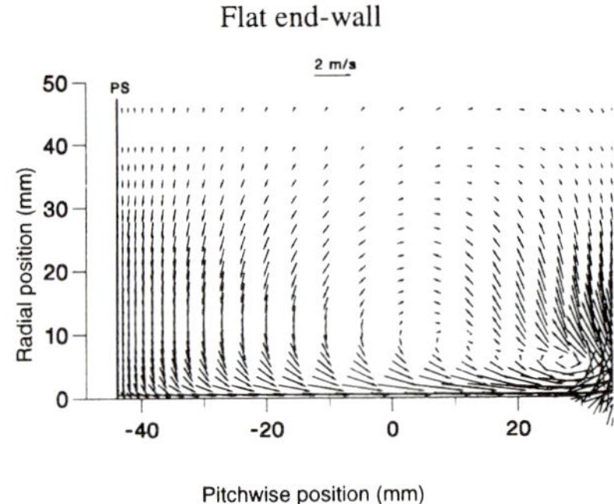

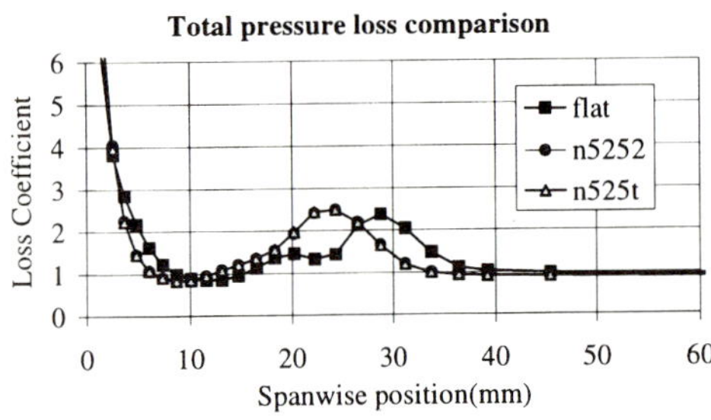

Figure 12: Pitch Averaged Values at Slot 10

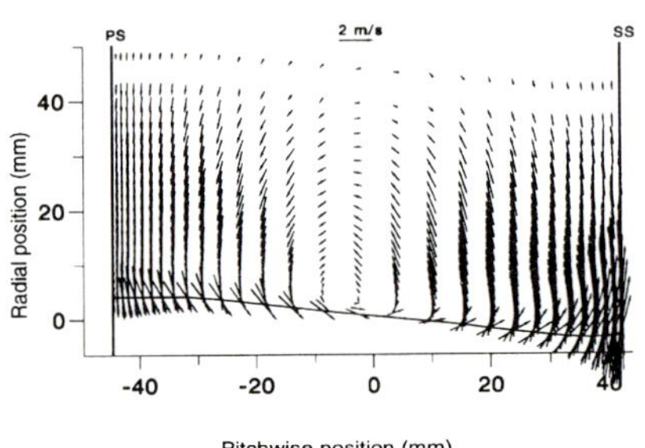

Figure 11: Secondary Flow at Slot 7

4.3 Design Evaluation

A detailed study was made of the flow with the chosen profile. The effect on the static pressure distribution around the blade close to the end-wall can be seen in Fig. 10. The effect of the end-wall shape is to reduce the pressure difference in the region of the peak, which is at 66% axial chord. So the cross-passage pressure gradient is reduced in the region of maximum turning. The loading is compensated around 80-90% axial chord, so that the overall loading is not changed, giving very nearly the same exit angle from the row. End-wall profiling provides an effective alternative to blade profile modification for adjusting the static pressure distribution near the end-wall.

Contour plots at various axial positions through the blade passage indicate the slower rolling up of the end-wall boundary layer and reduced secondary kinetic energy. This is illustrated in Figure 11 which shows the secondary vectors at slot 7, which is at 85% axial chord. The lower cross flow near the wall is very noticeable with the profiled end wall, and overall the secondary flow appears lower. The reduced secondary flow may be seen also in Fig. 12, where the loss core peak is closer to the end-wall having been convected less by the lower secondary velocities. It can be seen that the n525t and the n5252 profiles are almost identical, so that filling in the re-entrant corner on the pressure side has very little effect. The numerical results are given in Table 3. The loss coefficients are shown also based on isentropic exit velocity, which is a more familiar definition for turbine design, and which allows for the small changes in exit velocity caused by the slight changes in angle. The overall reduction is 6.6% of the loss and 14% of the secondary kinetic energy. The flat end-wall shows a small difference compared to Table 2, because Table 3 was obtained with an assumed inlet boundary layer before the actual one had been measured.

Table 3: End-wall Profile Results at Slot 10

Profile	ct (mm)	$2*at$ (mm)	YAW (degree)	Secondary KE based on inlet	Total loss based on inlet	Total loss based on exit
flat	none	none	77.35	0.0098	1.161	0.0546
n5252	52	10	77.28	0.0086	1.087	0.0511
n525t	52	10	77.26	0.0087	1.088	0.0510

5.0 CONCLUSION

The following conclusions from this work may be drawn.

- A computational code has been developed and validated against two sets of data of secondary flow in cascades. In the transitional case the loss is over-predicted at mid-span, due to regions of laminar flow which are ignored by the turbulence model. The nozzle blade, which had a boundary layer trip, is predicted better. The secondary flow is under-predicted for the high turning rotor blade, again probably due to the neglect of laminar flow regions, but it is over-predicted for the nozzle blade.

- A design philosophy was adopted to reduce the pressure on the end-wall near the pressure surface by convex curvature, and raise it near the suction surface by concave curvature. The intent was to reduce the cross-passage pressure gradient in the region of high turning and so reduce secondary flow.

- A profile was chosen with four parameters which could be varied to produce a range of profiles. A systematic study was carried out, giving an optimum profile with an amplitude of 10mm (compared to an axial chord of 152mm), with the peak positioned 52mm upstream of the trailing edge.

- A detailed evaluation of the flow given by the chosen profile showed the reduced secondary flow and less rolling up of the end-wall boundary layer to form the vortex core. An overall reduction of 6.6% of the loss was obtained.

The results of making and testing the chosen profile will be presented in a later paper.

ACKNOWLEDGEMENT

The authors thank ALSTOM Energy Ltd. for providing support for this work, and their permission to publish this paper.

REFERENCES:

1. Deich, M. E., and Zaryankin, A. E., et al, (1960). "Method of Increasing The Efficiency of Turbine Stages", *Teploenergetica* No. 2, Feb 1960, pp. 18-24. Translation in AEI Translation No. 2816.
2. Morris, A. W. H., and Hoare, R. G., (1975). "Secondary Losses Measurements in a Cascade of Turbine Blades with Meridional Wall Profiling", ASME paper 75-WA/GT-13.
3. Ewen J S, Huber F W, Mitchell J P, (1973) 'Investigation of the Aerodynamic Performance of Small Axial Turbines', *Journal of Engineering for Power*, pp.326-332
4. Boletis E, (1985) 'Effects of Tip Endwall Contouring on the Three Dimensional Flow-Field in an Annular Turbine Nozzle Guide Vane: Part 1 - Experimental Investigation', *Journal of Engineering for Gas Turbines and Power* Vol. 107, pp.983-990
5. Atkins M J, (1987) 'Secondary Losses and Endwall Profiling in a Turbine Cascade', *Proceedings of the I.MechE.*, Paper C255/87, pp.29-42
6. Rose, M.G., (1994). "Non-axisymmetric endwall profiling in the HP NGV's of an axial flow gas turbine." ASME Paper No. 94-GT-249
7. Hartland J.C., Gregory-Smith D.G., Rose M.G., (1998) "Non-axisymmetric endwall profiling in a turbine rotor blade." ASME Paper No. 98-GT-525
8. Patankar, S.V.(1980) "Numerical Heat transfer and Fluid Flow" Hemisphere Pub. Corp. New York.
9. Rhie, C.M and Chow, W.L. (1983), "Numerical Study of the Turbulent Flow Past an Airfoil with Trailing Edge Separation", *AIAA Journal.*, p125
10. Leonard, B.P. (1979) " A Stable and Accurate Convective Modelling Procedure Based on Quadratic Upstream Interpolation", *Comp. Meth. Appl. Mech. Eng.*, Vol 19, p59
11. Lien, F.S and Leschziner, M.A. (1994) "Upstream Monotonic Interpolation for Scalar Transport with Application to Complex Turbulent Flows", *Int. J. Num. Meth. Fluids,* Vol. 16, p527
12. Norris, L.H. and Reynolds, W.C. (1975) "Turbulent Channel Flow with a moving wavy Boundary', Rep. FM-10, Dept. of Mech. Engrg., Stanford University.
13. Kato, M. and Launder, B.E. (1993) 'The modelling of turbulent flow around stationary and vibrating square cylinders', *Proc. 9th Symp. Turbulent Shear Flows*, Kyoto, 10-4.
14. Gregory-Smith D.G., (1997) ERCOFTAC Seminar and Workshop on 3D Turbomachinery Flow Prediction, *ERCOFTAC Bulletin*, 18-22
15. Moore H., Gregory-Smith D.G., (1996) "Transition Effects on Secondary Flows in a Turbine Cascade." ASME Paper No. 96-GT-100
16. Gregory-Smith D.G., (1995) 'Calculations of the Secondary Flow in a Turbine Cascade.' *International Symposium on Computational Fluid Dynamics in Aeropropulsion,* ASME International ME'95 Congress and Exposition, AD-Vol.94, 77-88
17. Yan J., (1998) 'The effect of end wall profiling on secondary flow in nozzle guide vanes', PhD. Thesis, University of Durham, *to be submitted.*
18. Gregory-Smith D.G., Biesinger Th., (1992) 'Turbulence evaluation within the secondary flow region of a turbine cascade.' ASME Paper No. 92-GT-60

C557/144/99

Tip-clearance-affected flow fields in a turbine blade row

M SELL, M TREIBER, C CASCIARO, and **G GYARMATHY**
Turbomachinery Laboratory, Swiss Federal Institute of Technology, Zurich, Switzerland

SUMMARY

For effective validation of CFD codes for design, numerical simulations must be compared and contrasted with experimental data sets which are well posed and measured. This paper presents, with the emphasis placed primarily upon the description and explanation of the experimental results, a test case designed for CFD validation concerning the flow field generated by the presence of tip clearance in an annular cascade turbine blade row.

The test case is based upon a moderately loaded prismatic blade profile, operating with axial inflow and a nominal mid span exit Mach number of 0.5. In addition to the case with no clearance, three gap heights, representing 1, 3 and 5% of chord length have been selected. These tip clearances were chosen because they represent three conditions where the tip clearance is below, approximately equal to and substantially larger than the inlet boundary layer displacement thickness. At the leading edge this produces differing behaviours, with the smallest gap producing a stagnation point, equivalent to the case without tip clearance, whilst the largest clearance allows through flow over the leading edge into the tip gap.

In the presence of larger tip clearances, the tip leakage flow induces a stagnating flow in the trailing edge region, which results in a large area of blocked flow in the blade passage downstream of the trailing edge.

NOMENCLATURE

$$C_{pt} \quad [-] \quad = \frac{p_{oi} - p_o}{p_{oi} - p_{si}} \quad \text{Total pressure loss coefficient}$$

$$C_{ps} \quad [-] \quad = \frac{p_s - p_{oi}}{p_{oi} - p_{si}} \quad \text{Static pressure coefficient}$$

$$L_\varepsilon \quad [\text{mm}] \qquad\qquad \text{Eddy length scale}$$

M	[-]	Mach number
p	[pa]	Pressure
Tu	[%]	Turbulence intensity
α	[°]	Flow yaw angle, from tangential

Subscripts

o	Total conditions
s	Static conditions
i	Inlet conditions

Abbreviations

| CFD | Computational Fluid Dynamics |
| RMS | Root Mean Square radius |

1 INTRODUCTION

The interest in the experimental study of tip clearance affected flow fields, as measured in simple annular and linear cascades, falls into two distinct categories. Firstly there is the interest in understanding the flow physics of blading systems and secondly there is the interest in generating data sets for the validation of CFD codes for design. The work presented here is directed to the second case, however the tip physics revealed through this study have some distinct features which merit closer attention.

The flow over a free ended blade in an annular cascade is an attractive proposition for a CFD validation data set. In terms of design, the case is unrepresentative, due to the lack of relative blade to casing motion, and the fact that the inlet conditions within a cascade are far from matched to that of a real multi-stage machine, particularly with respect to the boundary layer height and flow unsteadiness. Nevertheless the case is valuable, as the geometry is simple to specify but produces a flow field with sufficient inherent complexity to address real issues in CFD validation such as meshing strategies, numerical smoothing, turbulence modelling etc. These topics are not part of the current work, but will be dealt with in later papers.

In effect, the test case is generic and as such does not have to incorporate tip clearance gaps representative of any particular machine. For this study, the values of the tip clearances have been selected based upon an assessment of the inlet boundary layer. In the development of the case the displacement thickness has been taken as one of the important identifying markers for the flow. The reason being that it can be reasonably assumed to be a controlling parameter for the near casing flow around the leading edge of the blade, because it will dictate the degree of blockage at the tip.

A structure that may be influential in the formation and development of the tip clearance flow is the leading-edge casing horseshoe vortex system. For a narrow ("aerodynamically closed") leading edge clearance it will be reasonable to assume that boundary layer will roll up and the horseshoe vortex system will form, with a strength and position related (amongst other factors) to the displacement thickness (see for example the correlations and results shown in (1) and (2)). For a wide open tip gap, it is reasonable to expect that the inlet boundary layer will not roll up and will flow more or less axially over the blade leading edge, and hence the horseshoe vortex system will not form. To allow for the possibility of differing flows in the leading edge region, three cases have been purposely developed corresponding to clearances less than, approximately equal to, and greater than the boundary layer displacement thickness. These choices add a subtle complexity to the CFD validation problem, as it requires that for all three cases to be calculated the complete flow field must be correctly (or systematically incorrectly) simulated from the leading edge onwards.

Extensive work has been performed in the realm of tip clearance studies in cascades. The majority of the idealised experimental work, such as (3) and (4), have been performed in linear cascades with the emphasis on the understanding of the physics. Studies with annular cascades are few, however notable is the work of (5) and (6). From the viewpoint of the current study the most relevant work in the literature pertains to that with large tip gaps, for example the work of (7) and (8).

2 EXPERIMENTAL AND NUMERICAL METHODOLOGY

2.1. Test Stand

The experiments have been conducted with an annular cascade test stand (9,10). The test stand operates at quasi-atmospheric conditions with the operating point of a nominal mid span exit Mach number of 0.5. The most important geometrical features of the test rig are given in Table [1].

Table [1]: Test stand geometry

Tip Radius	[mm]	400
Blade Height	[mm]	90
Blade Number	[-]	48
Axial Chord	[mm]	47.34
Tip-Hub Ratio	[-]	1.29

The experiments were performed with a moderately loaded generic steam turbine profile, which has a stagger angle of 54.5° relative to the tangential direction, and a pitch of 40.58 mm at the hub, figure [1]. The blading is prismatic, with the leading edge aligned radially. The inlet flow is axial, and the total turning in the undisturbed flow is approximately 68°.

In addition to the zero clearance case, the three tip clearances selected were 1, 3 and 5% of axial chord. The blade tip surface is cylindrical, that is to say the tip clearance is constant over the complete area of the blade profile.

2.2. Measurement Method and Accuracy

The experimental data presented here was derived exclusively using pneumatic pressure measurement techniques. These comprise probe data in the flow field and static pressures on the tip casing. The flow field data were collected in three tangential measurement planes downstream of the trailing edge, at 1/6, 1/3 and 1/2 axial chord. Three probe types were used. Firstly a simple round head Pitot with a diameter of 0.5 mm, secondly a small head diameter (0.9 mm) five hole probe, and finally a so-called "virtual four hole needle probe". The virtual four hole needle probe is an arrangement whereby two single hole probes are used sequentially in order to determine the full three dimensional flow field. The first probe is a normal needle probe, which has a slim 0.5 mm diameter transverse cylinder with a single hole bored radially. This probe is used to simulate a three hole probe, by measuring in sequence at any given measurement point three pressures corresponding to three rotational yaw positions of the probe, each separated by 40°. The fourth pressure is derived by measuring in the same location with a needle probe with a hole bored into a canted tip. With the two probes *a priori* calibrated in a free jet test stand as a quasi-four hole probe, it is possible to use the four measured pressures to derive the full three dimensional flow field information.

Because of the sequential nature of the measurements the virtual four hole needle probe is time consuming to use. Therefore its use was restricted to the tip region flow field, where the aerodynamic advantages of its small size were essential.

As the flow field in the tip region was shown to be subject to large yaw angle variations, it became necessary to optimise the probe behaviour by operating both the five hole probe and the virtual four hole needle probe in a nulling mode, i.e. turning the probe head into the local flow direction until the pressure difference between the two lateral (yaw) indicating probe holes was nominally zero. This was done by utilising a fuzzy logic controller (11). This controller is highly efficient, being capable of nulling a probe even from a reverse flow alignment in typically three or fewer steps, reducing iterations and thus making it ideal for use in combination with the virtual four hole needle probe.

Static pressures were measured on the tip casing in unusually fine detail. Typically over 2000 individual static pressure points were resolved per blade pitch. This was achieved by displacing the cascade blades in the circumferential direction in small steps relative to a series of fixed pressure taps drilled in the tip casing, as described in detail in (9).

The experimental accuracy is dependent upon the flow conditions, the quality of the probe, probe calibration and measurement chain (12). For the results shown here, typically an accuracy in flow angles of better than ±0.2° in the main flow field have been demonstrated, and in flow velocity to 0.1m/s. In the tip region, principally because of the presence of low local flow velocities and hence pressure head, the accuracy of the flow angles degrades to ±0.7°.

Extensive tip casing surface oil flow visualisation of the three main tip clearance cases were performed. The test stand is equipped with a plexi-glass window covering several blade passages over the full axial extent of the blade flow path. The flow visualisation method consists of introducing a light oil onto the casing surface which can be recorded photographically as it is swept downstream under the action of the surface shear stresses.

2.3. Numerical Methodology

The focus of this paper is strongly oriented to the description of the experimental results, however some supporting results from a CFD simulation are presented. The computations have been performed with the 3D Navier-Stokes solver CFX-TASCflow, V2.7, by AEA Technology Ltd. This uses a finite volume scheme, whose discretisation is based on upstream differencing, using a pure linear profile (equivalent to second order in space), with physical advection correction terms. The closure for the turbulent flows is achieved with a Kato-Launder model with a logarithmic law at the walls. A block-structured mesh has been used to describe the geometry. Extensive grid dependency studies have been performed acting on both the block topology (using pure H, H-O and H-C grid types) and on node distribution and density (from a coarse grid with 28,000 nodes to a fine one of 570,000 nodes). These studies showed that the results presented herein are mesh independent. The basic structure of the mesh at a mid span cut is shown in figure [2], where one of the medium node density meshes has been used for clarity.

For boundary conditions, at the inlet the code permits the application of either the total pressure or the velocity. The latter was prescribed using the measured velocity profile, with constant total temperature and constant free-stream turbulence characteristics (L_ε=5mm and Tu=2%). At the outlet the measured radial profile of static pressure was imposed from hub to the radial height (70% for the 5% clearance case results presented herein) where no influence of the leakage jet could be identified from the experiment; for the remaining area of the annulus to the casing the code was left to find a suitable distribution. For a discussion of the use of the exit static pressure distribution as a boundary condition see section 3.3. An in depth discussion of the issues relating to the validation of the code can be found in (13) and (14). The aim of the paper is to generate an experimental data set for CFD validation data, and it is outside the scope of the paper to discuss all the issues raised by the numerical simulation.

3. RESULTS

3.1. Presentation of the Principal Results

Figures [3.a-d] show contour plots of the tip static pressure coefficient distribution for the four clearances. Note that the pressures are non-dimensionalised with the dynamic head of the *inlet* flow (taken at the RMS radius). Consider first the leading edge region of the blade. The zero clearance and 1% clearance show similar distributions in the leading edge region and the forward portion of the blade passage, which are typical of a stagnating or low velocity flow. This supports the contention that for the front part of the blade the tip clearance gap could be blocked by the inlet boundary layer ("aerodynamically closed") and therefore could be sustaining a horseshoe vortex system. The most open tip clearance, the 5% case, shows no sharply stagnating flow at the leading edge. The 3% case shows no clear evidence of a leading edge stagnation.

Consider the rear portion of the blade passage. The 1% case [3.b] shows the characteristic pressure distribution of a tip leakage vortex formed in the blade passage at 70% axial chord, causing a pressure trough almost parallel to the suction side. Consider next the 5% clearance case [3.d], here the static pressure distribution has an interesting feature. Similar to the 1% case, the pressure depression due to the tip leakage vortex in the main passage is seen. However, the figure shows a "wasp-waist" type distribution of low static pressure, surrounded by two regions of higher pressure. An interpretation of this picture is that the strength of the leakage vortex is diminished locally shortly downstream of the trailing edge plane, and the flow is locally stagnating.

This region of stagnating flow is quite surprising, as it shows that there is, off the walls, a very substantial region of decelerating flow in the tip region. This deceleration is quite sharp, representing a static pressure recovery of approximately 6 inlet stagnation pressures in less than 10% of axial blade chord. The 3% case (see figure [3.c]) shows a similar trend to the 5% case, though the "wasp-waist" is less marked, and the pressure recovery is substantially weaker than in the 5% case, at approximately 2 inlet stagnation pressures over the same distance.

Considering the far downstream limit of the static pressure, the no tip clearance case shows an almost uniform static pressure distribution at one axial chord downstream (at a level of Cp_S of approximately -7.4), however all the tip clearance cases, and particularly the 5% case, show a persistence in the static pressure structure downstream, indicating the flow structures generated by the tip clearance are strong.

At the blade pressure side, the air inflow into the tip gap is made apparent (figures [3.b to d]) by the steep pressure gradient perpendicular to the blade contour. The sudden decrease in Cp_S by 6 to 8 units is a sign of a transverse velocity in the gap of the order of the nominal cascade exit velocity. No transverse recovery of pressure is seen behind the blade edges, indicating the absence of contraction effects due to edge separation.

The flow visualisation, through the position and character of the limiting wall streamlines, supported the experimental design assumption that the 1% tip clearance could be generating a horseshoe vortex like structure at the leading edge, whereby the 5% tip clearance case had no leading edge stagnation point, with flow moving axially at the tip over the leading edge of the blade.

The most important flow visualisation result is shown in figure [4]. This picture is a digitally enhanced view of tip casing wall surface oil flow focusing upon the rear portion of the blade for the 5% tip clearance case. The figure shows a limiting streamline and a distinct area near the suction side in the trailing edge region where the visualising oil has gathered in an apparently stagnating zone. It should be stressed that this flow feature and its extent is repeatable, periodic and independent of the volume of visualising material used. During the course of the experiment the thinning of the oil film due to the action of the applied surface shear stress could be observed; generally the flow feature shown in figure [4] thinned comparatively slowly, indicating that the shear stress in this local region is markedly lower than in the neighbouring areas.

Although it is, with a few exceptions, generally inadvisable to link surface oil flow visualisations with static pressure distributions, figure [4] is consistent in terms of the location of the stagnant flow field with the static pressure distribution shown in figure [3].

As support for this observation, consider figure [5]. This shows a cut from the CFD simulation at 94% radial height, which is just below the blade tip edge. The local flow vectors in the flow field can be seen, and superimposed upon them are the limiting streamlines observed from the simulation. The flow in this region is seen to be reversed, compatible with the flow visualisation and static pressure distribution. The importance of figure [5] is that it shows that numerically the flow field feature indicated by figure [4] can be captured and that given the depth into the flow field of the cut in figure [5], a simple separation bubble may not be a sufficient explanation.

Figure [6.a-d] shows the distribution of the total pressure coefficient for the four cases at 1/6 axial chord downstream of the trailing edge. The view shown is the upper half of the blade passage, with the mid span and tip radii indicated. For the no clearance case [6.a], the typical features associated with the secondary flow field are seen (see (13)). Because of the decision to use a constant contour step for all four plots of 0.5, some of the fine detail of the measured flow field for this case is lost in this figure. The fact that such a large contour step is required to plot the isolines for the tip clearance cases reflects perfectly the dominant nature of the loss core associated with the tip leakage flow, as compared to the zero-clearance secondary flows.

For the tip clearance cases, the area of high loss induced by the tip leakage flow becomes predominant, and is large in comparison to the blade passage pitch, the 1% case occupying about 50% of the blade pitch, the 3% case 75% and the 5% case almost entirely occupying the near casing part of the passage. The depth of the loss core is also a strong marker. Even though the measurement plane is very close to the trailing edge, the loss core has extended down some 20% into the blade passage. As the tip clearance opens, the strength and area of the loss core associated with the tip vortex progressively increases, but not proportionate to the tip clearance. There appears to be no evidence of any multiple loss cores in the measured data. Considering the low total and static pressure, figures [6] and [3] can be interpreted that the tip leakage flow is inducing a significant area of low momentum fluid. As the region of the tip leakage generated loss is large this implies a substantial radial redistribution of mass flow downstream of the trailing edge.

Figures [7.a-c] shows the separate development of total pressure loss system for the 5% clearance for the three measurement planes downstream of the trailing edge. The loss structure mixes out slowly, spreading a little in width, and hardly anything in radial depth, quite consistent with the persistence downstream of the patterns of static pressure distribution seen in figure [7.d].

The tip leakage flow has a much higher exit angle, referenced to the tangential, than the main flow (42° compared to 22°). Thus relative to the main flow wake the leakage flow is progressively sheared away, until at the half axial chord position the loss structure has worked its way (relative to the wakes) across almost a full pitch. Figure [7] shows the usual apparent geometric rotation of the main wake relative to the radial direction as flow proceeds downstream. The region of the tip leakage loss represents an area of localised vorticity, which induces an additional local rotation of the wake close to the tip leakage flow. This is seen most markedly in figure [7.c], where the wake has been deformed by the leakage vortex.

Figure [8] shows the exit flow angle, resulting from the mass averaging velocity components over the pitch, for 1/2 axial chord downstream for the 5% case, both measured and computed. The high angular variation in the pitch angle is striking, over the outer 10% of radial height the pitch averaged flow angle changes by 70°. Pitch averaging, naturally, masks the extremes of behaviour, and in fact locally at mid pitch a range of 120° in exit angle has been measured and computed, including an element of locally reversed flow.

The agreement between the CFD and experiment below 70% radial height is excellent, however for the simulation in the tip region the maximum overturning (lowest α) occurs higher in the channel, and its value is approximately 5 degrees too high. Outside this region towards the casing the agreement is once again very good. It should be noted that the region of maximum discrepancy between the CFD and experiment within a pitch coincides within the area of almost stagnant flow (typical Mach numbers are of the order 0.05). For the experimental values a consequence of this is that despite the very good accuracy of the flow angle itself, a mass averaging procedure will inevitably be less precise in the vortex core because of the low flow velocities.

3.2. Interpretation

It is clear that the flow field presented is completely dominated by the tip clearance induced flow. The most open clearance case shows several interesting features. One possible interpretation of the physics seen in this case is that a form of leakage vortex breakdown is occurring. Vortex breakdown, which can be broadly defined as a change in the structure of a swirling flow, is a phenomena observed in many flow situations, for example above delta wings.

The occurrence of vortex breakdown within turbomachinery flow fields is not unknown, however it is predominantly seen as a precursor to stall within compressor flow fields (15). Imperfectly understood, there are several explanations for the physics of vortex breakdown. One explanation for vortex breakdown in the tip region of compressor flows (given in (15)) is that there is a redistribution of vorticity from the axial into the circumferential direction. This arises from the fact that the positive axial pressure gradient of the core flow acts to reduce the axial velocity component, simultaneously the spanwise mass redistribution increases the radial component and reduces the circumferential component. The vortex strength axially decreases, making the centreline static pressure increase. Where this pressure makes the purely axial centreline flow to come to a halt, a stagnation point arises with back flow on the downstream side of it.

Although the deceleration ratio (defined as the ratio of peak to exit Mach number) in the blade passage under study is relatively benign (approximately 1.29 at mid span), and thus the positive axial pressure gradient over the rear portion of the blade passage is slight in comparison to that within an axial compressor. It is hypothesised that the tip leakage vortex induced by the large tip clearances have sufficient streamwise vorticity to trigger the breakdown phenomena described above.

A vortex breakdown phenomenon is also a plausible explanation of the area of reversed flow seen at some distance away from the tip casing within the CFD simulation, (figure [5]).

It has to be clearly stated, that the largest tip clearance under investigation here is greater than usually seen in the literature, arising from the present objective of validating CFD programs over a wide variety of flow configurations. However such undesirably large tip clearances may occur in damaged turbomachinery configurations.

3.3. Implications for CFD Computations and Validation Experiments

The measured flow fields have a clear implication for CFD computation and validation experiments. Figure [9] shows the distribution of static pressure coefficient at one third axial chord downstream of the trailing edge. The distribution shows a more or less uniform gradient, associated with the exit swirl, extending from the hub to about 70% radial height. From 70% radial height to the tip the picture of static pressure is more confused. No distinct radial stratification is revealed and structures corresponding to the positions of the tip blockage structure appear (compare figures [9] and [7.b]). The lack of stratification of the static pressure over the full height arises from the breakdown in the uniformity of the tangential (swirl) velocity component.

It is therefore open to debate as to the correct definition and application of the downstream boundary conditions for a computation of this case. The correct selection of the boundary conditions can significantly influence the convergence, stability and end accuracy of a numerical simulation. The usual procedure (16) would be to assume a fixed value for the static pressure, perhaps combined with a radial pressure gradient derived. This pressure gradient could be derived from an assumed value for the exit swirl, or from measured hub and static pressure values. For the case shown here, specification of a fixed radial distribution of the static pressure over the full annulus height is clearly inappropriate. This lead to the usage of a fixed radial distribution over 70% of the height described in section 2.3.

It is clear, that generally the ideal far field boundary condition for a CFD computation should be that the exit plane is sufficiently downstream for the fully mixed out values to be applied. However, with incentive of the effective use of computational resources "short" far field boundary conditions are often applied in combination with for example, non-reflecting boundary conditions (17), (18). Thus the non-uniform radial distribution of the exit conditions is of importance. A tightly prescribed definition of the static pressure has the potential to produce an incorrect computation for this case.

4. CONCLUSIONS

The flow over a moderately loaded turbine blade with zero to large tip clearances has been measured with the primary aim of developing a CFD validation test case. The case has achieved the objective of producing a complex and challenging flow field with a relatively simple geometrical arrangement.

The tip clearance induces a tip leakage vortex. Downstream of the trailing edge the flow field is characterised by a strong local flow blockage in the tip region which is extremely large, persistent and the dominant single source of loss within the blade passage.

At the largest tip clearance, 5% of chord, wall flow visualisation and the tip static pressure distributions show that the flow in the region of the trailing edge is producing a stagnating domain away from the walls. One possible explanation for this behaviour is the interaction of the tip leakage with the positive pressure gradient over the rear third of the blade to induce a vortex breakdown type flow.

The presented CFD simulation agrees well with the measurements, and fully confirms the hypothesis that a vortex breakdown is the cause of the unusual flow features measured.

The presented static pressure distribution indicates that care has to be taken in the selection of appropriate downstream boundary conditions to be applied for the computation of this, and analogous, cases.

5. ACKNOWLEDGEMENTS

The annular cascade test rig used here was sponsored by ABB Power generation, Ltd. and NEFF (Swiss National Energy Research Foundation). The third author acknowledges the financial support of the Swiss ERCOFTAC Pilot Centre. The assistance of Mr. P. Köppel with the reduction of the virtual four needle hole probe data is gratefully acknowledged. Mr. E. Pizza provided many helpful insights into the flow field through CFD simulation, which due to space constraints cannot be presented herein. A portion of the experimental data was gathered by Mr. K. Selvarajoo. Ernesto Casartelli is particularly thanked for his help in interpreting the results.

6. BLADE PROFILE CO-ORDINATES

The test stand geometry, blade co-ordinates and exact boundary conditions are available from the authors for those wishing to compute the case.

7. REFERENCES

(1) C.J. Baker, "The turbulent horseshoe vortex", Journal of Wind Engineering and Industrial Aerodynamics, 1980.
(2) C.J. Baker, "The laminar horseshoe vortex", Journal of Fluid Mechanics vol. 95, part 2, pp.347-367, 1979.
(3) M.I. Yaras and S.A. Sjolander, "Losses in the tip leakage flow of a planar cascade of a turbine blade", AGARD conference proceedings CP-469 "Secondary flows in turbomachines", 1989.
(4) J. Moore and J.S. Tilton, "Tip leakage flow in a linear turbine cascade", Journal of turbomachinery vol.110 pp18-26, 1988.
(5) G. Morphis and J.P. Bindon, "The effects of relative motion, blade edge radius and gap size on the blade tip pressure distribution in an annular turbine cascade with tip clearance", ASME Paper 88-GT-256, 1988.
(6) F.J.G. Heyes, H.P. Hodson and M. Bailey, "The effect of blade tip geometry on the tip leakage flow in axial turbine cascades", Journal of turbomachinery, vol. 114 pp643-651, 1992
(7) J.P. Bindon, "The measurement and formation of tip clearance loss" Journal of turbomachines vol. 111, 1987.
(8) J.K.K. Chan, M.I. Yaras and S.A. Sjolander, "Interaction between inlet boundary layer, tip leakage and secondary flow in a low speed turbine cascade", ASME paper 94-GT-250, 1991
(9) M. Sell, M. Treiber, P. Althaus, and G. Gyarmathy, "The design and construction of a new test stand for the study of basic turbine flow phenomena", Proceedings 2nd European turbomachinery conference, Antwerp, Belgium, 1997.
(10) M. Sell, P. Althaus and M. Treiber, "Data Acquisition and Control within the Zürich Annular Cascade test stand", Proceedings XIIIth symposium on measuring techniques for transonic and supersonic flows in cascades and turbomachines, Zürich, Switzerland, 1996.
(11) R. Biswas, "A fuzzy logic controller to balance pneumatic multi-hole probes", Proceedings XIIth symposium on measuring techniques for transonic and supersonic flows in cascades and turbomachines, Prague, Czechoslovakia, 1994.

(12) M. Treiber, P. Kupferschmeid and G. Gyarmathy, "Analysis of error propagation arising from measurements with a miniature pneumatic 5-hole probe", Proceedings XIVth symposium on measuring techniques for transonic and supersonic flows in cascades and turbomachines, Limerick, Ireland, 1998.

(13) C. Casciaro, M. Treiber, M. Sell, A.P. Saxer and G. Gyarmathy, "A comparison of experimental with computational results in an annular turbine cascade with emphasis on losses", ASME paper 98-GT-146 IGTI Congress, Stockholm, Sweden, 1998.

(14) C. Casciaro, M. Sell and G. Gyarmathy, "Towards reliable computations for a subsonic turbine", Proceedings of the symposium on verification of design methods by test and analysis, London, UK, 1998.

(15) S. Schlechtriem and M. Lötzerich, "Breakdown of tip leakage vortices in compressors at flow conditions close to stall", ASME paper 97-GT-41 IGTI Congress, Orlando, USA, 1997

(16) C. Hirsch, "Numerical computation of internal and external flows", J. Wiley & Sons, Chichester, 1988.

(17) A.P. Saxer and M.M. Giles, "Quasi-3D non-reflecting boundary conditions for Euler equations", AIAA paper 91-1603, 1991

(18) A. Fatsis and R. van de Braembussche, "Evaluation of an Euler code with non-reflecting boundary conditions for the analysis of 3D flow in compressors", International Journal of Turbo- and Jet Engines, vol. 12. 283-295, 1995.

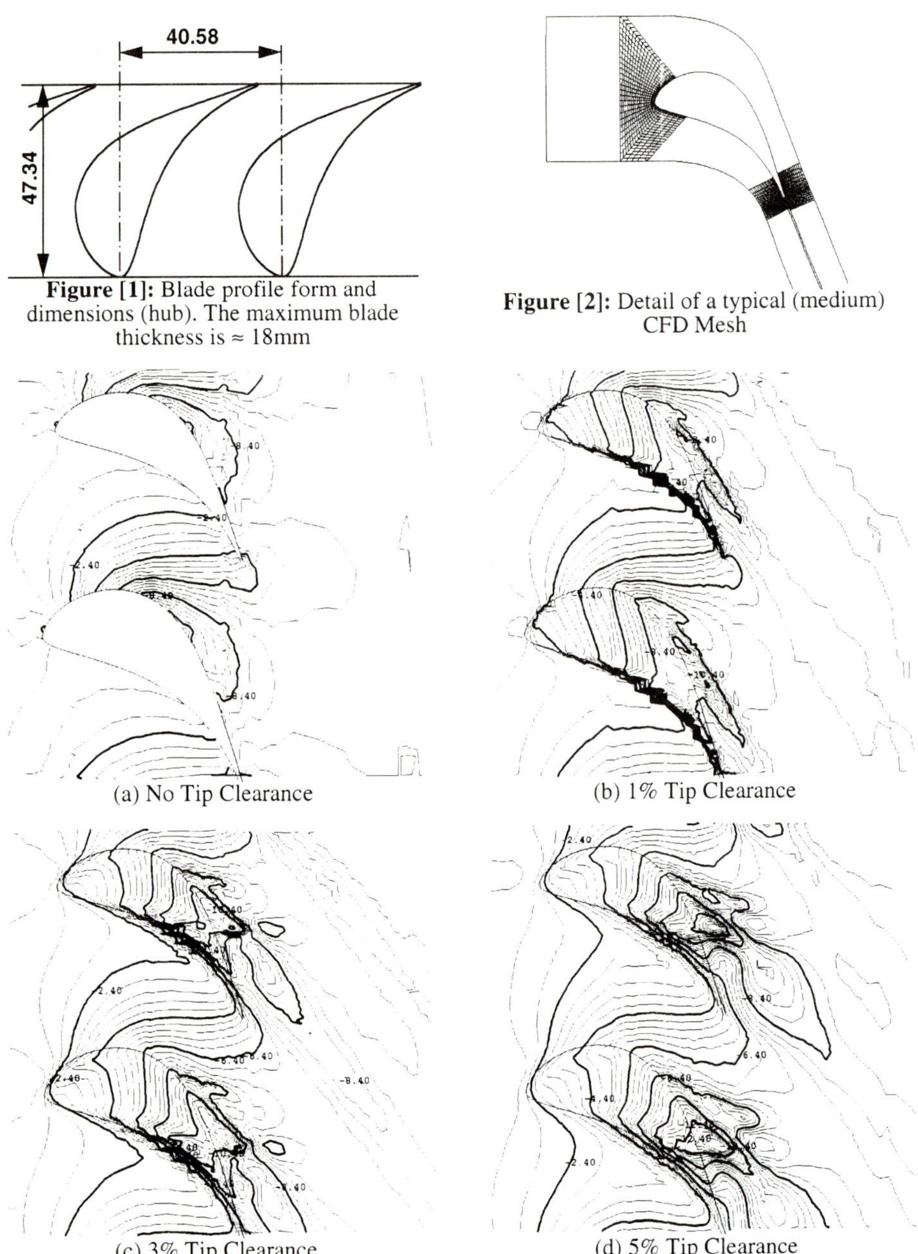

Figure [1]: Blade profile form and dimensions (hub). The maximum blade thickness is ≈ 18mm

Figure [2]: Detail of a typical (medium) CFD Mesh

(a) No Tip Clearance

(b) 1% Tip Clearance

(c) 3% Tip Clearance

(d) 5% Tip Clearance

Figure [3]: Measured Casing static pressure distribution (Cps) (ΔContour= 0.4)

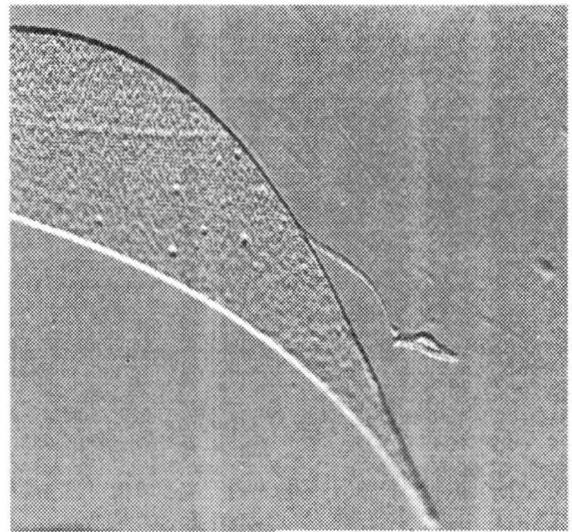

Figure [4]: Surface oil flow visualisation for the rear of the 5% case (digitally enhanced).

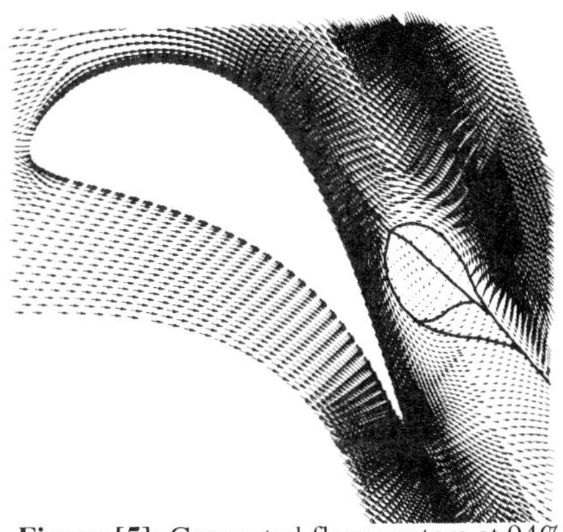

Figure [5]: Computed flow vectors at 94% radial height, with surface flow separation lines. (The blade ends at 97.4%)

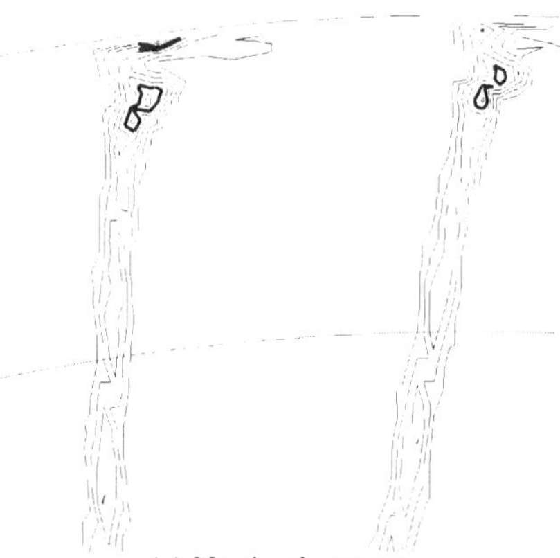

(a) No tip clearance

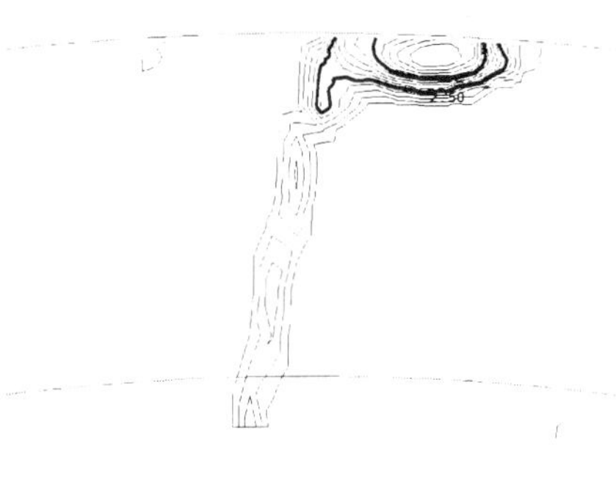

(b) 1% Tip Clearance

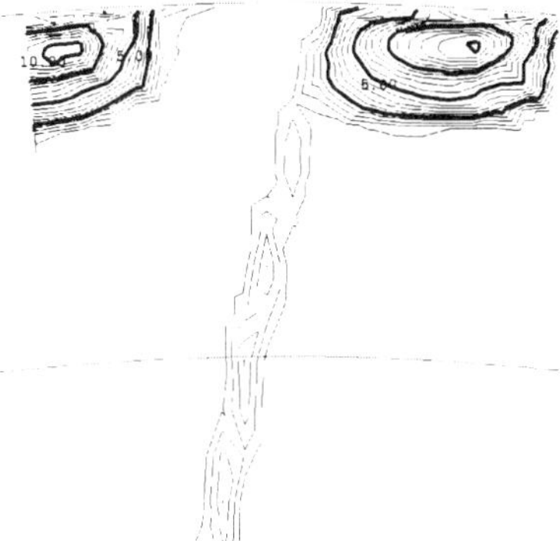

(c) 3% Tip Clearance

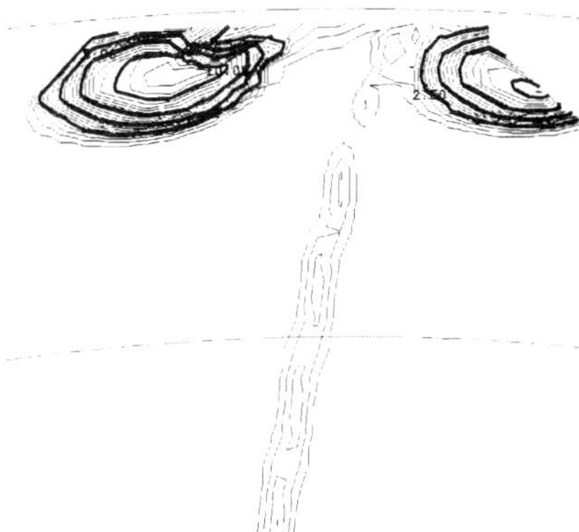

(d) 5% Tip Clearance

Figure [6]: Measured total pressure loss coefficient, 1/6th axial chord downstream of trailing edge. ΔContour=0.5, Midspan and Tip radial heights indicated

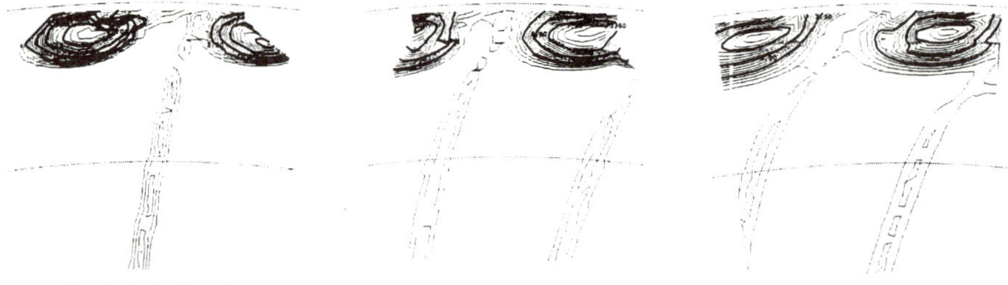

(a) 1/6 Axial Chord downstream   (b) 1/3 axial chord downstream   (c) 1/2 Axial Chord downstream

Figure [7]: Measured axial development of Total Pressure Coefficient for the 5% Tip clearance case downstream of the trailing edge. ΔContour = 0.5

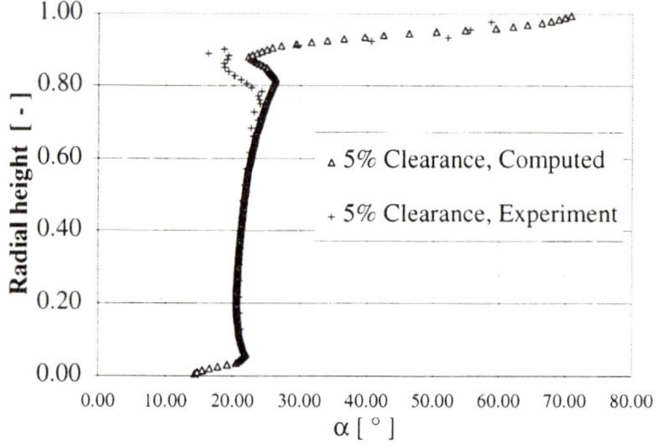

Figure [8]: Pitch-averaged exit flow angle distribution, 1/2 chord downstream of the trailing edge; comparison of CFD and measurement.

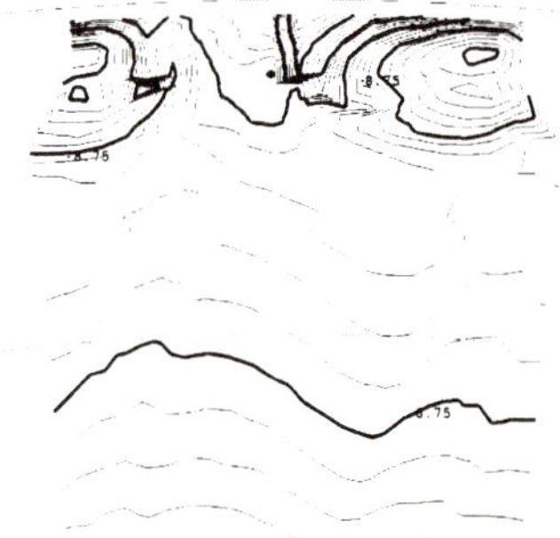

Figure [9]: Measured Static Pressure distribution, 1/3rd axial chord downstream, 5% tip clearance case ΔContour = 0.5

Experimental and numerical investigations on the influence of the tip leakage flow on the radial flow field in a four-stage turbine

D BOHN, K KUSTERER, and M LAMPING
Insitutute of Steam and Gas Turbines, RWTH Aachen, Germany

A test rig of a four stage axial turbine has been set up at the Institute of Steam and Gas Turbines, Aachen University of Technology, in order to investigate the flow characteristics in the axial gaps before and behind an intermediate stage. Therefore, extensive unsteady measurements have been carried out using a five hole pressure miniature probe. Furthermore, a numerical investigation of the 3-D flow field in the third stage of the test turbine is performed using a steady Navier-Stokes computer code. The calculations include the detailed simulation of the leakage flows using a high resolution grid in the near wall regions.

1 NOMENCLATURE

c	[m/s]	velocity	*Indices:*	r	relative
h	[m]	passage height		t	total
i	[kg/J]	specific dissipation work		φ	polytrophic
K	[kg/J]	specific kinetic energy			
n	[1/min]	rotor revolution number			
p	[N/m^2]	pressure	*Superscripts:*	–	irreversible
s	[J/kg/K]	entropy		~	reversible
sp	[mm]	gap size		$\rightarrow$	vector
T	[T]	temperature			
y	[kg/J]	specific flow work			
x	[m]	passage height coordinate			
τ	[-]	dimensionless time	*Abbreviation:*	PS	pressure side
ξ,η,ζ	[m]	curvilinear coordinates		SS	suction side
ξ,χ,ζ	[-]	equilibration coefficients		MP	measurement plane

2 INTRODUCTION

The flow in a multi-stage turbine is characterized by complex steady and unsteady 3-D flow phenomena such as rotor-stator interactions, secondary flow phenomena and leakage flows in radial gaps. These phenomena disturb the flow field around the vanes and blades leading to a loss in efficiency. The rotor-stator interactions can be divided into two parts: potential flow and wake interactions. The interaction of secondary flows and leakage flows in radial gaps lead to complex vortex systems responsible for a major part of the loss in efficiency. Furthermore, the inhomogeneity of the flow coming out of the previous stages heavily influences and disturbs the flow in the succeeding stages in a multi-stage turbine (1, 2).

A test rig of a four stage axial turbine has been set up at the Institute of Steam and Gas Turbines, Aachen University of Technology. According to the measurement task, miniature probes have been specially developed. They are used for unsteady flow measurements in the narrow axial gaps in front of and behind the 3rd stage of the turbine. Only few working groups have measured the unsteady flow field inside the turbine rotor. Binder (3) investigated the effects of secondary vortex cutting in a single stage axial cold air turbine with large axial gaps in order to exclude potential influences. Hodson (4) measured the flow field at mid-span of an single stage turbine. The work is focused on the stator wake transport through the turbine rotor passage. Flow visualization by application of the Laser Doppler Anemometry measurement technique in a 1 1/2 stage turbine have been presented by Walraevens and Gallus (5). They have shown that the stator secondary flow vortices have been transported through the rotor passages and can still be identified behind the stage. Therefore, they will significantly influence the flow field of a following stage.

The new test rig in combination with new miniaturized measurement techniques has been developed and set up in order to investigate the 3-D unsteady flow field in a multi-stage axial cold air turbine. The flow conditions in this test turbine are typical as in real turbines. Due to the narrow axial gaps, potential effects influence the flow field significantly. This paper contains exemplary results of the unsteady measurement campaign. The results give a solid data base for the detailed description of the flow phenomena described above and their mutual influences. In particular, the influence of the tip leakage flow in interaction with secondary flows on the radial flow field behind the third stage is investigated.

Furthermore, a numerical investigation of the flow field in the third stage of the test turbine is performed using a steady Navier-Stokes computer code in order to support the interpretation of the experimental data. Modern numerical codes are of high use in order to get more detailed information on the origin and development of complex 3-D flow structures as the tip clearance vortex and its interaction with secondary flows in the passage. The calculation includes the detailed simulation of the leakage flows using a high resolution grid in the near wall regions. By application of a balance-based averaging procedure (6, 7, 8, 9) reversible and irreversible averaging of the flow field is possible. The spatial inhomogeneity of the flow is given by equilibration coefficients derived from the transfer of the reversible averaged values to the irreversible averaged ones.

3 EXPERIMENTAL SETUP

3.1 Experimental Rig

Figure 1 shows a cut and a photo of the air turbine test rig "Regula". It is a four stage turbine, supplied with air from compressors at a maximum pressure of 6,3 bar with a mass flow of maximum 5,6 kg/sec. The geometrical data of the turbine and the operating conditions during the investigations are listed in Table 1 and Table 2. In order to homogenize the turbine inflow, coming through four inlet pipes arranged circumferential around the turbine, a new inlet casing has been built with a Honeycomb and a breaker plate. The power of the rotor is dissipated in a water break mounted in a momentum pedulum with hydrostatic bearings. Via these components the momentum of the rotor can be measured very sensitively.

The stator blades are fixed in an inner casing which is fixed in an outer casing. The inner casing can be turned around over an circumferential area of 12,5 degrees, that means more

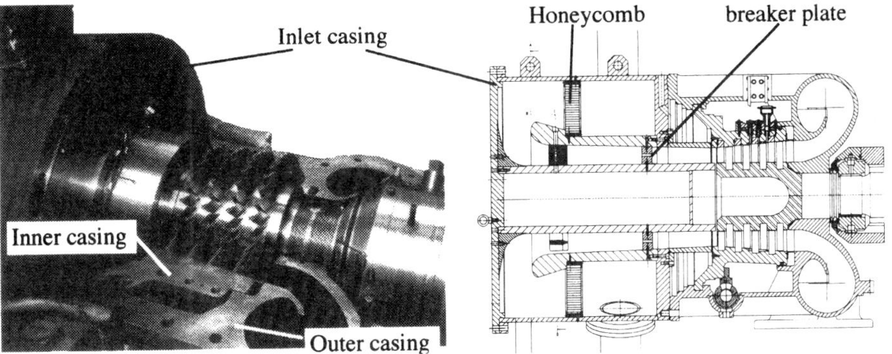

Figure 1: Photo and cut of the air turbine "Regula"

Table 1: Geometrical data of the test turbine

average diameter, MP30	$D_{a,30}$	[mm]	298,6
average diameter, MP32	$D_{a,32}$	[mm]	303,7
vane height, stator 3rd stage	L	[mm]	63,6
blade height, rotor 3rd stage	L	[mm]	68,7
axial chord length	b	[mm]	23,4
chord length	s	[mm]	32,1
average pitch	t	[mm]	23,5
number of rotor blades	z	[-]	33
number of stator blades	z	[-]	32
blade angles	α_1, β_2	[°]	36,3
tip clearance	sp	[mm]	0,6
axial gap rotor - stator	-	[mm]	6
degree of reaction	r	-	0,5

Table 2: Operating conditions during investigations

r.p.m.	4995
T_{Inlet}	353 K
T_{Outlet}	316 K
p_{Inlet}	1.5 bar
Π	1,5
mass flow	3,3 kg/s

than one stator vane spacing. While the inner casing is moved, the probes, mounted on a special dish with an air seal, remain at a fixed position. So the relative position of the probe to the stator blades is changed. This can be carried out at four planes: The inlet of the third stage (called Measurement Plane 30), behind the stator of the third stage (MP 31), at the outlet of the third stage (MP 32) and after the fourth stage and so the outlet (MP 42). Figure 3 shows the location of the measuring planes 30 and 32 and the probe positioning.

3.2 Measuring Probes

The measurements have been carried out with a five-hole pressure probe (Fig. 2) for unsteady measurements. This probe has been especially developed for unsteady investigations in small axial gaps. The extraordinary of this probe is, that the instationary pressure signal is transported via capillary tubes from the measurement point (head of the probe) to the transduers which have a diameter of 1.63 mm. This way of instationary measurement allows to build a small probe head with a diameter of 3 mm for 3-D flow measurements. Furthermore the height of the probe head amounts only 5 mm, which is necessary because the axial gap is approximately 6 mm between the rotor and the stator blades. The disadvantage of this probe is

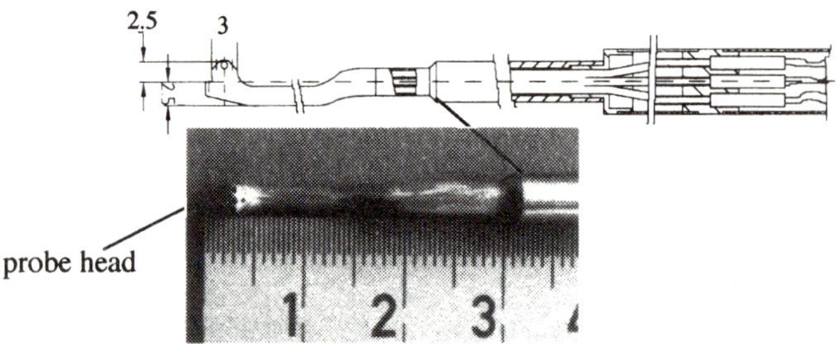

Figure 2: Five hole pressure probe for unsteady measurements

a damping and phase shift of the instationary pressure signal depending on the length and the diameter of the capillary tubes. To build a small probe head, capillary tubes of 0.6 mm diameter and a length of 145 mm are used. Further intensive investigations lead to a maximum frequency resolution of 20 kHz for this diameter and length (10). Above this maximum measure frequency the damping of the signal in the tube is too high and higher frequencies can't be reproduced. Due to the Nyquist criterion instationary flow phenomena with a resolution of 10 kHz can be investigated. The signal damping and phase shift of the signal is corrected with a transfer function. This function is found experimentally and numerically (10). Under the head of the probe a shielded thermocouple is integrated for measurements of the fluid temperature.

4   RESULTS

4.1 Experimental Results

Figure 3 shows an example of the unsteady experimental results, detected after the rotor blades of the second stage (MP 30). As shown in the pictures, the rotor blading passage is observed at a constant circumferential position of the probe. Here, the constant circumferential position of the probe no. 4 has been chosen as the potential influence of the succeding vanes is minimized. It will be the task of future work to complete the unsteady measurement campaign for a full spacing.

Concerning probe position no. 4, a dimensionless time axis (τ) is drawn against the relative height of the passage. τ is then the quotient of the absolute time interval and the time difference between two rotor blading passages. With a maximum discretization of 20 kHz , depending on the resolving power of the probe, 10 measurement points are represented per rotor blading division.

In locating the dissipating areas, the most effective method is to see the continuous course of the total pressure observed over an interval of time, so that this quantity is drawn as contour plots in dependency from the dimensionless passage heights and the dimensionless time. Figure 3 shows a total pressure loss area, consisting of tip leakage- and passage vortices, to be located near the casing. The graphical representation in Fig. 3 in combination with the numerical results presented later on shows very well how the leakage flow gets winded-up, through interaction with the passage vortices, and hence developing tip leakage vortices. In the region near the hub, as shown in Fig. 3, the loss area, arising from the hub passage vortex, is well resolved.

On the whole, the contour plots show that the unsteady effects with the probe, induced by the previous rotor blade can be satisfactory resolved. The continuous flow course is for every rotor blading passage different, for one due to the combination of the 33 rotor blades and 32 stator vanes, for the other, due to the fact that the blades do not posses 100% identical geometries. The flow pattern will repeat itself only after one full rotor rotation, on the pretext that these are periodical stationary processes.

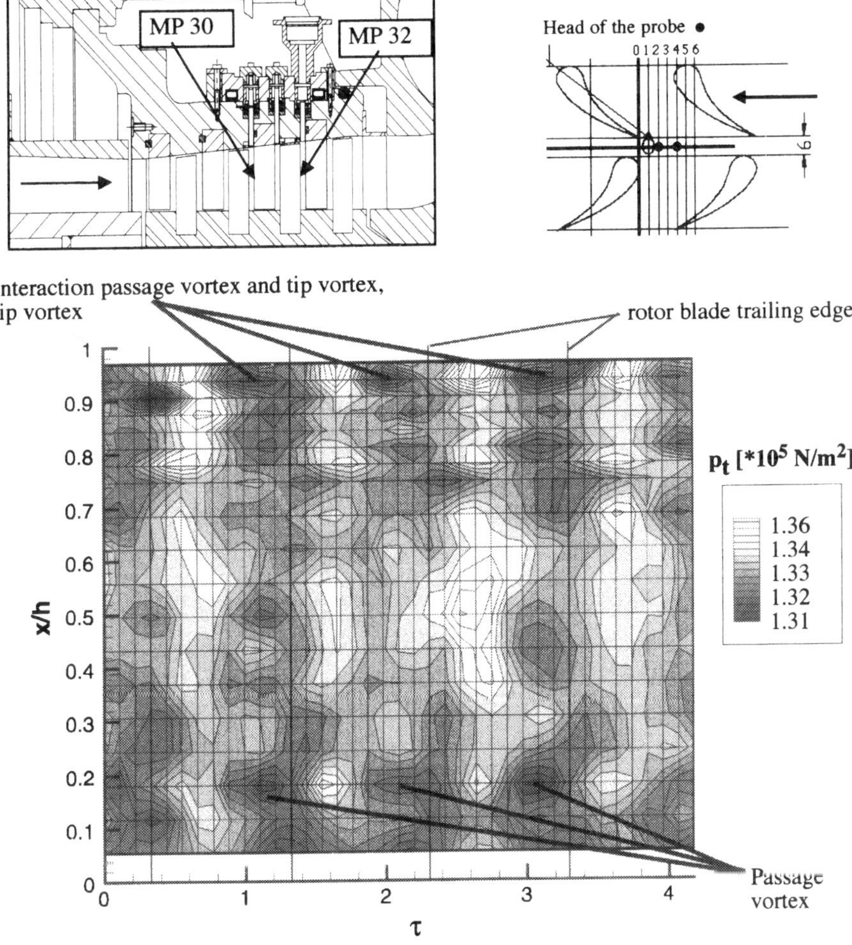

<u>Figure 3</u>: Instationary experimental results, MP 30, circumferential position 4

4.2 Numerical Method and Results

The numerical scheme for the simulation of the fluid flow works on the basis of an implicit finite volume method combined with a multi-block technique for structured grids. The physical domain is divided into separate blocks and the full, compressible, three-dimensional Navier-Stokes equations are solved in the fluid blocks. The governing equations for the conservative variables are developed in arbitrary, body-fitted coordinates ξ, η, ζ with the fluxes in normal directions to ξ, η, ζ = const.

The conservation equations are discretized implicitly in the first order in time using Newton´s method (11). Upwind discretization is used for the inviscid fluxes. Godunov-type flux-differencing is employed for the numerical diffusion. In order to achieve third order accuracy, van Leer's MUSCL-technique (12) is used. Since this Godunov flux is not sufficiently diffusive to guarantee stability in regions with high gradients, it is combined with

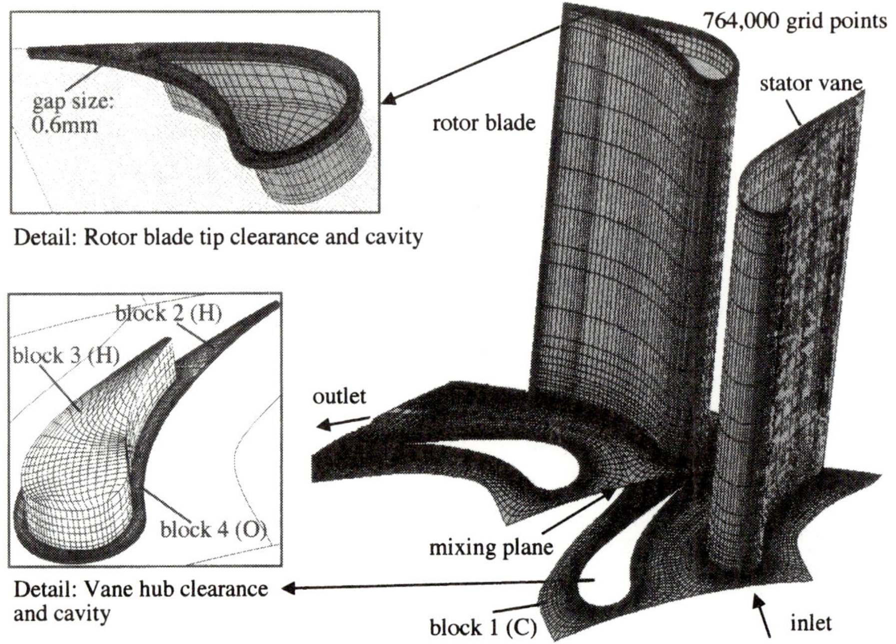

Figure 4: 3-D structured multi-block-grid for steady calculation of the 3rd stage inclusively radial clearances

a hyperdiffusive modified Steger-Warming flux (13). The viscous fluxes are approximated using central differences. The resulting system of linear equations is solved by a Gauss-Seidel point iteration scheme allowing high vectorization on present day computers. The closure of the conservation equations is provided by the algebraic eddy-viscosity turbulence model by Baldwin and Lomax (14). The performance of the code has been proven in a benchmark calculation (15) of a 1.5-stage cold air axial flow turbine (16).

Figure 4 gives an impression of the numerical grid. In order to discretize the narrow radial gaps and the cavities by the crown sharpening of the vanes and blades, several H-blocks and a O-block have been used. In the radial direction 18 calculation planes have been established in the radial clearances. Therefore, a detailed calculation of the fluid flow in a clearance becomes possible.The profile geometries of the vanes and blades are identical. The full size of the 3-D grid is about 764,000 grid points in 54 radial planes. The boundary conditions have been derived from the experimental data. A mixing plane is arranged between the vane and the blade using a flux coupling method.

The calculations have been performed on the vector-parallel-supercomputer SNI VPP300/8 with a theoretical computational performance of 2.2 Gigaflops and 2 Gigabyte RAM for one processor (total of 8 processors). The numerical solver needs approx. 400MB RAM for this kind of calculation. For the calculations, approx. 15,000 time steps (40 CPU-hours) were necessary to obtain a converged solution. The convergence of the calculation has been checked at different monitoring points.

Figure 5 gives an impression of the flow field in the stator vane row of the stage. The total pressure is displayed by isolines in axial cutting planes. At a position of approx. 80% axial vane chord length, the influence of the hub clearance vortex can be easily detected (marked A') in Fig. 5a. The vector plot detail shows the interaction of the hub clearance flow and the hub side wall flow leading to a dominant hub vortex. Near the casing, the influence of the casing side flow is leading to a weak passage vortex (marked B'). The flow field behind the

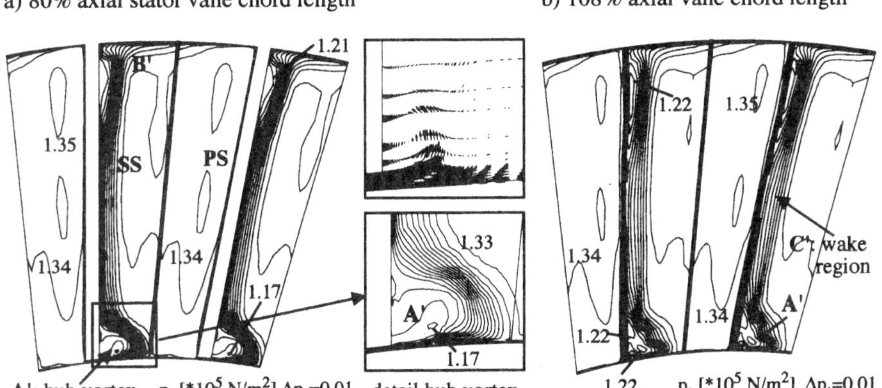

Figure 5: Total pressure field in axial cutting planes of the stator vane row

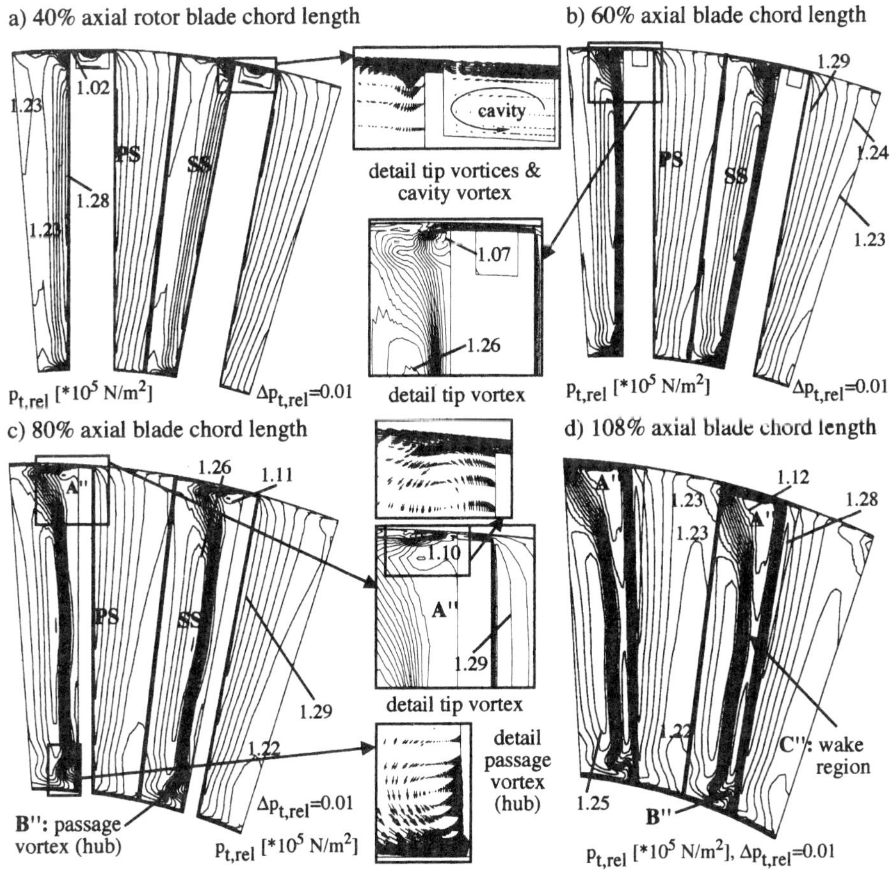

Figure 6: Relative total pressure field in axial cutting planes of the rotor blade row

stator in Fig. 5b reveals the total loss area created by the secondary flow interaction and the hub vortex still as the dominant flow feature beside the wake region (marked C').

In Fig. 6 the development of the relative total pressure field in the rotor blade row is shown in axial cutting planes. The position of the cutting plane in Fig. 6a is approx. at 40% axial chord length of the rotor blade. At the rotor blade tip a circumferential leakage flow from the pressure side to the suction side is visible. The leakage flow is reduced by viscosity effects induced by the relative motion of the casing and by supplying the cavity vortex with kinetic energy (Fig. 6a - detail). On the suction side, the tip clearance flow meets the opposing secondary flow in the casing boundary layer which is responsible for the development of the passage vortex. The two opposing secondary flows roll up to the counter-rotating tip clearance vortex and the casing passage vortex. Figure 6b shows the increase of the common loss region. The casing side wall and the tip leakage flow are supplying a huge loss area (marked A", Fig. 6c) with low energetic flow material. This phenomenon is supported by local flow separation on the suction side. On the hub side, the hub passage vortex (marked B") is the main feature in the flow field. The passage vortex transports high energetic flow material from the core flow near to the hub wall. On the suction side, the development of a thick boundary layer is obvious which is in danger of separation leading to further total pressure losses. The investigation of the flow field 8% axial chord length behind the trailing edge (Fig. 6d) reveals a huge region of total pressure losses at approx. 90% radial height (marked A"). As shown before, it is created by low energetic flow material supplied by the tip clearance vortex and the passage vortex and has also been detected in the experimental results. Furthermore, the radial flow field is characterized by a large wake region (marked C").

Altogether, the numerical simulation reveals a more or less well-structured flow field where the inhomogeneities can be referred to detected flow phenomena in the stage. This is in contrast to the experimental detections which have revealed a highly inhomogeneous flow field due to the large influence of the previous stages and the potential influence of the following stator vane row. These effects cannot be taken into account in the numerical simulations with a steady code. Nevertheless, both results have shown the development of the interaction zone and the passage vortex region. Furthermore, the numerical results help to understand the development process and origins of the main flow phenomena.

4.3 Application of a Balance-based Averaging Method

Reducing complex 3-d data to 2-d or 1-d values is necessary for the determination of the efficiency of a geometric configuration. Furthermore, it is helpful to understand the quantitative relevance of the calculated and measured results. By doing so, it is necessary, too, to keep as much information of space inhomogeneities in the averaging process as possible. The most common averaging methods such as area averaging and mass averaging have the disadvantage that they are not based on balance equations. Thus, the averaged values are not an equivalent substitute to the spatial flow through a free control surface. Furthermore, it is only possible to obtain statistically based information on the degree of spatial inhomogeneity of the flow through the free control surface.

Kreitmeier has introduced a balance-based time averaging (6) and space averaging procedure (7). Examples of the usage of the methods in turbomachinery flows are given in (8) and (9).The methods take into account all relevant conservation and non-conservation equations. Thus, the averaged values are physically based substitutes of an homogeneous flow through the free control surface, which means that the averaged flow values will lead to the same conservation fluxes (mass, total linear momentum, total angular momentum, total enthalpy) and non-conservation fluxes (enthalpy, kinetic energy, entropy). Kreitmeier's method is based on an equilibration model which uses a hypothetical equilibration chamber (Fig. 7). To conceive the model, the reader should imagine to hang a suitably deformed equilibration chamber on the observed cross-section (S) of the fluid flow. Such chamber could be made up of, say, from a Carnot diffuser and a nozzle together. The fluid flow shall be homogenized fully and extensively right in this hypothetical equilibration chamber.

Depending on the treatment of equations describing the equilibration process Kreitmeier has defined a reversible and a irreversible equilibration process. For an irreversible balance averaging, Kreitmeier develops the model as follow (17): As in the above exemplary

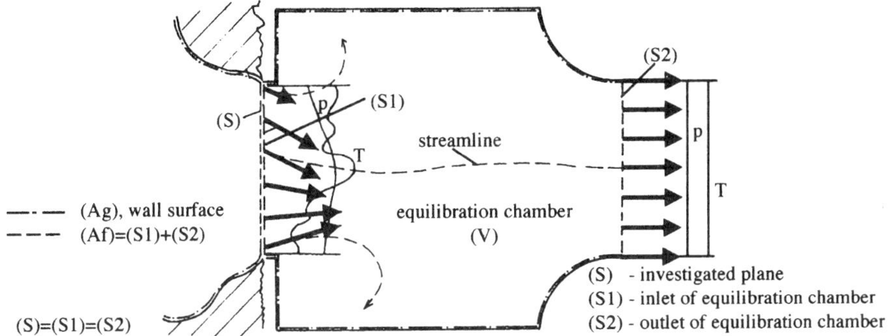

<u>Figure 7</u>: Equilibration chamber model of averaging procedure

equilibration chamber (<u>Fig. 7</u>), with the connected surface (Ag) with limits going to zero and both (S1) and (S2) infinitesimal neighbouring one another, all surface integrations will disappear in the equilibration process, which means the walls can exert no more influence. The fluxes through (S) must then, with exception of the fluxes of the enthalpie H, kinetic energy K, and entropy S, directly equal to those through (S2). For H, K, and S rest-fluxes will remain. The rest-entropy flux can be understood accordingly as source-fertility, which arises from the equilibration inside the equilibration chamber. It is derived from an abstract dissipation term, based on internal friction and a dissipation term resulting from the equalizing of the temperature. Due to the common occurrence of both source terms, Kreitmeier calls this type of averaging as "Irreversible averaging". For illustrating the reversible averaging process, it would be best to conceive the imagined equilibration process again: Kreitmeier derives it by having (Ag) -> 0 (no influence from the walls), Φ->0 (no internal friction), λ->0 (no internal heat transfer) and that all the internal output powers resulting from pressure difference from cross-section 1 to 2, must compensate themselves. In other words, it does not allow energy, like the kinetic energy, to transform from one form to another. For a more detailed description of the theory and the derivation of the theory equations the reader should refer to (7).

Thus, the differences between the reversible and irreversible averaged quantities are a measure for the flow inhomogeneity and equilibrations coefficients can be derived from the transfer from reversible averaged to irreversible averaged values (7). The definitions of the equilibration coefficients are given in <u>Fig. 8</u>. The loss coefficient describes a dissipative loss in the equilibration process from the reversible to the irreversible averaged values due to temperature inhomogeneities and inner friction during the process. It is normalized by the kinetic energy of the reversible averaged flow field. The kinetic energy coefficient is the quotient of kinetic energies in the averaged flow fields. The recovery coefficient is the quotient of the specific work by pressure changes and the reversible averaged kinetic energy meaning a loss of mechanical work during the equilibration process.

The spatial inhomogeneity of the flow coming out of a previous stage will lead to futher losses in succeeding stages. Optimization of the aerodynamic stage efficiency will desire reducing flow inhomogeneities, in particular those created by the tip clearance flow. The above described balance-based averaging method is a reliable numerical tool to quantify the positive or negative effects of a chosen geometric design on the inhomogeneity of the flow. In order to demonstrate the application of the method in the design process the reference configuration is compared to a configuration with a larger radial gap which obviously will lead to an increased inhomogeneity of the flow.

In <u>Fig. 8</u> circumferential reversible and irreversible averaging has been executed in the examination plane at 8% behind the stage for every radial height of the computational grid. Thus, the spatial inhomogeneity in circumferential direction can be given by the radial distribution of the equilibration coefficients. The method has been applied to three different

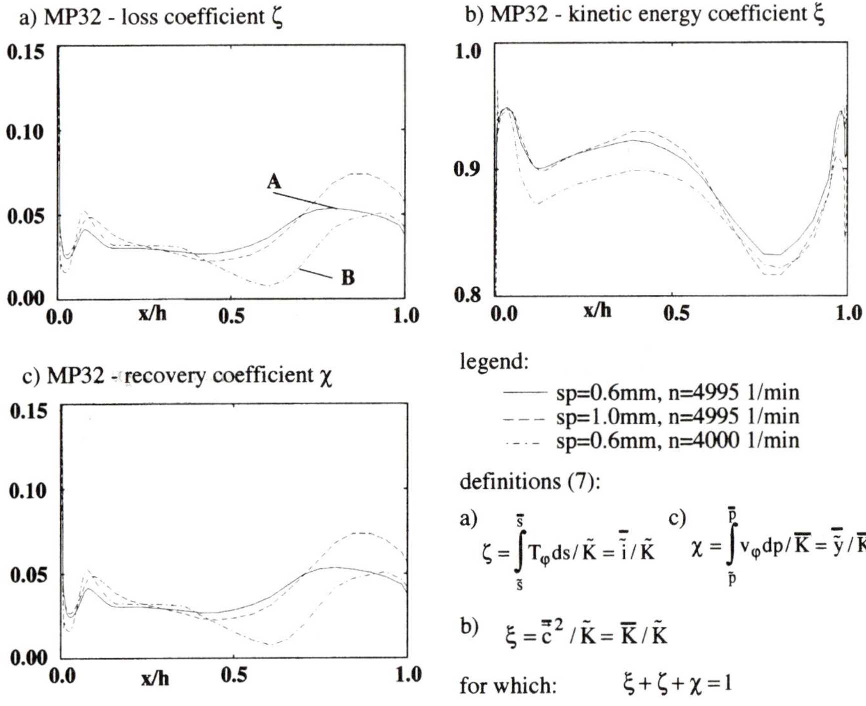

$$\text{a)} \quad \zeta = \int_{\underline{s}}^{\overline{s}} T_\varphi \, ds / \tilde{K} = \overline{\tilde{i}} / \tilde{K} \qquad \text{c)} \quad \chi = \int_{\underline{p}}^{\overline{p}} v_\varphi \, dp / \overline{K} = \overline{\tilde{y}} / \overline{K}$$

$$\text{b)} \quad \xi = \overline{\tilde{c}}^2 / \tilde{K} = \overline{K} / \tilde{K}$$

$$\xi + \zeta + \chi = 1$$

Figure 8: Radial distributions of equilibration coefficients

computational results:

1) the reference configuration with radial gap size sp=0.6 mm and rotor speed n=4995 1/min
2) configuration with increased radial gap size of sp=1.0 mm and rotor speed n=4995 1/min
3) the reference configuration with radial gap size sp=0.6 mm and with reduced rotor speed of n=4000 1/min

The radial distribution of the loss coefficient (Fig. 8a) for the reference configuration shows a loss plateau (A) near the casing which can be related to the loss region A" in Fig. 6d. The distribution of all three investigated cases is similar up to a radial height of h/H=0.5. The major differences can be detected in the upper part. Of course, the values of the loss coefficient increase for the configuration with the large radial gap due to the enforced tip clearance flow. In case of the reduced rotor speed, the inhomogeneity losses are smaller (B) as a local flow separation on the suction side is reduced. Nevertheless, the radial distributions of the kinetic loss coefficients (Fig. 8b) reveal losses for the second operation point. Therefore, this operation point is less efficient as the reference case. The reason can be found in the radial distribution of the recovery coefficient (Fig. 8c), which is on a distinctly higher level than in the reference case meaning that a large amount of mechanical work has not been used. The recovery coefficient distribution for the increased gap size does not differ distinctly from the reference case as operating conditions are the same. Differences in the distributions of the kinetic energy coefficient can be related mainly to the large gap size.

5   CONCLUSIONS

An experimental rig of a four stage axial cold air turbine has been built up at the Institute of Steam and Gas Turbines, Aachen University of Technology. Test runs have shown the working performance of the new test turbine. According to the measurement task, new developed five-hole measurement probes of miniaturized size have been successfully used in axial gaps of the machine. The experimental results have shown a great influence of instationary flow phenomena related to previous stages and potential effects on the flow field in the measurement planes. It will be the task of future work to separate the individual flow phenomena and effects. For this, the unsteady measurement campaign has to be completed for a full spacing.

The experimental work is accompanied by 3-d steady numerical calculations of the flow in the 3rd stage of the test turbine. Although instationary effects in general and stationary effects of previous stages and potential effects of the following stage are neglected in such a calculation, the numerical results give a good impression of the origin and development of several secondary flow phenomena and their interactions. Therefore, several of the experimental detected phenomena can be related to the numerical predictions. Within the numerical investigation a study concerning the influence of the operating point and the radial gap size on the spatial inhomogeneity of the flow field has been performed. The application of a balance-based averaging process allows one to quantify the flow losses related to the spatial inhomogeneities.

6   ACKNOWLEDGEMENTS

This report is the scientific result of a research project which was initiated by the Forschungsvereinigung Verbrennungskraftmaschinen e.V. (FVV, Frankfurt) and carried out at the Institute of Steam and Gas Turbines at the RWTH Aachen headed by Professor Dr.-Ing. Bohn and accompanied by a counselling circle headed by Dipl.-Ing. Kreitmeier, ABB Power Generation AG. The work was funded by the Bundesministerium für Wirtschaft (BMWi, Bonn) via the Arbeitsgemeinschaft industrieller Forschungsvereinigungen e.V. (AiF, Köln, AiF-No. 10527).

7   REFERENCES

(1) Zaccaria, M.A.   Unsteady Flow field Due to Nozzle Wake Interaction with the Rotor
     Lakshminarayana, D. in an Axial Flow Turbine: Part I   Rotor Passage Flow Field;
                          Journal of Turbomachinery, Vol. 119, pp. 201-212, 1997

(2) Zaccaria, M.A.   Unsteady Flow Field Due to Nozzle Wake Interaction with the
     Lakshminarayana, B. Rotor in an Axial Flow Turbine: Part II - Rotor Exit Flow Field;
                          Journal of Turbomachinery, Vol. 119, pp. 214-224, 1997

(3) Binder, A.   Turbulence Production Due to Secondary Vortex Cutting in a
                          Turbine Rotor;
                          *Journal of Engineering for Gas Turbines and Power, Vol. 107,
                          pp. 1039-1046, 1985*

(4) Hodson, H.P.   Measurement of Wake-Generated Unsteadiness in the Rotor
                          Passages of Axial flow Turbines;
                          *Journal of Engineering for Gas Turbines and Power, Vol. 107,
                          pp. 467-476, 1985*

(5) Walraevens, R.E.   Experimental Investigation of Three-dimensional Unsteady Flow
     Gallus, H.E.   Downstream the Rotor in a 1 1/2 stage Turbine;
                          95-IGTC-10, Yokohama, 1995

(6) Kreitmeier, F.   A New Time-Averaging Procedure for Compressible, Unsteady
                          Turbulent Flows;
                          ASME-paper 87-GT-83

(7) Kreitmeier, F.   Space-Averaging 3D Flows Using Strictly Formulated Balance
                  Equations in Turbomachinery;
                  IGTI-Vol. 7, ASME COGEN-TURBO, 1992

(8) Kreitmeier, F.   Numerical Estimation of the Balance Errors Made by Neglecting
     Juvet, P.J.   the Free Control-Surface Integrals that Contain the Shear-stress
     Benim, A.C.   Tensor and the Heat Flux Vector;
                  1st European Conference on Turbomachinery, Erlangen, 1995

(9) Kreitmeier, F.   Demonstration of a Balance-Based Procedure for the Time-
     Juvet, P.J.   averaging and Modelling of Compressible Three-dimensional
     Benim, A.C.   Unsteady Turbulent;
                  2nd European Conference on Turbomachinery, Antwerpen, 1997

(10) Bohn, D.   The Dynamic Response of Capillary Tubes for the Use in Miniature
     Schnittfeld, T.   Pressure Probes;
                  *Proc. of the 11th Symposium on Measuring Techniques for
                  Transonic and Supersonic Flow in Cascades and Turbomachines,
                  Munich, 1992*

(11) Eberle, A.   Generalized Flux Vectors for Hypersonic Shock-Capturing,
     Schmatz, M. A.   *AIAA -paper 90-0390, 1990*
     Bissinger, N.

(12) Anderson, W. K.   A Comparison of Finite Volume Flux Vector Splittings for the
     Thomas, J. L.   Euler Equations;
     van Leer, B.   *AIAA -paper 85-0122, 1985*

(13) Eberle, A.   Generalized Flux Vectors for Hypersonic Shock-Capturing;
     Schmatz, M. A.   *AIAA -paper 90-0390, 1990*
     Bissinger, N.

(14) Baldwin, B. S.,   Thin Layer Approximation and Algebraic Model for Separated
     Lomax, H.   Turbulent Flows, *AIAA-paper 78-257, 1978*

(15) Emunds, R.   The Computation of Adjacent Blade-Row Effects in a 1.5 Stage
     Jennions, I. K.   Axial Flow Turbine,
     Bohn, D.   *ASME -paper 97-GT-81, 1997*
     Gier, J.

(16) Walraevens, R.E. Testcase 6 - 1.5 Stage Axial Flow Turbine,
     Gallus, H.E.   *ERCOFTAC Testcase 6, IST-Report, RWTH Aachen, 1997*

(17) Kreitmeier, F.   Integralbehandlung inhomogener Strömungsfelder (Integral
                  Treatment of Inhomogeneous Flow Fields);
                  Brown Boveri report, 1977 (unpublished)

Parametric study of the flow in swirl brakes by means of a three-dimensional Navier–Stokes solver

K K NIELSEN* and C M MYLLERUP
Machinery Dynamics Group, Ødegaard and Danneskiold-Samsøe A/S, Copenhagen, Denmark
R A VAN DEN BRAEMBUSSCHE
Turbomachinery Department, von Karman Institute for Fluid Dynamics, Rhode-St-Genèse, Belgium

This paper describes a parametric study of the flow in swirl brakes at the inlet of the shroud seal of a centrifugal pump or compressor. After comparison of current and previously obtained results, the significant geometrical parameters have been modified over a relevant range and their influence on the effectiveness of the swirl braking vanes has been evaluated. The data presented in this paper allow an optimisation of the swirl brake geometry and an estimation of the performance degradation if the optimum dimensions are not possible.

1 INTRODUCTION

Rotordynamic stability is an important issue in turbomachinery design. A given rotor system can be classified as being rotordynamically stable if a free vibration decays exponentially in time. Equivalently, an exponentially increasing free vibration indicates rotordynamic instability.

The destabilising effect of the swirling flow in the narrow internal clearances of radial turbomachinery seals, operating at high pressure, is generally acknowledged by turbomachinery designers today (1). An increasing swirl velocity in the direction of rotor rotation creates a circumferential non-uniform pressure field that enhances the whirling motion of an eccentric rotor. Swirl braking vanes, to reduce the swirl at seal inlet, have a favourable effect on the destabilising rotordynamic forces. Vanes that facilitate even counter (negative) swirl act stabilising by counteracting the whirling motion of the rotor (1).

Although swirl brakes are now widely used, the results reported in the literature on performance and flow structures are very sparse. Their optimum aerodynamic design has not yet been studied in detail.

---

* Attending an industrial PhD study in collaboration with Aalborg University, Denmark.

The current paper is a follow-up on previous studies (2), (3) in which the real cavity inlet flow conditions have been defined and a better understanding of the flow structures in swirl brakes has been obtained. The studies also revealed which geometrical parameters have a major influence on the performance. The main purpose of the present study is the evaluation of performance variations as a function of the relevant parameters and to provide guidelines for an optimum design.

Whereas traditional bladings are designed with strong emphasis on low blade losses, a design of swirl braking vanes does not have low losses as a main priority. On the contrary, higher losses can even be beneficial in terms of an additional pressure drop, limiting the seal leakage flow rate.

The only performance criterion for swirl brakes is the ability to increase the rotordynamic stability by minimizing the swirl or even creating negative swirl at the inlet of the seal. The performance is quantified by the swirl ratio defined by the mass averaged swirl velocity at the seal inlet non-dimensionalised by the local peripheral velocity of the rotor surface. The rotordynamic stability improves with a decreasing swirl ratio.

Previous studies have shown that the flow between the vanes is mainly determined by a vortical structure resulting from a flow separation at the vane leading edge. A different flow solver (CFX-TASCflow (4) instead of TRAF3D (5)) was used in the present study. A first question one tries to answer is how much the predicted flow structure is influenced by the Navier-Stokes solver and the models employed. After comparison of the results, CFX-TASCflow is used for a parametric study in order to define the most effective geometry.

2 GEOMETRY AND BOUNDARY CONDITIONS

The axisymmetric pump shroud cavity in which the vanes are introduced is presented in figure 1. Except for the shape of the vanes the geometry is the same as in previous studies (2), (3).

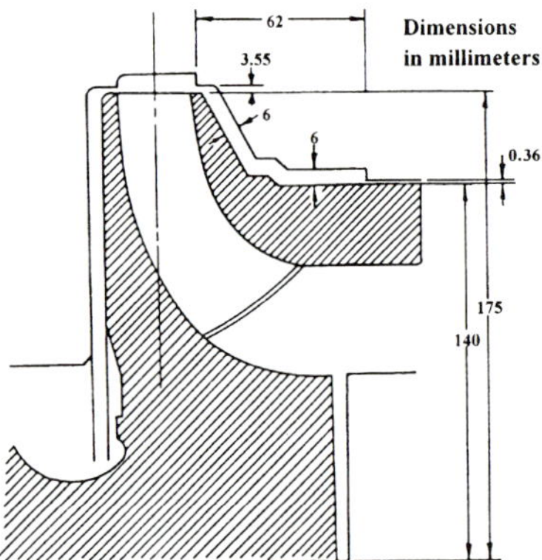

Figure 1. Detailed sketch of pump impeller and cavity geometry.

A schematic view on the shroud cavity and swirl braking vanes is shown in Figure 2. The baseline geometry has 40 vanes that are straight in both axial and radial direction, with a pitch to chord ratio of 1.33 at midspan. The vanes are fixed on the stator wall, only contour shown on this figure for clarity, and are separated from the rotating shroud by tip clearance. Note that the grid lines included do not represent the actual cell number or density.

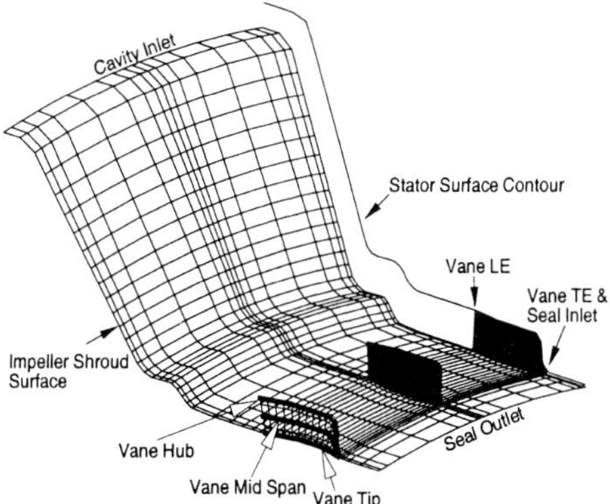

Figure 2. 3D representation of cavity geometry including vanes.

Seals are designed to limit the leakage mass flow by means of a small radial gap whereas shroud cavities are much wider to reduce the disk friction losses. As a consequence the flow entering the seal experiences a strong convergence which in the present case has a ratio of 17 to 1. This results in a low meridional velocity in the leakage cavity, where the cross section is much larger than in the seal.

The circumferential velocity component in the cavity, which is proportional to the shroud peripheral speed, is much larger than the axial one and flow angles, measured against the axial direction, are very high. A typical example of the pitchwise averaged flow angle distribution just upstream of the vane leading edge is shown in Figure 3.

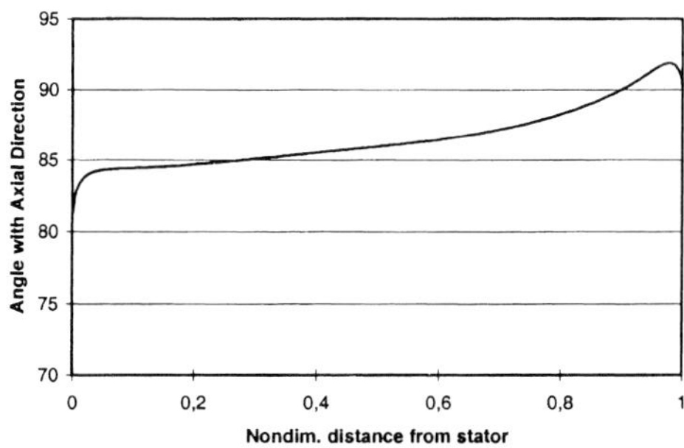

Figure 3. Angle distribution just upstream vane leading.

Angles exceeding 90 deg. near the shroud, indicate local return flow. It is quite obvious that for this highly swirling flow, it is impossible to design vanes that do not cause considerable flow separation without introducing an unacceptable large number of vanes, creating excessively high blockage when using vanes of finite thickness.

The large meridional contraction and flow deflection result in large areas of 3D separated flows interacting with the large leakage flows between the shroud and vane tip. Swirl braking vanes therefore differ significantly from traditional bladings in several aspects.

Previous studies have shown that straight axial vanes are as efficient as profiled or curved ones because massive flow separation can hardly be avoided and has even a favourable effect. The vane thickness distribution is only of secondary importance when considering flows with leading edge separation. The very thin vanes with wedge-shaped leading- and trailing edges, used in previous calculations, have been replaced by straight vanes of zero thickness.

The vanes extend over half of the space between the seal entrance and the first upstream cavity bend. The aspect ratio (height/chord) is 0.31. The clearance between the vane tip and the impeller shroud is twice the seal clearance, except close to the seal, where the tip clearance decreases to the seal clearance.

The inlet flow angles and leakage mass flow specified at the cavity inlet for the vane calculations are those reported in (6), (7) and (8). The leakage mass flow is 3.85 kg/s and results from the seal geometry and a total impeller head of 68 m. The last one is obtained with an impeller speed of 2000 rpm at a tip Reynolds number of $8.02 \cdot 10^6$. At the seal outlet fully developed flow is assumed i.e. Neumann conditions are applied. Working medium is water at 30°C.

3 SWIRL BRAKE CALCULATIONS

3.1 Flow solver

The flow simulations presented in this paper have been made with the 3D Navier-Stokes solver CFX-TASCflow from AEA Technology. This is a general purpose CFD code able to deal with complex 3D flows. It includes many features that make it suitable for turbomachinery flows.

The code uses a pressure based finite volume approach in colocated (i.e. unstaggered) grids. The discretisation in the code is strongly conservative. The algebraic multigrid method and the time stepping techniques included in the code to speed up convergence were used in the present study. Likewise, the second order accurate Linear Profile Skew scheme including physical advection correction (PAC) was used.

The code provides three types of k-ε turbulence models. The wall boundary conditions can be handled by using a wall function approach or a two-layer model. The latter facilitates resolving the wall boundary layer with a one-equation model (4). In the current study the k-ε RNG model was employed together with both wall boundary treatments. The two-layer approach requires typically 10 to 20 extra cells in the boundary layer and is therefore computationally more expensive to use. Consequently, the two-layer model was primarily used for test calculations. Additional information on the flow solver is available in (4).

3.2 Calculation methodology and comparison

All calculations reported in this paper are incompressible. The same mass flow was specified for all vane designs considered. The pressure drop was only slightly affected by vane design. It was assumed that the rotor is concentric with respect to the stator. The circumferential extent of the calculation domain was limited to one pitch by applying periodicity conditions.

H grids have been used for all calculations, as they facilitate excellent grid orthogonality in the almost rectangular flow channels between the sharp nosed vanes. Typical dimensions of the grid are: 148·40·32 respectively in the streamwise, radial and pitchwise direction.

The grid independence of the solutions was verified by doubling the number of grid points in the three directions. The resulting change of seal inlet swirl ratio was less than 1%.

The use of boundary layer resolution is appropriate when considering separated flows. In the separated flow region the wall function assumptions fail. However, applying the two-layer model increases the demand for grid points considerably, resulting in long calculation times. The test calculations performed with the two-layer model did not show a significant influence on the seal inlet swirl ratio.

The grid shows significant variations of the grid size between the shroud cavity and the seal resulting in large variations of the maximum allowable time step. The maximum time step for the meridional transport from the cavity inlet to the vane leading edge is of the order of 100 times larger than the one allowed by the seal and tip clearance flow. Local time stepping is appropriate under these conditions.

Low convergence rates were observed and the convergence to a stable circumferential velocity distribution was rather time consuming. These findings are similar to the experiences reported in (9) where leakage flows have been calculated using the same program. They explained the long computer times by the non-linearity of the wall function.

The majority of the present calculations were carried out on an IBM RS/6000 43P-140 with a 332 MHz processor and 256 MB of RAM memory. A typical calculation lasted 8 hours, equivalent to $1.5 \cdot 10^{-3}$ sec/node/iteration.

Calculations by means of CFX-TASCflow on the baseline geometry predict a seal inlet swirl ratio that is 5% higher than the one obtained previously with TRAF3D. The differences must in the first place be attributed to a difference in vane geometry. Only a secondary part is due to the flow solvers that differ from each other in terms of turbulence models (TRAF3D uses the Baldwin-Lomax model), wall boundary layer treatment, advection schemes, tip region handling etc. The relatively good agreement in swirl ratio and the similarity in the flow structure confirm that the separation of the flow on the vane leading edge is indeed the driving force for the vortices in the vane passage. The conclusions made in (3) therefore remain valid and no further comparisons have been made.

4 SWIRL BRAKE FLOW FIELDS

4.1 Vane-to-vane flow field

Figure 4 shows the vane-to-vane flow field upstream of the vanes, inside the vanes and at the seal inlet for two spanwise positions. Plot 4A corresponds to a position 25% from the stator surface whereas plot 4B corresponds to a position 25% from the rotating shroud surface (12% from the vane tip). The vane-to-vane flow structure shown in plot 4A is representative for the upper 2/3 of the cavity height.

As pointed out previously the flow is almost tangential at the vane leading edges (LE). Considering the characteristic plot 4A, the very high incidence on the straight axial vanes results in a complete separation of the flow from the suction side. The flow impinges on the pressure side, creating a very large vortex that more or less controls the whole vane-to-vane flow. At the vane suction side the flow moves backward from the trailing edge (TE) area towards the LE. The vane-to-vane vortex is very beneficial from the point of view of creating strong counter swirl at the seal inlet.

The strength of the vane-to-vane vortex decreases as the distance from the stator increases. Plot 4B shows how the vortex weakens near the vane tip due to the strong shear stress imposed by the proximity of the rotating shroud. The central part of the vortex structure has thus disappeared for this position. At the seal inlet, the region of counter swirl at the vane pressure side is due to radial inflow of counterswirling fluid.

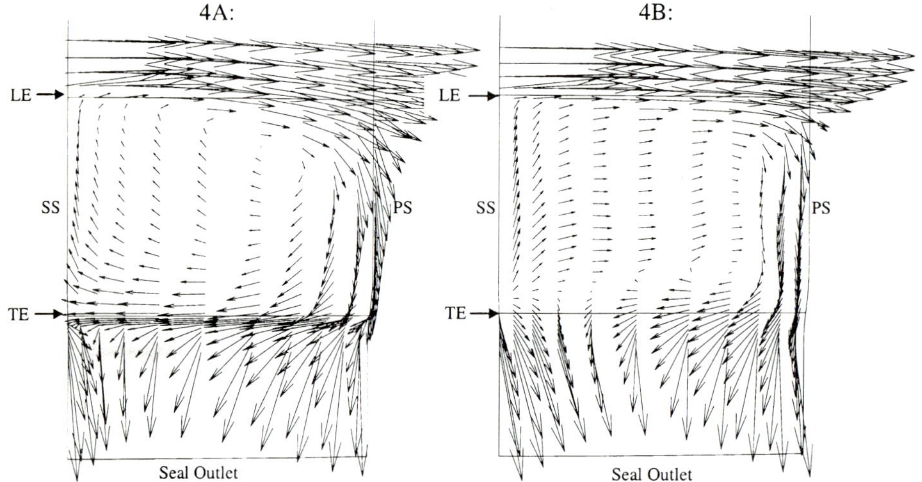

Figure 4. Baseline flow field 25% from stator (4A) and 25% from rotor (4B).

The meridional flow acceleration at the seal inlet is large due the local reduction in radial extend of the flow channel. A smaller increase of the tangential velocity is due to the conservation of tangential momentum when decreasing the radial position from the vane root towards the seal inlet.

The flow in the (vaneless) seal is strongly influenced by the large tangential velocity of the impeller shroud surface and therefore experiences a deflection from counter swirling at the inlet to swirling with the rotor at the outlet. The average seal inlet flow of the baseline design is counter swirling, with an inlet swirl ratio of −5%. This value should be compared with a swirl ratio of +45% in the geometry without swirl braking vanes.

4.2 Meridional flow field

The large spanwise variation of the flow in the vane-to-vane section is due to the highly three-dimensional flow structure and can be further clarified by considering velocity distributions in the meridional plane shown in Figure 5. Vector plots of the meridional velocity in planes at 16% of the pitch from the vane pressure side (PS), 5A, and 16% of pitch from the vane suction side (SS), 5B, are presented. The projections of the vane contour on the plots are included.

Considering plot 5A, the flow enters the blading mainly in the upper half of the cavity where the swirl velocity and hence the incidence is the smallest. Approaching the seal the flow first moves outwards before it is deflected downwards along the radial surface to enter the seal. This explains the existence of counterswirl at the vane pressure side near the vane tip in plot 4B.

A large vortex in plot 5A is observed adjacent to the rotating shroud just upstream of the seal inlet. This seal inlet vortex was also identified by (3) and has an important influence on the swirl at the seal inlet. It prevents the highly swirling fluid adjacent to the rotor surface from entering the seal by "pumping" it upstream and allows the counter swirling fluid, that comes from the outer part of the blading, to enter the seal. The upstream moving leg of the vortex is mainly in the clearance whereas the downstream flowing part is within the blading.

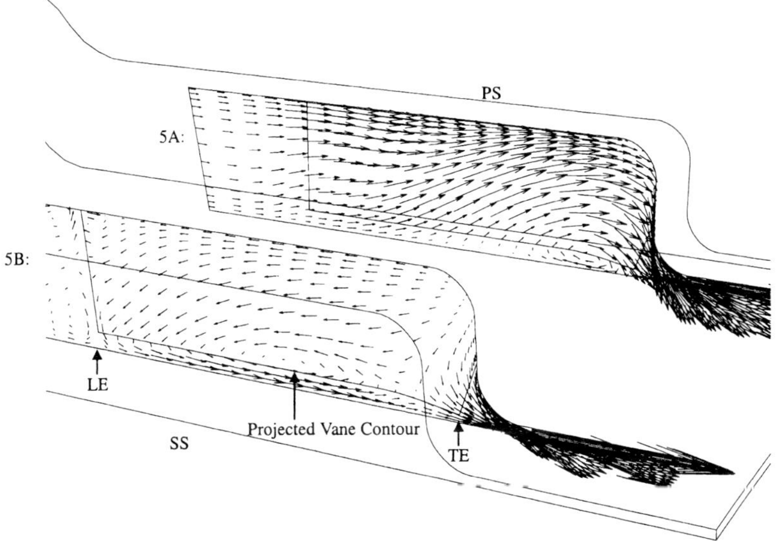

Figure 5. Meridional flow fields at 16% of pitch from PS (5A) and SS (5B).

The flow field near the suction side (5B) differs fundamentally from the pressure side flow. In agreement with plots 4A and 4B the flow generally moves from the TE to the LE. The velocity magnitude is also generally smaller. The seal inlet vortex does not exist close to the suction side. Instead, a vortex with opposite direction of rotation is found close to the shroud surface at an upstream position. The downstream moving leg of this vortex is in the tip clearance and the upstream leg is within the blading. The flow in the tip clearance has strong swirl with the shroud rotation. Fortunately, this flow is generally deflected into the blading and only a small part enters the seal. The deceleration of the flow close to the shroud is facilitated by the local pressure field imposed by the circular stator contour at the seal inlet.

5 PARAMETRIC STUDY

5.1 Pitch to chord ratio
Figure 6 shows the influence of the vane pitch to chord ratio on the swirl brake performance. The ratios investigated correspond to 10, 20, 40, 50, 60, 80 and 100 vanes in the cavity. All other parameters remained unchanged.

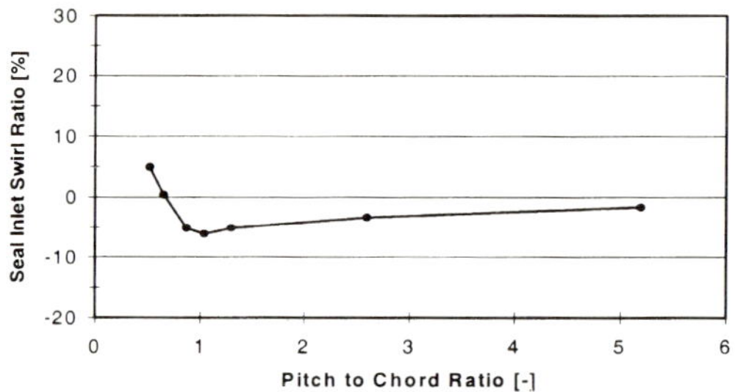

Figure 6. Performance for varying pitch to chord ratio.

The optimum pitch to chord ratio for the geometry investigated, is slightly larger than 1.0. The performance degrades rapidly for values of the pitch to chord ratio less than one. The performance degradation is less severe for values larger than 1. Generalising from this, one can conclude that a designer who is not sure about the exact performance of a given geometry should rather choose a pitch to chord ratio slightly larger than the optimum one. Less than optimum number of vanes is better than too many vanes and in addition it is also cheaper.

5.2 Clearance variation

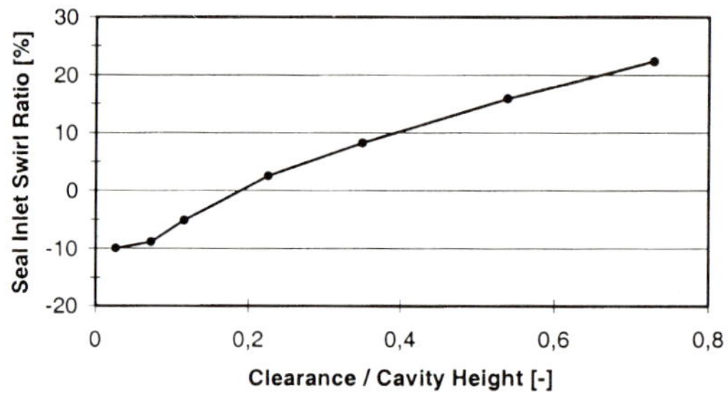

Figure 7. Performance for varying clearance.

Figure 7 shows the influence of the variation of vane clearance on the swirl brake performance. The clearance has been non-dimensionalised by the cavity height. The ratio varies between 0 for no clearance and 1 for maximum clearance (i.e. no vanes).

The seal inlet swirl ratio varies almost linearly with the clearance for most of the range considered. The flow at seal inlet changes from counter-swirl to co-swirl when the clearance exceeds approximately 20% of the cavity height. The decrease of swirl brake performance with increasing clearance is not surprising because increasing clearance results in less flow control. When increasing the clearance the seal inlet vortex strength decreases and for large clearance the vortex disappears.

Manufacturing considerations may necessitate a large vane tip clearance and it should be realised that low seal inlet swirl ratios are still possible even with a rather large clearance. A clearance of 50% of the cavity height will thus provide a seal inlet swirl ratio less than 15% whereas the ratio without vanes is 45%.

A clearance of 5% of the cavity height provides a seal inlet swirl ratio that is almost identical to the one with zero clearance.

5.3 Mass flow variation

It is of interest to study also how the seal inlet swirl ratio changes as a function of the leakage mass flow. The leakage mass flow will change when varying the operating point of the compressor or pump as it will alter the static pressure difference over the impeller and hence over the seal.

The influence on seal swirl ratio of varying the mass flow in the seal is shown in Figure 8. It should be pointed out that the cavity inlet swirl was kept constant when varying the mass flow in the calculations. This will not necessarily be the case when changing operating point.

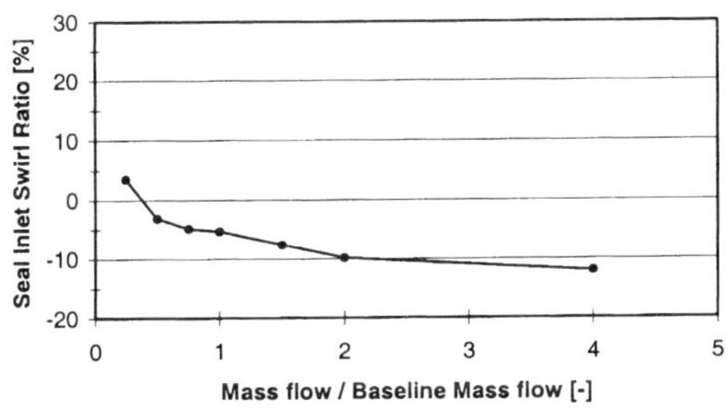

Figure 8. Performance for varying leakage mass flow.

Figure 8 indicates an improvement of the swirl brake performance with increasing leakage mass flow. A possible explanation may be that an increasing mass flow reduces the incidence angle at the vane leading edge, resulting in a stronger vane-to-vane vortex. Certainly, there must be an optimum vane inlet angle in terms of counter swirl creation. However, there is no interest to define it for the current design because the corresponding leakage flow will be very large and of no practical interest.

The magnitude of the mass flow variations shown in Figure 8 is large when compared to the typical variations that occur in a given machine. However, this range may be interesting in terms of swirl brake design. Widening the seal gap, to allow more leakage flow, allows improving the rotordynamic stability however at the cost of aerodynamic performance of the pump or compressor.

6 CONCLUSION

The investigations described in this paper clearly show that the flow structure in swirl braking vanes is governed by vortices. They are due to flow separation, and not resulting from conventional flow guidance by the vanes. This, together with the favourable effect of high losses, results in optimum designs which are completely different from what is normally seen in turbomachinery.

The highly three-dimensional flow is characterized by a large vane-to-vane vortex, covering the whole vane passage, and a smaller vortex, with axis in the peripheral direction, upstream of the seal. The first one contributes to the creation of counter swirl at the seal inlet by giving the flow a negative tangential velocity whereas the second one prevents the fluid with positive swirl from entering the seal.

The parametric study revealed the existence of an optimum pitch to chord ratio of approximately 1.0. Decreasing the pitch to chord ratio below the optimum value, by increasing the vane number or increasing the vane chord, leads to rapid performance degradation whereas an increasing ratio, by reducing the vane number, has only a very weak effect.

The swirl brake performance was found to decrease with clearance. The investigation showed an almost linear dependence with almost no gain when decreasing the clearance below 10% of the cavity height. Clearances that are easily achievable allow good results.

The variation of leakage mass flow revealed an improving swirl brake performance with increasing leakage mass flow. However seals are installed to limit the leakage flows, because they have a detrimental influence on the aerodynamic performance. Increasing the seal gap to enhance the rotordynamic stability, can therefore not be recommended.

ACKNOWLEDGEMENTS

The discussions with professor K. E. Widell, Aalborg University, and Dr. F. Pløger, HV-Turbo A/S, are appreciated. We also wish to thank the Academy of Technical Sciences in Denmark for financial support.

REFERENCES

(1) Childs, D., 1993, "Turbomachinery Rotordynamics", John Wiley & Sons, Inc., New York.
(2) Krogh Nielsen, K., 1997, "Rotordynamic Impact of Swirl Brakes", Diploma Course Report 1997-28, von Karman Institute for Fluid Dynamics, Rhode-St-Genèse, Belgium.
(3) Krogh Nielsen, K., Van den Braembusssche, R. A., Myllerup, C. M., 1998, "Optimization of Swirl Brakes by Means of a 3D Navier-Stokes Solver", 98-GT-328
(4) AEA Technology / ASC Engineering, 1997, CFX-TASCflow User Documentation.
(5) Arnone, A., 1992, "Notes on the Use of the TRAF Codes", Technical Report, University of Florence, Italy.
(6) Baskharone, E. A., Daniel, A. S., and Hensel, S. J., 1994, "Rotordynamic Effects of the Shroud-to-Housing Leakage Flow in Centrifugal Pumps", ASME Journal of Fluids Engineering, Vol. 116, pp. 558-563.
(7) Baskharone, E. A., and Hensel, S. J., 1993, "Flow Field in the Secondary, Seal-Containing Passages of Centrifugal Pumps", ASME Journal of Fluids Engineering, Vol. 115, pp. 709.
(8) Bolleter, U., Leibundgut, E., and Sturchler, R., 1989, "Hydraulic Interaction and Excitation Forces of High Head Pump Impellers", Presented at the Third Joint ASCE/ASME Mechanics Conference, University of California, La Jolla, CA.
(9) Mack, R., Casey, M. V., Schmied, J., Krause, M., 1997, "The Use of a Three-Dimensional Navier-Stokes Code for Sealing and Leakage Flows in Turbomachinery Applications", Transactions of the Second European Conference on Turbomachinery, pp. 513-520.

Additional Paper

C557/138/99

Through-flow model for design and analysis integrated in a three-dimensional Navier–Stokes solver

A STURMAYR and **Ch HIRSCH**
Department of Fluid Mechanics, Vrije Universiteit Brussel, Belgium

Abstract

A multiblock, multigrid Navier-Stokes solver has been extended to include a through-flow model for the design and analysis of turbomachines. The presence of the blades in the inviscid axisymmetric flow is modelled in the classical way through a distributed blade force to produce the desired turning, a blockage factor that accounts for the reduced area due to blade thickness, and a distributed friction force representing the entropy increase due to viscous stresses and heat conduction. The exact blade geometry is not required. All features of the 3D code concerning the physical fluid model, boundary conditions, spatial and time discretization, convergence acceleration techniques, and data visualization are available to the through-flow module. This includes the capability to treat the entire range of relevant Mach numbers, from strictly incompressible (through a preconditioning technique) to supersonic, as well as any number of blade rows in any configuration, including, e.g., bypass engines. Selected elements comprising the through-flow model are discussed, with special emphasis on the blade force and its discretization. The properties of analysis and design mode with respect to shocks and the associated losses are investigated. The methodology is demonstrated on a transonic compressor rotor and a four-stage low speed turbine.

1 Introduction

To use the Euler equations instead of the traditional streamfunction or streamline curvature methods as the basis for an axisymmetric throughflow model is appealing from several perspectives: the massflow is a result of the computation, choking of blade rows can be predicted, certain types of shocks can be captured and the well-developed numerical solution techniques for the Euler equations can be applied. Throughflow models based on the Euler equations have been proposed by Spurr (1980), Nigmatullin and Ivanov (1994), Yao and Hirsch (1995), Boure and Gillant (1995), Baralon et al. (1997) and Damle et al. (1997) and unsteady results reported by Vuillez and Petot (1994). Additional benefits arise from integration with a 3D Navier-Stokes solver, namely the possibility to set up computational domains comprised of both throughflow and 3D blade rows, and the utilization of the 3D code's infrastructure. Dawes (1992) has adapted

Nomenclature

Latin symbols

b	blockage factor $(1 - d/s)$
c	axial chord
c_p	specific heat at constant pressure
d	tangential blade thickness
E	total energy per unit mass
F	multigrid forcing function
$\vec{f}_B$	blade force per unit mass
$\vec{f}_F$	friction force per unit mass
H	total enthalpy per unit mass
I	unit tensor
K	blade force correction operator
L	residual operator
M	Mach number
m	meridional coordinate
$\dot{m}$	massflow
p	pressure
R	restriction operator for variables
$\hat{R}$	restriction operator for residuals
$\vec{R}$	radius vector $\perp \vec{\Omega}$
S	surface of volume V
s	tangential blade spacing
T	temperature
t	time
U	pseudo-vector of conservative variables, entrainment velocity $(|\vec{\Omega} \times \vec{R}|)$
V	volume
$\vec{W}$	relative velocity vector

Greek symbols

α	absolute flow angle
β	relative flow angle
γ	ratio of specific heats
δ	deviation angle
η	efficiency
κ	time step for the blade force
Π	pressure ratio
ρ	density
τ	pseudo-time
ψ	loss coefficient
$\vec{\Omega}$	vector of angular velocity

Subscripts

is	isentropic
m	meridional vector component
ref	reference dynamic quantity
rel	relative
rot	rotary
t	total quantity
TE	trailing edge
θ	azimuthal vector component

Superscripts

(k)	Runge-Kutta stage
(n)	grid level
t	target value

his 3D Navier-Stokes code in this sense, allowing blades with a given geometry to be modelled either as 3D or as throughflow rows. In the same spirit, a genuine throughflow model has been implemented in the state-of-the-art 3D Navier-Stokes solver Euranus, Hirsch et al. (1991). This paper first presents some key elements of the numerical implementation, then discusses the shock capturing properties of the Euler throughflow equations, and finally illustrates the potential of the method through selected results.

2 The Euler Throughflow Model

The dominant aspect of an Euler throughflow method, in particular when trans- and supersonic applications are considered, is modelling of the turning and the associated distributed blade force. Most authors so far have opted for a straightforward algebraic approach, in which the blade force is effectively determined from the tangential projection of the momentum equation in conjunction with the imposed flow direction or swirl, leaving four differential equations to be solved. This can be viewed as spatial discretization of the blade force (more precisely, its θ-component). Alternatively, it may be treated as an additional time-dependent unknown, Baralon et al. (1997), that is, discretized in time. Here, a time dependent approach is presented. The following system of equations is solved in blade passages:

$$\int_V \frac{\partial}{\partial t} \left| \begin{array}{c} \rho \\ \rho\vec{W} \\ \rho E \end{array} \right| dV + \frac{1}{b} \oint_S \left| \begin{array}{c} b\rho\vec{W} \\ b(\rho\vec{W} \otimes \vec{W} + p\boldsymbol{I}) \\ b\rho\vec{W}H \end{array} \right| \cdot d\vec{S} =$$

$$\int_V \left| \begin{array}{c} 0 \\ \rho(\Omega^2 \vec{R} - 2\vec{\Omega} \times \vec{W}) + \rho(\vec{f}_{\mathrm{B}} + \vec{f}_{\mathrm{F}}) + p\dfrac{\vec{\nabla}b}{b} \\ \rho\Omega^2 \vec{R} \cdot \vec{W} \end{array} \right| dV \tag{1}$$

$$\frac{\partial}{\partial \tau}(\rho f_{\mathrm{B}\theta}) = \kappa(\widetilde{\rho f_{\mathrm{B}\theta}}) \frac{W_\theta^{\mathrm{t}} - W_\theta}{\widetilde{W}_\theta} \tag{2}$$

The throughflow model is introduced through the blockage factor due to tangential blade thickness, b, the blade force, $\vec{f}_{\mathrm{B}}$, and the friction force, $\vec{f}_{\mathrm{F}}$. In accordance with the basic code, Eq. (1) is formulated in cartesian coordinates and solved in the relative system. The finite volume spatial discretization employs either the central or upwind TVD schemes, time integration is explicit through a Runge-Kutta scheme with implicit residual smoothing and multigrid. The mesh consists of a single cell in the tangential direction and does not reflect the blade geometry; axisymmetry is expressed through a periodic boundary condition.

The θ-component of the blade force is treated as an additional unknown for which Eq. (2) is solved. The other two components are obtained from the conditions that the blade force be orthogonal to the relative velocity vector and tangential to the local axisymmetric stream surface. For stability reasons, this second condition is approximated by orthogonality to spanwise mesh lines. τ is a pseudo-time and κ a constant coefficient of the order of 1. Variables with a tilde are mean absolute values over the blade passage. The target relative tangential velocity, W_θ^{t}, is either direct external input (the so-called design mode) or locally calculated from the prescribed relative flow angle, β^{t}, and the current meridional velocity (the so-called analysis mode),

$$W_\theta^{\mathrm{t}} = W_m \tan \beta^{\mathrm{t}} \tag{3}$$

W_θ^{t} (in design mode) or β^{t} (in analysis mode) are given as explicit input only at the blade trailing edge (TE). The respective distribution inside the blade passage is recalculated at each Runge-Kutta stage by applying a normalized power function bewteen the current value at the blade leading edge (LE), determined by the oncoming flow, and the imposed TE value.

The discretization and multigrid implementation of Eq. (2) are crucial for the efficiency of the following global algorithm. $\rho f_{\mathrm{B}\theta}$ is explicitly corrected at each Runge-Kutta stage,

$$(\rho f_{\mathrm{B}\theta})^{(k+1)} - (\rho f_{\mathrm{B}\theta})^{(k)} = \kappa \left((\widetilde{\rho f_{\mathrm{B}\theta}}) \frac{W_\theta^{\mathrm{t}} - W_\theta}{\widetilde{W}_\theta} \right)^{(k)} \tag{4}$$

The multigrid procedure applied to the blade force differs from the standard full approximation scheme (Brandt; 1977; Wesseling; 1991). For the main variables, the first coarse-grid residual is the restricted fine-grid residual. In symbolic notation for two successive grid levels n and $n+1$, U representing the pseudo-vector of conservative variables, L the residual operator, R and $\hat{R}$ restriction operators for variables and residuals, F the forcing function,

$$\partial_t U^{(n)} + L(U^{(n)}, f_{\mathrm{B}\theta}^{(n)}) = F^{(n)}$$
$$\partial_t U^{(n+1)} + L(U^{(n+1)}, f_{\mathrm{B}\theta}^{(n+1)}) = \underbrace{L(R(U^{(n)}), R(f_{\mathrm{B}\theta}^{(n)})) + \hat{R}(F^{(n)} - L(U^{(n)}, f_{\mathrm{B}\theta}^{(n)}))}_{F^{(n+1)}} \tag{5}$$

For the blade force, the residual is directly visible on the coarse grid as the difference between the restricted current solution and the restricted target solution, which tends to zero at convergence. Use of a forcing function and the restricted fine-grid residual is therefore not necessary. Eq. (2) is applied on the coarse grids unchanged, writing K for the right-hand-side,

$$
\begin{aligned}
\partial_\tau(\rho f_{\mathrm{B}\theta})^{(n)} &= K(U^{(n)}, f_{\mathrm{B}\theta}^{(n)}) \\
\partial_\tau(\rho f_{\mathrm{B}\theta})^{(n+1)} &= K(U^{(n+1)}, f_{\mathrm{B}\theta}^{(n+1)})
\end{aligned}
\tag{6}
$$

Restriction and prolongation applied to the blade force are the same as those for the main variables. If shocks are captured in analysis mode, only linear prolongation, as opposed to the slightly less expensive constant prolongation, permits satisfactory multigrid convergence. The target on grid level $n + 1$ is obtained by restricting W_θ^t from grid level n. In analysis mode, $\tan\beta^t$ is calculated immediately after restriction,

$$
(\tan\beta^t)^{(n+1)} = \frac{R((W_\theta^t)^{(n)})}{W_m(R(\vec{W}^{(n)}))}
\tag{7}
$$

and then kept constant.

The friction force $\vec{f}_\mathrm{F}$ introduces losses in the inviscid flow according to the distributed loss model, Hirsch (1989). It acts in the direction opposite to the relative velocity vector and for a perfect gas its magnitude is given by

$$
\rho f_\mathrm{F} = \frac{p}{p_{\mathrm{t\,rot}}} p_{\mathrm{ref}} \partial_m \psi
\tag{8}
$$

This formulation ensures that a loss coefficient of zero will give exactly zero friction force. The imposed loss coefficient ψ is defined as

$$
\psi = \frac{p_{\mathrm{t\,rot\,LE}} - p_{\mathrm{t\,rot}}}{p_{\mathrm{ref}}}
\tag{9}
$$

where $p_{\mathrm{t\,rot}}$ is the total pressure associated with rothalpy,

$$
p_{\mathrm{t\,rot}} = p\left(\frac{T_{\mathrm{t\,rot}}}{T}\right)^{\frac{\gamma}{\gamma-1}} \qquad T_{\mathrm{t\,rot}} = T + \frac{W^2 - U^2}{2c_p}
\tag{10}
$$

and p_{ref} a reference dynamic pressure, taken at the leading edge (for compressors) or trailing edge (for turbines) on each streamwise mesh line.

For output purposes, true streamlines and spanwise profiles of efficiency, pressure ratio and calculated loss coefficient are defined by connecting points of equal massflow, integrated along spanwise mesh lines.

3 The Relevance of Captured Throughflow Shocks

An analysis has been performed about the different shock patterns as seen by the design mode, by the analysis mode and in 3D. This will be discussed in a following paper. We briefly state the main results: In design mode, captured shocks are axisymmetric. Figure 1c shows such a shock (unrealistically) placed in the context of a blade-to-blade section. In analysis mode, captured shocks are quasi-normal on axisymmetric streamsurfaces, obliquity being admitted in the meridional plane. They can be thought of as achieving in axisymmetric flow the effect of a blade-to-blade normal shock, Fig. 1a or shock B in Fig. 1b. Oblique shocks such as shock A in Fig. 1b cannot be represented.

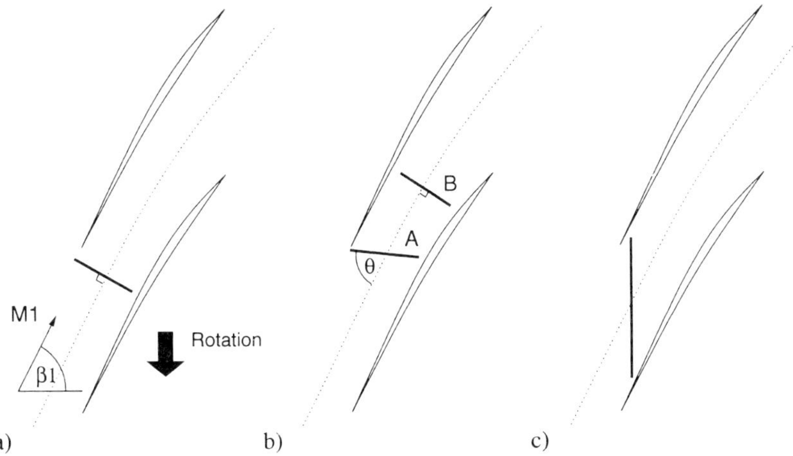

Figure 1: 2D compressor blade passage with supersonic relative inflow ($M_1 > 1$): (a) near-stall, (b) choked, (c) axisymmetric shock admitted in design mode if $M_1 \cos \beta_1 > 1$.

Having established the shock capturing properties of the Euler throughflow model, it can now be examined how they are best brought to bear with the important application of transonic axial compressor rotors. The supersonic sections of these are characterized by thin blades with low camber, and subsonic outflow. Information on the shock structure is best gleaned from Laser anemometer measurements, e.g. by Dunker (DLR Single-Stage Compressor) or by Strazisar et al. (1989) (NASA Rotor 67), both documented in (Fottner; 1990). A complete picture, consistent with earlier, more punctual experimental results, of the shock structure as well as the choking and principal loss mechanisms is painted in two papers by Freeman and Cumpsty (1992) and by Bloch et al. (1996). Bloch has identified a continuous variation between three qualitatively distinct patterns:

- Near stall, a single normal shock stands ahead of the closed part of the blade passage.

- Around peak efficiency, this first shock turns oblique and is trailed by a normal shock. Initially (i.e., for higher back pressures) these two shocks coalesce near the suction surface.

- For choked flow, the first, oblique shock is not influenced by the downstream flow anymore. The second, normal shock, by analogy with quasi-1D nozzle flow, positions itself in the diverging blade passage to match the imposed back pressure.

Since the flow remains axially subsonic, no shock can be captured in design mode; the imposed or modelled losses must include the shock losses.

The captured normal shock in analysis mode may replace or complement a shock loss model. It might automatically yield the correct shock obliquity in the hub-to-tip plane. With the exception of Wennerstrom's model (Wennerstrom and Puterbaugh; 1984) this is normally neglected in even the most sophisticated shock loss models, e.g., König et al. (1996), constructed from purely 2D considerations in the blade-to-blade plane, but was demonstrated, amongst others by Strazisar (1985) and by Wood et al. (1987), to have decisive influence on shock strength. Preliminary experience seems to indicate, however, that the structure and upstream Mach number of shocks captured in analysis mode are difficult to control. Although their response to changes in back pressure is similar to Bloch's scheme in certain cases, an unequivocal association with the actual 3D shocks over the complete design speed operating range remains yet to be established.

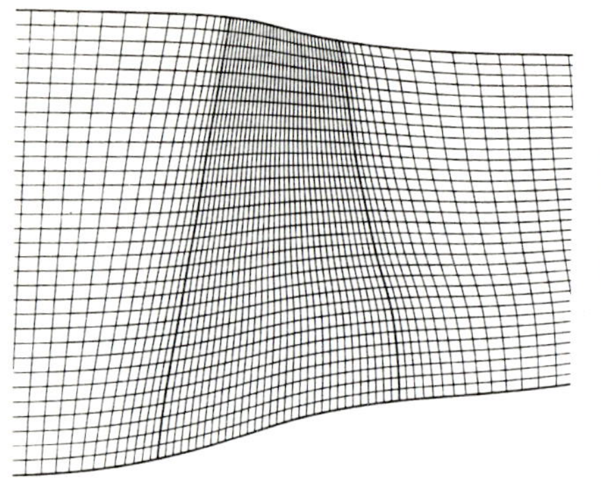

Figure 2: NASA Rotor 67: 32 × 72 = 2304-cell mesh.

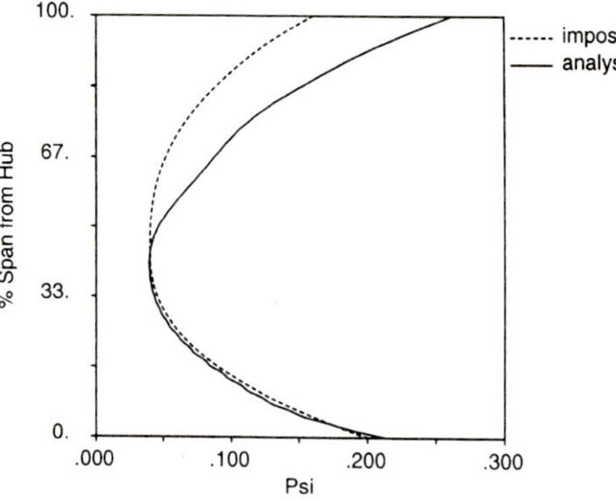

Figure 3: NASA Rotor 67: viscous loss profile imposed in analysis run (dashed line) and calculated loss profile (solid line).

Table 1: Comparison of analysis mode and design mode.

Mode	$\dot{m}$ [kg/s]	Π_t	η_{is} [%]
analysis	30.25	1.831	89.87
design	30.37	1.844	89.51

4   Results

4.1   Transonic compressor rotor

The different shock capturing properties of design and analysis mode are illustrated with the NASA rotor 67 (design pressure ratio 1.63, inlet tip relative Mach number 1.38, 22 blades). Figure 2 shows the mesh. To get the same results in a design run as in an analysis run, the losses imposed in the design run must be the losses imposed in the analysis run plus the shock losses captured in the analysis run:

$$\psi_{\text{design}}^{\text{imposed}} = \psi_{\text{analysis}}^{\text{imposed}} + \text{captured shock losses} \tag{11}$$

We first performed an analysis run with a viscous loss profile, shown in Fig. 3. The captured shock appears clearly in the isolines of relative Mach number, Fig. 4a; the bold isoline marks $M_{\text{rel}} = 1$. The computed trailing edge values for swirl and loss coefficient were then imposed in an otherwise identical design run, Fig. 4b. Although — in conformance with theory — no shock is captured, nearly the same massflow, pressure ratio and efficiency are predicted as in the analysis run, table 1.

Conceptually, solutions calculated with an axisymmetric throughflow method are an approximation to the pitchwise averaged 3D flow. Their accuracy should therefore be assessed by comparison with the latter. 3D Navier-Stokes calculations have been performed with the same flow solver on a mesh of 550,000 cells including tip clearance and using a two-equation turbulence model. Figure 5 shows isolines of relative Mach number in the meridional average, to be compared with Figs. 4a and b. Globally, the design solution represents the pitch averaged 3D flow field more faithfully than the analysis solution. The 3D solution corresponds to the peak efficiency operating point on the calculated characteristic ($\dot{m} = 33.65$ kg/s, $\Pi_t = 1.58$, $\eta_{is} = 89.0\,\%$).

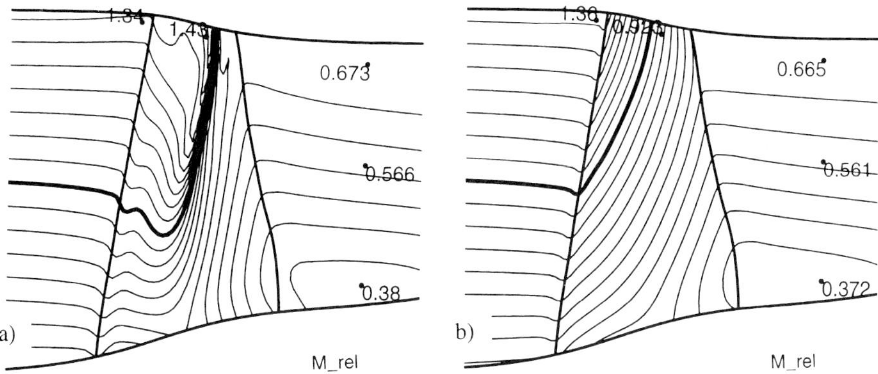

Figure 4: NASA Rotor 67: isolines of relative Mach number: (a) analysis mode, (b) design mode (bold line marks $M_{rel} = 1$, increment 0.05).

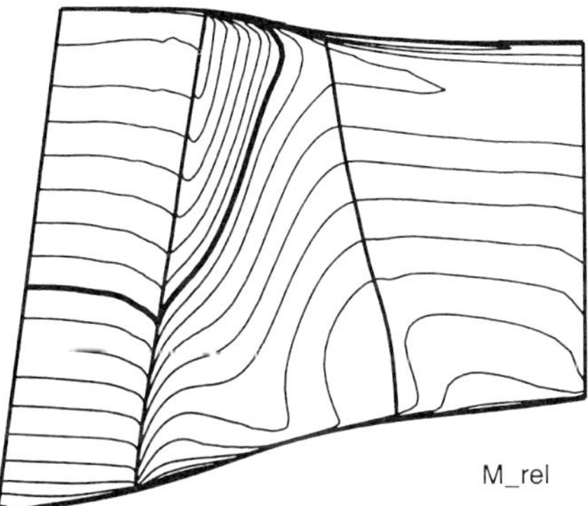

Figure 5: NASA Rotor 67: meridional average of 3D Navier-Stokes calculation (isolines of relative Mach number, bold line marks $M_{rel} = 1$, increment 0.05).

Because the primary goal of the throughflow calculations had been the comparison between design and analysis mode, simple choices have been made for the different parameters that model the blade (e.g., linear distribution of flow angle and losses, standard parabolic thickness distribution). In particular, the choice of a deviation angle of only 5° over the entire span constitutes a considerable underestimation. The pressure ratio is therefore too high ($\Pi_t = 1.83$). The throughflow operating point should nonetheless occupy a similar position on the characteristic, shifted to a higher pressure ratio; the massflow of 30.25 kg/s equals 99.15 % of the throughflow choking massflow, versus 98.25 % for the 3D calculation, efficiency is 89.9 % versus 89.0 % in 3D. The 3D calculation determines a slightly oblique shock at the blade passage entrance which appears as a smooth deceleration in the meridional average. This is realized in design mode, whereas the analysis solution first shows substantial acceleration upstream of the captured normal shock. Only in the subsonic region near the blade root does the analysis solution appear to be of higher fidelity than the design solution: the weaker deceleration (and even reacceleration at the hub), leading to an inflection point in the isolines, is present, albeit overestimated in its severity, whereas the design solution misses this effect. The obliquity of the analysis shock provokes considerable deflection of the meridional streamlines. Figure 6a compares analysis mode (solid bold lines) and design mode (dashed bold lines), shown superimposed on streamwise mesh lines.

Table 2: Sensitivity of the choking massflow to modelling parameters.

Case	Modification	$\dot{m}_{\text{choke}}$ [kg/s]	Π_t	η_{is} [%]
1		**30.8**	**1.53**	78.
2	d_{max} at 0.75 c	32.7	1.52	86.
3	d_{max} at 0.25 c	27.2	1.53	66.
4	no smoothing	28.9	1.53	72.
5	$\delta - 5°$	32.6	**1.63**	78.
6	$\delta + 5°$	28.6	**1.44**	79.
7	β-exponent 2	**34.8**	1.55	89.
8	β-exponent -2	**24.7**	1.52	63.
9	ψ-exponent -5	31.0	1.53	79.
10	ψ-exponent 5	30.1	1.52	76.
11	$d_{\text{max}} - 20\,\%$	32.2	1.54	82.
12	$d_{\text{max}} + 20\,\%$	29.3	1.51	75.
13	d_{max}-exponent 1.5	31.8	1.53	81.
14	d_{max}-exponent 3	29.1	1.52	75.

The reference 3D solution, Fig. 6b, confirms the smooth design mode streamlines. Despite the difference inside the blade passage, the downstream flow fields are nearly identical, witnessed by the profiles of β and absolute total pressure ratio, Figs. 7a and b.

An advantage of the analysis mode over the design mode is the ability to run performance curves by mere variation of exit pressure, including the prediction of choke. (Simultaneous variation of exit swirl would be required in design mode to maintain the correct exit flow angle, which is not practicable.) The choking massflow turns out to be highly sensitive to some of the blade modelling parameters. The blade thickness is defined through a spanwise profile of maximum tangential thickness (d_{max}), the location of d_{max} as a fraction of chord, an exponent for the streamwise power function distribution and the number of cells over which to smooth b at LE and TE. The turning is prescribed through a spanwise profile at the trailing edge of W_θ or β and an exponent for the streamwise power function distribution. Finally, a spanwise profile at the trailing edge for ψ as well as an exponent for the streamwise power function distribution define the imposed losses.

The sensitivity of the choking massflow to these parameters has been investigated by applying sensible variations to each in turn, keeping a constant exit pressure of 100 kPa at $r = 14\,\text{cm}$. The results are summarized in table 2 (Π_t = total pressure ratio, $\delta - 5°$ indicates a reduction of the reference deviation angle by $5°$). In terms of the blade modelling parameters, the reference case, 1, is defined as follows: d_{max}, rather than being deduced exactly from the blade geometry, is assumed to be co-located with the directly available maximum normal blade thickness, $d_{n\,\text{max}}$, and thus approximated as $d_{\text{max}} \approx d_{n\,\text{max}} / \cos \beta_{n\,\text{max}}$, where $\beta_{n\,\text{max}}$ is the blade angle at the location of $d_{n\,\text{max}}$. The streamwise thickness distribution seen by the throughflow computation is quadratic (exponent 2) with d_{max} at mid-chord (0.5 c) and the resulting b smoothed over ± 4 cells at LE and TE. The imposed trailing edge flow angle and the loss coefficient are those measured at peak efficiency at a station downstream of the blade row. Both are distributed linearly (exponent 1), β_{TE} is corrected by the calculated variation between the measurement station and the TE. Maximum variations of choking massflow and total pressure ratio are highlighted by bold print. The greatest influence on choking massflow is exerted by the flow angle distribution. Figure 8 compares the actual distributions, extracted from the meridional average of the 3D solution at 13 %, 52 % and 88 % span from the hub, with the power function distributions assumed in the throughflow module. At the hub and near mid-span the simple power function, with a suitable exponent, fits

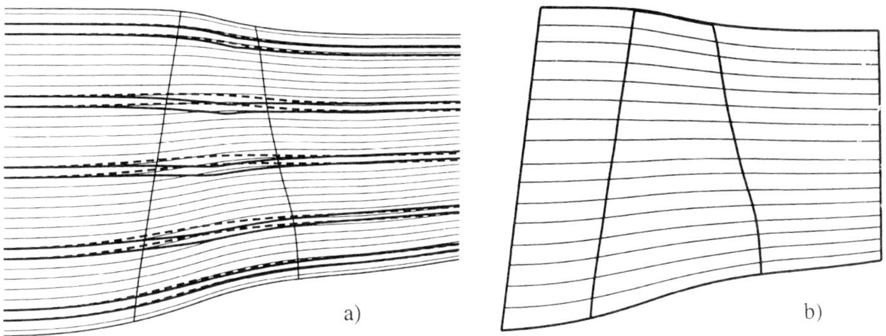

Figure 6: NASA Rotor 67: streamlines: (a) throughflow (solid bold lines = analysis mode, dashed bold lines = design mode, light lines = streamwise mesh lines), (b) meridional average of 3D Navier-Stokes calculation.

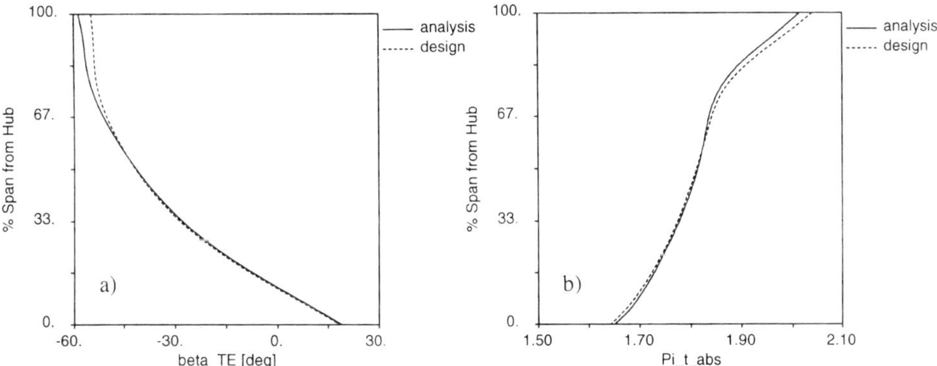

Figure 7: NASA Rotor 67: downstream profiles of (a) relative flow angle and (b) absolute total pressure ratio (solid line = analysis mode, dashed line = design mode).

the actual distributions remarkably well. The tip distribution follows the same general trend of more turning in the forward region, but shows a defect around mid-chord. The location of d_{max} and the β-exponent have also been modified either only for the tip-section or only for the hub-section, with a linear variation along the span towards no modification at the respective other end of the blade. The choking massflow turns out to be more sensitive to modification of the tip section, but only slightly so (by 15 % for the β-exponent and by 9 % for the location of d_{max}).

No control or limiting of the leading edge flow angle has been performed in these calculations. The choking mechanism for all blade sections is thus passage choking, as opposed to choking at the unique incidence angle. At the experimental choking massflow (35.0 kg/s, adjusted with, e.g., d_{max} at 0.6 c, a d_{max}-exponent of 1.6 and a β-exponent of 1.5) the throughflow computation indicates zero incidence ($\pm 0.5°$) with respect to the suction side blade angle for the outer 40 % span.

The multigrid algorithm described in section 2 is highly effective in the absence of shocks. This is demonstrated by an analysis run at 50 % speed, Fig. 9, for which a convergence level of 4 orders, sufficient for practical purposes, is reached 14 times faster (in terms of CPU time) with the multigrid algorithm than on a single grid. If shocks are captured, multigrid still provides an advantage both in terms of robustness and convergence rate for all operating points

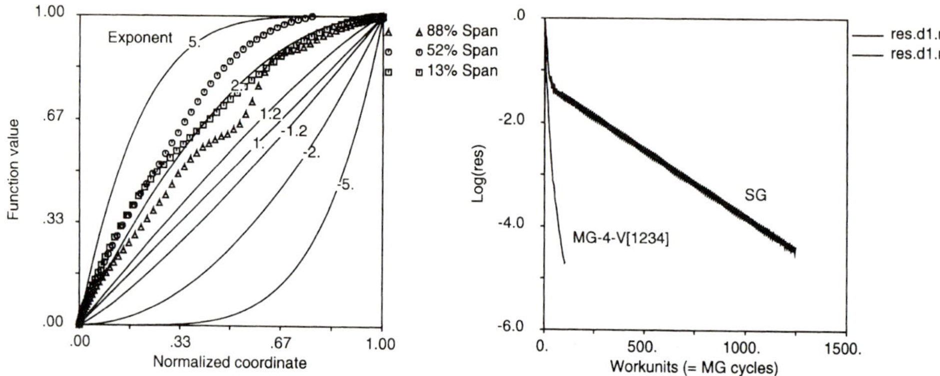

<table>
<tr><td>

Figure 8: NASA Rotor 67: actual flow angle distributions (meridional average of 3D Navier-Stokes calculation, symbols) and power functions for selected exponents.

</td><td>

Figure 9: NASA Rotor 67: comparison of multigrid (MG) and single grid (SG) convergence rates in analysis mode at 50 % speed.

</td></tr>
</table>

Table 3: Influence of the spatial discretization.

Scheme	$\dot{m}$ [kg/s]	Π_t	η_{is} [%]
central	30.25	1.831	89.87
FDSTVD	29.67	1.830	88.65
1st OU	29.63	1.804	88.36

except choked ones. In the latter case, a slight slowdown compared with single grid is observed in terms of CPU time. Single grid then only requires around 300 iterations for engineering accuracy (residual reduction of 5 orders), however. Multi-stage machines might favour multigrid even in those cases.

The computations discussed so far use the second-order central scheme with a Jameson-type blend of fourth- and second-order artificial dissipation (Jameson et al.; 1981; Hirsch; 1990), in combination with a four-stage Runge-Kutta scheme and time-step dependent residual smoothing (Zhu; 1996) with a coefficient of 2. The CFL number is 6 and the coefficient κ in Eq. (4) is 5. The analysis run has been repeated with a first-order upwind scheme (Roe; 1981) (1st OU, with simplifications such as variable-averaging) and a second-order TVD extension of this scheme (Hirsch; 1990) (FDSTVD for flux difference splitting TVD scheme) using the minmod limiter. All three schemes yield comparable convergence rates, giving a residual drop of 4 orders in about 500 iterations. Although, apart from pressure ratio, the first-order upwind scheme appears to agree well with the FDSTVD scheme globally, table 3, inspection of the radial profiles of the calculated loss coefficient, Fig. 10a, reveals that this agreement is fortuitous: insufficient strength of the strongly diffused shock in the outboard region, Fig. 11, is balanced by excessive numerical dissipation in the inboard region; at the hub, the numerically generated losses reach 10 %. The downstream absolute flow angle, α_{TE}, is another sensitive variable, Fig. 10b. Here, the maximum difference, occurring at the casing, is 4.5°. The two second-order schemes predict practically the same pressure ratio, but differ by 2 % in massflow and 1.2 percentage points in efficiency. Referring again to the spanwise profiles, this divergence is easily traced to the different shock capturing properties: the solutions are nearly identical in the subsonic region below about 40 % span, but differ by up to 2.5 % in loss coefficient and 1.5° in α_{TE} in the supersonic

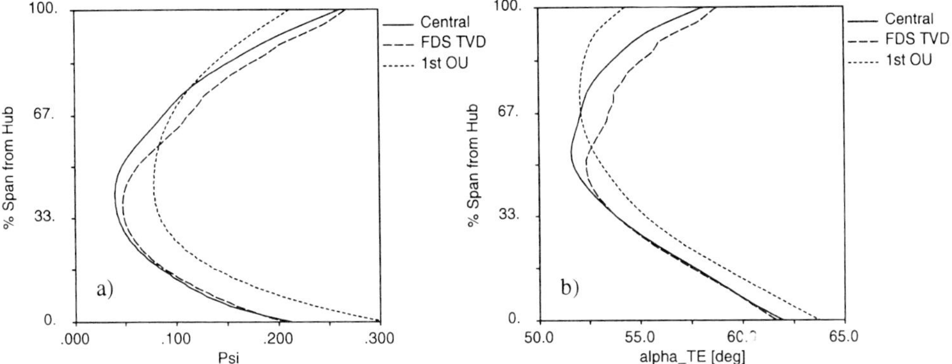

Figure 10: NASA Rotor 67: analysis mode: comparison of (a) calculated loss coefficient and (b) downstream absolute flow angle for the central scheme (solid line), the FDSTVD scheme (dashed line) and the first-order upwind scheme (dotted line).

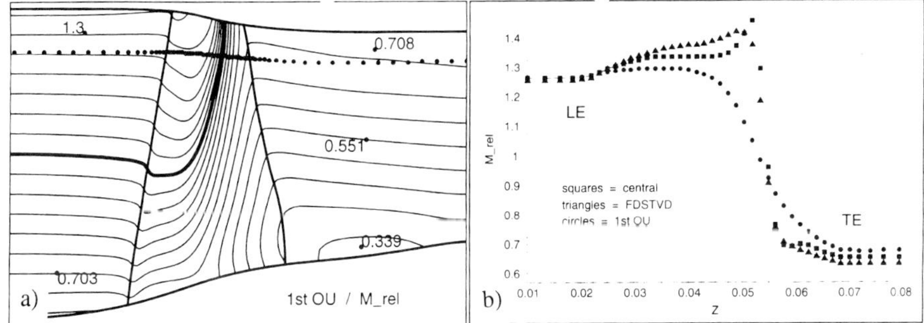

Figure 11: NASA Rotor 67: analysis mode: (a) isolines of relative Mach number computed with the first-order upwind scheme, (b) relative Mach number at 85 % span (squares = central scheme, triangles = FDSTVD scheme, circles = first-order upwind scheme).

region.

A mesh dependence study performed in analysis mode with the central scheme indicated negligible dependence on the spanwise direction (16, 32 and 64 cells tested), but considerable dependence on the streamwise direction (16, 32 and 64 cells between LE and TE tested). In the latter case, the differences were again limited to the supersonic outboard region and similar in magnitude to those between the central and FDSTVD scheme. The observed mesh and scheme dependence is clearly shock-related and not present in design mode.

4.2 Four-stage low speed turbine

The multistage capability of the throughflow module is demonstrated with test case E/TU-4 of Fottner (1990), a four-stage low speed turbine (nominal power 703 kW, 7500 rpm, massflow 7.8 kg/s, pressure ratio 2.5, efficiency 91 %, degree of reaction 0.5, hub ratio at exit 0.525, constant hub radius 135 mm). Figure 12 shows the mesh.

An important benefit of the multigrid technique is the mesh size-independent convergence rate. Applied to multi-stage configurations this means that even for machines with a large number

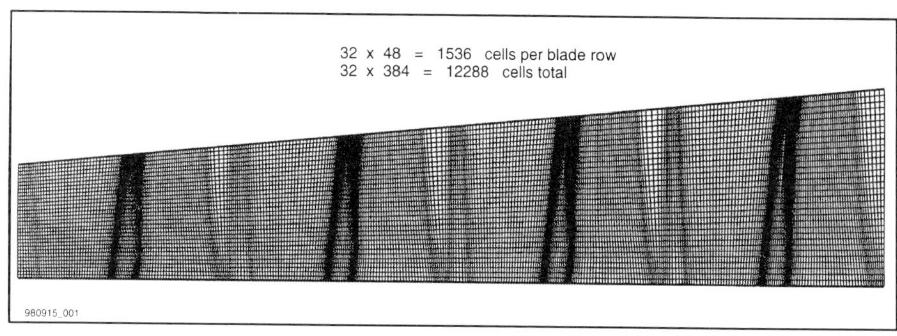

Figure 12: Turbine E/TU-4: computational grid.

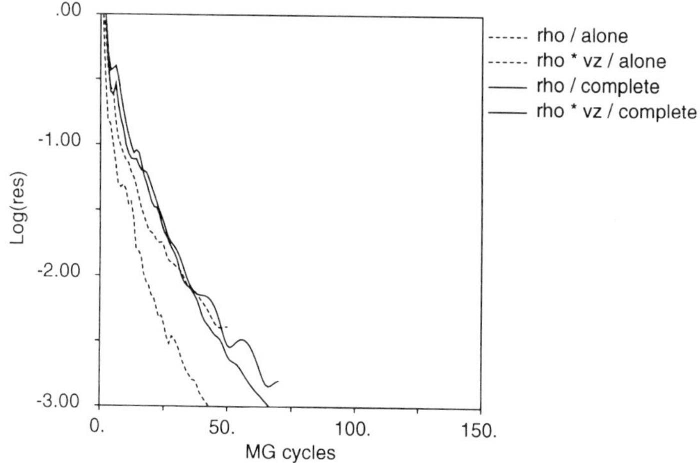

Figure 13: Turbine E/TU-4: comparison of convergence of rotor 1 if computed alone (dashed lines) and in a computation of the complete machine (solid lines), curves are for density and axial momentum.

of stages there is no significant slow-down compared with individual components. Figure 13 shows the convergence histories of density (the lower curves) and axial momentum for rotor 1 computed alone (dashed lines) and embedded in a computation of the complete machine (solid lines). For this test case, a residual drop of 3 to 4 orders and stabilization of the massflow are reached after 30 to 70 multigrid cycles, depending on the initial solution and on the operating point. Four-level multigrid in a V-cycle with 1, 2, 3 and 4 smoothing sweeps on respectively grid levels 1, 2, 3 and 4 were used. The computations were performed with the central scheme, the coefficient κ in Eq. (4) is 1 and the CFL number is 6.

Figure 14 shows contours of relative Mach number at the design point. Contrary to transonic test cases, blade thickness and the flow angle distribution are not critical parameters. Deviation angles and loss coefficients are the same for all stages and were adjusted to match massflow and efficiency at the design operating point with the experimental boundary conditions at machine inlet (total pressure and temperature) and exit (static pressure) and kept constant. Figure 15 shows the resulting performance curves.

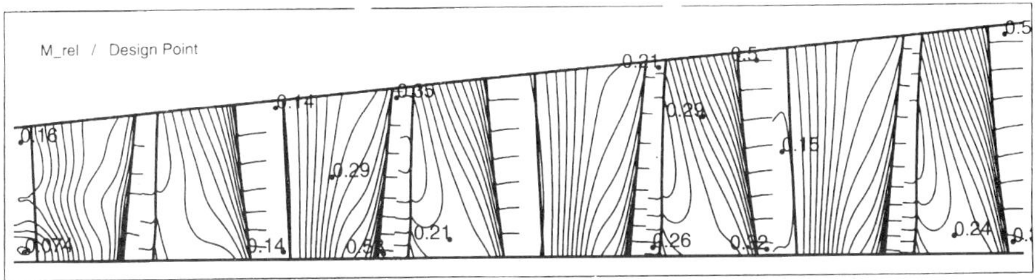

Figure 14: Turbine E/TU-4: contours of relative Mach number at design point.

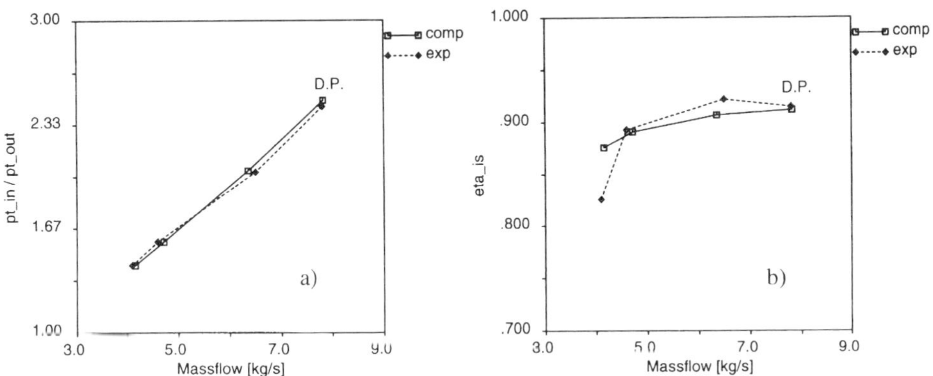

Figure 15: Turbine E/TU-4: performance curves: (a) absolute total pressure ratio, (b) isentropic efficiency (solid lines = computed, dashed lines = experiment, D.P. = design point).

5 Conclusions

A 3D Navier-Stokes solver has been extended to include a throughflow model based on the Euler equations. Blades are described through distributions and profiles for blade thickness, flow angle or swirl, and losses. The efficiency of the multigrid-accelerated Euler throughflow algorithm with a time-dependent blade force was demonstrated for multi-stage machines. It allows the use of an Euler through-flow solver in the design cycle at acceptable turn-around times of a few minutes. The algorithm is equally effective for subsonic and for supersonic flows. An investigation of the shock capturing properties of the Euler throughflow equations showed that quasi-normal shocks are captured in analysis mode and axisymmetric shocks in design mode. The implications for the important class of applications of transonic compressor rotors have been pointed out. The choking massflow in analysis mode was found to be highly sensitive to the distribution of thickness and, especially, flow angle across the blade. Analysis of a 3D Navier-Stokes solution in the meridional (pitchwise) average shows that the actual evolution of the flow angle is well described by simple power functions. The throughflow model may thus serve as a tool to predict the effect of design changes in those parameters. Although due to the captured shock the analysis mode flow field can differ from the design mode flow field, which more closely resembles the pitch-averaged 3D flow field, the differences downstream of the blade row appear to be small.

Acknowledgement

Dr. Shun Kang and Dr. Andrei Khodak are acknowledged with performing the 3D calculations of NASA Rotor 67.

References

Baralon, S., Eriksson, L.-E. and Hall, U. (1997). Viscous Throughflow Modelling of Transonic Compressors Using a Time-Marching Finite Volume Solver, 13th International Symposium on Airbreathing Engines (ISABE), Chattanooga, USA.

Bloch, G. S., Copenhaver, W. W. and O'Brian, W. F. (1996). Development of an Off-Design Loss Model for Transonic Compressor Design, *Loss Mechanisms and Unsteady Flows in Turbomachines*, AGARD-CP-571.

Boure, G. and Gillant, P. (1995). Aerodynamic Modelling of a Transonic Radial Equilibrium of a Multistage Compressor, *Turbomachinery — Fluid Dynamic and Thermodynamic Aspects*, VDI Berichte 1185, pp. 143–155.

Brandt, A. (1977). Multi-Level Adaptive Solutions to Boundary-Value Problems, *Mathematics of Computation* **31**: 333–390.

Damle, S. V., Dang, T. Q. and Reddy, D. R. (1997). Throughflow Method for Turbomachines Applicable for All Flow Regimes, *Transactions of the ASME, Journal of Turbomachinery* **119**(2): 256–262.

Dawes, W. N. (1992). Toward Inproved Throughflow Capability: The Use of Three-Dimensional Viscous Flow Solvers in a Multistage Environment, *Transactions of the ASME, Journal of Turbomachinery* **114**: 8–17.

Fottner, L. (ed.) (1990). *Test Cases for Computation of Internal Flows in Aero Engine Components*, AGARD-AR-275.

Freeman, C. and Cumpsty, N. A. (1992). Method for the Prediction of Supersonic Compressor Blade Performance, *AIAA Journal of Propulsion and Power* **8**(1): 199–208.

Hirsch, C. (1989). *Numerical Computation of Internal and External Flows*, Vol. 1: Fundamentals of Numerical Discretization, John Wiley & Sons.

Hirsch, C. (1990). *Numerical Computation of Internal and External Flows*, Vol. 2: Computational Methods for Inviscid and Viscous Flows, John Wiley & Sons.

Hirsch, C., Lacor, C., Rizzi, A., Eliasson, P., Lindblad, I. and Häuser, J. (1991). A Multiblock/Multigrid Code for the Efficient Solution of Complex 3D Navier-Stokes Flows, Presented at the First European Symposium on Aerothermodynamics for Space Vehicles.

Jameson, A., Schmidt, W. and Turkel, E. (1981). Numerical Solutions of the Euler Equations by Finite Volume Methods Using Runge-Kutta Time-Stepping Schemes, AIAA Paper 81-1259, AIAA 5th Computational Fluid Dynamics Conference.

König, W. M., Hennecke, D. K. and Fottner, L. (1996). Improved Blade Profile Loss and Deviation Angle Models for Advanced Transonic Compressor Bladings: Part II — A Model for Supersonic Flow, *Transactions of the ASME, Journal of Turbomachinery* **118**(1): 81–87.

Nigmatullin, R. Z. and Ivanov, M. J. (1994). The mathematical models of flow passage for gas turbine engines and their components, *Mathematical Models of Gas Turbine Engines and their Components*, AGARD-LS-198, chapter 4.

Roe, P. L. (1981). Approximate Riemann Solvers, Parameter Vectors, and Difference Schemes, *Journal of Computational Physics* **43**(2): 357–372.

Spurr, A. (1980). The Prediction of 3D Transonic Flow in Turbomachinery Using a Combined Throughflow and Blade-to-Blade Time Marching Method, *International Journal of Heat and Fluid Flow* **2**(4): 189–199.

Strazisar, A. J. (1985). Investigation of Flow Phenomena in a Transonic Fan Rotor Using Laser Anemometry, *Transactions of the ASME, Journal of Engineering for Gas Turbines and Power* **107**(2): 427–435.

Strazisar, A. J., Wood, J. R., Hathaway, M. D. and Suder, K. L. (1989). Laser Anemometer Measurements in a Transonic Axial-Flow Fan Rotor, NASA Technical Paper 2879.

Vuillez, C. and Petot, B. (1994). New Methods, New Methodology — Advanced CFD in the SNECMA Turbomachinery Design Process, *in* A. S. Üçer (ed.), *Turbomachinery Design Using CFD*, AGARD-LS-195, chapter 7.

Wennerstrom, A. J. and Puterbaugh, S. L. (1984). A Three-Dimensional Model for the Prediction of Shock Losses in Compressor Blade Rows, *Transactions of the ASME, Journal of Engineering for Gas Turbines and Power* **106**(2): 295–299.

Wesseling, P. (1991). *An Introduction to Multigrid Methods*, John Wiley & Sons.

Wood, J. R., Strazisar, A. J. and Simonyi, S. (1987). Shock Structure Measured in a Transonic Fan Using Laser Anemometry, *Transonic and Supersonic Phenomena in Turbomachines*, AGARD-CP-401.

Yao, Z. and Hirsch, C. (1995). Throughflow Model Using 3D Euler or Navier-Stokes Solvers, *Turbomachinery — Fluid Dynamic and Thermodynamic Aspects*, VDI Berichte 1185.

Zhu, Z. W. (1996). *Multigrid Operation and Analysis for Complex Aerodynamics*, PhD thesis, Vrije Universiteit Brussel, Belgium.

UNIVERSITY COLLEGE
LONDON